Conversion Factors

Quantity	U.S. Customary Units	SI Units
Length	1 in.	= 2.54 cm
—	1 ft	= 0.3048 m
Force	1 lb	= 4.448 N (kg·m/s^2)
Mass	1 slug	= 14.59 kg
Pressure	1 lb/in^2	= 6.895 kPa (N/m^2)
Energy	1 ft·lb	= 1.356 J (N·m)
Power	1 ft·lb/sec	= 1.356 W

Other Useful Conversions

Area

1 in^2 = 6.45 cm^2
1 ft^2 = 0.093 m^2
1 cm^2 = 0.155 in^2
1 m^2 = 10.76 ft^2

Volume

1 in^3 = 16.39 cm^3
1 m^3 = 35.29 ft^3
1 m^3 = 1000 liter (L)
1 ft^3 = 0.028 m^3 = 28.3 L
1 gal = 231 in^3
1 gal = 3.785 L
1 gal = 0.135 ft^3

Volume Flow Rate

1 ft^3/sec = 449 gpm
1 ft^3/sec = 0.028 m^3/s
1 gpm = 6.309 × 10^{-5} m^3/s
1 Lpm = 16.67 × 10^{-6} m^3/s

Temperature

°C = 5/9(°F − 32)
°F = 9/5 °C + 32
°K = °C + 273
°R = °F + 460

Energy

1 Btu = 778 ft·lb
1 ft·lb = 1.356 J

Weight (not mass)

1 lb = 4.448 N
1 kg$_f$ = 2.2 lb
1 kg$_f$ = 9.81 N

Power

1 HP = 550 ft·lb/sec
1 HP = 746 W
1 HP = 42.4 Btu/min
1 kW = 56.8 Btu/min

Fluid Power Technology

Fluid Power Technology

F. Don Norvelle

Oklahoma State University
Stillwater, Oklahoma

West Publishing Company

Minneapolis/St. Paul   New York   Los Angeles   San Francisco

West Legal Studies Staff:

Compositor: Monotype Composition Company

Cover Design: Paul Konsterlie

Artist: Paul Cary

Text Design: John Rokusek

Copyediting: Margaret Jarpey

Production, Prepress, Printing and Binding by West Publishing Company.

For more information, contact Delmar, 3 Columbia Circle, PO Box 15015, Albany, NY 12212-0515; or find us on the World Wide Web at http://www.westlegalstudies.com

Library of Congress Cataloging-in-Publication Data

Norvelle, F. Don.
    Fluid power technology / F. Don Norvelle.
        p. cm.
    Includes index.
    ISBN 0-314-01218-4
    1. Fluid power technology. I. Title.
TJ843.N58. 1994   93-46736
620.1'06—dc20   CIP

Dedication

To Evelyn for her loving encouragement and patience.
To Cery and Debbie for long hours of typing.
To my students for not letting me get by with anything.

Contents

Chapter 3

Hydraulic Fluids, 51

Chapter 4

Hydraulic Pumps, 85

Contents

Preface to the Instructor

This textbook, *Fluid Power Technology*, has been written with the needs of both the students and the instructor as our primary consideration. We have presented the fundamentals of fluid power in a style and depth suitable for community colleges, technical colleges and technical institutes as well as undergraduate courses in engineering and engineering technology.

In this textbook we have attempted to use a writing style and a technical approach that the student will find easy to read and understand. The chapters begin with easily grasped descriptive material supported by artwork that has been prepared to illustrate the component or concept in a simple, yet technically correct, manner. This descriptive material is followed by analytical material using algebra-based mathematics to support operation and performance analyses.

In addition to covering the basic concepts and components—which every fluid power textbook addresses—this text includes complete chapters on filtration, fluid conduits, and ancillary devices (with emphasis on reservoirs and accumulators) which are not given much attention in most textbooks. Two full chapters are devoted to pneumatics. Throughout the text, the importance of understanding graphic symbols and of familiarization with fluid power standards is stressed. Both U.S. customary units and S.I. units are used in the text.

Significantly this textbook addresses virtually all of the technical material covered on the Fluid Power Specialist Certification Examination administered by the Fluid Power Society. As such, it can be used as a single-source technical reference for persons preparing for that examination.

Text Organization

The text contains fifteen chapters. This lends itself nicely to the normal one-semester basic fluid power course. The detail of the coverage of each chapter can easily be adjusted because of the organization of the material. Chapters 1 and 2 are introductory in nature, covering the basic concepts of fluid power—pressure and force, flow and speed, power, and basic system requirements. Chapter 3 is devoted to fluids. This chapter discusses fluid properties in some detail. It also includes sections on fluid analysis as an important aspect of system maintenance as well as a discussion of fire-resistant fluids.

The standard fluid power hardware—pumps, motors, cylinders, and valves—are discussed in Chapters 4 through 9. In each chapter, the components are described, their operation is explained, and their relationship with other system components is explored. Analytical material is then presented to allow the student to determine power inputs and outputs, pressures, flow rates, speeds, forces, heat generation, and other parameters that are important in component selection and system design. System concepts, rather than individual components, are emphasized.

Chapters 10 through 13 expand on the system theme. Chapter 10 introduces electrical and electronic control of the valves of Chapter 7, 8, 9. Chapter 11 presents filtration and contamination control as the single-most important factor in the life and reliability of fluid power systems. Chapter 12—Fluid Conduits—discusses in detail pipes, hoses, and tubing and their associated fittings and connectors. This chapter also includes a discussion of pressure losses due to flow-through conduits. Chapter 13 covers the so-called ancillary devices—reservoirs, accumulators, heat exchangers, and other components that are important to system design and operation, but never seem to get much attention in fluid power texts. An effort was made here to give them better coverage than they usually receive.

Chapters 14 and 15 are devoted to pneumatics. These two chapters provide a much more in-depth coverage of the subject than is found in most fluid power textbooks. Chapter 14 deals with the basic gas laws and with air preparation—the most critical aspect of pneumatic system life and reliability. Compressors, compressor sizing, dryers, filtration, and pressure regulation are among the topics discussed. Chapter 15 describes those pneumatic components that are different in concept from their hydraulic counterparts. It also discusses pneumatic system design concepts and addresses the unexpectedly high costs of system leakage and of operating pneumatic tools at less than their rated pressure. Pressure losses in pneumatic piping is also addressed. Considerable space in Chapter 15 is devoted to the control of pneumatic systems. Pilot operation of valves is presented, and pneumatic logic is discussed in detail.

Chapter Organization

Each chapter begins with a brief introduction of the chapter and a list of learning objectives. The body of each chapter begins with descriptive information followed by analytical material. The analytical material is exceptionally well supported by examples which begin with simple concepts and become progressively more difficult.

At the end of each chapter there is a summary of key equations and a section containing troubleshooting tips. Each chapter also features an extensive selection of review questions and problems. The questions are to be answered by listing, explaining, and describing components and concepts. The problems are divided by topic and become progressively more difficult within each grouping. Each grouping includes problems using U.S. customary units and problems using S.I. units.

Pedagogy

The writing style, organization, generous use of examples, and the review problems enhance the learning process. This textbook is very practical. It attempts to answer four questions for the student:

1. What is it? Components and concepts are described in practical terms.

2. What does it do? The function of each type of component in the system is stressed.
3. How does it work? Component operation is described in practical terms. The detail of component design is not deemed important for students at this stage of their fluid power education.
4. How does it relate to the system? Nothing in a fluid power system operates independently of other system components. These component interactions are continuously explored.

The writing style is conversational and less formal than that found in many technical textbooks. Explanations are made in the same tone as used by classroom instructors, thereby extending the lecture to the students' home study.

End-of-Text Materials

The textbook includes six appendexes, including the most comprehensive listing of fluid power standards available in any textbook of this type. Appendix B is a complete reprinting of the ISO fluid power graphic symbols. Answers to selected review problems follow the appendixes.

Acknowledgments

The efforts of many people brought this text from an idea to a reality. The entire staff at West Educational Publishing has been most helpful, courteous, and supportive throughout the long process of writing and producing this textbook. My special thanks go to acquisition editors Chris Conty and Tom Tucker, to developmental editor Liz Riedel, and to the several production editors who helped make this book ready for printing.

As author, I gratefully acknowledge the support of my family, particularly daughters, Cery and Debbie, and my wife, Evelyn, who so patiently and helpfully typed the manuscript. I also make a special acknowledgment of the help I received from the fluid power students at Oklahoma State University, who struggled through the early days of the manuscript. There are two people whose contribution to every page of this book deserve special recognition: artist Paul Cary and Professor Donald Miller. Their special attention to accuracy will be unknowingly appreciated by many students who use this text in the years to come.

Finally, this text is much better because of the conscientiousness of the many reviewers of my manuscript. Therefore I thank:

Larry Biddle
Oklahoma State University–Okmulgee

James A. Collier
University of Wisconsin–Stout

Ed Dekker
Cincinnati Technical College, retired

Charles Drake
Ferris State University (MI)

Barry Halliwell
Southern Alberta Institute of Technology

Roy Hartman
Texas A & M University

Charles Hermach
Oregon Institute of Technology

Dan Koonts
Central Piedmont Community College (NC)

Jim Kramer
Schoolcraft College (MI)

Frederick G. Krauss
Delta College (MI)

Al Manore
Macomb Community College (MI)

Edward Messal
Indiana-Purdue University-Fort Wayne

David C. Miller
Clackamas Community College (OR)

Donald J. Miller
Northern Alberta Institute of Technology

Randy Nobles
Womack Machine Supply (OK)

George Nordenholt
Oakland Community College (MI)

Howard Olson
State Technical Institute at Memphis (TN)

John J. Pippenger
Fluid Power Educational Foundation, retired

Patrick Rubino
Triton College (IL)

John G. Slater, Ph.D.
Milwaukee School of Engineering

Ken Stoll
Cincinnati Technical College

James Weber
Hennepin Technical College (MN)

Steve Widmer
Purdue University

F.D. Norvelle
Stillwater, Oklahoma

Fluid Power Technology

Chapter 1

Introduction

Outline

1.1 A Brief History of Fluid Power

Even before the written history of mankind, the power of air and water were put to use in an effort to make life and labor a little easier. The currents of flowing rivers and the wind were probably the first power sources harnessed by early man, primarily for transportation.

The first machines to utilize fluid power were water mills. These very large hydraulic motors used flowing water to turn the rotor (the water wheel). This rotated the shaft, which used gears, belts, pulleys, etc, to rotate the millstone to grind grains. Recorded history indicates that water mills were used as early as 100 BC, so they were probably around much earlier than that.

Hero of Alexandria was a scientist of the first century AD who is credited with the development of a number of early machines, including a reaction jet steam engine called an *aeolipile* (Figure 1.1). Steam was heated in the large container and piped into the sphere. The steam escaping through the two nozzles caused the sphere to rotate. This device was not used to do work. It was simply a novelty enjoyed by the nobility.

Another of Hero's inventions was a clever device for opening the doors of the temple. Air was heated in a closed container. The expansion of the air forced water through tubes to suspended containers that were attached by ropes and pulleys to the doors. As the containers filled, their weight slowly, quietly, and mysteriously opened the doors.

Hero is only one of the early scientists to make contributions to what we now call *fluid power*. Archimedes developed the screw pump (Figure 1.2) in the third century BC. Greeks and Romans used water power in aqueducts and piping systems in which lead pipes carried water for both power and domestic use. The water clock—invented by Ktesbios around 250 AD—provided the most accurate method of keeping time until the Swiss movements were invented in the fourteenth century.

1

During the so-called Dark Ages, little technical progress of any kind occurred. Fluid power stagnated, as did all other scientific fields. The seventeenth century saw a new surge of scientific endeavors, and fluid power began to be seen as a major field of study. During that century, Sir Isaac Newton developed the concepts of *viscosity* and *drag,* along with his *Laws of Motion.* Torricelli did his work in fluid motion about the same time.

In the eighteenth century, Daniel Bernoulli, a Swiss scientist and scholar, studied energy in fluid streams and pipelines and developed the mathematical expressions for the transmission of fluid energy. A French scientist and mathematician, Blaise Pascal, developed the theories of hydrostatic pressure and force known as *Pascal's Law* (which we'll examine later). An Englishman, Joseph Bramah, used the principals of Pascal's Law and the related concepts of multiplication of force to develop the first hydraulic presses as the eighteenth century drew to a close.

The Industrial Revolution began in Britain in 1850. Interest in the use of fluid power to operate machines of all types expanded rapidly during this period. The forerunners of many of today's important hydraulically operated manufacturing tools were developed during that time. This was accompanied by the development of many different types of pumps—driven by hand- or foot-operated mechanisms, horse-operated turnstiles or treadmills, water wheels, windmills, steam engines, and other such devices—to provide hydraulic power for the machines.

One of the most remarkable developments was the pressurized water main system in the city of London.[1] This system, which originated in 1848, was

Figure 1.1. Hero's aeolipile was a reaction jet steam engine designed for entertainment, not work.

Source: *Aircraft Gas Turbine Technology,* 2nd ed., 1979, by I. Treager. Reprinted by permission of Glencoe Division of the Macmillan/McGraw-Hill School Publishing Company.

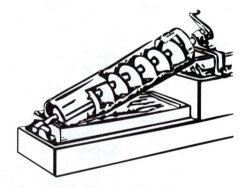

Figure 1.2. The Archimedean screw pump was invented in third century BC.

Source: From *The American Peoples Encyclopedia,* 1976 Edition, Copyright 1976, permission of Grolier Incorporated.

intended initially to supply hydraulic power to operate the dock elevators along the Thames River to ease the problems of moving cargo from the ships to the docks and then into upper floors of the warehouses. Initial progress was slow, but by 1883 there were 12 miles of pipeline laid out along the Thames supplying pressurized water at 700 psi. By 1920 this system had grown to 160 miles of piping providing hydraulic power for over 5000 freight elevators (with capacities to more than 40 tons), cranes and capstans, 1000 industrial presses, and 1500 passenger elevators. In addition, many thousands of small service elevators were in use by London merchants. The Great Stage at the Theatre Royal in London was powered from this main for many years.

It was also during the nineteenth century that Navier expanded on Bernoulli's work by developing the equations for fluid flow that included the effects of friction. In the latter part of the century, Sir George Stokes, an Irish physicist and mathematician, independently confirmed Navier's work and continued to explore related topics. Osborne Reynolds, an English physicist, presented his work on the resistance to fluid flow and his definitions of laminar and turbulent flow late in the century.

The concept of *pneumatic fluid power*—the use of compressed air to do work—also blossomed in the nineteenth century. Early compressors had a maximum pressure of about 90 psi. One of the most important early applications of pneumatics was its use in the construction of an 8½ mile tunnel through Mt. Cenis in the French Alps. The project took 14 years to complete, which was less than half the original estimate based on the use of manual rock drills instead of pneumatic drills.

While the city of London was progressing with its pressurized water mains, the city of Paris was looking to compressed air mains. In 1888, 4 miles of mains with 30 miles of branch lines were established in an old sewer system. The 90 psi air was provided by a 65 HP compressor. In three years, the demand was so great that the compressors required over 25,000 HP. As with London's water mains, the compressed air service found many applications in both commercial and private sectors. Both the hydraulic and pneumatic mains gave way to electricity early in the twentieth century.

Fluid power became more sophisticated as the early twentieth century brought new materials and concepts. Water was replaced by petroleum-based oils with their improved viscosity and lubricity and their improved low-temperature characteristics. Higher pressures were possible because of improved sealing capabilities. Better machining processes provided closer tolerances and smaller clearances to reduce internal leakage, again allowing the use of higher pressures.

Strangely, while electricity was displacing pressurized water and compressed air as power distribution utilities, hydraulics and pneumatics were displacing electricity for heavy industrial and mobile applications. In 1906 the electric gun turret drives on the battleship USS *Virginia* were replaced with hydraulic drives. The results were sufficiently impressive that virtually all subsequent ships' systems—propellor pitch control, steering systems, elevators, winches, etc—were (and still are) hydraulic.

In 1926 the first packaged hydraulic power unit—containing a reservoir, pump, control valves, actuator, etc—was developed in the United States. This led to major innovations in industrial fluid power applications, because these self-contained, relatively small units could be easily repositioned and operated independently of other plant systems (Figure 1.3).

The Second World War was the catalyst for rapid advances in both hydraulic and pneumatic fluid power. The major impetus was provided by the aircraft industry. The advances in aircraft—larger, heavier airframes, higher speeds, and higher altitudes—meant that the earlier cable-operated flight control systems were no longer adequate because the flight crew simply didn't have the physical strength to operate them. Auxiliary systems such as retractable landing gear, bomb bay doors, cargo doors, and the like required power systems for their operation. Aviation applications also prompted the development of the early servo systems—both mechanical and electrohydraulic—to provide rapid and accurate response to pilot input. These systems also improved flight stability by providing accurate and near instantaneous control corrections to counteract external disturbances from wind gusts, up- and down-drafts, and buffeting (Figure 1.4). Many devices that were originally developed for aircraft systems are presently used in the automotive industry. The most notable of these are flow limiters and automatic braking systems.

Fluid power technology continues to grow. Higher pressures, improved hardware, better fluids, improved control capabilities, and ingenious design innova-

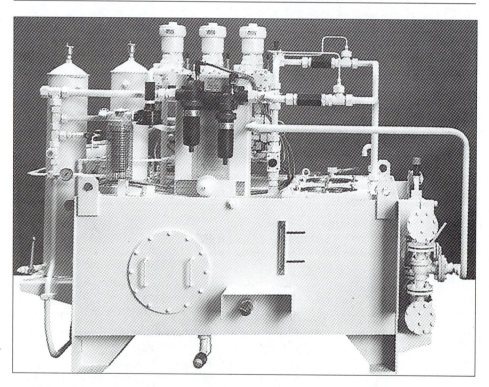

Figure 1.3. The hydraulic power pack has allowed major innovations in industrial fluid power applications.
Source: The Rexroth Corporation

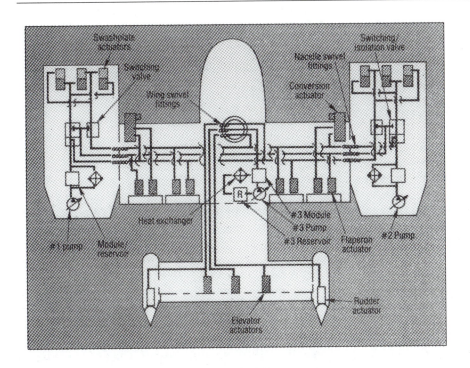

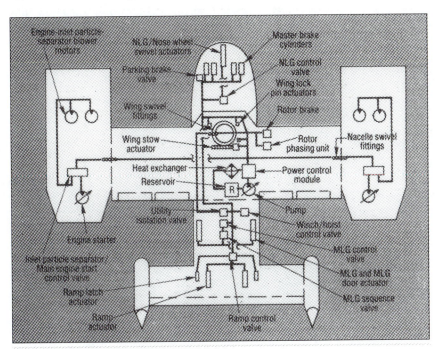

Figure 1.4. The V-22 Osprey's hydraulic systems are an example of an aviation application.

Source: *Hydraulics & Pneumatics*, December 1990.

tions, along with improved filtration and contamination control, continue to expand its applications and improve the performance and reliability of its systems and components.

1.2 Fluid Power Applications

We are often reminded of our dependence on electrical power. Just recall the inconvenience of the last brief power outage, with no light, no air conditioning (in summer), factories and businesses—or at least their computers—shut down. Few people, however, realize that fluid power plays just as important a role in our lives. Other than the brakes, automatic transmissions, and power steering on our cars, hydraulic jacks, and occasionally an elevator, few of us ever come into contact with hydraulic systems in our everyday home life. But there are very few segments of industry that do not depend to some extent on fluid power; and, of course, industry influences the daily lives of us all. Some applications are in industries where we might expect to find them—foundries, rolling mills, large cranes, bulldozers, etc. Other applications are less well known—in textiles, food processing, automated monsters at theme parks, glass manufacturing, plastic fabricating, medical examination tables, garbage trucks, oil drilling rigs—the list goes on and on. Barbers' and dentists' chairs, for example, are raised and lowered by fluid power. In fact, almost 90 percent of American industry uses fluid power in some way. Figures 1.5 through 1.9 show some typical fluid power applications.

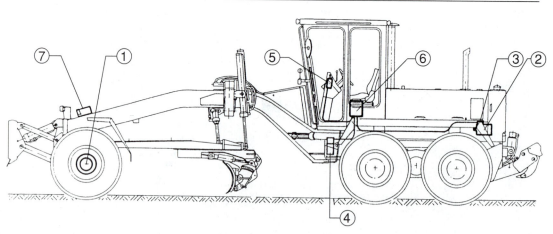

Figure 1.5. A modern road grader has several hydraulic components. ① Hydraulic wheel motors, ② variable displacement pump, ③ electronic proportional control valve, ④ solenoid valve bank, ⑤ control panel, ⑥ electronics, ⑦ flow divider.
Source: The Rexroth Corporation.

1.3 The Fluid Power Industry

The fluid power industry can be divided into six broad categories:

- Fluid manufacturers
- Component manufacturers
- Original equipment manufacturers (OEMs)
- Distributors
- Users of fluid power–driven machines
- Component repair facilities

The first three categories represent the "hard core" of the industry. It is within this group that most of the trained fluid power engineers and technologists are currently found. According to recent information from the National Fluid Power Association, this group consists of approximately 800 companies with 1200 plants employing some 340,000 people in the United States. The industry accounts for over $8 billion in sales annually.

Fluid power distributors provide the interface between those who manufacture fluid power products and those who use the products. While many people consider the distributors to be hardware salesmen, they are actually very talented and competent systems engineers and designers. They provide assistance in the design of fluid power systems as well as the modification and modernization of existing equipment. While they do sell and service components, the demands of their customers require that a large percentage of their

Figure 1.6. This large CNC lathe is hydraulically operated.
Source: The Rexroth Corporation.

time be spent in the engineering design function. Most fluid power distributors are members of the Fluid Power Distributors Association (FPDA).

The users group includes a major proportion of industry, but it actually employs a relatively small number of fluid power engineers and technologists. Rather, it employs a large number of mechanics and pipefitters to maintain its fluid power systems. These maintainers are often relatively untrained in fluid power, which poses huge problems in keeping today's sophisticated, high-pressure, often electronically controlled systems operating properly.

The last category—component repair facilities—can provide an invaluable service to the user group. In many cases, hydraulic components that have been removed from service can be repaired and put back into service for a fraction of the cost of a new component. It is often not practical, however, for users to maintain their own repair facilities, so they may choose to use the services of a *rebuild house* for overhaul. Many of these exist, offering factory-trained technicians, state-of-the-art test equipment, and shop facilities to properly service and rebuild components at a significant saving for the user.

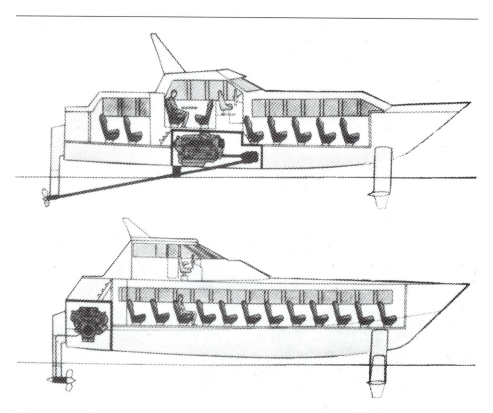

Figure 1.7. The use of a hydraulic-motor-driven propellor on this modern hydrofoil (bottom) gives a 25% increase in carrying capacity and a 9% increase in speed over the traditional configuration (top). Both boats use the same engine.

Source: The Rexroth Corporation.

1.4   **Future of Fluid Power**

Recessions have changed the face of the fluid power industry. Many small companies have closed their doors permanently because of these business cycles. Others have merged or been acquired by larger, more stable companies. Even some of the giants of the industry have been forced to reduce their staffs and eliminate marginal product lines. The progressive companies, however, have also taken advantage of the lull in activity to regroup, reevaluate their futures, and even do some serious research and development that the earlier rush to keep up with production had not allowed. The result has been a technologically stronger industry with more progressive leadership and a better vision of what will be required to compete with other modes of machine power (primarily electricity) and with international competition. For example, significant

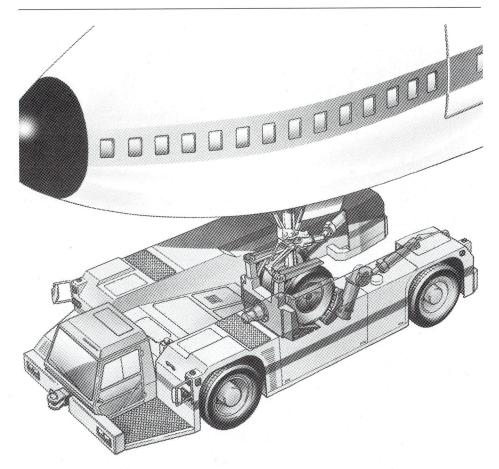

Figure 1.8. A ground mover system for large aircraft uses hydraulic cylinders to lift and clamp the aircraft nosewheel. Hydraulic motors drive the vehicle's wheels.

Source: The Rexroth Corporation.

research has been done in the marriage of microprocessor controls and fluid power. Herein lies the real future of fluid power. The precision of electronic control applied to the mechanical speed and power of hydraulics and pneumatics will provide a versatility never before available to industry.

The prospects are exciting! Imagine huge manufacturing robots able to place and hold large loads weighing hundreds of pounds with a precision measured in thousandths of an inch. Consider programming an excavator to cut a trench to an exact depth and along a path that requires numerous changes in direction and do it all automatically. Lasers have already been used to control a motor grader to ensure precision leveling of an area.

The future of fluid power belongs to those engineers and technologists who are able to look past the traditional ways of designing systems and imagine

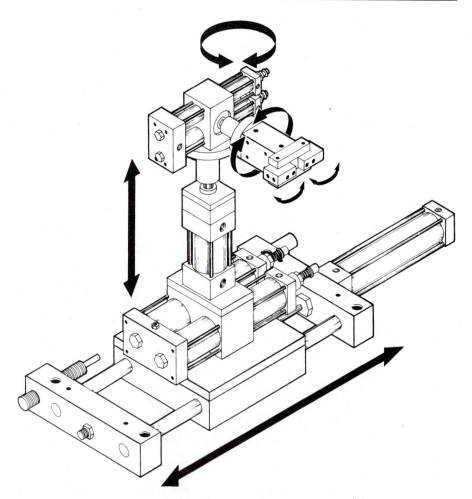

Figure 1.9. This multi-axis, multi-rotation pick-and-place mechanism uses fluid power.

Source: PHD Inc., Fort Wayne, IN.

new approaches to control, materials, reliability, energy conservation, filtration, health monitoring, maintainability, and all other such aspects. The future of fluid power depends on the ability—and inclination—of the people in the industry to stay on the leading edge of technology.

1.5 Fluid Power Standards

Throughout this text, reference will be made to the various standards applicable to specific aspects of fluid power. Standardization has made important inroads into the fluid power industry, although not to the extent that it has penetrated other fields. Mountings for pumps, motors, and cylinders, shaft sizes, seals, sampling ports, sampling procedures, particle counter calibration, and many other areas have been subjected to standards produced by one or more recognized professional organizations. Unfortunately, the development of a standard does not necessarily mean the acceptance and application of that standard. In the United States the standards are not mandatory for any portion of the industry. The reader is strongly encouraged to seek out standards and apply them wherever possible. If a standard has shortcomings, work to get them corrected. If there is no standard where one is needed, work to get it developed. Most major companies participate in standards-making groups and encourage their employees to take part in their areas of expertise.

In the United States, several professional organizations produce standards that either directly or indirectly influence the fluid power industry. The most predominant of these groups are:

- The National Fluid Power Association (NFPA)
- Society of Automotive Engineers (SAE)
- American Society for Testing and Materials (ASTM)
- American Petroleum Institute (API)

Other groups, including component manufacturers, OEMs, government agencies, and insurance companies, initiate standards in areas of specific interest to them.

Because there is some overlap in the areas of responsibility of some of these standards bodies, there are often several (sometimes conflicting) standards or procedures on the same subject. The American National Standards Institute (ANSI) acts as an arbitor to help resolve any differences and to develop a truly national standard. Representatives from the many concerned groups participate in the ANSI process. Committee B93 is the ANSI group primarily concerned with fluid power standards.

As one would expect, a similar situation exists on an international scale. The major industrial countries of the world have their own standardization bodies similar to ANSI. It is the task of the International Organization for Standardization (ISO) to develop internationally accepted standards for products other than electrical and electronic devices. This procedure is often long and laborious and usually requires considerable diplomacy, but the result is a document that allows fluid power people from different countries to "speak the same language" about a specific area of common interest.

Despite continued progress toward standardization in component sizing, mounting, description, assessment, and so forth, surprises often cause problems in the industry as a result of many companies' ignoring standards, either from ignorance or by deliberate choice.

As a fluid power technologist or engineer, you should accept, as part of your job, the responsibility to be aware of the standards that apply to the industry and encourage your company to work with those standards. Appendix A is a listing of fluid power standards.

1.6   Fluid Power Symbols

An important part of the standardization effort has been the development of a system of graphic symbols by which individual components and entire circuits can be represented. Symbols may be of three types—pictorial, cutaway, and graphic.

Pictorial symbols are drawings that represent the actual physical appearance of the components (Figure 1.10). They are very useful for illustrating the interconnections between individual components, but they are very difficult to standardize, especially concerning their functioning.

Cutaway drawings (Figure 1.11) are particularly helpful in showing the construction of a specific component. Since cutaway drawings are peculiar to a specific component design, standardization is virtually impossible. In addition, they are very complex to draw and do not readily illustrate the functioning of the component.

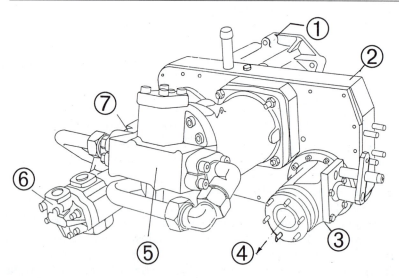

Figure 1.10. Pictorial drawings such as this representation of a hydrostatic gearbox show the actual physical appearance of the components. ① Bell housing, ② intermediate gearbox, ③ central differential, ④ to the front axle, ⑤ variable displacement motor, ⑥ pump for service hydraulics, ⑦ variable displacement pump.
Source: The Rexroth Corporation

Graphic symbols depict the function and operation of components. They do not represent any particular design or method of construction. For example, it is impossible to determine from the graphic symbol representing a hydraulic pump whether the unit in question is a gear, vane, piston, or screw pump. Graphic symbols are commonly used and have been standardized on an international basis by ISO Standard 1219 "Fluid Power Systems and Components— Graphic Symbols." Figure 1.12 is a simple circuit diagram employing ISO graphic symbols. Appropriate ISO symbols will be used throughout this text in association with component discussions. Appendix B contains the ISO graphic symbols for fluid power.

The relatively few rules concerning the use of graphic symbols may be stated as follows:

- Graphic symbols indicate only the function of the components, including connections, flow paths, actuators, etc., necessary to fully describe the function.
- Graphic symbols do not indicate component construction, system parameters (pressure, flow rate, etc), spatial relationships, or the physical layout of a circuit or component.
- The physical orientation of a graphic symbol in a drawing does not change its function nor indicate its actual orientation.
- Symbols are normally shown in their unactuated (normal) positions.
- The graphic symbol for a hydraulic reservoir is analogous to the ground

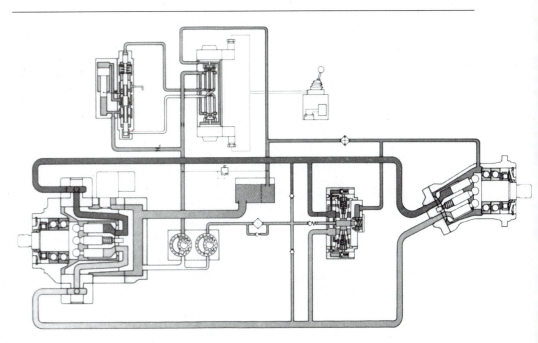

Figure 1.11. Cutaway drawings such as this representation of a closed-loop hydrostatic transmission are peculiar to a specific component design.
Source: The Rexroth Corporation

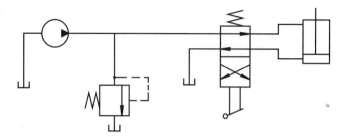

Figure 1.12. ISO graphic symbols such as those used here for a hydraulic circuit diagram depict the function and operation of components and are not peculiar to a specific design.

---

symbol in an electrical circuit diagram. Consequently, several such symbols may be used for convenience or clarity in a circuit diagram and still represent a single reservoir.

Recognizing and understanding graphic symbols, so that you can read circuit diagrams, is essential to circuit design, analysis, and troubleshooting. Therefore, one of the primary objectives of this book is to help you learn to read circuit diagrams. Once you learn the basic concepts, you will find that it is relatively simple, similar to reading written text.

1.7 Hydraulics and Pneumatics

Fluid power is divided into two categories—hydraulics and pneumatics. The distinguishing feature of these categories is the power transmission medium. *Hydraulic* fluid power uses liquid, while *pneumatic* fluid power uses a gas (most commonly air). The primary differences between these two fluids are summarized in Table 1.1.

Table 1.1 Comparison of Liquids and Gases

Parameter	Liquid	Gas
Compressibility	Incompressible (essentially)	Readily compressible
Mass	Has its own mass	Has its own mass
Volume	Has its own volume	Volume determined by container
Shape	Takes shape of container, but only to its own volume	Expands to completely fill and take the shape of the container

Note that we consider liquids to be incompressible, but that's not really the case. Hydraulic fluids are compressible to about 0.4 percent per 1000 psi. For most applications, this is insignificant. It becomes significant when dealing with high-frequency cycling or load applications such as are experienced in some electrohydraulic servovalve applications. In this text, we'll ignore liquid compressibility unless specifically stated otherwise.

We can't get by with ignoring gas compressibility, though, so our approach will be to deal with the incompressible liquid (hydraulics) in the first part of the text. Then we'll devote separate chapters to the compressible gases (pneumatics). Most of the hydraulic concepts will relate directly to pneumatics, so the later chapters will emphasize only those areas that are significantly different.

Reference

1. "U.K. Diary: Editor's Odessy," *Elevator World,* 30, no. 30 (October 1982): 56.

Review Questions

1. List at least ten areas in which fluid power influences your life.
2. Define the following abbreviations:

    a. NFPA
    b. SAE
    c. ANSI
    d. ASTM
    e. ISO
    f. FPDA

3. What is meant by the terms *hydraulics* and *pneumatics?* What is the primary difference between the two?
4. Are the liquids used for hydraulic fluids really incompressible? Explain your answer.
5. Where is the compressibility of a hydraulic fluid an important consideration?

Chapter 2

Basic Hydraulic Concepts

Outline

2.1 Introduction

In this chapter, we'll explore many of the basic concepts that govern and define the operation and performance of hydraulic systems. Much of the information in this chapter is common to both hydraulic and pneumatic systems, and we'll cover their application to pneumatics in later chapters.

Objectives

When you have completed this chapter, you will be able to

- Explain the difference between hydrostatics and hydrodynamics and how each provides power.
- State Pascal's Law and explain its significance.
- Explain the advantages and disadvantages of hydraulic fluid power.
- Apply the equation expressing the pressure-force-area relationship.
- Solve problems concerning the multiplication of force.
- Calculate work, power, and horsepower.
- Explain the basic concept of a fluid power system.
- Calculate input, hydraulic, and output horsepower.
- Apply the equation expressing the flow-velocity-area relationship and the continuity equation.
- Trace fluid power flow through a hydraulic fluid power system.

17

2.2   Hydrodynamics versus Hydrostatics

Fluids can be used to transmit useful energy in two ways—hydrodynamically and hydrostatically. Webster's *New World Dictionary of the American Language* defines *hydrodynamic* as "derived from, or operated by, the action of water, etc., in motion." In a hydrodynamic system, the kinetic energy or the impact of the moving fluid is converted to mechanical energy (usually rotational) and is thus used to produce work. Water wheels (Figure 2.1) and water turbines (Figure 2.2) such as those used in hydroelectric generators are examples of hydrodynamic applications.

Webster defines *hydrostatics* as "having to do with the pressure and equilibrium of water and other liquids." In a hydrostatic system, energy is transferred through a confined fluid by the pressure that is created by the application of a force to that confined fluid. Thus, to compare, in a hydrodynamic system, energy is transmitted by the motion of the fluid itself, while in a hydrostatic system, energy is transmitted through the confined fluid by pressure.

Most hydraulic systems are considered to be hydrostatic, because the pressure energy in the fluid is the primary source of force. Fluid motion is required, but not as a force producer.

2.3   Pascal's Law

In the midseventeenth century, the noted French scientist and mathematician, Blaise Pascal, demonstrated that in a confined liquid at rest, the pressure at

Figure 2.1 A water wheel is an example of a hydrodynamic system.
Source: War Eagle Mill, Rogers AR

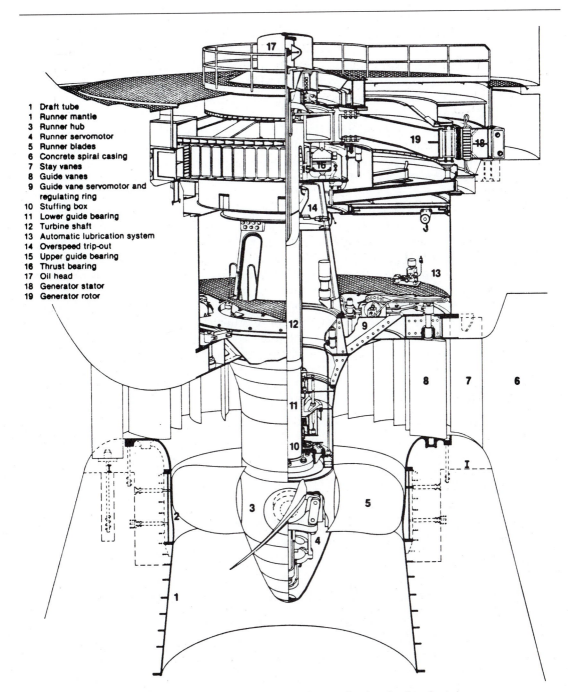

1   Draft tube
1   Runner mantle
3   Runner hub
4   Runner servomotor
5   Runner blades
6   Concrete spiral casing
7   Stay vanes
8   Guide vanes
9   Guide vane servomotor and
    regulating ring
10 Stuffing box
11 Lower guide bearing
12 Turbine shaft
13 Automatic lubrication system
14 Overspeed trip-out
15 Upper guide bearing
16 Thrust bearing
17 Oil head
18 Generator stator
19 Generator rotor

Figure 2.2 Cross section of a Kaplan Turbine for producing hydroelectric power.
Source: J. George Wills, *Lubrication Fundamentals* © 1980 Mobil Oil Corporation (New York: Marcel Dekker).

any point in the system was the same as at any other point. This discovery, which has become known as *Pascal's Law*, can be stated as follows:

> Pressure exerted on a confined liquid at rest is transmitted equally in all directions, is the same at any point in the liquid, and acts at right angles to the surfaces of the container.

It is also important to note that any change in the exerted pressure is seen almost instantly throughout the liquid. Figure 2.3 illustrates Pascal's Law.

When considered in light of some of the basic properties of fluids, the significance of Pascal's Law can be better understood. For instance, we know that a fluid will conform to the shape of the vessel in which it is confined (Figure 2.4). This is an important concept in fluid power because the "vessel" need not conform to our usual concepts of a bottle or tank, but, in fact, can include an entire network of tanks, cylinders, and so forth, connected by any number of fluid conduits (Figure 2.5). In this case, just as in the case of a simple container, Pascal's Law applies, so the pressure exerted at the plunger exists throughout the entire system and is available as a force at any point in the system. The shape of the vessel does not diminish the validity of Pascal's Law.

If we carry the fluid power concept another step and consider that liquids are virtually incompressible at the more commonly used pressures (2000 psi or less), we find still another important aspect of fluid power. Because of this

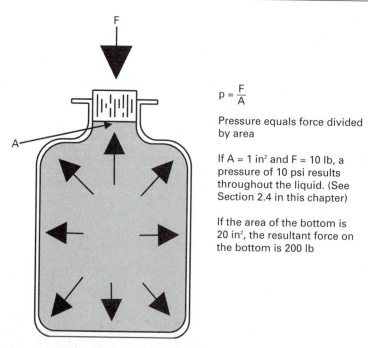

$$p = \frac{F}{A}$$

Pressure equals force divided by area

If A = 1 in² and F = 10 lb, a pressure of 10 psi results throughout the liquid. (See Section 2.4 in this chapter)

If the area of the bottom is 20 in², the resultant force on the bottom is 200 lb

Figure 2.3 Illustration of Pascal's Law using a bottle filled with liquid. F = force, and A = area. Pressure (p) is indicated by the arrows.

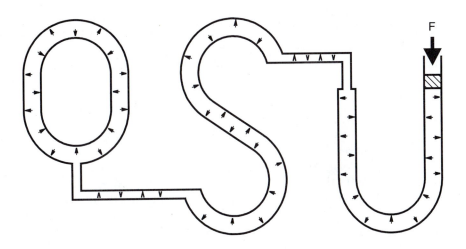

Figure 2.4 Pascal's Law applies regardless of the shape of the container.

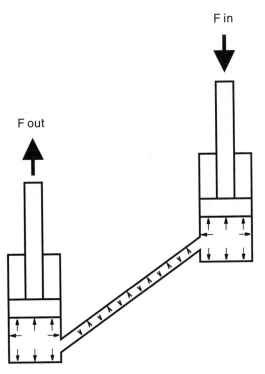

Figure 2.5 Pascal's Law applies as well to a network of vessels connected by fluid conduits.

property, the fluid in a system pipeline forms a column for transmitting power that is essentially as rigid as a steel rod. It has the advantage over steel, however, in that its rigidity is undiminished by bends, loops, and other changes in direction. Even the use of flexible tubes and hoses does not reduce its capabilities as long as the burst pressure of the conduit is not exceeded.

There are several important advantages of hydraulic power systems:

1. They provide high levels of readily regulated torque and force.
2. They offer infinitely variable linear or rotary speed over a wide range.
3. They are instantly reversible, eliminating the need to come to a gradual stop.
4. They can be stalled without damage and without the necessity to restart the prime mover when the stall-producing load is removed.
5. High power output is possible from relatively small, lightweight packages.
6. High accuracy and extreme stiffness facilitates the positioning and holding of heavy loads.
7. They are readily automated without electronics.
8. They are fully adaptable to electrical or electronic controls, including programmable logic controllers.
9. They can provide cushioning to reduce the mechanical effects of impact or shock loads.
10. The fluid itself provides lubrication.

Along with these advantages, one must also consider certain disadvantages of hydraulic power systems:

1. The hazards associated with any high-pressure system exist.
2. A fire hazard exists for hydraulic fluids; almost all of them (including the so-called fire-resistant fluids) are flammable to some extent.
3. Hydraulic fluid leakage is possible.
4. Disposal of used fluids and the cleanup of large spills must be conducted carefully.
5. Accelerated component wear is caused by particulate contamination of the fluid, especially on high-pressure systems, unless adequate filtration is provided.

From these lists, it is evident that hydraulic power may not be ideal for every application. In fact, for many applications, hydraulic power makes no sense at all. The alternative power systems—mechanical, electrical, and pneumatic—perform these tasks far better than hydraulics. On the other hand, there are many applications for which hydraulic fluid power provides the only practical answer. These include conditions requiring high forces for heavy loads, long-stroke linear motion, rigidity and accuracy with heavy loads, high-stall torque, and so on.

2.4 Pressure-Force-Area Relationship

Pressure is the result of resistance to fluid flow. In the most simplistic sense, pressure in a fluid power system is the result of a force being applied to confined

Table 2.1 Commonly Used Units

	U.S. Customary	**SI Metric**
Force (F)	Pounds force (lb$_f$)	Newtons (N)
Distance (d)	Inches (in) or feet (ft)	Meters (m)
Area (A)	Square inches (in)^2	Square meters (m)^2
Pressure (p)	Pounds per square inch (lb/in^2 or psi)	Newtons per square meter (N/m^2) or pascals (Pa) (Note: Bar is used in some European countries. A bar is 14.5 psi or 100 kPa.)
Work (W)	Foot-pounds force (ft·lb)	Newton-meters (N·m) or joules (J)
Power (P)	Foot-pounds/second (ft·lb/sec)	Newton-meters/second = watt (N·m/s = W = J/s)
Volume Flow Rate (Q)	Gallon/minute (gpm)	Liter/minute (Lpm)

Note: The abbreviation for second is "sec" in U.S. Customary and "s" in SI metric.

fluid. The resulting pressure is directly proportional to the area over which the force is applied and is defined mathematically as

$$\text{Pressure} = \frac{\text{Force}}{\text{Area}}$$

or   **(2.1)**

$$p = \frac{F}{A}$$

It is important here, as with all calculations, to be consistent with our units. Table 2.1 shows the units that are commonly used for pressure, force, and area in both the U.S. Customary and metric systems. Note that in the U.S. Customary System, pressure is in *pounds per square inch* (lb/in^2 or psi). In System International (SI) units, *newtons per square meter* (N/m^2) is used. This has been designated the *pascal* in honor of Blaise Pascal; thus, we most often see pressure in the SI system expressed as Pa or kPa.

The ability to convert between the U.S. Customary and SI systems is very important, since both systems are in use. The more important conversion factors are listed inside the covers of this text. The conversion from psi to Pa is made as follows using the conversions

$$1 \text{ lb} = 4.448 \text{ N}$$

$$1 \text{ m} = 39.37 \text{ in}$$

$$1 \text{ m}^2 = 1550 \text{ in}^2$$

$$\frac{1 \text{ lb}}{\text{in}^2} = \frac{1 \text{ lb} \times 4.448 \text{ N/lb}}{\text{in}^2 \times \text{m}^2/1550 \text{ in}^2} = 6895 \text{ N/m}^2$$

Since a N/m² is a pascal, 1 psi is 6895 Pa. Obviously, a pascal is a very small unit of pressure, so the numbers involved in considering high pressures become very large. Therefore, it is more common to see pressure expressed in thousands of pascals, termed *kilopascals* (kPa) rather than pascals. For example, 1 psi would be expressed as 6.895 kPa.

It is important to note here that there are those in Europe who prefer to use the *bar* as the unit for pressure. The bar is approximately one atmosphere and is defined as 14.5 psi, or approximately 10^5 Pa (100 kPa).

Example 2.1

Fluid is confined in a vessel as shown in the illustration for this example. A force of 100 lb is applied to the plunger. Find the resultant pressure.

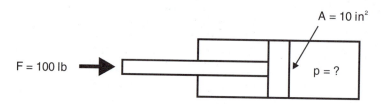

Solution

$$p = \frac{F}{A} = \frac{100 \text{ lb}}{10 \text{ in}^2} = 10 \text{ psi}$$

Note: The units are an important part of the answer.

Example 2.2

In the hydraulic cylinder shown here, the pressure is 250 psi. If the diameter of the piston is 2 in, find the resultant force.

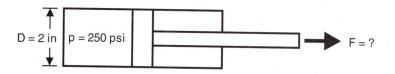

Solution

$$p = \frac{F}{A}$$

$$F = p \times A$$

The piston area is

$$A = \frac{\pi D^2}{4} = \frac{\pi \times (2 \text{ in})^2}{4} = 3.14 \text{ in}^2$$

Therefore, the force is

$$F = 250 \text{ lb/in}^2 \times 3.14 \text{ in}^2 = 785 \text{ lb}$$

This example assumes that there is actually a 785 lb resistance opposing the cylinder motion. The cylinder pressure will never be greater than the pressure required to overcome the resistance.

Example 2.3

How much pressure would be required to raise a 1000 kg load at a constant velocity if the cylinder has a diameter of 10 cm as shown?

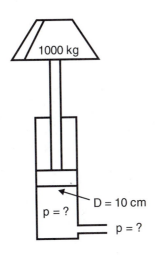

Solution

In this problem, as in the remainder of this text, we will ignore static friction and inertia. We will also assume steady (constant speed) motion so that the forces on both sides of the piston are equal. Thus, we again use the equation for pressure, as in Example 2.1:

$$p = \frac{F}{A}$$

$$F = 1000 \text{ kg} \times 9.81 \text{ N/kg} = 9.81 \text{ kN}$$

$$A = \frac{\pi D^2}{4} = \frac{\pi \times (10 \text{ cm})^2}{4} = (78.54 \text{ cm}^2)\left(\frac{\text{m}^2}{10{,}000 \text{ cm}^2}\right) = 0.00785 \text{ m}^2$$

$$p = 9810 \text{ kN}/0.00785 \text{m}^2 = 1.25 \text{ MPa}$$

2.5   Multiplication of Force

One of the most attractive features of fluid power concerns the concept of multiplication of force. Simply stated, this means that a small input force can be multiplied to produce a large output force by using different sized input and output pistons, as demonstrated in Figure 2.6. Let us trace the forces and pressures through the system shown here.

First consider the force exerted on piston A. Since that force works on an area of 1 in², we can calculate the pressure generated in cylinder A using Equation 2.1. Thus,

$$p = \frac{F}{A} = 100 \text{ lb/1 in}^2 = 100 \text{ psi}$$

Now, what is the pressure generated in cylinder B? Since we have a confined fluid, Pascal's Law applies and tells us that the pressure throughout the vessel is the same. That is, the pressure in cylinder B (as well as the pressure in the connecting pipe) is the same as that in cylinder A, or

$$p_B = p_A = 100 \text{ psi}$$

Knowing p_B and having A_B given, we can solve for F_B as in the earlier examples. Thus,

$$F_B = p_B \times A_B = 100 \text{ lb/in}^2 \times 10 \text{ in}^2 = 1000 \text{ lb}$$

From this example, we can derive a mathematical relationship between the input force (F_A) and the output force (F_B). Pascal's Law makes this derivation a straightforward algebraic exercise. From Pascal's Law,

$$p_B = p_A$$

From Equation 2.1,

$$p_B = \frac{F_B}{A_B}$$

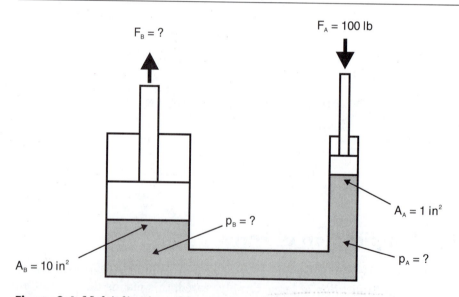

Figure 2.6 Multiplication of force means that a small input force can be multiplied to produce a large output force. F-force, A-area, and p-pressure. The subscripts refer to piston A (right) and piston B (left).

and

$$p_A = \frac{F_A}{A_A}$$

so that

$$\frac{F_B}{A_B} = \frac{F_A}{A_A}$$

Solving for F_B, we see that

$$F_B = F_A \times \frac{A_B}{A_A} \qquad (2.2)$$

So we see that the output force is equal to the input force multiplied by the ratio of the areas of the output piston to the input piston. Observe also that

$$A = \frac{\pi D^2}{4}$$

so that

$$\frac{A_B}{A_A} = \frac{\dfrac{\pi D_B^2}{4}}{\dfrac{\pi D_A^2}{4}} = \frac{D_B^2}{D_A^2}$$

Thus,

$$F_B = F_A \times \frac{D_B^2}{D_A^2} \qquad (2.3)$$

We see, then, that multiplication of force simply means that by use of a large output area, a small input force can be increased to a larger output force. Note that the word "force" is key here, because only force can be increased. As will see later, a fluid power system cannot increase energy or power.

Example 2.4

A mechanic using a hand-operated hydraulic jack applies a force of 85 lb to a 1 in² input piston. If the output piston has an area of 4 in², how much weight can he lift?

Solution

From Equation 2.2,

$$F_{OUT} = F_{IN} \times \frac{A_{OUT}}{A_{IN}} = 85 \text{ lb} \times \frac{4 \text{ in}^2}{1 \text{ in}^2} = 340 \text{ lb}$$

Example 2.5

In the system shown, how much force must be applied to piston A to raise the load on piston B?

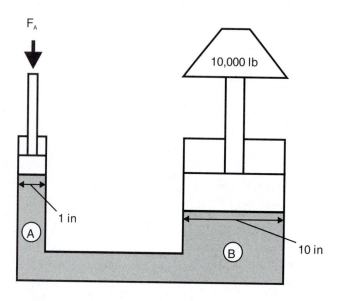

Solution

From Equation 2.3,

$$F_B = F_A \times \frac{D_B^2}{D_A^2}$$

Solving for F_A gives

$$F_A = F_B \times \frac{D_A^2}{D_B^2} = 10{,}000 \text{ lb} \times \frac{(1 \text{ in})^2}{(10 \text{ in})^2}$$

$$= 10{,}000 \text{ lb} \times \frac{1 \text{ in}^2}{100 \text{ in}^2} = 100 \text{ lb}$$

Example 2.6

A force of 20 N is exerted on a 5 cm² input piston that is connected to a 50 cm² output piston. What is the output force capability of the system?

Solution

Again using Equation 2.2,

$$F_{OUT} = F_{IN} \times \frac{A_{OUT}}{A_{In}} = 20 \text{ N} \times \frac{50 \text{ cm}^2}{5 \text{ cm}^2} = 200 \text{ N}$$

2.6   Work

Work is defined as a force acting over a distance. Mathematically, it is expressed as

$$\text{Work} = \text{Force} \times \text{Distance}$$

or

$$W = F \times d \qquad (2.4)$$

In the U.S. Customary system of units, where force is expressed in pounds and distance is in feet, we see that the units of work are foot-pounds (ft·lb). In the metric system, force is in newtons and distance is in meters, so work is expressed in newton-meters (N·m) or joules (J).

Example 2.7

How much work is required to raise a 2000 lb load by 4 feet?

Solution

$$W = F \times d = 2000 \text{ lb} \times 4 \text{ ft} = 8000 \text{ ft·lb}$$

In simple fluid power systems, if we ignore such things as friction losses, we can generalize with reasonable accuracy that the work input and the work output are equal. Mathematically, we can write

$$W_{\text{IN}} = W_{\text{OUT}}$$

From our definition of work (Equation 2.4), we can rewrite this as

$$(F \times d)_{\text{IN}} = (F \times d)_{\text{OUT}} \qquad (2.5)$$

Letting subscript A denote "in" and B denote "out," we have

$$F_A \times d_A = F_B \times d_B \qquad (2.5a)$$

This equation gives us another method for determining the output force based on a known input force. In this case, rather than being related by piston diameters, the relationship is based on the relative movements of the pistons.

Example 2.8

In the system shown here, an input force of 100 lb is applied to piston A and pushed downward 5 in. How far will piston B move?

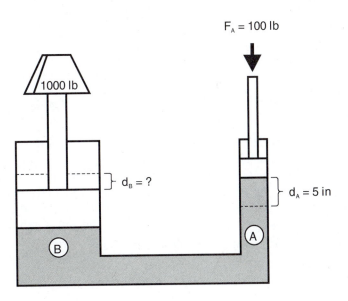

Solution

From Equation 2.5

$$F_A \times d_A = F_B \times d_B$$

so that

$$d_B = \frac{F_A \times d_A}{F_B} = \frac{100 \text{ lb} \times 5 \text{ in}}{1000 \text{ lb}} = 0.5 \text{ in}$$

Example 2.9

How much force is exerted on piston A in this illustration?

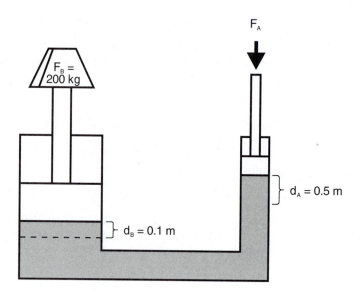

Solution

From Equation 2.5,

$$F_A \times d_A = F_B \times d_B$$

so that

$$F_B = 200 \text{ kg} \times 9.81 \text{ N/kg} = 1.96 \text{ kN}$$

$$F_A = \frac{F_B \times d_B}{d_A} = \frac{1.96 \text{ kN} \times 0.1 \text{ m}}{0.5 \text{ m}} = 0.392 \text{ kN}$$

Actually, we could eliminate the somewhat shaky assumption of "work in" equals "work out" by considering the volumetric relationships of the two cylinders. Once we have sufficient pressure to move the output cylinder, the distance we move it depends strictly on the amount of fluid transferred.

Referring to the figure for Example 2.8, assume that no fluid leaks or gets lost along the way, and that all of the fluid from the input cylinder (A) goes into the output cylinder (B). Denoting the volume of fluid leaving cylinder A as Vol_A, and the volume of fluid entering cylinder B as Vol_B, we can write

$$Vol_A = Vol_B$$

But

$$Vol_A = A_A \times d_A$$

and

$$Vol_B = A_B \times d_B$$

where d_A and d_B are the distance moved by piston A and piston B, respectively. Therefore,

$$A_A \times d_A = A_B \times d_B \qquad (2.6)$$

From this, we can see that the ratio of the distances moved is proportional to the ratio of the piston areas and is independent of the forces on the pistons.

Example 2.10

The area of cylinder A in this figure is 2 in². The area of cylinder B is 7 in². If the piston of cylinder A is pushed down 12 in, how far will the piston of cylinder B be pushed upward?

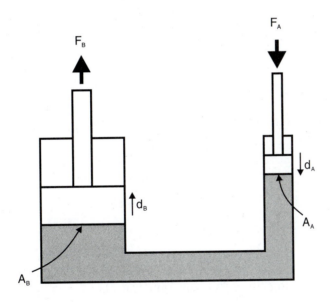

Solution

From Equation 2.6,

$$A_A \times d_A = A_B \times d_B$$

Therefore,

$$d_B = d_A \times \frac{A_A}{A_B}$$

$$= 12 \text{ in} \times \frac{2 \text{ in}^2}{7 \text{ in}^2}$$

$$d_B = 3.43 \text{ in}$$

2.7 Power and Horsepower

The terms *power* and *horsepower* are frequently misunderstood and misused. *Power* refers to the rate at which work is done, that is, how fast work is accomplished. Consequently, it can be expressed as

$$\text{Power} = \frac{\text{Work}}{\text{Unit Time}}$$

Incorporating our definition of work (Equation 2.4), we can express this as

$$p = \frac{W}{t} = \frac{F \times d}{t} \qquad\qquad (2.7)$$

Dimensionally, we can see that in the U.S. Customary system, power is given in ft·lb/sec. In the SI system, we use N·m/s or J/s which has been defined as a watt (W). (Note that "second" is abbreviated "sec" in U.S. Customary and "s" in SI metric.)

Example 2.11

How much power is required to move a 2000 lb load 4 feet vertically in 10 seconds at a constant velocity?

Solution

From Equation 2.7,

$$p = \frac{F \times d}{t} = \frac{2000 \text{ lb} \times 4 \text{ ft}}{10 \text{ sec}} = 800 \frac{\text{ft·lb}}{\text{sec}}$$

In Equation 2.7, we have distance divided by time. Since distance divided by time is velocity, we can also express power as

$$\text{Power} = \text{Force} \times \text{Velocity}$$

$$p = F \times v \qquad\qquad (2.8)$$

Example 2.12

A 1000 kg load is being lifted at a constant velocity of 0.75 m/s. How much power is required?

Solution

Using Equation 2.8, we get

$$F = 1000 \text{ kg} \times 9.81 \text{ N/kg} = 9.81 \text{ kN}$$

$$P = F \times v = 9.81 \text{ kN} \times 0.75 \text{ m/s}$$

$$= 7.36 \text{ kN·m/s} = 7.36 \text{ kW}$$

In these examples, we have used the term *load* to mean "resistance". If the load were being moved horizontally instead of vertically, we could still use *load* to mean resistance to movement, and the resulting answer would be the same.

In contrast to this, note that the terms *load* and *weight* can be used interchangeably when speaking of vertical motion to which there is no resistance other than gravity; however, if a *weight* is being moved horizontally, the actual *load* will be a function of the friction between the weight and the surface over which it is moving. If we also consider starting and stopping the load, then acceleration forces must be included in the total load.

While power is a valid measure of the relationship between work and time, it is more common in the fluid power discipline to discuss a pump or motor's capabilities in terms of *horsepower*. This term was originated by James Watt in the latter part of the nineteenth century when he encountered difficulty in comparing the capabilities of his steam engines with the work done by the horses used to pull wagons in mines. It was logical for the mine owners to be interested in the number of horses that could be replaced by a single steam engine. Watt, therefore, devised a test to determine how a horse would perform when lifting weights using a block-and-tackle type of arrangement. His testing showed that the horse used could raise a 150 lb weight at an average rate of 3.67 ft/sec or 550 ft·lb/sec (33,000 ft·lb/min), which he defined as 1 horsepower (1 HP). Although it is generally agreed that Watt's horse was perhaps a little stronger than the average horse, this definition became the standard for comparing the output of machinery.

To convert the power of Equation 2.7 to the more convenient horsepower, it is necessary to divide the power by the horsepower constant. Thus,

$$HP = \frac{P}{550} = \frac{W}{550\,t} = \frac{F \times d}{550\,t} \tag{2.9}$$

Dimensionally, we have

$$\frac{F(\text{lb}) \times d(\text{ft})}{\left(550\,\dfrac{\text{ft·lb}}{\text{sec}}/\text{HP}\right)\text{sec}}$$

leaving horsepower as the only dimension.

Example 2.13

Express 800 ft·lb/sec in terms of horsepower.

Solution

$$HP = \frac{P}{550} = \frac{800\,\dfrac{\text{ft·lb}}{\text{sec}}}{550\,\dfrac{\text{ft·lb}}{\text{sec}}/\text{HP}} = 1.45\ \text{HP}$$

It is essential to note that the use of the horsepower constant (550 ft·lb/sec) forces us to be extremely careful concerning the units used to express power. For example, it would be completely valid (although somewhat unusual) to express power in mile-tons per year; however, it would be necessary to convert those strange units to ft·lb/sec in order to find the equivalent horsepower.

Example 2.14

Express a power of 2,389 mile-tons per year (mi·ton/yr) as horsepower.

Solution

$$HP = \frac{\left(2389 \frac{mi \cdot ton}{yr}\right)\left(\frac{5280 \text{ ft}}{mi}\right)\left(\frac{2000 \text{ lb}}{ton}\right)\left(\frac{yr}{365 \text{ day}}\right)\left(\frac{day}{24 \text{ hr}}\right)\left(\frac{hr}{3600 \text{ sec}}\right)}{\frac{550 \text{ ft} \cdot lb}{sec}/HP}$$

$$= 1.45 \text{ HP}$$

Notice that all units except horsepower cancel top and bottom. Therefore, our answer is 1.45 HP.

Horsepower can also be related to the international system of units (SI). It is found the same way as when U.S. Customary units are used. The derivation of the SI horsepower constant will be left as an exercise for the reader. Please note, though, that it is more common to use watts or kilowatts to express power in SI units.

In fluid power systems, we are concerned with power in several different areas:

1. Input to pumps.
2. Output from pumps.
3. Input to hydraulic motors.
4. Output from hydraulic motors.
5. Input to hydraulic cylinders.
6. Output from hydraulic cylinders.
7. Losses across valves.
8. Losses resulting from the fluid flowing through the system.

Before we can thoroughly understand the significance of these different powers, we need to define what is meant by a fluid power system.

2.8 Fluid Power Systems

A *fluid power system* is simply a mechanism for converting mechanical energy into kinetic and pressure energy in a fluid, transmitting that energy through a conduit, and then reconverting it to mechanical energy in order to do work. The system consists basically of three segments—the power-input segment, the control segment, and the power-output segment (Figure 2.7).

The *power-input segment* consists of a prime mover and a pump. The *prime mover* is normally either an internal combustion engine (usually on mobile equipment) or an electric motor (usually on stationary equipment). Obviously, both of these sources produce rotational energy and are coupled to the rotating shaft of a hydraulic pump (see Chapter 4). The pump converts the rotational energy into kinetic energy (due to flow) and pressure energy (due to the resistance to flow) in the system fluid.

The *control segment* is the "inner part" of the hydraulic system and consists primarily of a series of valves that control (or are operated by) the system pressure, the flow rate in the system, and the direction of flow to and from components, actuators, the reservoir, and so forth. The pump may be included in this control group if it is a variable-displacement unit. Likewise , a hydraulic motor or cylinder can be included if it incorporates a device (an orifice or variable-displacement capability) that allows it to exert some type of influence over the fluid energy.

The *power-output segment* is the whole reason for the existence of the system, since, after all, it is designed to do work. The task of this segment is to convert the kinetic and pressure energy of the fluid into mechanical energy. The mechanical energy may be rotational, as in the case of a hydraulic motor or rotary actuator, or linear if the device is a hydraulic cylinder.

Now we begin to see the significance of horsepower to the total system.

2.9   Types of Power in the Fluid Power System

Keeping in mind the segments of a complete system, let us follow a "chunk" of horsepower through each of them. This very brief look will be followed by a more detailed treatment in the chapters concerning components that affect horsepower.

Input Power

Input horsepower is that which is transmitted from the prime mover to the hydraulic pump. It is alternately termed *brake horsepower* or *power input*. It is a function of the torque and the rotational speed of the input device.

Torque is defined as any force or combination of forces that produces a rotation or twisting motion. A familiar example of torque for most of us is the twisting motion exerted on a nut by a wrench (Figure 2.8). Note that it is not necessary to produce any motion in order to produce torque. In our example of the wrench, even if the nut won't budge, you are still exerting a torque on it with the wrench. The amount of torque produced is a function of the force

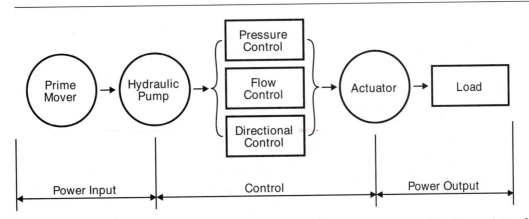

Figure 2.7 This block diagram of a basic fluid power system shows that it consists of three segments: power input, control, and power output.

applied to the lever (the wrench) multiplied by the distance from the center of rotation (the nut) to the point of the application of the force. Thus,

$$Torque = Force \times Wrench\ Length$$

or, more generally,

$$Torque = Force \times Distance$$

or

$$T = F \times d \tag{2.10}$$

Dimensionally, we see that in the U.S. Customary system we have a force in pounds multiplied by a distance so that the units of torque are lb·in or lb·ft. In SI units, torque is in newton-meters (N·m). Notice that these are the same units as used for work, although torque and work are not the same thing.

In physical systems, rotational speed (N) is correctly measured in terms of *radians per unit time*. (There are 2π radians in 360 degrees, or one complete revolution.) In the case of engines and motors, however, tradition has us use *revolutions per minute* (rpm). Thus, for calculating the input horsepower (*IHP*) to a hydraulic pump, we multiply rpm by 2π radians. This way we finally arrive at an expression for calculating the input horsepower:

$$IHP = \frac{T(\text{lb·in}) \times 2\pi N(\text{rev/min})\ (\text{ft/12 in})}{(60\ \text{sec/min}) \times (550\ \text{ft·lb/sec/HP})}$$

or   (2.11)

$$IHP = \frac{T \times N}{63,025}$$

A word of caution is needed concerning this equation. Because of the constant (63,025) in the denominator, you have no choice but to use torque in pound-

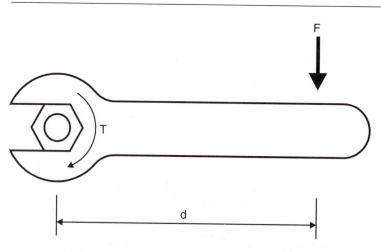

Figure 2.8 Torque (T) equals force (F) times distance (d), as when a wrench is applied to a nut.

inches and rotational speed in rpm. If torque is in pound-feet, the constant is 5252. Any other units will give the wrong answer, so they must be converted to fit the equation.

The equivalent calculation in SI units would be in terms of power (P_{IN}) instead of horsepower, with torque expressed in N·m and speed in rpm. This would give

$$P_{IN} = \frac{T \times N}{9550} \tag{2.12}$$

yielding kN·m/s, or kW, as the units for input power.

Hydraulic Power

The power output from a pump is commonly termed *hydraulic horsepower (HHP)*. This is the horsepower put into the fluid and is a function of the fluid pressure and flow rate. The standard equation for calculating hydraulic horsepower is

$$\text{Hydraulic Horsepower} = \frac{\text{Pressure (psi)} \times \text{Flow Rate (gpm)}}{1714}$$

or
$$\tag{2.13}$$

$$HHP = \frac{p \times Q}{1714}$$

where

$$p = \text{pressure in psi.}$$
$$Q = \text{flow rate in gpm.}$$

The constant (1714) incorporates the conversions from psi and gpm to ft·lb/s so that the horsepower constant (550) applies. Again you are cautioned to be careful about units. Only if pressure is in psi and flow rate is in gpm can they be substituted directly into this equation; otherwise, conversion will be necessary.

To express the power possessed by the fluid in watts (or kilowatts), we use Equation 2.14.

$$P_{HYD} = \frac{p \times Q}{60,000} \tag{2.14}$$

where p is in kilopascals (kPa) and Q is in liters per minute (Lpm) giving P_{HYD} in kilowatts (kW).

If, and *only* if, we assume that we have a pump that is 100 percent efficient—meaning that there are no losses due to mechanical friction, fluid leakage, and so forth—then the input horsepower and the hydraulic horsepower are equal. (As we'll see later, pumps are never 100 percent efficient.) Thus, we can write

$$IHP = HHP$$

or
$$\tag{2.15}$$

$$\frac{T \times N}{63,025} = \frac{p \times Q}{1714}$$

Example 2.15

A hydraulic pump operating at 2200 rpm requires an input of 10 HP. How much torque is required to turn the pump?

Solution

Using Equation 2.11, we write

$$IHP = \frac{T \times N}{63,025}$$

Solving for T gives

$$T = 63,025 \times IHP/N$$

$$= 63,025 \times 10 \text{ HP}/2200 \text{ rpm} = 286.5 \text{ lb·in}$$

Because of the derivation of Equation 2.11, torque will be in lb·in if the proper units are used in solving the equation.

Example 2.16

A hydraulic pump produces a flow rate of 30 gpm at 3000 psi. What is the horsepower output of the pump?

Solution

From Equation 2.13, we know that

$$HHP = \frac{p \times Q}{1714}$$

Substituting from the problem statement gives

$$HHP = \frac{3000 \text{ psi} \times 30 \text{ gpm}}{1714} = 52.5 \text{ HP}$$

Example 2.17

A hydraulic pump is delivering 25 Lpm at 13,000 kPa. How much hydraulic power is available from the fluid?

Solution

We use Equation 2.14 to get

$$P_{\text{HYD}} = \frac{p \times Q}{60,000}$$

$$= \frac{13,000 \text{ kPa} \times 25 \text{ Lpm}}{60,000}$$

$$= 5.42 \text{ kW}$$

Example 2.18

A 10 HP electric motor is used to drive a hydraulic pump. If the pump produces a flow rate of 15 gpm, what is the maximum pressure at which the pump can operate? Assume no losses.

Solution

Assuming no losses, Equation 2.15 applies. Therefore, we have

$$IHP = HHP = \frac{p \times Q}{1714}$$

Solving for pressure gives

$$p = \frac{1714 \times IHP}{Q} = \frac{1714 \times 10\text{ HP}}{15\text{ gpm}} = 1143\text{ psi}$$

This is the *maximum* pressure we can have based on the given parameters. The actual pressure will be determined by system parameters such as the load and other resistances. Likewise, the actual horsepower output of the electric motor will be determined by the pressure. The pressure will not exceed the system requirements, and the motor horsepower will not exceed that actually required to turn the pump.

As we will see in Chapter 4, there is no such thing as a 100 percent efficient pump. It is assumed here only to illustrate the input-output relationship. In reality, the prime mover must have a higher horsepower output than the pump.

Hydraulic Power and Actuators

The *actuators* (hydraulic motors and cylinders) in a hydraulic system utilize the pump output horsepower as their input horsepower. In all systems, there are losses due to pipe friction, resistances in valves, and so forth, which can be significant. For now, we will ignore these losses—although in the real world, they simply cannot be ignored—and will deal with some of them in Chapter 9.

We can start to see a method for system analysis coming together. Consider, for example, a hydraulic motor as our actuator. The output from a hydraulic motor is calculated in the same way as the input to a hydraulic pump:

$$\text{Output Horsepower} = \frac{\text{Torque (in·lb)} \times \text{Speed (rpm)}}{63,025}$$

or   **(2.16)**

$$OHP = \frac{T \times N}{63,025}$$

In SI units, we would calculate output power from

$$P_{\text{OUT}} = \frac{T \times N}{9550}$$ **(2.17)**

The input horsepower to the motor is the same as the output horsepower (*HHP*) from the pump (ignoring system losses). So we see that the horsepower calculations for a motor are just reversed from those for a pump, and that the power output of the motor equals the hydraulic (input) horsepower if, *and only if*, we assume 100 percent efficiency.

Example 2.19

Calculate the torque available from a hydraulic motor if the motor speed is 1700 rpm. The hydraulic horsepower available at its inlet is 15 HP. Assume no losses.

Solution

Assuming no losses means that the output horsepower is equal to the hydraulic horsepower, or, from Equation 2.16,

$$OHP = \frac{T \times N}{63{,}025} = HHP$$

Solving for torque and inserting the given values, we have

$$T = \frac{63{,}025 \times HHP}{N}$$

$$= \frac{63{,}025 \times 15 \text{ HP}}{1700 \text{ rpm}} = 556 \text{ lb·in}$$

2.10 Fluid Flow

We have previously touched on the concept of fluid flow and volume flow rate in a fluid power circuit. We mentioned that fluid power systems are generally considered to be neither fluid sources nor fluid sinks as far as fluid flow through piping is concerned. In other words, fluid is not manufactured, nor is it absorbed by the circuit; it flows around a closed circuit. (An exception to this is an accumulator, which we'll discuss in Chapter 13.) Thus, we can consider that the flow rate past any given point in a circuit is the same as that past any other point in the same circuit as long as the flow does not divide into two or more branches.

A very simple equation is used to determine the flow rate at any given point:

Flow Rate = Pipe Area × Fluid Velocity

or   **(2.18)**

$$Q = A \times v$$

where

$$Q = \text{volume flow rate.}$$
$$A = \text{flow area.}$$
$$v = \text{fluid velocity.}$$

Please observe and remember the difference between Q and v. The *volume flow rate*, Q, is the volume of fluid flowing past a given point per unit time. Thus,

$$\text{Volume Flow Rate} = \frac{\text{Volume}}{\text{Time}}$$

$$Q = \frac{Vol}{t}$$

and is usually expressed in gallons per minute or liters per minute for fluid power applications.

Velocity, on the other hand, is the speed with which a "chunk" of fluid is moving past a given point. Velocity is usually measured in feet per second, feet per minute, or meters per second for fluid power applications.

We assume that the fluid completely fills the pipe.

Example 2.20

Determine the volume flow rate if the fluid is flowing at 15 feet per second (fps) in this pipe, which has a 2 in inside diameter (ID).

Solution

From Equation 2.18

$$Q = A \times v$$

The area of a 2 in ID pipe is 3.14 in². Thus,

$$Q = 3.14 \text{ in}^2 \times 15 \text{ ft/sec} \times 12 \text{ in/ft} = 565.2 \text{ in}^3/\text{sec}$$

This is a valid answer; however, since in the United States flow rate is normally expressed in gallons per minute, it is best converted. Here we introduce an extremely important conversion factor that you should commit to memory because it will be used over and over:

$$1 \text{ gallon} = 231 \text{ in}^3$$

Note: The reference here is to a *U.S. gallon*. An *imperial gallon* is equal to 1.20 U.S. gallons, or 277 in³. All references to gallons throughout this text will be to U.S. gallons unless imperial gallons are specified.

We use this to convert our answer to gpm thus:

$$Q = 565.2 \text{ in}^3/\text{sec} \times (1 \text{ gal}/231 \text{ in}^3) \times 60 \text{ sec/min}$$

$$= 146.8 \text{ gpm}$$

Example 2.21

Fluid is flowing through this 2.5 cm ID pipe at 30 Lpm. Find the fluid velocity.

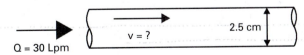

Solution

Again using Equation 2.18 and solving for velocity, we have

$$Q = A \times v$$

or

$$v = \frac{Q}{A}$$

The flow area is

$$A = \frac{\pi D^2}{4} = \frac{\pi \times (2.5 \text{ cm})^2}{4} = 4.91 \text{ cm}^2 = .000491 \text{ m}^2$$

Therefore,

$$v = \frac{(30 \text{ L/min} \times (\text{m}^3/1000 \text{ L}) \times (\text{min}/60 \text{ s})}{0.00049 \text{ m}^2}$$

$$= 1.02 \text{ m/s}$$

2.11 Continuity Equation

Another way to say that the fluid flow rate remains constant through a given portion of a circuit is provided by the *continuity equation:*

$$Q_1 = Q_2 = Q_3 = \ldots \qquad \textbf{(2.19)}$$

This equation is illustrated in Figure 2.9. We can expand the usefulness of this expression by utilizing Equation 2.17 so that

$$A_1 v_1 = A_2 v_2 = A_3 v_3 = \ldots \qquad \textbf{(2.20)}$$

where the subscripts denote different points in the circuit.

It is crucial to remember that the flow rate, Q, is the parameter that stays constant. The velocity will change if the inside diameter of the pipe changes.

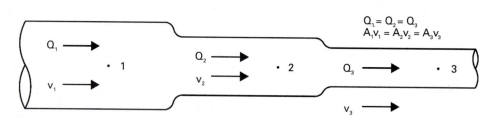

Figure 2.9 The continuity equation says that the fluid flow rate remains constant through a given portion of a circuit.

Example 2.22

In this partial circuit, calculate the flow velocities at points B and C if the velocity at point A is 15 ft/s.

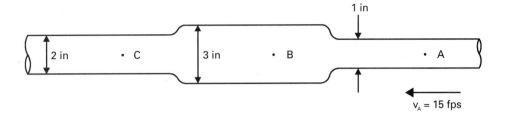

Solution

From the continuity equation (Equation 2.18 and Equation 2.19), we know that

$$A_A v_A = A_B v_B = A_C v_C$$

Using points A and B, we can solve this equation for the velocity at point B to get

$$v_B = \frac{A_A}{A_B} \times v_A$$

The areas are

$$A_A = 0.785 \text{ in}^2$$

$$A_B = 7.069 \text{ in}^2$$

Therefore,

$$v_B = \frac{0.785 \text{ in}^2}{7.07 \text{ in}^2} \times 15 \text{ ft/sec} = 1.7 \text{ ft/sec}$$

Moving to point C, we find

$$A_C = 3.14 \text{ in}^2$$

Thus,

$$V_C = \frac{A_A}{A_C} \cdot v_A = \frac{0.785 \text{ in}^2}{3.14 \text{ in}^2} \times 15 \text{ ft/sec} = 3.8 \text{ ft/sec}$$

We can also find the flow rate through the system by using Equation 2.17.

$$Q = A_A v_A$$

$$= 0.785 \text{ in}^2 \times 15 \text{ ft/sec} \times 12 \text{ in/ft} \times \text{gal}/231 \text{ in}^3 \times 60 \text{ sec/min}$$

$$= 36.7 \text{ gpm}$$

In these examples, we have very deliberately kept track of the units throughout each equation. This should serve as a pattern for you. *The importance of units cannot be overemphasized.*

This is a good time to discuss the myth of multiplication of power. As seen earlier, force can be multiplied by hydraulic systems. In fact, that is one of the great advantages of fluid power. We will see in Chapter 13 that pressure can be multiplied by a device known as an intensifier. Power, however, *cannot* be multiplied in a fluid power system. We will always have losses, so that HP_{OUT} = HP_{IN} − Losses.

This can easily be shown by considering an idealized 100 percent efficient hydraulic system in which there are no losses. Beginning at the pump, assume an input from the prime mover of, say 20 HP. Our idealized pump converts all of this input horsepower to fluid horsepower, which is defined by Equation 2.13 in terms of pressure and flow rate. Since the pump output horsepower is equal to the pump input horsepower, we still have 20 HP (Figure 2.10). The 20 hydraulic horsepower (*HHP*) arrives at the motor (or other output device) and, at 100 percent efficiency, is converted to 20 horsepower of mechanical output at the output shaft.

As we have already mentioned, a fluid power circuit cannot add to or subtract from the total system flow rate. Also, since we are assuming an idealized situation where there are no pressure losses, we see that the product of p and Q (and, consequently, fluid horsepower) remains constant. Therefore, there can be no increase in the horsepower delivered to the actuator.

Whether the actuator is a hydraulic cylinder or a hydraulic motor makes no difference, but since we've already seen that in a motor that is 100 percent efficient, the output horsepower is equal to (but not greater then) the input horsepower, let's consider a hydraulic cylinder. As we have seen, a hydraulic cylinder has a force output that moves a load over a distance. If we consider the time element, we see, as in Equation 2.7, that the power output is given by

$$\text{Power} = \frac{\text{Force} \times \text{Distance}}{\text{Time}}$$

$$P = \frac{F \times d}{t}$$

In this equation, if we separate the terms, we have

$$P = F \times \frac{d}{t}$$

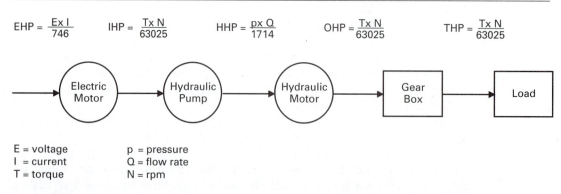

$$EHP = \frac{E \times I}{746} \qquad IHP = \frac{T \times N}{63025} \qquad HHP = \frac{p \times Q}{1714} \qquad OHP = \frac{T \times N}{63025} \qquad THP = \frac{T \times N}{63025}$$

Electric Motor → Hydraulic Pump → Hydraulic Motor → Gear Box → Load

E = voltage　　　p = pressure
I = current　　　Q = flow rate
T = torque　　　N = rpm

Figure 2.10 This diagram shows the horsepower flow through a hydraulic motor circuit.

Distance (d) divided by time (t) is simply velocity (v), so the equation becomes

$$P = F \times v$$

As we will see in Chapter 6, the velocity with which a cylinder (actually, the piston) moves is a function of the system flow rate and the area of the piston, or

$$v = \frac{Q}{A}$$

Since force is simply pressure times the area of the piston ($F = p \times A$) we can write

$$P = (p \times A) \times \frac{Q}{A} = p \times Q$$

Again, in our idealized system, neither p or Q changes, so the power output from the cylinder can be no more than the power put into it by the fluid. Because horsepower and power differ only by the constant, 550, we can substitute the term *power* for *horsepower* in the first part of this discussion. Doing that, we can see that even in this idealized system, the output from the actuator can be no more than the power input to the pump. Thus, there is no multiplication of power.

Horsepower losses due to friction losses in valves will be discussed in detail in Chapters 7, 8, and 9. Horsepower losses due to friction in pipes, fittings, and so on, will be discussed briefly in Chapter 12.

2.12 Summary

We see throughout this chapter the importance of Pascal's Law in hydraulic fluid power, because it defines the basis for all hydrostatic system operations. Both the production of force and the multiplication of force demonstrate this basic concept. Pascal's Law, however, defines only a static situation such as when there is no movement of a cylinder. If there is to be any movement of the load (whether we're using a cylinder or a hydraulic motor), flow is required. The flow rate will determine how fast any given actuator will move.

Thus, we see that in order to move a load, we must first have sufficient pressure to generate the required force. Only then can the flow produce the desired motion. As we'll see later, although flow and pressure are discussed separately, they are actually closely related and interdependent to some degree in system operation.

The calculations of the various horsepowers throughout the system provide us with important information concerning system input and output power. We'll use these concepts of horsepower in later chapters to determine efficiencies, operating costs, heat generation rates, and temperature rises through the system.

Beginning with Chapter 3, we'll look at the components that make up a fluid power system. These chapters will first describe the component hardware, then explain how the components function as an integral part of the system and how they interact with other components as the system is operated.

2.13   Key Equations

Pressure:

$$p = \frac{F}{A} \qquad \textbf{(2.1)}$$

Force:

$$F = p \times A \qquad \textbf{(2.1a)}$$

Multiplication of force:

$$F_B = F_A \times \frac{A_B}{A_A} \qquad \textbf{(2.2)}$$

$$F_B = F_A \times \frac{D_B^2}{D_A^2} \qquad \textbf{(2.3)}$$

Work:

$$W = F \times d \qquad \textbf{(2.4)}$$

$$(F \times d)_{IN} = (F \times d)_{OUT} \qquad \textbf{(2.5)}$$

Piston movement:

$$F_A \times d_A = F_B \times d_B \qquad \textbf{(2.5a)}$$

$$A_A \times d_A = A_B \times d_B \qquad \textbf{(2.6)}$$

Power:

$$P = \frac{W}{t} = \frac{F \times d}{t} \qquad \textbf{(2.7)}$$

$$P = F \times v \qquad \textbf{(2.8)}$$

Cylinder output power:
   U.S. Customary

$$HP = \frac{F \times d}{550\,t} = \frac{F \times v}{550} \qquad \textbf{(2.9)}$$

   SI

$$P = \frac{F \times d}{t} = F \times v \qquad \textbf{(2.8)}$$

Torque

$$T = F \times d \qquad \textbf{(2.10)}$$

Pump input power
   U.S. Customary

$$IHP = \frac{T \times N}{63{,}025} \qquad \textbf{(2.11)}$$

$$(T = \text{lb·in}, N = \text{rpm})$$

   SI

$$P_{IN} = \frac{T \times N}{9550} \qquad \textbf{(2.12)}$$

$$(T = \text{N·m}, N = \text{rpm})$$

Hydraulic power:
   U.S. Customary

$$HHP = \frac{p \times Q}{1714} \qquad \textbf{(2.13)}$$

$$(p = \text{psi}, Q = \text{gpm})$$

   SI

$$P_{HYD} = \frac{p \times Q}{60{,}000} \qquad \textbf{(2.14)}$$

$$(p = \text{kPa}, Q = \text{Lpm})$$

Motor (both electric and hydraulic)
output power:
U.S. Customary

$$OHP = \frac{T \times N}{63{,}025} \qquad \textbf{(2.16)}$$

$(T = \text{lb·in}, N = \text{rpm})$

SI

$$P_{\text{OUT}} = \frac{T \times N}{9550} \qquad \textbf{(2.17)}$$

$(T = \text{N·m}, N = \text{rpm})$

Flow rate:

$$Q = A \times v \qquad \textbf{(2.18)}$$

Continuity equation:

$$Q_A = Q_B = Q_C = \ldots \qquad \textbf{(2.19)}$$

$$A_A v_A = A_B v_B = A_C v_C = \ldots \qquad \textbf{(2.20)}$$

Review Problems

Note: The letter "S" after a problem number indicates SI units are used.

General

1. Define hydrodynamics.
2. Define hydrostatics.
3. State Pascal's Law.
4. List the advantages of hydraulic fluid power.
5. List the disadvantages of hydraulic fluid power.
6. Define multiplication of force.
7. Define work.
8. Define power.
9. Define horsepower.
10. What are the three basic sections of a fluid power system?
11. Can power be multiplied in a hydraulic fluid power system? Why?
12. What is the difference between flow rate and flow velocity?

Force, Pressure, and Area

13. A closed cylinder with a piston area of 10 in^2 supports a load of 4000 lb. What is the pressure in the cylinder?
14. A cylinder with a 2 in diameter piston produces 5500 lb of force. What is the cylinder pressure?
15. Find the force output of a cylinder that has a 5 in^2 piston and a pressure of 3000 psi.
16. A cylinder has a 3 in diameter piston. If the maximum system pressure is 2500 psi, what is the maximum load it can move?
17. A broaching machine used to cut the keyway in gear blanks uses a 4 in cylinder to produce 20,000 lb of force. How much pressure is required?
18S. A cylinder with a 5 cm piston operates with a pressure of 20,000 kPa. What is its force output capability?
19S. An automobile weighing 2500 kg is lifted by a lift that has a 25 cm piston. What pressure is required?

Multiplication of Force

20. In a fluid power system consisting of two hydraulic cylinders, a force input of 10 lb to one cylinder results in a force output of 1000 lb from the other cylinder. Find the multiplication of force.
21. Two fluid power cylinders are connected by a flexible hose. Cylinder A is 2 in in diameter. Cylinder B is 8 in in diameter. A force of 200 lb is input to cylinder A. Find the

    a. Pressure in cylinder A.
    b. Pressure in cylinder B.
    c. Pressure in the connecting hose.
    d. Force output of cylinder B.

22. How much work is required to move a 50,000 lb load 6 feet?
23. Two cylinders are connected like those in Figure 2.6. The load on one cylinder is 50 lb, while the load on the other is 600 lb. When the 50 lb load moves downward 6 inches, how far will the 600 lb load move?
24. In Review Problem 23, do the diameters of the pistons have anything to do with the operation? Justify your answer.
25. Two cylinders are connected as shown in Figure 2.5. Assume that cylinder A is a single-acting hand pump with a 0.5 in diameter and a 3 in stroke. An operator is using the hand pump to lift a 1000 lb load. He needs to raise the load 2 feet. He exerts a force of 50 lb on the piston of cylinder A. Find the

    a. Number of hand-pump strokes required.
    b. Multiplication of force.
    c. Diameter of cylinder B.

26S. A hydraulic pallet jack has a 2.5 cm input piston and a 7.5 cm output piston. What input force on the small piston is required to produce an output force of 12,000 N?
27S. A hydraulic jack has a 2 cm input piston and a 5 cm output piston. How much can the jack lift if a 500 N force is applied to the input piston?
28S. In a system similar to that shown in Figure 2.5, a 200 N force is applied to a 1.5 cm piston. The output from the large piston is 1000 N. Find the

    a. Multiplication of force.
    b. Diameter of the output piston.
    c. Pressure in the system.

Power and Horsepower

29. A hydraulic system moves a 30,000 lb load 4 feet in 12 seconds. How much power is expended in the effort? Work the problem in both U.S. Customary and SI units.
30. How much power is produced by the system in Review Problem 29? Work the problem in both U.S. Customary and SI units.
31. Derive the SI equivalent of the U.S. Customary definition of horsepower (550 ft·lb/sec).
32. The deck elevators on aircraft carriers are required to raise a fully loaded and armed aircraft weighing 215,000 lb from the hanger deck 22 feet to the flight deck in 8 seconds. If the elevator itself weighs 96,000 lb, how much horsepower is required?
33. A torque of 240 lb·ft is required to lock the companion nut on a vehicle rear end. If a torque wrench 18 inches long is used, how much force must be applied to the wrench?

34. An electric motor is used to drive a hydraulic pump. If the motor turns at 1800 rpm and produces 50 lb·in of torque, how much horsepower is required to operate it? How many watts?

35. A hydraulic pump requires 25 horsepower to drive it at 2200 rpm. How much torque is needed to turn the pump?

36. How much hydraulic horsepower is produced by a pump operating at 1500 psi with a 12 gpm flow rate?

37. Derive the constant 1714 that appears in the denominator of the hydraulic horsepower (*HHP*) equation.

38. Find the torque required to drive a hydraulic pump at 2000 rpm. The pump has a flow rate of 20 gpm and operates at 2500 psi. Assume 100 percent efficiency.

39. A 10 HP electric motor drives a hydraulic pump at 2200 rpm. What is the maximum pressure at which the pump can be operated if its output is 16 gpm? Assume 100% efficiency.

40S. A hydraulic press is used to insert hinge pins in a large portable bridge structure. A force of 10,000 N is required. The press moves the pins 18 cm in 5 seconds. What is the power output of the press?

41S. A log splitter uses a 10 cm cylinder with a 1 m stroke. To split oak logs, a 20,000 N force is required. The cylinder extends completely in 6 seconds. How much pressure is required? How much power?

42S. The drum on a ball-pulverizer turns at 60 rpm. It requires 300 N·m of torque. How many watts are required?

43S. A hydraulic pump produces 100 Lpm at 10,000 kPa. It turns at 1800 rpm. Assume 100% efficiency. Find the

    a. Hydraulic power produced.
    b. Input torque required.
    c. Power required.

44S. A hydraulic motor produces 1200 N·m of torque at 300 rpm. The pressure at the motor inlet is 12,500 kPa. Find the

    a. Power output.
    b. Flow rate.

Fluid Flow

45. Fluid is flowing through a 2 in ID pipe at 25 ft/sec. What is the flow rate?

46. Fluid is flowing through a 1.5 in ID pipe at 30 gpm. What is the flow velocity?

47. The flow rate of a fluid in a 1 in ID pipe is 346.5 in³/s. What is the flow velocity?

48. A system contains 1, 1.5, and 2 in ID pipe sections. If the system flow rate is 25 gpm, what is the fluid velocity in each section?

49. A circuit uses pipes with 2, 2.5, and 3 in diameters. The fluid velocity in the 2 in pipe is 20 fps. Find the flow rate in the circuit and the velocities in the other two pipe sections.

50. A pump is operating at 200 psi. The fluid flow velocity is 20 ft/sec through a 1 in ID pipe. What is the horsepower of the electric motor driving the pump? What assumption did you make to get your answer?

51S. Fluid is flowing at 3 m/s through a 2.5 cm ID pipe. What is the volume flow rate?

52S. A 20,000 L tank is being drained through a 7.5 cm ID pipe. If the tank drains completely in 30 minutes, what is the average flow rate through the pipe? What is the average velocity?

53S. Fluid is flowing at 200 Lpm through a pipe with a 2.5 cm ID. That pipe is connected to a 4 cm pipe. What is the velocity in each pipe?

54S. A pump is delivering a flow rate of 125 Lpm. If the flow rate in the pressure line must not exceed 8 m/s, what is the smallest acceptable ID?

55S. The flow velocity in a return line in a hydraulic system should not exceed 3 m/s. What is the maximum flow rate that can be used with a 5 cm return line?

Hydraulic Fluids

Outline

3.1   Introduction

In mechanical power systems, force is transmitted by means of solid components—rods, cables, gears, belts, and so on. In fluid power systems, force is transmitted through a column of fluid. While this sounds relatively simple and uncomplicated, it is, in fact, an extremely complex subject, because there are many factors that must be considered in selecting the proper fluid for a particular application.

At one time, this fluid-selection problem did not exist. Originally, fluid power systems all used water as their power-transmission medium, posing some severe limitations on application. One immediately obvious problem was that of the environment in which the systems could be used—exposure to freezing temperatures, for example, could result in extreme maintenance problems. In addition, the low viscosity of water contributed to high internal leakage of pumps and motors, while its poor lubricating characteristics contributed to rapid wear, particularly in components that experienced high mechanical loading. These two factors—low viscosity and poor lubricity—meant that system operating pressures had to be relatively low, with the consequence that actuators had to be large in order to develop the necessary forces to move large loads.

In spite of the limitations, some very successful systems were operated on straight water. Recall the discussion of the pressurized water mains in London. Many other cities, including Paris, also had pressurized water as a utility. The great stage of the Theatre Royal in London and the elevators in the Eiffel Tower in Paris still use water as the hydraulic fluid.

51

With the proliferation of electrical power and improved distribution capabilities, there was a shift away from the pressurized water mains. The companies that operated these mains were eventually forced to cease operations because they were no longer economical to run. When this happened, many of their customers installed their own pumping units and continued to operate the systems on water. In 1975 there were still over 600 known systems in London using water as the power-transmitting medium.

Late in the nineteenth century, the advantages of petroleum-based hydraulic fluids brought about a revolution throughout the fluid power industry. The higher viscosities and far better lubricating characteristics of these fluids enabled operation at higher system pressures, allowing significant reductions in the sizes of components required to produce very large forces. Today many industrial systems operate at 2000 to 5000 psi or higher, while mobile construction equipment frequently uses 3000 psi or higher in the United States (up to 4500 psi in Europe), and the U.S. Navy now has a major research program under way in which some aircraft are flying with 8000 psi systems. Even petroleum-based fluids have their limitations and problems, though, as will be discussed.

Objectives

When you have completed this chapter, you will be able to

- List the important functions of hydraulic fluids.
- Explain basic fluid characteristics such as mass, weight, density, specific gravity, and specific weight.
- Explain the basic differences between liquids and gases.
- List the basic hydraulic fluid properties.
- Explain the differences between flash point, fire point, and autogeneous ignition temperature.
- Calculate the compression of a liquid based on its bulk modulus.
- Explain viscosity, viscosity index, and pour point.
- Explain the three lubrication regimes.
- Know which fluid analyses are needed for a successful fluid-condition monitoring program.
- Give the definition of a fire-resistant fluid.
- List and describe the four categories of fire-resistant fluids.

3.2 Fluid Functions

The fluid in a hydraulic system must do more than simply transmit a force. It must serve many functions, the most important of which are to

   a. Transmit energy.
   b. Lubricate components.
   c. Transfer heat from heat-generating components to the walls of fluid conductors, reservoirs, or heat exchangers, where convection removes the heat.

    d. Carry wear-generated particles to filters.

    e. Prevent rust and corrosion.

    f. Provide electrical insulation for certain applications.

    g. Seal clearances.

In addition, it must

    a. Be fully compatible with all system hardware, seals, and hoses.

    b. Have low foaming tendencies.

    c. Release entrained and dissolved air rapidly.

    d. Allow water to separate rapidly (see Section 3.3).

    e. Have the proper viscosity and viscosity index (see Section 3.4).

    f. Have a high shear stability.

    g. Have good performance over a wide temperature range.

    h. Have a high bulk modulus (see Section 3.5).

    i. Have a low specific gravity (Section 3.4).

Unfortunately, no single fluid is outstanding in its ability to meet all of these requirements. Therefore, it becomes the responsibility of the system designer to decide which of the large number of available fluids will best meet a particular need. Only with a sound understanding of fluid properties and characteristics can the proper choices be made.

In the remainder of this chapter, we will look at various types of fluids, their properties and characteristics, and methods for assessing the suitability of fluids for use in a specific fluid power system.

3.3   Physical Characteristics and Properties of Liquids versus Gases

The term *fluid* refers to both liquids and gases. In discussing hydraulic fluid power, it is important to remember that the "fluid" of concern is a liquid. As discussed in Chapter 1, a liquid is distinguished from a gas by two important characteristics. The first of these is that a liquid is relatively incompressible—to the point of being considered incompressible for most fluid power applications. In reality, most hydraulic fluids will compress up to 0.4 percent per 1000 psi. Gases, on the other hand, are readily compressible and tend to have a springy or spongy operation in power systems as opposed to the stiff fluid column in a liquid system.

The second distinguishing characteristic involves the actions of liquids and gases in enclosed vessels. A liquid has a mass and a relatively fixed volume, but has no shape of its own. This means that it will conform to the shape of the vessel up to the volume of the liquid. A gas, however, while it has a mass, and like a liquid has no shape of its own, changes volume with changes in pressure and temperature. Consequently, a gas may be considered to completely fill its containing vessel and conform to the shape of the entire vessel.

For convenience, in the early part of this book, we will use the word *fluid* to refer to liquids only; air and gases will be expressly termed air or gas.

3.4   Fluid Properties

The basic properties of a fluid include

- Mass
- Weight
- Density
- Specific weight
- Specific gravity

Mass is the property associated with a fluid's inertia. It is a fixed property, and is not affected by its location in space. Mass is denoted by the symbol m and, in the U.S. Customary system, has units of lb·sec²/ft, defined as a *slug*. In the SI system, mass is expressed in kilograms. When discussing mass, some authors prefer to use *pounds mass* rather than slugs. This is a perfectly acceptable approach. At standard gravity (on the earth's surface), 1 pound mass equals 1 pound force, indicating that, on the earth's surface, mass and weight are equal. This author feels, however, that the limitation to earth gravity is unduly restrictive in the space age in which we live and therefore chooses to express mass in slugs.

The *weight* of a fluid is the result of its mass being acted upon by gravity. The relationship between weight and gravity is defined by Newton's Second Law of Motion,

$$F = m \times a \qquad\qquad (3.1)$$

where

$$F = \text{force.}$$
$$m = \text{mass.}$$
$$a = \text{acceleration.}$$

In this discussion, the force in the equation is the weight of the fluid, while the acceleration term is the acceleration of gravity. Thus, we can write that

$$\text{Weight} = \text{Mass} \times \text{Gravity}$$

or

$$(3.2)$$

$$w = m \times g$$

For our purposes, taking g to be 32.2 ft/sec² (386.4 in/sec²) or 9.81 m/s² will give satisfactory answers as long as we are considering earthbound systems. In other gravitational fields, or if the fluid were being accelerated in an elevator, an aircraft, or a piece of mobile equipment, we would need a different constant.

Dimensionally, if we use mass in slugs (or lb·sec²/ft) and gravity in ft/sec², we see that weight is given simply in pounds (lb). In the SI system, where mass is in kilogram (kg) and gravity is in meters per second squared (m/s²), weight has the units of kilogram-meters per second squared (kg·m/s²), which is defined as a newton (N).

Example 3.1

Find the weight of a body with a mass of 5 slugs (a) on the earth's surface, and (b) on a planet where the acceleration of gravity is 50 ft/sec².

Solution

a. From Equation 3.2, substituting lb·sec²/ft for slugs,

$$w = m \times g = (5 \text{ lb·sec}^2/\text{ft}) \times (32.2 \text{ ft/sec}^2)$$
$$w = 161 \text{ lb}$$

b. Again, using Equation 3.2 and substituting for slugs,

$$w = m \times g = (5 \text{ lb·sec}^2/\text{ft}) \times (50 \text{ ft/sec}^2)$$
$$w = 250 \text{ lb}$$

The *density* of a material is defined as its mass per unit volume. Mathematically, we express this as

$$\text{Density} = \frac{\text{Mass}}{\text{Volume}}$$

or   (3.3)

$$\rho = \frac{m}{V}$$

where

$$\rho \text{ (rho)} = \text{density.}$$
$$V = \text{volume.}$$

If we express mass in slugs (lb·sec²/ft) and volume in ft³, we find that density is given in terms of lb·sec²/ft⁴. In the SI system, density has the units of kg/m³.

Example 3.2

If the body of Example 3.1 has a volume of 1.5 ft³, find its density.

Solution

From Equation 3.3,

$$\rho = \frac{m}{V} = \frac{5 \text{ lb·sec}^2/\text{ft}}{1.5 \text{ ft}^3} = 3.33 \text{ lb·sec}^2/\text{ft}^4 = 3.33 \frac{\text{slug}}{\text{ft}^3}$$

A fluid's *specific weight* is found by dividing its weight (*not* mass) by its volume. In the U.S. Customary system, we express specific weight in lb/ft³, while the SI system uses N/m³. Mathematically, we have

$$\text{Specific Weight} = \frac{\text{Weight}}{\text{Volume}} = \frac{\text{Mass} \times \text{Gravity}}{\text{Volume}}$$

or   (3.4)

$$\gamma = \frac{W}{V} = \frac{m \times g}{V}$$

where

$$\gamma \text{ (gamma)} = \text{specific weight}$$

Example 3.3

Find the specific weight on earth of the body of Example 3.1 and 3.2.

Solution

Using Equation 3.4,

$$\gamma = \frac{W}{V} = \frac{m \times g}{V} = \frac{(5 \text{ lb·sec}^2/\text{ft}) \times (32.2 \text{ ft/sec}^2)}{1.5 \text{ ft}^3}$$

$$= 107.33 \text{ lb/ft}^3$$

Specific gravity (Sg) is the ratio of the density of the liquid (or any substance) in question to the density of water at 39°F (4°C), so we can write

$$\text{Specific Gravity} = \frac{\text{Density of Liquid}}{\text{Density of Water at 39°F}}$$

or

$$Sg = \frac{\rho}{\rho_{water}}$$

We can make this equation a little more flexible if we note that the equation for specific weight (γ) can be written

$$\gamma = \frac{m \times g}{V}$$

Since $m/V = \rho$, we have

$$\gamma = \rho g$$

Using this relationship, we see that

$$Sg = \frac{\rho}{\rho_{water}} = \frac{\gamma}{\gamma_{water}} \qquad\qquad \textbf{(3.5)}$$

The values of ρ and γ in the denominators are constants and are not affected by the temperature of the liquid in question. In the U.S Customary system, the density of water at 104°F is 1.94 slug/ft³, and the specific weight is 62.4 lb/ft³. In the SI, the density is 1000 kg/m³, while the specific weight is 9.81 kN/m³.

Example 3.4

Find the specific gravity of the body of Example 3.1.

Solution

From Equation 3.5, we know that

$$Sg = \frac{\rho}{\rho_{water}} = \frac{\gamma}{\gamma_{water}}$$

In Example 3.2, we found the density of the body to be 3.33 slug/ft³; thus,

$$Sg = \frac{3.33 \text{ slug/ft}^3}{1.94 \text{ slug/ft}^3} = 1.72$$

Also, in Example 3.3, we found that the specific weight of the body was 107.33 lb/ft³. So we can also say that

$$Sg = \frac{107.33 \text{ lb/ft}^3}{62.4 \text{ lb/ft}^3} = 1.72$$

Note that specific gravity is dimensionless.

In many fluid power applications, fluid density, weight, specific weight, and specific gravity are of little significance. Since most petroleum-based hydraulic fluids have a specific gravity of 0.85 to 0.9, there is no great difference in these properties. Some alternate fluids—for instance, high water-base fluids or some synthetic oils—have very different properties. Water-based fluids have specific gravities very near 1.0, while that of some synthetic fluids may approach 2.

This high specific gravity carries several negative connotations. First, a high specific gravity means a high weight relative to petroleum-based oils. This can be a major disadvantage in aircraft and space vehicles where weight is a major consideration.

A second disadvantage, we'll see in Chapter 12, is that the higher the specific gravity, the higher the pressure loss will be as the fluid flows through the system piping. This can be costly in heat generation and energy waste. It may also cause cavitation in the pump suction line and lead to pump damage.

The final disadvantage of a high specific gravity shows up in shock-absorbing devices, specifically in gun recoil mechanisms. In any such device in which the fluid is forced through small orifices at very high velocities, the higher the specific gravity of the fluid, the higher the pressure will be in the shock absorber (for a given orifice size).

Several chemical and physical properties that must be considered when evaluating a hydraulic fluid for potential use in a system follow:

- Oxidation resistance
- Corrosion and rust protection
- Anti-wear and lubrication characteristics
- Emulsibility/demulsibility
- Flash and fire point
- Bulk modulus
- Viscosity (absolute and kinematic)
- Viscosity index
- Pour point

Oxidation

Fluid oxidation pertains to a degradation of the fluid itself. It is a chemical reaction between most hydraulic fluids and oxygen that frequently results in the formation of resins and sludges in the fluid. Oxidation begins almost as soon as the fluid is put into use. Its rate depends on several factors, the most important being the chemistry of the fluid itself and the amount of oxygen and the temperatures to which the fluid is exposed. The amounts and types of particulate contamination in the fluid also play an important role. The materials of which the components are constructed—especially alloys containing cadmium, zinc, and copper—may also act as catalysts. Most hydraulic fluids contain oxidation inhibitors in an attempt to counteract these factors.

Obviously, since oxidation is the result of exposure to oxygen, it follows that the more oxygen there is in the fluid, either as free air or combined with water, the higher the oxidation rate will be. When the presence of oxygen is accompanied by high temperatures, whether it be a high bulk temperature or a localized hot spot, the situation is aggravated. As a general rule, you can consider the rate of oxidation to double with every 18°F (10°C) increase in fluid temperature above 140°F (60°C). Doubling the oxidation rate cuts the useful life of the oil in half. As a result of this critical temperature limit, most hydraulic fluid manufacturers recommend a maximum operating termperature of 130°F (55°C).

There are two major problems that result from oxidation. The first is that the resins and sludges formed can mechanically degrade system components by clogging orifices and by causing components to become sluggish—or even jam completely due to the sticky deposits. The second problem is that many of the oxidation products are acidic and therefore may cause etching or corrosion of metallic surfaces.

Because of the many factors involved in oxidation and the multitude of possible variations of these factors, it is very difficult to determine by a standard laboratory test how a given fluid will resist oxidation in use. A standard test does, however, exist—ASTM D 943—which is commonly used for this purpose. While its results are not infallible, they can indicate the probable performance of a fluid. A user of fluid power systems can increase system reliability by (1) using fluids that have good results in this test, (2) regularly monitoring the oxidation levels in the fluid, and (3) taking steps to reduce the tendency for oxidation to occur. To monitor oxidation levels, visually check the fluid and any removed components for evidence of discoloration due to oxidation products and also check the fluid acid level to reduce the tendency toward oxidation.

1. Keep air and water out of the system.
2. Reduce the operating temperature of the system to eliminate hot spots.
3. Remove particulate contaminants to eliminate their catalytic effects.
4. Avoid the use of cadmium, zinc, and copper in contact with the fluid. These metals act as catalysts that promote oxidation.

Corrosion and Rust Protection

Corrosion and rust protection pertain to the materials with which the fluid comes in contact. In fluid power terms, *corrosion* can generally be considered to be any deterioration of a metal surface due to a chemical reaction with a fluid. Frequently, this reaction is the result of acids formed by oxidation products. *Rusting*, on the other hand, is generally defined as the surface oxidation of a ferrous metal.

The visible evidences of these processes are very different. In corrosion, metal is actually dissolved and removed from the surface. This results in the formation of surface pits or voids that are often filled with a dark-colored substance—usually consisting of the oxidation products and acids that caused the corrosion. Because metal is actually removed, a corroded part loses weight.

This means that the part becomes weaker, the fluid is contaminated by the metal that is removed, and leakage across the part may be increased.

Rusting, on the other hand, is the formation of ferrous oxides on the surface. The reddish color of rust is due to the oxides that have built up on the surface. Because the oxides are heavier than the parent metal, a rusting component actually gains weight, at least in the initial stages.

Neither of these situations is acceptable in a fluid power system. The degradation of metallic surfaces results in internal and external leakage, lowered efficiencies, poor controllability, and an overall increase in the particulate contamination level, which further aggravates the oxidation and wear problem.

Two laboratory tests can be used to evaluate the rust-preventing characteristics of hydraulic fluids—ASTM D 665 and ASTM D 3603. Again, these tests are not foolproof, but they can provide an indication of how a fluid might perform in service.

As with oxidation, fluid power system users can take certain steps to reduce the problems of corrosion and rusting:

1. Select fluids with good rust and oxidation inhibiting properties (termed *R and O oils*).
2. Regularly check fluids and removed components for signs of corrosion or rusting.
3. Provide good filtration to remove the by-products of rusting and corrosion.
4. Limit, as much as possible, the introduction of air, water, and chemicals into the hydraulic fluid.

Anti-wear and Lubrication Characteristics

While anti-wear and lubrication characteristics are related, they are not actually the same. We will discuss anti-wear characteristics in this section and devote a later, major section to the concept of lubrication.

While all hydraulic fluids have as one of their major functions the lubrication of components, some oils are specially formulated for use in systems subjected to high temperatures, high surface loading, or high levels of particulate contamination. These oils, often referred to as *extreme-pressure (EP) oils*, are usually fortified with one or more anti-wear additives. These materials protect moving surfaces by reacting with the metals to form a protective surface film. The formation of this film requires freshly exposed, chemically reactive metal surfaces, as well as time and heat. Once the film is formed, contact is actually between the layers of the surface films rather than the metals. As the film is worn off, new film is formed very rapidly so that little metal wear occurs.

Extreme pressure additives usually contain sulfur, chlorine, phosphorus, or combinations of these elements. Zinc dithialphosphate is the most commonly used additive for hydraulic fluids.

There are several tests for rating the anti-wear characteristics of fluids, including the Four Ball Test (ASTM D 2266), Falex (ASTM D 2670), Vickers Vane Pump (ASTM D 2882), the Timkin (ASTM D 2782), and the recently developed Gamma Test from the Fluid Power Research Center (FPRC) at Oklahoma State University.

Emulsibility/Demulsibility

Emulsibility refers to an oil's ability to disperse water droplets and hold them in suspension as tiny droplets in the oil. *Demulsibility*, on the other hand, defines an oil's ability to separate from water. The particular application and environment determine which property is more important.

Oils with good emulsibility are formulated with detergents that cause water droplets to disperse. These droplets are then surrounded by a film of detergent that encapsulates the water and prevents it from agglomerating with other water droplets. (You have probably seen this demonstrated in dishwashing detergent commercials on TV, except in those commercials, the grease droplet is dispersed in water.) This prevents the settling of water on metal surfaces when the system is not operating. Two situations are especially suited for the use of these detergent oils. One is when there is little dwell time of the oil in the reservoir or sump where the water could settle out. The second is where there is less than 2 percent water in the oil.

Automotive brake fluids (DOT Type 3, DOT Type 5, etc)—which are a special type of hydraulic fluid—are required to contain emulsifiers. This is necessary in order to prevent water from collecting at low points in the brake lines where it could freeze and cause brake failure.

Oils from which water separates readily are usually formulated with chemicals termed *demulsifiers*. They help water to settle out rapidly in reservoirs and sumps so that it can be drained off and are used where high levels of water are expected.

Three tests are widely used to determine the emulsibility/demulsibility characteristics of oils: ASTM D 1401, ASTM D 2711, and Institute of Petroleum IP 19. Most reputable formulators can provide data sheets showing the results of these tests for their fluids.

Because the chemicals that promote emulsibility and demulsibility can be depleted during system operation, the fluid should be checked periodically for additive depletion. Most fluid-analysis laboratories can do this using infrared analysis techniques.

Flash Point, Fire Point, and Autogenous Ignition Temperature (AIT)

Most people mistakenly believe that the *flash point* is the temperature at which a fluid will burst into flame. Actually, the flash point is the lowest temperature at which the fluid releases enough vapor at its surface to ignite in the presence of an open flame. Only the vapors above the fluid are ignited. Since the release of vapors is not rapid enough to sustain combustion, the flame dies immediately—hence the term "flash" point. The most common test methods for determining the flash point of a liquid are ASTM D 92 (Cleveland Open Cup), ASTM D 93 (Pensky-Martens Closed Cup), and ASTM D 3828 (Setaflash Closed Tester).

If the oil temperature is increased, a point will eventually be reached where enough vapors are released to support combustion for 5 seconds in the presence of an open flame. This temperature is termed the *fire point*. It is usually at least 50°F higher than the flash point. The fire point is determined using the Cleveland Open Cup test (ASTM D 92).

The temperature at which a fluid will "burst into flame" in the absence of an open flame or other ignition source is called the *autogenous ignition temperature* (AIT). The AIT is usually much higher than the fire point for most oils. The standard test method for the AIT of oils is ASTM E 659.

Other groups, including the Society of Automotive Engineers (SAE), the Mining Safety and Health Agency (MSHA), and the military have tests for flash point, fire point, and AIT that are similar to the ASTM procedures.

None of these points is of great importance to the average hydraulic system user. In certain cases, however, where high temperatures are encountered, they can be significant. This is especially true in the foundry and die-casting industries, where extremely hot metals and open flames can be found near hydraulic systems. Distributors of hydraulic fluids will gladly offer assistance in selecting the best fluid for these applications.

It is important to know that the data derived from these tests do not necessarily define the flammability of a fluid. There are literally hundreds of such test methods in use today, but, unfortunately, most of them are virtually useless because of their lack of definition of test parameters or the manner in which results are reported. The reader would be well advised to remember that, while some hydraulic fluids are less flammable than others, all of them (with the exception of pure water and some highly sophisticated synthetic fluids) will burn under some conditions. This includes the 95-5 high-water-content fluids. Consequently, one should never treat a fluid as if it were nonflammable. (See Section 3.1 for a discussion of fire-resistant fluids.)

Bulk Modulus

Earlier, in our comparison of liquids and gases, we discussed the concept of liquids being almost, but not quite, incompressible. The degree to which liquids are compressible is expressed as the *bulk modulus* of the liquid. It is actually the reciprocal of compressibility. We can express bulk modulus mathematically as

$$E = \frac{-\Delta p}{(\Delta V / V)} \tag{3.6}$$

where

E = bulk modulus.
Δp = change in pressure.
ΔV = change in volume due to Δp.
V = original volume.

The units of E are the same as pressure (psi or kPa). As you can see, the denominator is actually the percent change of volume that results from a change of pressure. The minus sign takes care of the fact that the volume decreases when pressure increases.

Example 3.5

Determine the percent decrease in volume of an oil that has a bulk modulus of 2×10^5 psi when a pressure of 3000 psi is applied.

Solution

Rearranging Equation 3.6 to solve for the percent change in volume gives

$$\frac{\Delta V}{V} = \frac{-\Delta p}{E}$$

$$= \frac{-3000 \text{ psi}}{2 \times 10^5 \text{psi}}$$

$$= -0.015$$

which is a 1.5 percent decrease in volume.

The value 2×10^5 psi (approximately 1.4 MPa) is a typical bulk modulus for petroleum-based fluids, while water has a bulk modulus of around 3×10^5 psi (approximately 2.1 MPa).

In practice, we really don't become concerned with bulk modulus unless we are interested in high cycling rates. In these applications, the fluid compressiblity causes the liquid to behave similarly to a gas and can significantly reduce the useful cycle rate of the system.

Liquids that are oversaturated with air show drastically reduced bulk moduli because the gas itself compresses.

3.5   Viscosity, Viscosity Index, and Pour Point

The viscosity, viscosity index, and pour point of a fluid are three very important fluid properties, because both system performance and component wear rate depend on them to a large extent.

Viscosity

Viscosity is defined as a fluid's resistance to flow, or internal resistance to shear, at a given temperature. When a layperson refers to the "weight" of an oil, or talks about a "thick" or "thin" oil, he or she is actually talking about its viscosity. It is probably the single most important property of an oil. Among the areas affected by viscosity are the formation and maintenance of a lubricating film, the pressure losses due to fluid friction in pipes, and the ease with which the fluid can be pumped. Another important aspect of viscosity is the destructive effects of cavitation on the pump that result from the inability of atmospheric pressure to push a highly viscous fluid into the pump inlet port.

The basic concept of viscosity is illustrated in Figure 3.1, where a plate is being moved across a fluid film that separates it from a stationary plate. Because fluids tend to cling to the surfaces with which they are in contact, the fluid adjacent to the stationary surface will remain stationary, while that adjacent to the moving surface will be moving at velocity v. The oil between the two surfaces can be considered a series of molecule-thick shear planes that are moving relative to one another at a velocity that is somewhere between zero and v. A shearing stress is created in the fluid that was defined by Sir Isacc Newton as

$$\text{Shear Stress} = \text{Constant} \times \frac{\text{Change in Velocity}}{\text{Change in distance}}$$

or   (3.7)

$$\tau\text{(tau)} = \mu\text{(mu)}\frac{\Delta v}{\Delta y}$$

where τ is shear stress and μ is a proportionality constant that is defined as the *dynamic* or *absolute* viscosity of the fluid. In the U.S. Customary system of units, dynamic viscosity has the units lb·sec/ft² or slug/ft·sec. In the SI system, its units are N·s/m² or Pa·s.

If we consider Equation 3.6 on a macro scale, that is, let $\Delta v = v$ and $\Delta y = y$, we can write

$$\tau = \mu\frac{v}{y}$$

Another way of expressing shear stress is

$$\text{Shear Stress} = \frac{\text{Force}}{\text{Area}}$$

or

$$\tau = \frac{F}{A}$$

where

F = force required to move the plate at velocity v.

A = area of the plate.

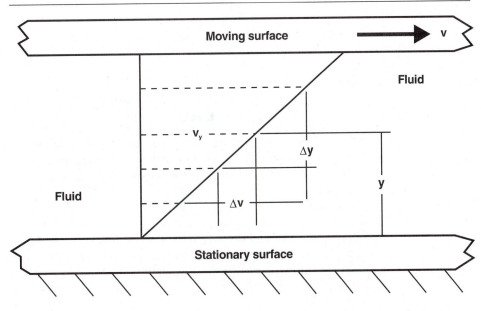

Figure 3.1 Fluid velocity profile between stationary and moving plates.

Equating these two and solving for the dynamic viscosity gives us

$$\mu\frac{v}{y} = \frac{F}{A} \tag{3.8}$$

$$\mu = \frac{F \times y}{v \times A}$$

Example 3.6

In Figure 3.1, the moving plate is 2 ft on a side and the fluid film is 0.1 in thick. A 28.8 lb force is required to move the plate at 2 ft/sec. Find the dynamic viscosity of the fluid.

Solution

Using Equation 3.8,

$$\mu = \frac{F \times y}{v \times A} = \frac{28.8 \text{ lb} \times 0.1 \text{ in} \times (1 \text{ ft}/12 \text{ in})}{(2 \text{ ft/sec})(4 \text{ ft}^2)}$$

so that

$$\mu = 0.03 \text{ lb·sec/ft}^2$$

Another common unit for dynamic viscosity is the *centipoise*, which is shorthand for g/(cm·s)/100.

The *kinematic* viscosity of a fluid is simply its dynamic viscosity divided by the fluid density at the temperature at which the dynamic viscosity was determined. thus, we have

$$\text{Kinematic Viscosity} = \frac{\text{Dynamic Viscosity}}{\text{Density}}$$

or

$$\tag{3.9}$$

$$\nu(\text{nu}) = \frac{\mu}{\rho}$$

where

$$\nu = \text{kinematic viscosity.}$$

If we use lb·sec/ft^2 for the dynamic viscosity and slug/ft^3 for the density, we can express kinematic viscosity in ft^2/sec. In the SI system we would use m^2/s. While these units are convenient for calculations, it is very common in the fluid power field to see kinematic viscosity expressed in *Saybolt Seconds Universal* (SSU) or in *centiStokes* (cSt).

The measurement of viscosity in SSU (or SUS, as it is also commonly expressed) is accomplished by using a test apparatus called a Saybolt Viscometer in accordance with ASTM D 88. This standard has been rescinded by ASTM in deference to other, more accurate viscometers; however, the traditional use of the SSU terminology continues. The change from SSU to cSt is facilitated by ASTM D 2161, which contains both formulae and tables for converting from one to the other.

When converting SSU to cSt, different equations are required, depending upon whether the SSU is greater or less than 100. If the viscosity is between 32 and 100 SSU, the viscosity in cSt is found from

$$\nu = 0.226t - \frac{195}{t} \qquad (3.10)$$

where

$$\nu = \text{cSt.}$$

$$t = \text{time in seconds (SSU at } 100°\text{F).}$$

For viscosities above 100 SSU,

$$\nu = 0.226t - \frac{135}{t} \qquad (3.11)$$

(handwritten: 0.220t)

Note: These equations provide *estimates* of the equivalent values. For exact conversions, refer to ASTM 2161.

Example 3.7

Convert the following SSU viscosities to centistokes:
a. 55 SSU
b. 200 SSU

Solution

a. Since 55 < 100, we use Equation 3.10. Thus,

$$\nu = 0.226t - \frac{195}{t}$$

$$= (0.226 \times 55) - \frac{195}{55}$$

$$= 8.9 \text{ cSt}$$

b. For the 200 SSU, we must use Equation 3.11.

$$\nu = 0.226t - \frac{135}{t}$$

$$= 0.226 \times 200 - \frac{135}{200}$$

$$= 44.5 \text{ cSt}$$

Most methods for determining viscosity require that the tests be conducted at specific temperatures. Commonly 40°C and 100°C. The oil is then classified into a viscosity grade normally based on the viscosity at 100°C. In the United States, and in fact in a large part of the world, the SAE viscosity grades have been standard for many years, especially for crankcase oils. As seen in Figure 3.2, these SAE grades define fairly large ranges of viscosity. In recent years, however, a classification system has been developed jointly by ASTM and

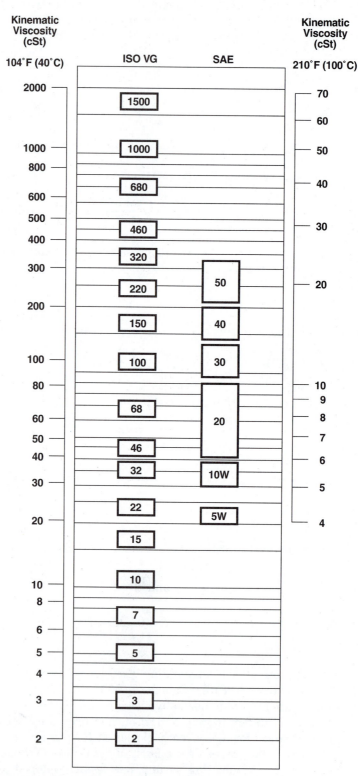

Figure 3.2 Viscosity grade comparisons.

the Society of Tribologists and Lubrication Engineers (STLE) [formerly the American Society of Lubrication Engineers (ASLE)], which grades fluids according to their kinematic viscosities in cSt at 100°F. This standard has been adopted worldwide with the minor change that it uses the viscosity at 40°C instead of 100°F. It has been designated ASTM D 2422 and ISO 3448. Fluids rated by the ISO scale are specified by ISO viscosity grade (VG). Most fluid power systems use ISO VG 32, ISO VG 46, or ISO VG68. Pump manufacturers will usually specify the viscosity range in which their pumps will operate properly.

Figure 3.3 shows some typical viscosity-temperature curves that illustrate the change of viscosity with temperature for several fluids. On this standard ASTM chart, which has a linear scale for temperature and a specially derived scale for viscosity, the curves for most fluids are straight lines.

The chart shows us what we have already known intuitively and from experience, if not from actual testing; that is, that the viscosity of a liquid changes with temperature—as the temperature increases, the viscosity decreases; as the temperature decreases, the viscosity increases. We're all familiar with this phenomenon from observing how fluids pour slowly when they are cold but fast (sometimes too fast) when they are hot. The degree to which the viscosity of a fluid is affected by temperature is termed its *Viscosity index*, discussed next.

Viscosity Index (VI)

As explained, the *viscosity index* (VI) is an indication of a fluid's rate of change of viscosity with a change in temperature. From Figure 3.3, we see that not all fluids respond to temperature changes at the same rate. The VI rating was originated to account for this and to provide a standard method for describing this fluid characteristic. Unfortunately, the VI rating system can cause a little confusion at the first encounter, because it does not, as one might expect, refer to the slope of the line of the ASTM temperature-viscosity chart. In fact, VI is simply a dimensionless number that indicates the relative viscosity of a fluid between two arbitrary limits. These limits were originally set at zero (based on oils from the fields along the coast of the Gulf of Mexico that had a very high rate of viscosity change with temperature) and 100 (based on Pennsylvania oils with a very low rate of change). The originators felt that these two oils represented the extremes and that all other fluids would fall between these limits. While this may have been the case at that time, today we have many oils with VIs far in excess of 100 and others (usually specialty oils) with VIs of far less than zero. In any case, the main source of confusion arises from the fact that a *lower* VI indicates a *higher* rate of change of viscosity with temperature, while a *higher* VI indicates a *lower* rate of change. Many fluid formulations include long chain polymers as VI improvers; that is, they increase the VI.

The viscosity index of a fluid can be calculated, using information contained in ASTM D 2270, as follows:

$$VI = \frac{L - U}{L - H} \times 100$$

(3.12)

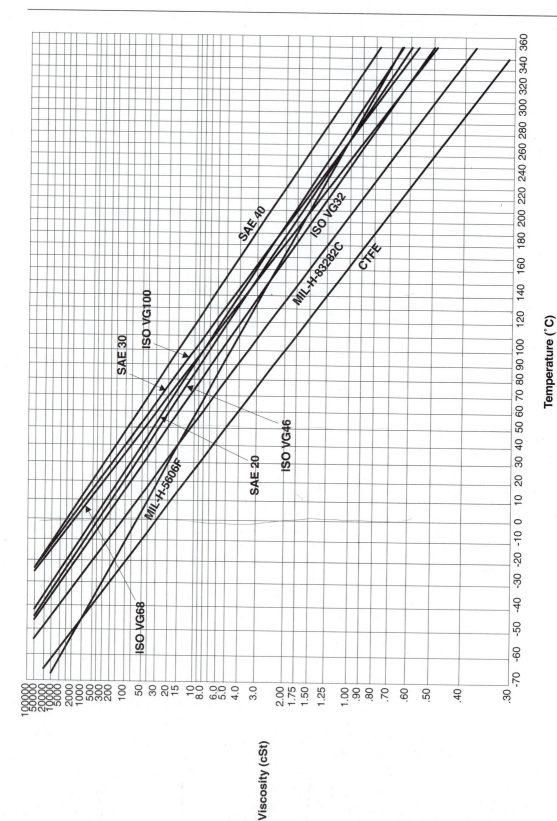

Figure 3.3 ASTM Standard viscosity temperature charts for liquid petroleum products.

where

L = the viscosity, in cSt, at 40°C of an oil that has a VI of zero and the
same viscosity as the test oil at 100°C.
H = the viscosity, in cSt, at 40°C of an oil that has a VI of 100 and the
same viscosity as the test oil at 100°C.
U = the viscosity, in cSt, of the test oil at 40°C

Example 3.8

Find the VI of an oil that has a 48 cSt viscosity at 40°C. The 0 VI reference
oil has a viscosity of 86 cSt at 40°C, while the 100 VI reference has a 32 cSt
viscosity at 40°C.

Solution
From Equation 3.12,

$$VI = \frac{L - U}{L - H} \times 100$$

$$= \frac{86 - 48}{86 - 32} \times 100$$

so that

$$VI = 70.4$$

Pour Point

The *pour point* of an oil is defined as the lowest temperature at which movement
of the oil is observed when it is chilled without disturbance under controlled
conditions. This is a very important point when the system is to be operated
under extreme environmental conditions, but it becomes less of a problem for
systems that are not exposed to outside conditions. If cold temperatures might
be a consideration for a specific application, the fluid selected should have a
pour point 15°F to 20°F below the lowest expected environmental temperature.
Pump manufacturers normally indicate the highest acceptable viscosity for
operating their pumps.

Viscosity Considerations in Fluid Selection

Since a system design calls for a specific fluid viscosity for optimum operation,
it is obvious that there are two important viscosity considerations in selecting
the initial or a replacement fluid. The first of these is the fluid viscosity at the
expected (or actual) operating temperature. The second is the viscosity index
(VI).

The use of an oil with too low a viscosity can lead to several system problems,
including

1. Loss of pump efficiency due to internal leakage.
2. Loss of system static pressure due to internal and external leakage.

3. Reduction of actuator speed due to internal leakage.
4. Increased component wear due to breakdown of the lubrication film.
5. Loss of overall system efficiency due to leakage.

If the viscosity of the fluid is too high, the problems may include

1. Loss of the pump's ability to move the fluid (pumpability).
2. Pump cavitation (starvation).
3. System overheating due to excessive fluid friction in the system piping.
4. Sluggish operation of cylinders and motors.
5. High pressure drops due to excessive fluid friction in the system piping.

A casual review of fluid manufacturers' catalogs will reveal numerous fluids that have a given viscosity at a given temperature. If we could be assured that the fluid would always be at that temperature, any of those fluids would be acceptable, at least as far as viscosity is concerned. As the temperature moves away from that temperature, the fluid viscosity will change. In this case, the VI becomes important, because we would like to keep the viscosity as close to the optimum operating point as possible to preclude viscosity-related problems. Again, a review of manufacturers' data sheets reveals many fluids available to meet the VI requirements.

The extremes of temperatures to which the system will be exposed determine the significance of VI and pour point. In aerospace vehicles, construction equipment, and systems that might see wide fluctuations in temperature, it is critical to choose a fluid with a very high VI and a very low pour point. For an industrial system operating in a relatively stable and protected environment, other considerations might be far more important than VI and pour point.

3.6 Lubrication

Many volumes have been written on the subject of lubrication. It is not the purpose of this small section to make you an expert on the subject, but to introduce you to some of the basic concepts especially as they are related to fluid power and fluid viscosity.[1]

In fluid power systems, as in most mechanical systems, we are concerned with three basic lubrication problems—linearly sliding surfaces, plain (or journal) bearings, and antifriction bearings. One of the major functions of the hydraulic fluid is to provide a lubricating film to reduce friction and wear in these situations.

Regardless of the surfaces involved, there are three basic regimes of lubrication, as illustrated by Figure 3.4:

1. Full film lubrication.
2. Mixed film lubrication.
3. Boundary lubrication.

In spite of appearances, no surface is absolutely flat. Microscopic examination will reveal that the surface is completely covered by rough peaks called *asperities*. When two surfaces move relative to one another, contact between these asperities creates friction, generates high local temperatures, and causes the peaks to break off and contaminate the system. Good lubrication will eliminate or at least significantly reduce the effects of this asperity contact.

Full film lubrication is generally the most desirable regime, because during normal operation, the lubricating film is thick enough to prevent any asperity contact. As a result, there is almost no wear of the surfaces, and friction is reduced to the shearing of the fluid.

Mixed, or thin film lubrication occurs when there is not sufficient film thickness to completely separate the surfaces. In this case, the fluid that is present carries part of the load, and the surface asperities carry the remainder. There is more wear than in the full film mode because of the asperity contact; however, many hydraulic fluids are formulated with extreme-pressure (EP) additives to minimize this problem. If this regime can be carefully balanced to virtually eliminate asperity contact, friction is actually reduced to a minimum because of the very thin fluid film that must be sheared. Any very small disturbance can upset this critical balance, however, so it is the practice of most designers to move well out of this regime and into the full film regime.

In the boundary lubrication regime, there is virtually no oil film to separate the surfaces. In this situation, friction is high, and heavy wear is likely to occur. Extreme-pressure fluids offer the only solution to this problem short of solid film (greases), and this solution is only partial at best.

A common way to display these lubrication regimes for rotating devices is by use of the Stribeck diagram shown in Figure 3.5. This graph is often termed the *ZN/P diagram* because it depicts the effects of viscosity (Z), speed (N), and pressure (P) on the coefficient of friction (f). When consistent units are used, the term ZN/P is dimensionless.

From a coefficient-of-friction consideration, the boundary lubrication portion of the ZN/P diagram should be avoided. The diagram shows that the mixed film zone has the lowest coefficient of friction. It is attractive from that viewpoint; however, disturbances such as shocks or changes in speed or pressure could cause the lubricating film to rupture and allow the surfaces to be damaged by unlubricated contact. As a result, most designers attempt to ensure that their equipment operates in the full film zone slightly to the right of the mixed film zone.

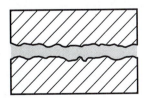
Full Film Lubrication, R ≥ 4

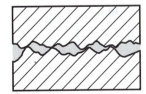
Mixed Film Lubrication, 4 ≥ R ≥ 1

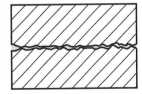
Boundry Lubrication, R ≤ 1

$$R = \frac{\text{oil film thickness}}{\text{surface roughness}}$$

Figure 3.4 There are three basic regimes of lubrication: full film, mixed film, and boundary lubrication.

Most lubricated surfaces experience all three lubricating regimes, at least on startup if not during normal operation. This can be easily demonstrated by looking at the development of the lubricating film in a journal bearing during startup. This type of bearing is commonly used in hydraulic pumps and motors.

In Figure 3.6a, the shaft is stationary, and metal-to-metal contact exists between the shaft and the bearing. When the machine is started, there is a short time of unlubricated rotation. During this time, friction is high and the shaft tends to "climb" the bearing in the direction opposite the direction of rotation. As it moves upward, it comes in contact with the lubricating oil and moves from the unlubricated state into the boundary regime. As it continues to rotate (Figure 3.6b), the tendency of the oil to adhere to the shaft and move with the moving surface allows a wedge of oil to be drawn into the space between the shaft and the bearing, developing the mixed film regime. As the shaft speed increases, the amount of oil drawn in increases until the full film regime is developed. At this time, the shaft will essentially be centered in the fluid film that is nearly concentric with the shaft and bearing (Figure 3.6c). Some eccentricity will exist because of the load, but in a well designed system using the proper fluid, full film lubrication will completely surround the shaft.[1]

The film thickness is a directly proportional function of viscosity for any given load and shaft speed. Therefore, as the viscosity decreases, the film

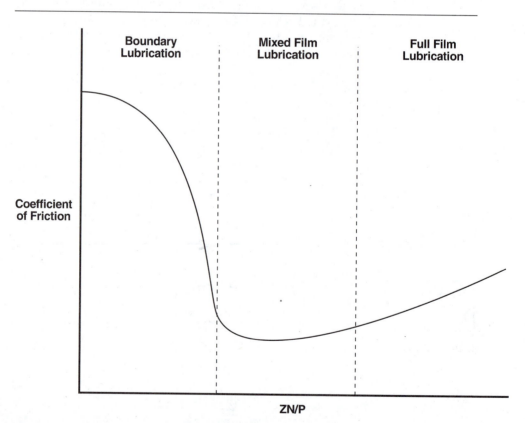

Figure 3.5 The Stribeck ZN/P diagram shows the effect of viscosity (Z), speed (N), and load (or pressure, P) on bearing friction.

thickness decreases until eventually the lubricating film can no longer maintain the separation between the surfaces. On the other hand, the coefficient of friction in the full film regime is also a direct function of viscosity, so as the fluid viscosity increases, fluid friction losses also increase.

You can see, then, how important viscosity is for both system performance and component lubrication. Remember that both the viscosity at the normal operating temperature and the viscosity index must be carefully considered in selecting hydraulic fluids.

3.7   Fluid Analysis

Selecting the right fluid for a particular application is only the first step in preventing fluid-related problems. Also vital to the health and reliability of fluid power systems and components, frequently omitted by the user, is *fluid analysis*—periodically taking a look at the condition of the fluid. Since hydraulic fluids degrade in service due to solid contaminants, oxidation, heat, and so forth, it stands to reason that there is a point at which it will become unusable. Without periodic analysis, the user will not know when that point has been reached until it is too late.

There is, however, an even more important reason for fluid analysis: Except for structural failure, fatigue, and fluid leaks, almost all hydraulic system failures are fluid related—contamination, viscosity loss, additive depletion, and so on. If these fluid-related problems can be detected, in many cases they can be corrected and the fluid restored to satisfactory condition. For example, it has been estimated that 75 percent of all hydraulic system failures are the result of particulate contamination in the fluid. Contrary to popular belief, contamination problems are not hard to solve. Consequently, a well designed

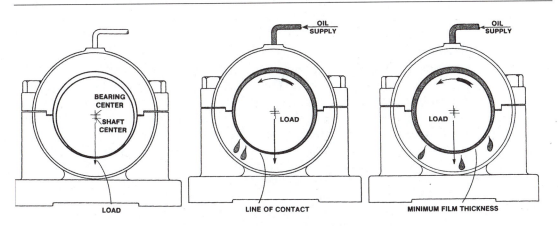

Figure 3.6 Development of hydrodynamic film in a full journal bearing with downward load.

Source: J. George Wills, *Lubrication Fundamentals* © 1980 Mobil Oil Corporation (New York: Marcel Dekker), p. 92.

and rigorously applied contamination-control program could eliminate 75 percent of industry's hydraulic problems! The basis of that contamination-control program is necessarily a good fluid-analysis program.

The subject of contamination control could (and indeed does) fill many volumes. Chapter 11 contains some basic information on the subject. Unfortunately, there is not much space in this text to treat the subject in detail, but the reader is encouraged to pursue it further before suffering too many catastrophic failures.

Important Analysis Parameters

Of the hundreds of analyses that can be conducted on hydraulic fluids, most are significant only to the fluid manufacturer and perhaps to those responsible for the initial specification of fluid for hydraulic systems. There are, however, a few tests that a user should employ to check the condition of the hydraulic system fluid. A fluid-condition monitoring program should include these analyses on a periodic basis.

Particulate

Particulate analysis should include both a contamination-level analysis (such as a particle count) and a materials analysis (visual microscope, etc.) to identify the types of particles present. Because of the almost infinite variety of systems, component designs, and operation parameters, it is virtually impossible to establish a general contamination limit other than to say that the cleaner a fluid is, the longer the components are likely to last. Only experience can lead to setting a limit that is acceptable from both the reliability and economical points of view.

Water Content

Water can be damaging to a system for several reasons, among them, rusting, oxidation, microbial growth, depletion of additives and loss of lubrication. As a general rule, a water content of 0.2 percent in the commonly used petroleum-based fluids indicates that corrective action is required.

Viscosity

The viscosity of a hydraulic fluid may either increase or decrease with fluid life, depending on the reason for the change. Normally, the viscosity of petroleum-based fluids will increase with use because of chemical breakdown, evaporation of lighter fractions of the oil, oxidation, chemical contamination, and so on. Many fluids, especially those whose formulations include long-chain polymers, will show a decrease in viscosity due to the shearing of the fluid as it passes through pumps and valves. Certain chemical contamination can also cause a decrease in viscosity. In either case, a change of more than 10 percent usually indicates a problem.

Acidity

Corrosion in fluid power components is often the result of a high acid level in the fluid. Acidity is normally measured by determining the amount of potassium hydroxide (KOH) required to neutralize the acid in one gram of the hydraulic fluid. It is commonly reported in milligrams of KOH per milligram of oil (mg KOH/mg). Many hydraulic fluids contain additives that are somewhat acidic, so the change in acid level is usually of more significance than the absolute level. A change of 1.0 mg KOH/mg from the original level is an indication of significant degradation. Since acid level increases at an exponential rate, the problem must be corrected quickly to prevent disaster, because a high acid level can cause corrosion, surface degradation, increased leakage through components, and even valve jamming.

Techniques of Fluid Analysis

Fluid analyses may be done either in-house or by outside laboratories depending on the size and policy of the company. If the samples are sent out to a laboratory, they will no doubt benefit from recent technology and skilled technicians using the latest standardized procedures. When seeking a laboratory to conduct such analyses, one should be familiar with the applicable standards so as to question technical personnel (the laboratory manager, for instance) about their adherence to these procedures.

Most commercial laboratories offer an analysis program consisting of several different tests. The laboratories can usually provide a particle count that will indicate the number and sizes of solid particles in a given volume of the fluid. They may also analyze the materials (iron, copper, aluminum, etc.) of the solid particles using *spectrographic analysis* techniques. In this method, a sample of the oil is burned in an electric arc. The spectrum of colors given off as the sample burns is analyzed electronically. Since each element produces a distinct and unique color spectrum when it is burned, the presence of the element can be detected by the presence of its spectrum. However, this analysis technique has not been highly successful for fluid power systems.

A very useful technique for analyzing fluid power system (in regard to the presence of iron) is *ferrography*, in which a sample of the oil is allowed to flow slowly across a glass slide positioned over a strong magnet. The magnetic attraction causes all particles that contain iron to stay on the slide. The slide is later inspected under a high-powered microscope to determine the amount and severity of the wear that has occurred in the system components.

Other analyses, as mentioned, involve water content, acid level, and viscosity. Infrared analyses may be conducted to determine if the fluid additives have been depleted and to check for changes in the chemical structure of the oil due to overheating or contamination by other fluids or chemicals. If the testing is done in-house, the same formal laboratory techniques employed by outside laboratories can be used, or some less formal field tests can be utilized with good results.

For instance, particle analysis can be accomplished readily and inexpensively on-site using sampling and analysis systems that provide a comparative means for determining contamination levels as well as allowing microscopic viewing of the particles (Figure 3.7). This visual microscopic analysis is important because it can provide much information concerning the type and source of the particles.

While it is generally expensive to obtain accurate viscosity measurements on-site using laboratory-grade viscometers, reasonably accurate results can be obtained using devices that compare the viscosity of the system fluid with that of a known reference. The Visgage (Figure 3.8) is one of several viscosity comparator units that can be used on-site with reasonable accuracy by relatively unskilled personnel.

Acidity can be determined on-site through the use of kits such as one from the Gerin Corporation that uses pre-measured, sealed containers of potassium hydroxide (KOH). The KOH is added to a test solution one container at a time until the solution changes color. The number of milligrams of KOH added indicates the acidity of the fluid sample.

Quantitive water content testing is difficult to do on-site. The most common check is termed the *crackle* test, in which a hotplate or a heat-resistant vessel is used. A sample of the oil is either dropped on the heated hotplate or heated in the vessel to about 250°F. If there is water in the sample, it will crackle and pop at that temperature. This technique indicates only that water is present; it gives little or no indication of the quantity.

The appearance and smell of a fluid sample can also give important information in on-site analysis. A milky appearance usually indicates water contamination, while dark color indicates either gross particulate contamination or oxida-

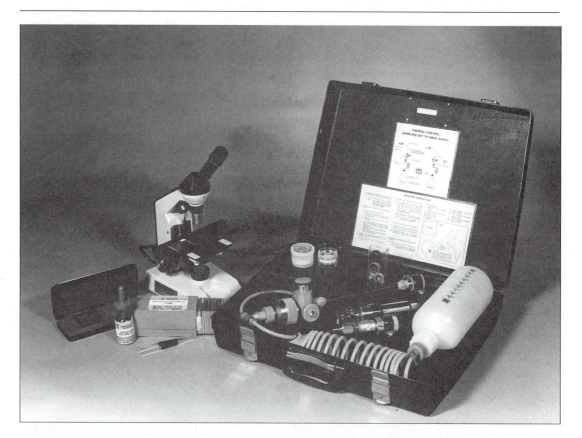

Figure 3.7 CONPAR fluid contamination analysis equipment for on-site analysis.

tion. A burnt or acrid smell indicates that the fluid has been overheated and may be severely oxidized, especially if it is discolored.

Having come from a maintenance background, the author well realizes the importance of having on-site analysis capability. In many cases, it is important to have immediate results rather than waiting for samples to be sent to a laboratory for analysis. This is especially true in commissioning and cleaning systems after repairs or overhauls. While analyses done in the field or on-site might not provide the accuracy and sophistication of those done in a well-equipped laboratory, they can be used to great advantage as a first, quick look, supplemented if necessary by formal laboratory analyses later.

A fluid-monitoring program can be only as good as the sampling technique, the analysis procedures, and the interpretations of results. A poorly taken sample may not be representative of the system fluid. A bad analysis makes the effort useless. Failure to correctly interpret the analysis results makes the whole program an exercise in futility.

3.8   Fluid Handling, Storage, and Disposal

One of the major sources of contamination—especially water and particulate—in a hydraulic system is the new fluid that is put into it. Most fluid manufacturers make efforts to ensure that the hydraulic fluid that comes out of the pipe at the end of the process is reasonably clean. Unfortunately, due to packaging, handling and storage, its condition usually goes downhill from there. Hydraulic fluid purchased in 1 to 5 gallon cans is usually relatively uncontaminated but the fluid from 55 gallon drums and bulk storage containers is often highly contaminated. A study conducted at the Fluid Power Research Center at Oklahoma State University[2] showed that one sample of new fluid contained well over 100,000 particles per milliliter larger than 5 μm. The average contamination level of new fluid was nearly 5,000 particles per milliliter. That's dirtier than many operating systems!

Maintaining the cleanliness of fluid in bulk storage is a problem because of the initial condition of the container itself, the process and equipment used for filling it, and the environment to which it is exposed. The cleaning of the container is very difficult, and the filling process is often practically uncontrolla-

Figure 3.8 The Visgage is a viscosity comparator unit that can be used with relative ease for on-site viscosity checks.
Source: Louis C. Eitzen Co., Inc., Glenwood Springs, Colo.

ble; therefore, a method for cleaning the fluid and keeping it clean while in storage is needed. The answer is a filtration system that continuously circulates the fluid through a good set of filters. In addition to this, exposure to the environment can be minimized by ensuring that all openings are sealed and that breathers contain good filters along with desiccants to remove moisture from any air that enters the vessel.

Transferring fluid from the storage tank to the machine is also a major source of contamination. The dirty bucket and the siphon hose have been the death of many good pumps and valves. Fluid servicing equipment must be maintained at almost surgically clean conditions and used only for servicing hydraulic fluids. Ideally, fluid should be pumped directly into the system through a very fine polishing filter.

One of the most critical issues facing all of us is the protection and restoration of the environment. A special problem to users of hydraulic systems is the handling of spills and the disposal of fluids. The Environmental Protection Agency (EPA) will no longer allow users to simply flush them down the drain or dump them in the pit out behind the plant. As responsible citizens, we should view such actions as unacceptable even if they were allowed.

Environmental concerns focus on two broad categories. The first is preventing spills and leaks. Systems should be properly designed to allow components to be changed without spilling significant quantities of fluid. Tapered pipe threads should be avoided wherever possible because of their tendency to leak. Fluid leakage should not be tolerated by either management or maintenance personnel. Conduits (hoses, pipes, tubing) should be routed so that failure due to abrasion, stress, or fatigue is not likely to occur.

The second category is disposal. All fluids and fluid-contaminated items must be disposed of in accordance with current EPA guidelines. This usually involves contracting with an EPA-approved hazardous-waste handler to collect waste materials for proper disposal or burning the oil in an EPA-approved unit (usually a heater or oil-fired boiler or furnace). Recent EPA rulings make the user responsible for hazardous waste "from the cradle to the grave"—from the time it is unloaded on the dock until the disposal operation is complete. This means that simply turning the waste over to an EPA-approved handler is not sufficient. *You* must ensure that the handler disposes of the waste properly. If the driver decides to take the load out somewhere and dump it on a dirt road, you can be held responsible.

There are a variety of systems available for reclaiming or reconditioning used oils. While there is some merit to these techniques, they usually result in a product that is inferior to the oil normally recommended for fluid power systems. It is much more effective to take care of the oil in the system to prolong its usual life. This will eliminate the disposal problem and save the money that would be spent in replacing the oil. A popular addage in the fluid power community is, "Keep it clean; keep it cool." Following this advice will increase the life of the oil and the components.

3.9 Fire-Resistant Fluids

In many applications, hydraulic fluids may be exposed to high-temperature materials and even open flames. In these applications, fire-resistant fluids are usually utilized. (It is important to caution you that although these fluids are less flammable than straight petroleum products, they are not nonflammable.) These fluids are formulated to reduce their tendency to ignite and their ability to support combustion in the absence of an ignition source, while still performing their basic functions as hydraulic fluids.

Defining Fire Resistance

A fire-resistant fluid can be defined, at least for now, as any fluid that shows less tendency to ignite and/or support combustion than a petroleum hydraulic fluid. This definition is very loose and far from absolute, because there are fluids that have better characteristics than petroleum fluids when exposed to some hazards, but worse characteristics when exposed to others.

Evaluating the fire resistance of a fluid is even more difficult than defining it. The International Organization for Standardization (ISO) has said that a minimum of 12 different tests are necessary to evaluate the fire-resistance characteristics of a fluid over the full range of expected hazards. Unfortunately, there are no accepted procedures for five of these tests, and many of the others are poorly defined. A tremendous amount of work will be required to standardize this area and allow the user to really understand the fluids in his system.

Fire-Resistant Fluid Categories

The categories of fire-resistant fluids as defined by ISO 6743/4 are as follows:

1. **HFA: Oil-in-water fluids containing in excess of 80 percent water, with remainder made up of petroleum or synthetic oils and additives**. These are commonly referred to as *high water base* or *high water content* fluids. They are broken down into two subcategories.

   a. **HFAE** is an emulsion of petroleum oil droplets suspended in water. The emulsion is maintained by surrounding the oil droplets with a film of an emulsifying agent, usually a soap.

   b. **HFAS** is a true solution that uses water soluble synthetic rather than petroleum oils.

   The use of HFA fluids rose dramatically in the mid-1970s when petroleum was in short supply. Unfortunately, many people who switched to these fluids at that time had very bad experiences because of bad advice by salespeople and because of the very poor performance and lubricating ability of the fluids that were available at that time. Their popularity suffered tremendously as a result. In spite of this, a great deal of research has been done to improve these fluids and the components, especially pumps, with which they are to be used. The result is that there are HFA fluids available today that are comparable in both performance and component life to many conventional fluids.

2. **HFB: Water-in-oil emulsions consisting of up to 60 percent water, with the remainder made up of petroleum oil and additives**. These fluids are often termed *invert emulsions* because they contain water suspended in oil rather than oil suspended in water like the HFA fluids. They have the unusual characteristic of

having a viscosity that is higher than either of the main constituents. Strangely enough, when water is added to an invert emulsion, the viscosity actually increases. Because of their high viscosities, these fluids can often be used where an HFA fluid would be completely unacceptable.

3. **HFC: Water-glycol fluids consisting of 50 percent water and 50 percent alcohol-derivative (commonly a polyglycol of some type).** These fluids have very good low-temperature characteristics and can be used at pressures up to 3000 psi. A major problem with the water glycols is their incompatibility with most paints and many seal materials. They can also cause damage to zinc- and cadmium-containing metals.

4. **HFD: synthetic hydraulic fluids.** While phosphate ester fluids make up the bulk of this group, the HFDs also include synthetic hydrocarbons, synthetic and petroleum blends, and many specialty formulations. These fluids usually have excellent performance characteristics, but they are very expensive and frequently have compatibility problems.

All of these categories of fire-resistant fluids have significant advantages and disadvantages. In addition, the individual fluids included within each category can differ greatly from other fluids in the same category. While fluids from different categories may perform well in a given application, the decision to change from one fluid to another must not be taken lightly. Such a decision usually involves more than simply draining the system and refilling it with a new fluid. Major system redesign is sometimes required. Likewise, care must be exercised in switching from one brand to another. Study the manufacturers' data sheets carefully and, if possible, talk with past and current users of the product. Don't allow your purchasing people to push you into using a less expensive fluid unless you are certain it will do the job at least as well as your current fluid. A bad choice could lead to repair and downtime costs that will far outweigh the savings.

3.10　Summary

The hydraulic fluid is the single most important component in the system. Therefore, selection of the proper fluid for each specific application is critical to ensure maximum life and performance from the system hardware. In most cases, the fluid requirements of the system pump will be the most stringent of any component and will dictate the fluid to be used.

The characteristics of hydraulic fluids are enhanced by the use of additives. Most hydraulic fluids contain additives to improve their viscosity index, rust and oxidation protection, emulsibility/demulsibility characteristics, lubricity, and extreme-pressure capabilities. These additives can be depleted during the life of the fluid, and depletion can be accelerated by particle contamination, overheating, and contamination by water, air, and chemicals.

Periodic fluid analyses are useful in detecting degradation of the fluid before serious damage to the system components occurs. These analyses—primarily for particle contamination, acid level, water content, and viscosity—can be conducted on-site or by a commercial laboratory. Infrared analyses can be conducted periodically to check for additive depletion.

Fire-resistant fluids should be used where there is a significant fire hazard associated with the application. The four basic types—high water-base, water-in-oil emulsions, water glycols, and synthetic fluids—are all more difficult to ignite than petroleum-base oils, but all will burn under certain circumstances. Their characteristics—viscosity, lubricity, specific weight, temperature and pressure operating ranges, and so on,—can differ significantly from those of petroleum-base oils. Therefore, they may require special precautions or even specially built components in order to perform optimally.

3.11 Key Equations

Weight:
$$w = m \times g \tag{3.2}$$

Density:
$$\rho = \frac{m}{V} \tag{3.3}$$

Specific weight:
$$\gamma = \frac{w}{V} \tag{3.4}$$

Specific gravity:
$$Sg = \frac{\rho}{\rho_{\text{water}}} = \frac{\gamma}{\gamma_{\text{water}}} \tag{3.5}$$

Bulk modulus:
$$E = \frac{-\Delta p}{(\Delta V/V)} \tag{3.6}$$

Kinematic viscosity:
$$v = \frac{\mu}{\rho} \tag{3.9}$$

Conversions from SSU to cSt:
$$v = \left(0.226t - \frac{195}{t}\right) \text{SSU} < 100 \tag{3.10}$$

$$v = \left(0.226t - \frac{135}{t}\right) \text{SSU} > 100 \tag{3.11}$$

Viscosity index:
$$VI = \frac{L - U}{L - H} \times 100 \tag{3.12}$$

References

[1] For an excellent treatment of lubrication, see *Lubrication Fundamentals,* J. George Wills, (New York: Marcel Dekker, Inc., 1980.)

[2] L.C. Moore, "The Cleanliness of New Hydraulic Fluids," *BFPR Journal* 11 (1978): 71–75.

Review Problems

Note: The letter "S" after a problem number indicates SI **units** are used.

General

1. List the most important functions of a hydraulic fluid.
2. What is the primary difference between a liquid and a gas?
3. Explain the differences between a liquid and a gas in a closed vessel.

4. Define mass. What are the units of mass?
5. Define weight. What are the units of weight?
6. What is a newton? How does it compare to a pound?
7. Define density. What are the units of density?
8. Define specific weight. What are the units of specific weight?
9. Define specific gravity. What are the units of specific gravity?
10. Compare density and specific weight.
11. Define fluid oxidation. Why is it a problem?
12. The rate of oxidation doubles for every _____ above _____ °F. How does oxidation affect the useful life of the oil?
13. What is the difference between emulsibility and demulsibility?
14. Define flash point.
15. Define fire point.
16. Define autogeneous ignition temperature.
17. Define bulk modulus.
18. Define viscosity.
19. Define viscosity index.
20. List and discuss the problems caused by low fluid viscosity.
21. List and discuss the problems caused by high fluid viscosity.
22. List and describe the three lubrication regimes.
23. What four types of analysis should be included in a fluid-condition monitoring program?
24. List and define the categories of fire-resistant fluids.

Fluid Properties

25. Oil has a specific gravity of 0.87. Find its density and specific weight.
26. Find the specific gravity of a fluid that has a specific weight of 54.75 lb/ft^3.
27. A fluid has a density of 1.65 slug/ft^3. Find its specific gravity.
28. Find the specific gravity of a fluid that has a density of 2.06 slug/ft^3.
29. A body has a mass of 12.3 slugs. Find its weight.
30. How much does the body of Problem 29 weigh where the acceleration of gravity is 23 ft/sec^2?
31. Find the volume in in^3, ft^3, and gallons of 5000 lb of a fluid that has a specific gravity of 0.89.
32. How long does it take to transfer 1,500,000 lb of a fluid with a specific gravity of 0.92 through a 4 in pipe if the flow velocity is 25 fps?
33. A body has a mass of 30 slugs. Find its weight

    a. On the earth's surface.
    b. On a planet where the acceleration of gravity is 15.6 ft/sec^2.

34. One cubic foot of a substance has a mass of 7 slugs. Find its density.
35. One gallon of a substance has a mass of 3 slugs. Find its density.
36. One cubic inch of a substance has a mass of 0.002 slugs. Find its specific gravity.
37. A liquid has a specific gravity of 0.96. Find its density and specific weight.
38. A fluid has a specific gravity of 0.90. Find its

    a. Density.
    b. Specific weight.
    c. Weight of 50 gallons of the fluid.

39. Find the density and specific gravity of a fluid that has a specific weight of 58.7 lb/ft^3.

40. A tank car holds 33,000 gallons of fluid. If the fluid has a specific gravity of 1.07, find the fluid's density and specific weight, as well as the weight of the load.

41. A tank contains 4000 lb of a fluid that has a specific gravity of 0.90. Find the volume of fluid in gallons.

42S. A body has a mass of 100 kg. Find its weight

    a. On the earth's surface.
    b. On a planet where the acceleration of gravity is 4.6 m/s^2.

43S. Find the specific weight and density of a fluid with a specific gravity of 0.92.

44S. Find the specific gravity of a fluid that has a specific weight of 8.62 kN/m^3.

45S. A liquid has a density of 1125 kg/m^3. Find its specific weight and specific gravity.

46S. Find the volume (liters) of 3000 kg of a fluid that has a specific gravity of 0.87.

47S. One liter of a liquid has a mass of 3 kg. Find its density, specific weight, and specific gravity.

48S. A tank truck is carrying 100,000 L of a fluid with a specific gravity of 0.85. Find the fluid's density and specific weight, as well as the weight of the load.

Bulk Modulus

49. By what percent will the volume of water change with a pressure increase of 10,000 psi?

50. How large a pressure increase is required to change the volume of a petroleum oil by 5 percent?

51S. A hydraulic fluid has a bulk modulus of 1.7×10^6 kPa. By what percent will its volume change with a pressure increase of 100,000 kPa?

Viscosity

52. Convert the following viscosities from SSU to cSt:

    a. 40 SSU
    b. 150 SSU
    c. 225 SSU
    d. 80 SSU

Viscosity Index

53. Find the VI of a fluid that has a viscosity of 60 cSt at 100°F. The 0 VI and 100 VI reference oils have viscosities of 80 cSt and 40 cSt, respectively, at 100°F.

54. Find the VI of an oil that has a 35 cSt viscosity at 40°C. The zero-VI reference oil has a viscosity of 90 cSt at 40°C, while the 100-VI reference oil has a 42 cSt viscosity at 40°C.

Hydraulic Pumps

Outline

4.1   Introduction

The old cliché that the hydraulic pump is the heart of the system has some validity. In fact, a hydraulic system is similar to the human body in many ways, and the pump, like the heart, causes fluid flow. The hydraulic fluid, like the blood, is the transporter of energy, although in a different form, through the piping (circulatory system). Hydraulic cylinders can be compared to the muscles of the arms and legs, while filters function as the kidney. Of course, there are big differences, too, so let's not get carried away with the analogy.

Objectives

When you have completed this chapter, you will be able to

- Explain the relationship between pump flow and system pressure.
- Explain the purpose of a pump.
- Describe the sources of pressure.
- Explain the difference between positive displacement and non-positive displacement pumps.
- Describe the operating principle of all positive displacement pumps.
- Describe the basic pump types.
- Discuss the difference between fixed and variable displacement pumps and list the pump types in each grouping.
- Recognize and draw the graphic symbols representing the various pump types.
- Calculate pump efficiencies.

- Discuss the effects of air and gas in the fluid on pump life and performance.
- Describe the difference between cavitation and entrained air.
- List and discuss the parameters important in the selection of a hydraulic pump.

4.2   Some Notes on Flow and Pressure

As you begin this chapter, it is very important that you understand three basic points about fluid power systems. First, regardless of the type pump used in a fluid power system, the purpose of that pump is to generate fluid flow. Many people incorrectly think that pumps generate pressure, which brings us to the second important point: pressure is the result of resistance to the flow that the pump is trying to produce. In Figure 4.1, assume that we are using a *fixed displacement pump* (that is, one that produces the same output flow regardless

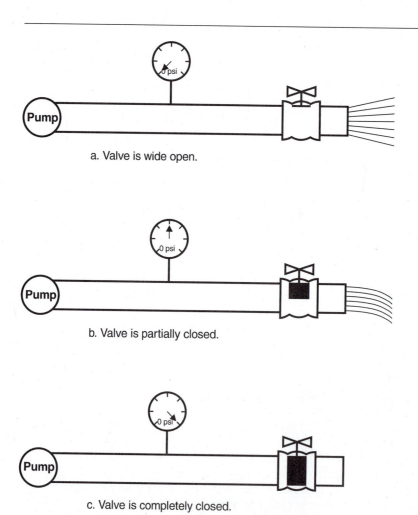

a. Valve is wide open.

b. Valve is partially closed.

c. Valve is completely closed.

Figure 4.1 Pressure is the result of resistance to flow.

of system pressure) being driven at a constant speed. In Figure 4.1a, the valve is wide open, allowing unrestricted flow of the fluid to atmospheric pressure. Other than a small friction resistance in the pipe, there is no resistance to the flow. Consequently, the pressure in the pipe is essentially zero. In Figure 4.1b, the valve has been partially closed, causing a restriction of the fluid flow. As a result of this restriction, the pressure in the pipe has increased. The increase will be proportional to the resistance to the flow.

We can relate this information to pump performance as follows: when a fixed displacement pump is described as a 30 gpm, 3000 psi pump, it actually means that the pump has a rated flow rate of 30 gpm and that it can continue to pump fluid at approximately that rate against a resistance of 3000 psi when driven at a constant speed. (As we will see later, if lower flow rates are required, we would be much wiser to use either a smaller pump or a variable displacement pump than to use a fixed displacement pump and a valve to reduce the flow.)

Finally, in Figure 4.1c, the valve has been closed completely, allowing no fluid flow. In this case, the pressure in the pipe will increase until something happens to arrest that increase—internal leakage in the pump increases, something breaks, a relief valve opens (our example doesn't have one), the drive unit stalls, and so on. The resistance to flow may be provided by a cylinder or hydraulic motor rather than a valve, but the resultant pressure rise will be the same. (Remember $p = F/A$.)

The third important point is that the maximum pressure in a hydraulic system depends on the least resistance to flow. Hydraulic fluid, like electricity, water, and many people, takes the path of least resistance. This can readily be demonstrated by a simple example. Look at the system shown in Figure 4.2. This is a *single-acting* hand pump, meaning that fluid is pushed into the pump from the tank by atmospheric pressure as the piston moves upward and is pushed into the system as the piston moves downward. Cylinder A has a piston area of 8 in², cylinder B is 3 in², and cylinder C is 2 in². Assume that the lines are all empty initially, and that all the pistons are at the bottom of the cylinders. The piston in the handpump has a ½ in² area. As the operator

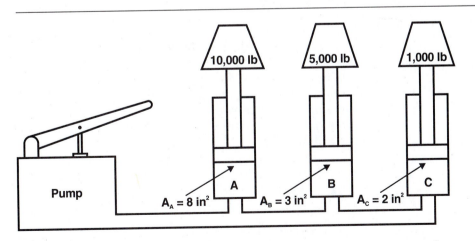

Figure 4.2 System pressure is determined by the least resistance to flow.

begins to stroke the handle, very little force is needed because he or she is simply transferring fluid from the tank to the empty lines. (The total operator effort simply compresses the air trapped in these lines.) As soon as all the lines are filled, however, the job gets a little tougher, because to push any more fluid into the system, enough force must be generated to move one of the loaded pistons. The easiest piston to move will be the one offering the least resistance to the fluid flow, that is, the one requiring the least pressure.

The pressures required to move the loads are found from Equation 2.1.

$$P_A = \frac{F_A}{A_A} = \frac{10,000 \text{ lb}}{8 \text{ in}^2} = 1250 \text{ psi}$$

$$P_B = \frac{F_B}{A_A} = \frac{5000 \text{ lb}}{3 \text{ in}^2} = 1667 \text{ psi}$$

$$P_C = \frac{F_C}{A_C} = \frac{1000 \text{ lb}}{2 \text{ in}^2} = 500 \text{ psi}$$

Cylinder C, at 500 psi, represents the minimum resistance to flow in the system. Therefore, as soon as the system pressure reaches 500 psi, the 1000 lb load can be lifted.

Now, consider what happens when piston C reaches the end of its stroke and contacts the end of the cylinder. At this point, the load on that piston becomes whatever force is required to stretch the cylinder material (in other words, the tensile strength of the cylinder). The operator, however, is still trying to force fluid into the system. Again, something must move to make room for it. As before, the cylinder with least resistance—now the 10,000 lb load on cylinder A—is the one that will move. As we have seen, a 1250 psi pressure on piston A will result in enough force to lift the load but will not be sufficient to cause piston B to move. As long as piston A will move, 1250 psi will be the maximum system pressure.

Finally, when piston A reaches the end of its stroke, the 5,000 lb load on piston B represents the least resistance to fluid flow. The result is a system pressure of 1667 psi if the load is to move.

To pull this all together, as you consider the role of hydraulic pumps, keep in mind these three facts:

1. Pumps generate flow, not pressure.
2. Pressure is the result of resistance to fluid flow. (We are ignoring static head pressure here because it is seldom a factor in fluid power systems because of their size.)
3. The maximum pressure in a fluid power system is the result of the minimum resistance to fluid flow at that time. (Another way to say this is that the actuator that requires the least pressure is the one that will move first.)

4.3 Types of Pumps

There are two very broad categories of pumps—non-positive displacement (sometimes termed *kinetic* or *hydrodynamic*) and positive displacement (sometimes referred to as *hydrostatic*). The design and use of the categories differ significantly.

Non-Positive Displacement Pumps

When we speak of the displacement of a pump, we are referring the amount of fluid moved during each rotation of the pumping mechanism. A *non-positive displacement pump* may be a centrifugal, axial propellor, or turbine pump. The movement of the fluid is the result of an impellor or propellor imparting momentum to that fluid (Figure 4.3). Jet pumps (sometimes called *eductors*) are also in this category. Because the internal design of these pumps does not include seals or sealing surfaces to ensure that the fluid cannot leak past the pumping mechanism, any slight increase in downstream pressure (resistance to flow!) causes an increase in internal leakage (called *slippage*) and reduces the pump output. Thus, in a non-positive displacement pump, the pump output is a function of pump design, rotation speed, fluid density, and downstream pressure. In fact, in some applications, the outlet of such a pump is completely blocked off during part of the operating cycle so that the pump output flow is zero. During this time, the rotating mechanism simply "churns" the fluid in the pump housing. The only result of this churning is heating of the fluid. Figure 4.4 shows a typical pressure-flow curve.

Another important characteristic of non-positive displacement pumps is that

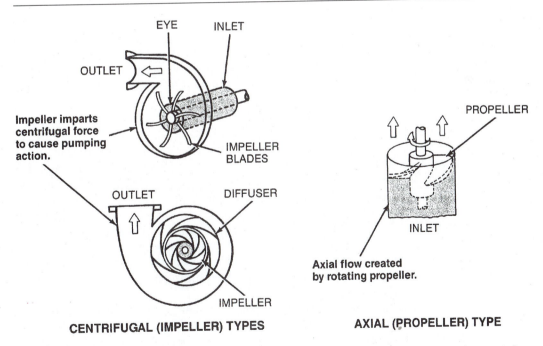

Figure 4.3 Non-positive displacement pumps may be centrifugal (impeller) or axial (propellor) types.
Source: Courtesy of Vickers, Inc.

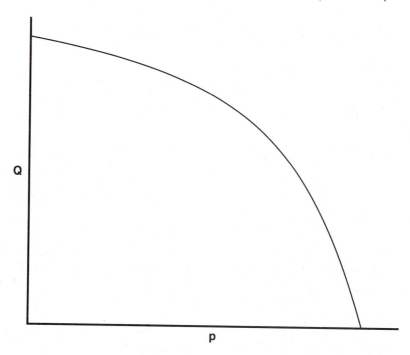

Figure 4.4 Typical pressure-flow curve for a non-positive displacement pump.

they do not prime themselves well, nor do they hold the prime. This means that the pump case and inlet lines must somehow be filled with liquid before they will start to pump. They will lose their prime again as soon as they are shut down unless a check valve is used in the inlet line to hold the liquid in the pump.

A common example of a non-positive displacement pump is the water pump on an automobile engine or a washing machine. In the engine application, the pump causes the water to circulate through the engine and radiator by imparting a high centrifugal force to it. The water is directed into the center of the impeller and is then thrown to the outside of the curved blades, into the diffuser portion of the housing, and subsequently out through the outlet port.

Non-positive displacement pumps are almost never used as fluid power pumps. Their role is normally that of transferring fluids where the primary resistance to flow is the result of flow friction through piping, valves, and process equipment and the specific weight of the fluid. These pumps can produce flows of up to several thousands of gallons per minute, but they work at pressure usually no higher than 200 psi.

This is a good place to clear up another problem in fluid power terminology. Quite often, when we see a stream of water emerging from an outlet (the water coming out of a fire hose, for instance), we utter exclamations about the "high pressure" of that flow stream. Such statements are inaccurate. The water, being unconfined by any sort of vessel, suddenly finds itself at atmospheric pressure. It is moving at a high velocity, however, so it possesses what is termed *velocity head* due to its kinetic energy. The ability to travel a long distance or knock a man down is due to the momentum of the water. When the fast-moving volume of water strikes an object, a force is created per Newton's Second Law ($F = m \times a$), because of the velocity head, not because of the static pressure in the system. We will discuss velocity head in more detail in Chapter 12 when the Bernoulli equation is presented.

Positive Displacement Pumps

The vast majority of pumps used in fluid power are *positive displacement pumps*. These units are designed so that the internal surfaces form seals to prevent the fluid being pumped from leaking from the high-pressure side of the pump to the low-pressure side. As a result, the pump takes in discrete "chunks" of fluid through the inlet port and expels them through the outlet port. Theoretically, all the fluid taken in on one revolution is discharged through the outlet port, hence the term "positive displacement." The amount of fluid moved in one revolution, which is a function only of the internal design of the pump, is termed the *pump displacement*. It is commonly expressed in cubic inches or cubic centimeters per revolution.

While the output of a positive displacement pump is a function of its internal configuration, it is possible to change the displacement of some pumps by changing the internal configuration during operation. These pumps are called *variable displacement pumps*. If it is not possible to change the displacement, the pump is termed a *fixed displacement pump*.

There are basically four families of positive displacement pumps:

  a. Vane
  b. Gear
  c. Piston
  d. Screw

The screw type is often considered a type of gear pump, but it is really a separate category. Only vane and piston pumps are generally available as variable displacement devices. Figure 4.5 shows the "family tree" of positive displacement pumps.

The principle on which all positive displacement pumps operate is basically very simple. Let's use a simple syringe as an example. In Figure 4.6a, the plunger of the syringe is pushed in. As the plunger is pulled back (Figure 4.6b), two things happen. First, the volume in the needle end of the syringe increases. Second, as a consequence, a low-pressure area, or vacuum, is created as the volume increases. If the end of the syringe is immersed in a liquid, atmospheric pressure on the surface of the liquid will push the liquid into the syringe.

In Figure 4.6c, the plunger is being pushed into the cylinder. This results in a decrease in volume in that area. Since the liquid is not compressible, it must be pushed out of the syringe.

Positive displacement pumps, regardless of their design, operate on the same concept. They contain pumping chambers that alternately increase and

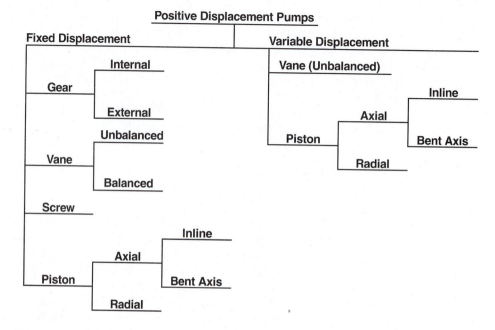

Figure 4.5 The family tree of positive displacement pumps has two main branches, fixed and variable displacement.

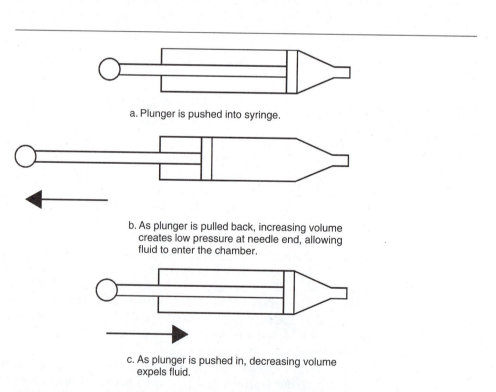

Figure 4.6 The principle on which all positive displacement pumps is based can be illustrated with a simple syringe.

decrease in volume. Fluid enters the chamber through an inlet port as the volume increases. It is pushed out of the chamber through the outlet port as the volume decreases. The exact mechanism by which this occurs varies greatly among the pump types, but the concept is the same for all pump types.

The principle we've discussed here concerns how the pump produces flow. Remember that pressure is the result of resistance to that flow. Therefore, the pressure at the outlet of the pump can vary dramatically. The flow rate, however, remains *essentially* constant for a positive displacement pump, regardless of the outlet pressure variations as long as the rpm remains constant (Figure 4.7).

Although the flow rate remains *essentially* constant, it does normally decrease somewhat as the pressure increases. This is due to internal leakage, often called *slippage*, inside the pump. Since the pumping mechanism must move inside the pump, there must be internal clearances. There is also a pressure difference across the mechanism—vacuum on the inlet side, system pressure on the outlet side. Any time there are clearances and a pressure difference, there will be leakage. The amount of leakage will depend on the size of the clearance and the pressure difference (as well as the fluid viscosity, as we'll see later). Thus, as the outlet pressure increases, so will the internal leakage in any given pump. We'll discuss this further in the section dealing with volumetric efficiency.

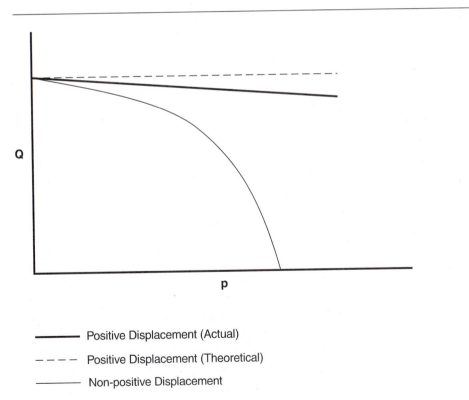

—————— Positive Displacement (Actual)

— — — — Positive Displacement (Theoretical)

—————— Non-positive Displacement

Figure 4.7 Pressure-flow characteristics of a positive displacement pump.

4.4 Vane Pumps

Vane pumps are categorized as *unbalanced* (or *simple*) and *balanced.*

Unbalanced Vane Pumps

One of the simplest of all positive displacement pumps is the basic sliding vane pump illustrated in Figure 4.8. Each pair of vanes form a pumping chamber that varies in volume as the pump rotates. As the chamber passes the inlet port, it is enlarging due to the eccentricity of the cam ring and the rotor. This creates a partial vacuum that allows fluid to be pushed into the cavity by the pressure (usually atmospheric) in the inlet line. As the rotation continues, the

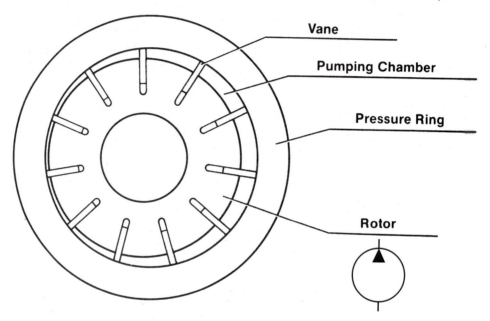

Figure 4.8 The basic sliding vane pump is one of the simplest of all positive displacement pumps. It is considered an unbalanced pump.
Source: Courtesy of Dana Fluid Power.

volume of the cavity is reduced. When the pumping chamber reaches the outlet port, the fluid is pushed out of the pump and into the system.

As the pump begins to rotate, the vanes are forced outward and held against the cam ring surface by centrifugal force. During operation, pressurized fluid that is ported into the slots in the rotor underneath the vanes may be used to supplement the centrifugal force and hold the vanes in their extended position. In some pumps, the vane pressure is monitored and controlled to prevent excessive wear due to high extension forces. Fluid leakage around the sides of the vanes and rotor is minimized by spring-loading and/or pressure-loading the side plates to seal in the fluid.

The pump illustrated in Figure 4.8 is often called an *unbalanced vane pump* because there is high pressure on the outlet side of the rotor and low pressure on the inlet side. This pressure differential causes a net force that pushes the rotor toward the inlet port (Figure 4.9). Although the actual movement of the rotor is minute, the result is an unbalanced loading of the rotor bearings. Recalling the principles of lubrication from Chapter 3, you will recognize that this situation can lead to a reduced lubrication film thickness, resulting in metal-to-metal contact and thus heavy wear on the bearings.

While bearing wear is inherent to the unbalanced vane pump design, it is

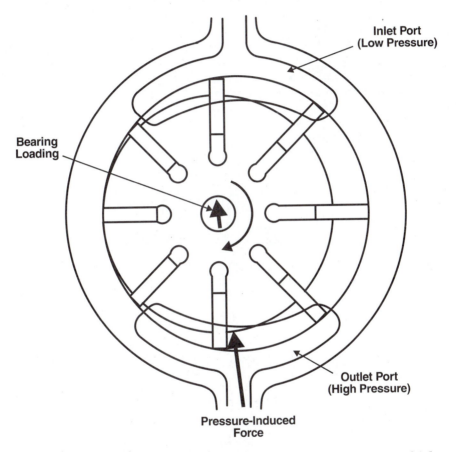

Figure 4.9 Force imbalance in an unbalanced vane pump causes high bearing loads.

often tolerated for one or more reasons. One reason is the relatively low cost of the pump due to the simplicity of the design. Another is that an unbalanced vane pump can incorporate a variable displacement feature by changing the position of the cam ring with respect to the rotor.

Figure 4.10 shows a *variable displacement vane pump*. The displacement of the pump is changed by changing the eccentricity of the cam ring with respect to the rotor. As the eccentricity is reduced, the rotor becomes centered in the cam ring (Figure 4.10a). This causes the differential volume experienced by the pumping chamber at the inlet and outlet ports to be eliminated. As a result, the fluid is not expelled through the outlet port, but is simply carried in the pumping chamber as it rotates. At this point, the pump has zero displacement and is said to be "off stroke." No fluid passes through it in this position.

Maximum pump displacement occurs when the greatest eccentricity exists between the rotor and the cam ring, as in Figure 4.10b. This position maximizes the differential volumes in the pumping chambers as the pump rotates and provides the largest volume at the inlet port of which the pump is capable.

The displacement of a vane pump can be varied manually by an external control such as a handwheel or automatically by a pressure compensator as in Figure 4.11. Such pressure compensators are devices that sense the pressure in the system and adjust the pump displacement (and consequently, the out-

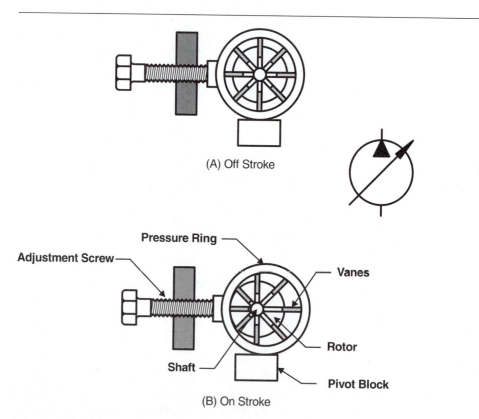

Figure 4.10 The flow of a variable displacement vane pump can be varied by using the adjustment screw.

Source: Courtesy of Dana Fluid Power.

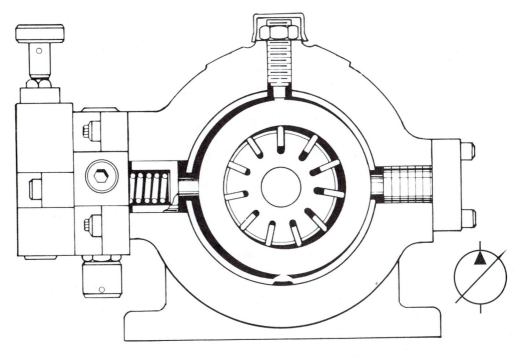

Figure 4.11 In a pressure compensated vane pump, displacement is automatically varied.
Source: Courtesy of Dana Fluid Power.

put)as necessary to maintain a preset pressure level. Remember that pressure is the result of resistance to flow, so in order to use a variable displacement pump to prevent the system pressure from exceeding the preset level, the flow rate must be reduced. In a vane pump, this is done by reducing the rotor/cam ring eccentricity.

The pump in Figure 4.11 has a compensator spring that applies a force to the cam ring and moves it to the right to obtain the maximum eccentricity. As the system pressure begins to increase, so does the pressure in the pressure side of the pump. This pressure exerted on the surfaces of the rotor and cam ring results in a force that opposes the compensator spring force. When the internal (pressure-generated) force exceeds the spring force, the cam ring moves to the left, the pump output is reduced, and the system pressure is reduced accordingly. The cam ring will continue to move to the left until the system pressure and resulting force are reduced to equal the spring force. At this point, the pump will provide just enough flow to make up any system leakage. The ring will remain in this balanced position until a subsequent action causes another change in the system pressure. The compensator can cut off the pump flow completely in some circumstances, but usually there is some minimum flow to provide lubrication and cooling.

This automatic pressure compensation accomplishes two major purposes:

1. It reduces the possibility of overpressurizing the system, thus averting damage or failure of system components. **Note:** While the pressure-compensated pump provides pressure control, it should never be considered to be a safety device. If the compensator should fail, there would be no protection for the system. *Always include a pressure-relief valve in the circuit, even with a pressure-compensated pump.*

2. It saves energy. Recall that the horsepower required to operate a pump is the product of the pressure and the flow rate. Reducing the flow rate reduces the horsepower as well.

Example 4.1

A hydraulic system uses a *fixed displacement vane pump* to provide 10 gpm at 900 psi to operate a hydraulic cylinder. When the cylinder reaches the end of its stroke, there is an internal leakage rate of 0.5 gpm. The system relief valve is set at 1000 psi. How much horsepower could be saved by using a pressure-compensated pump that would reduce the pump output to 0.5 gpm to make up the internal leakage while maintaining the 1000 psi system pressure (which would be the compensator setting for the pump)?

Solution

In the first case, when the cylinder reaches the end of its stroke, no further work output is produced by the energy put into the fluid by the pump. While the fixed displacement pump continues to produce 10 gpm, the relief valve maintains the system pressure at 1000 psi by "dumping" 9.5 gpm, while the other 0.5 gpm is leaked through system components. None of this fluid does any work, so all of the energy is lost. Thus, from Equation 2.11.

$$HHP = \frac{p \times Q}{1714} = \frac{1000 \times 10}{1714} = 5.83 \text{ HP(lost)}$$

Note: The pump doesn't care where the fluid goes. Whether it dumps over the relief valve or leaks through component clearances, it is still lost.

With the compensated pump, the fluid horsepower output of the pump is reduced by reducing the flow rate to only 0.5 gpm to make up the internal leakages to maintain 1000 psi. In this case,

$$HHP = \frac{p \times Q}{1714} = \frac{1000 \times 0.5}{1714} = 0.29 \text{ HP(lost)}$$

Assuming 100 percent efficiency, so that our input horsepower is equal to our output horsepower, we see that we are saving 5.54 horsepower by using the compensated pump. This saving occurs only when the cylinder is at the end of its stroke. While the cylinder is moving, the pressure is only 900 psi, so no fluid is flowing over the relief valve.

Balanced Vane pumps

Balanced vane pumps are designed to reduce the bearing loading experienced by the unbalanced pump as well as to provide a somewhat smaller pump unit. Figure 4.12 illustrates a typical balanced vane pump design. It is distinguished from the unbalanced design by the two internal inlet ports and two internal outlet ports. With the two high-pressure areas of the pump directly opposite one another, the resultant radial forces are equal and opposite. This tends to center the rotor in the bearings and virtually eliminates bearing wear due to loading. The result is that these pumps tend to have exceptionally long lives. Lambeck[1] states that lives exceeding 24,000 hours in industrial applications is not unusual. Balanced vane pumps are available in fixed displacement models only.

Vane pumps are reasonably rugged and can tolerate fluid contaminated with

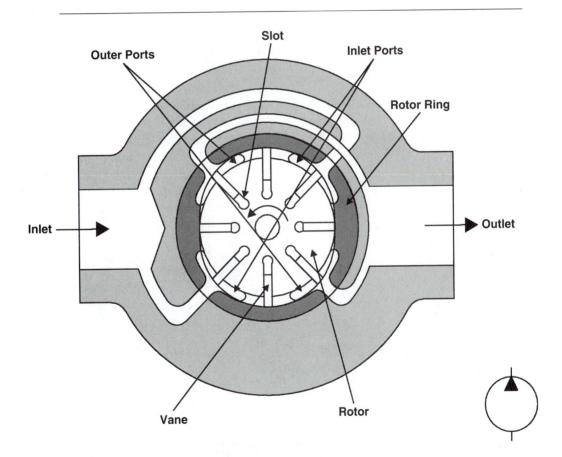

Figure 4.12 The balanced vane pump is distinguished from the unbalanced design by two internal inlet ports and two internal outlet ports. It is a fixed displacement pump.

fine particles fairly well. Because the vanes extend and load against the cam ring, vane wear does not generally result in a gradual loss of efficiency or reduction in output. Unbalanced vane pumps are normally considered to be low-pressure devices (less than 2000 psi), with 2500 psi considered the absolute maximum pressure. Balanced vane pumps, on the other hand, are capable of higher pressures. Industrial versions range up to 3000 psi, while some units designed for mobile applications can operate at 4000 psi. In general, the larger the pump displacement, the lower its maximum pressure.

4.5   Gear Pumps

It is generally agreed that gear pumps are the most robust and rugged of the fluid power pumps, although the bearings in modern, high-pressure pumps used in many mobile applications are very sensitive to contaminated fluid. There are several gear-type pumps, but we will limit our discussion to the two most common types—external gear and internal gear pumps.

External Gear Pumps

By far the most frequently used gear-type pump is the external gear pump (Figure 4.13). In this type pump, as the gears rotate, a partial vacuum is created at the inlet port by the increasing volume as the gear teeth separate. This allows fluid to be forced into the pump by the pressure in the inlet line.

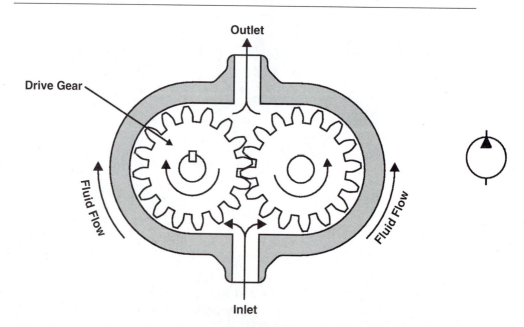

Figure 4.13 The external gear pump is the most frequently used type of gear pump.

The fluid is pushed into the spaces between the gear teeth and swept around the housing in the pumping chambers formed by the teeth, the pump housing, and the side plates that are on each side of the gears. As the teeth move past the outlet port and begin to mesh, the fluid is forced out through the outlet port.

Internal Gear Pumps

Internal gear pumps get their name from the fact that the fluid is carried from the inlet port to the outlet port in the spaces between the gear teeth; that is, internal to the gear set. There are several types of internal gear pumps. The

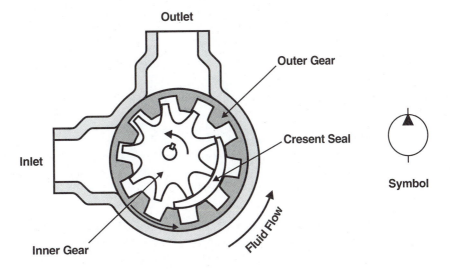

Figure 4.14 The crescent pump is a type of internal gear pump. Note the gear rotating within a gear.

most common are the crescent gear pump and the gerotor pump. These two types are discussed here.

A crescent (internal) gear pump is shown in Figure 4.14. This pump features a gear rotating within a gear. As the gear teeth unmesh near the inlet port, a partial vacuum results, and fluid is forced into the pumping chamber. The fluid is carried through the pump in the pumping chambers formed between the gear teeth, the crescent seal, and the side plates. As the gear teeth begin to remesh, fluid is forced out through the outlet port. Both gears rotate, but the inner gear has one less tooth than the outer gear, so they rotate at different speeds.

Figure 4.15 shows a gerotor pump, which is another internal design. Like the crescent gear pump, it contains one gear that rotates inside another. The two differ in that there is no crescent seal in the gerotor, so the gears remain

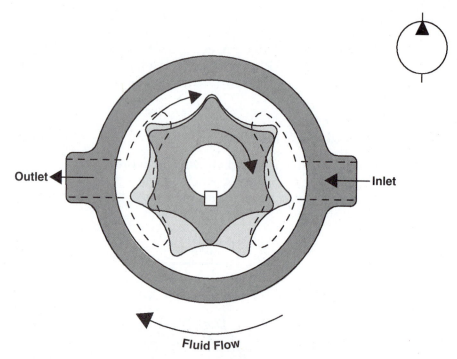

Figure 4.15 The gerotor pump is another type of internal gear pump. It differs from the crescent pump in that it has no crescent seal.

in contact, with the teeth in the inside gear riding over those of the outside gear. The outside gear of the gerotor rotates within the housing at a rate determined by the number of teeth on each gear. The outside gear always contains one more tooth than the inside gear and, therefore, rotates more slowly.

The pumping mechanism is basically the same as in other gear pumps. Fluid is forced into the pump by atmospheric pressure when a partial vacuum is created at the inlet port by the unmeshing of the gear teeth. As the teeth begin to mesh again at the outlet port, fluid is forced out of the pump.

Gear pumps, particularly the external gear type, are very rugged and robust and are more generally tolerant of particle contamination than other types of pumps. Consequently, they are commonly used in road and farm equipment, in industrial applications such as conveyor systems, and other applications that are subject to rough handling and dirty environments.

As mentioned, gear pumps are always fixed displacement. Since none of the hardware inside a gear pump can be repositioned to change the pump displacement, the only way to vary the pump output is to vary the pump speed. The maximum pressure of the gear-type pumps can be as high as 4000 psi, although gear pumps are normally not chosen for applications requiring more than 2000 to 3000 psi because their efficiency tends to drop off rather rapidly at higher pressures.

4.6 Piston Pumps

By far the most complex and expensive of the fluid power pumps is the piston pump family. This family includes radial piston and axial piston pumps. The axial pumps are of two types—in-line piston and bent axis. In all cases, the pumping action results from the reciprocation of pistons in enclosed cylinders.

Radial Piston Pumps

Radial piston pumps get their name from the fact that their pistons reciprocate radially rather than along the axis of the drive shaft. There are two different basic design concepts used for these pumps—a rotating cam and rotating pistons.

A rotating-cam pump is shown in Figure 4.16. In this design the pistons reciprocate in precision bores machined into the stationary cylinder block. Their reciprocating action is caused by the rotating eccentric cam that is part of the drive shaft. The fluid enters the pump through the inlet port and is distributed through an annular passage (or gallery) that connects to the inlet valves of each piston. As the cam rotates, the motion of the pistons causes them to alternately fill with fluid through the inlet valve, then discharge it through the discharge valve. The discharged fluid flows through the discharge gallery to the outlet port.

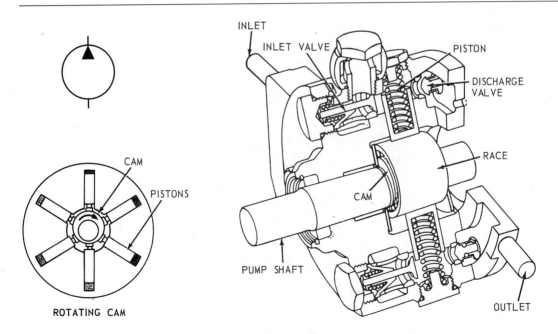

Figure 4.16 Rotating-cam type of radial piston pump.

The rotating-piston pump, shown in Figure 4.17, uses a rotating cylinder block that contains the pistons. The centerline of the cylinder block is offset from the centerline of the bearing ring on which the piston shoes ride (similar to an unbalanced vane pump). The eccentricity of the cylinder block causes the pistons to reciprocate as the block rotates. During a portion of the rotation, the pistons are thrown outward by centrifugal force, and fluid enters the piston through the inlet valves. As the block continues to rotate, the pistons are pushed back toward the center, and the fluid is forced out through the outlet ports. Both the inlet and outlet ports are located in a fixed (nonrotating) shaft termed a *pintle*.

Both types of radial piston pumps can be either fixed or variable displacement. In the rotating-cam type, variable displacement is achieved using a stroke-control mechanism that limits the stroke of the pistons hydraulically. In the rotating-piston type, the eccentricity of the cylinder block and the bearing ring are changed in much the same way as in an unbalanced vane pump.

Radial piston pumps are capable of high pressures and very high volume-flow rates. Most applications of this type pump are in mobile equipment, particularly in agricultural and mining machines.

Axial Piston Pumps

Axial piston pumps are the most complex of all the pump designs. Their name refers to the action of the pistons, which move more or less parallel to the axis

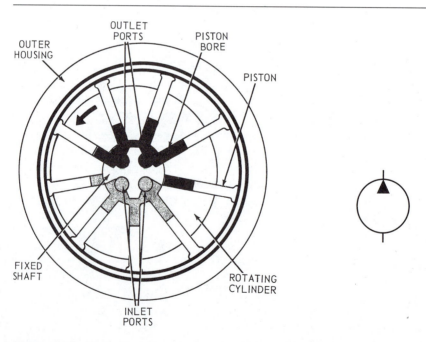

Figure 4.17 Rotating-piston type of radial piston pump.
Source: Reproduced by permission of Deere and Company, © 1987 Deere and Company. All rights reserved.

of rotation of the cylinder block. Axial piston pumps may use either in-line action, in which the axis of rotation is the drive shaft, or bent-axis, in which the axis of rotation is at an angle to the drive shaft.

The pumping mechanism of an in-line piston pump, as shown in Figure 4.18, consists primarily of a cylinder block, pistons that reciprocate in the block, an angled swashplate that causes the pistons to reciprocate, and a port plate that directs fluid into and out of the cylinder block. All of this is contained inside the pump housing. Depending on the pump design, the drive shaft may turn the swashplate while the cylinder block remains stationary, or it may turn the cylinder block while the swashplate remains stationary. The second case is the more common, so let's use it to describe the pumping action.

Each piston in the pump has a slipper bearing, usually called the *piston shoe*, that rides on the swashplate. As the cylinder block rotates, the piston shoes follow the angle of the swashplate. During the first 180 degrees of rotation, the angle causes the piston to pull out of the cylinder block. This gives us the increase in volume (like the syringe) that allows fluid to enter the cylinder bore through the inlet port and the port plate. As the cylinder block comes over the top of the rotation, the swashplate angle starts to push the piston back into the cylinder block. This reduces the volume of the pumping chamber, so the fluid is pushed out through the outlet port of the port plate and into the system.

The pistons in all piston pumps are hollow to reduce their mass and improve their power-to-weight ratios and mechanical efficiencies. They also have a lubrication port that runs through the ball joint on which the piston shoe is attached and right through the piston shoe itself. Fluid passing through this passage serves to lubricate the piston shoe/swashplate interface and balance the forces on the piston. As the piston moves through the pressure portion of the stroke, high-pressure fluid is pushed through this port to form a hydrodynamic lubrication film between the piston shoe and the swashplate. As a result, the piston shoe actually floats over the swashplate rather than riding on its surface. The machined grooves on the bottoms of the shoes are designed to counteract the pressure-generated force on the pistons and provide a controlled force against the swashplate. If the lubrication film is lost, metal-to-metal contact occurs, and pump failure will follow very quickly.

The most common reason for the loss of the lubricating film is particle-contaminated fluid. A large particle may jam the lubricating port, preventing flow through to the swashplate, or more commonly, particles may cause scoring of the surface of the piston shoe or the swashplate. Both of these surfaces are machined to exact flatness and smoothness specifications. If solid particles get caught between these surfaces and cut or erode them, leakage paths are created that allow the lubricating fluid to jet out of the bearing area. In either case, the high force exerted by the piston shoe on the swashplate due to the pressure on the outlet side of the pump will destroy both as the shoe wears into the swashplate surface.

The fluid that lubricates the swashplate is removed from the flow path through the pump. From the swashplate, it flows into the pump case (housing). The pump housings are very low-pressure units (all the high pressure is contained within the cylinder block) and are not capable of withstanding internal pressures above 5 to 10 psi in most cases. Therefore, a drain port called the *case drain* is provided to give the lubricating fluid a way to get out of the

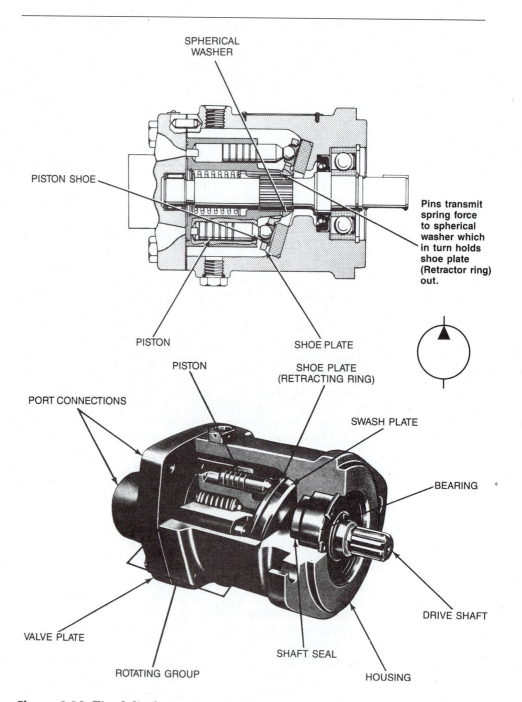

SPHERICAL
WASHER

PISTON SHOE

**Pins transmit
spring force
to spherical
washer which
in turn holds
shoe plate
(Retractor ring)
out.**

PISTON   SHOE PLATE

PISTON   SHOE PLATE
                                (RETRACTING RING)

PORT CONNECTIONS   SWASH PLATE

                                          BEARING

                                          DRIVE SHAFT

VALVE PLATE   SHAFT SEAL

        ROTATING GROUP   HOUSING

Figure 4.18 Fixed displacement axial piston pump with in-line pumping
mechanism.
Source: Courtesy of Vickers, Inc.

housing. The pump must always be installed with the case drain on the top. This will ensure that the case remains full of fluid so that there is some degree of lubrication on startup. The case drain line from the pump to the reservoir must be sized so that the back pressure will not exceed the pump case limits. Normally, the drain line terminates below the surface of the fluid in the reservoir to maintain a low positive pressure in the pump case.

In a bent-axis pump (Figure 4.19), the pistons reciprocate within a cylinder block. In this pump, however, the action is not the result of the pistons riding against a swashplate. Rather, the pistons are fixed to a rotating shaft that is mounted at an angle to the cylinder block. The cylinder block is connected to the same shaft by a universal joint. As the entire assembly rotates, the angle between the shaft and the cylinder block causes the pistons to move in and out within the cylinder block, alternately increasing and decreasing the pumping chamber volume. A stationary port plate allows fluid to flow into and out of the pump as the chamber volume changes with the rotation of the mechanism.

Both in-line and bent-axis piston pumps can be either fixed or variable-displacement. In an in-line pump, the displacement is varied by changing the angle of the swashplate (Figure 4.20). In a bent-axis pump, the angle between the rotating shaft and the cylinder block is changed. In both cases, the pump output approaches zero as the angle approaches zero. In some in-line piston pump designs, the swashplate can be pivoted past the zero position, causing the flow direction through the pump to be reversed. Manual as well as automatic control can be used to change piston pump displacement.

Piston pumps are much more complex than either gear or vane pumps. Because of this complexity and the close tolerances required in manufacturing, they are also more expensive than other types of pumps. In general, however, piston pumps have a higher volumetric efficiency (up to 98%) and a better

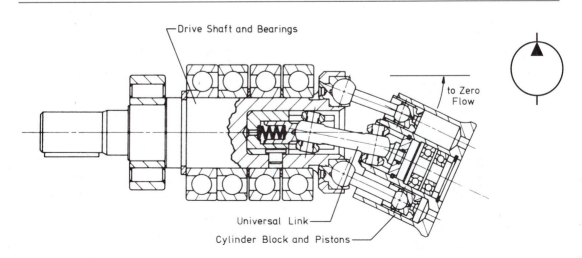

Figure 4.19 Bent-axis axial piston pump.
Source: Courtesy of Vickers, Inc.

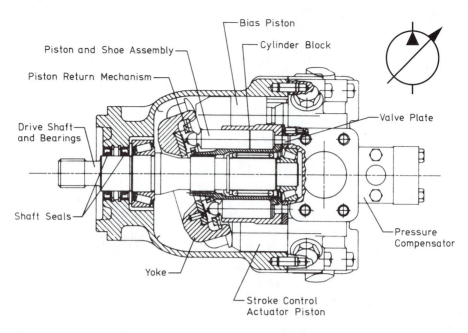

Figure 4.20 Variable displacement in-line piston pump.
Source: Reprinted from Ref. 1, p. 31 by courtesy of Marcel Dekker, Inc.

power-to-weight ratio, and offer a greater range of flows and pressures than other types. Piston pumps are far more sensitive to contamination than other pump types, though, primarily because of the high loading and small film thickness on the swashplate/piston shoe arrangement.

The power-to-weight ratios of piston pumps makes them the "pump of choice" for aerospace applications. They are also popular for industrial applications and for hydrostatic transmissions.

4.7 Screw Pumps

Screw pumps such as the one shown in Figure 4.21 are not widely used in industrial or mobile fluid power systems, but are used in submarine applications virtually to the exclusion of other pump types. The reason for their popularity in this application is that they are, as a group, among the quietest pumps made. Their lack of popularity in other applications is due to their high cost.

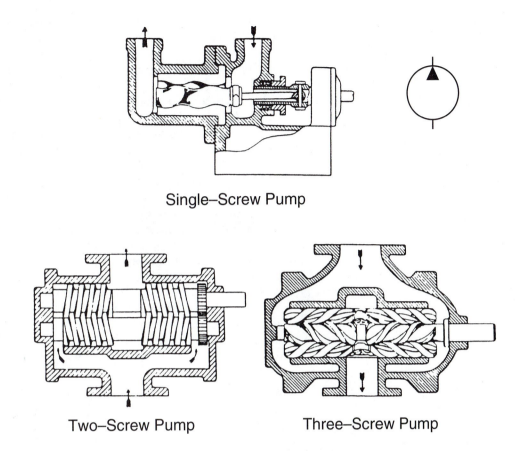

Single–Screw Pump

Two–Screw Pump   Three–Screw Pump

Figure 4.21 Screw Pumps may have one, two, or three screws.
Source: Courtesy IMO Pump Division, IMO Industries, Inc.

As seen in the figure, the name "screw" pump stems from the fact that the pumping mechanism is in the shape of a helical screw that literally feeds fluid through the pump in the same way that a feed auger moves feed from a bin to a hopper. There may be only one screw, or as many as three, rotating in a housing with very small clearances on the outsides of the screw to prevent the leakage of fluid back to the low-pressure side.

4.8   Graphic Symbols

The basic symbols representing fluid power pumps are shown in Figure 4.22. The symbol in 4.22a is the most basic of all pump symbols, so let's look at it in detail. First, we see that it consists of two primary geometric shapes—a circle and an equilateral triangle. The circle is often used in symbology to represent rotating devices. In fluid power, the circle is usually used to indicate an energy-conversion unit or a measuring instrument. Thus, for pumps the circle indicates a rotating energy-conversion device. Note that even if a hand pump or a plunger pump is used, the circle is still used to denote the pumping element.

The equilateral triangle denotes energy flow. There are two important aspects of the triangle used in the hydraulic pump symbol. One is that the apex of the triangle points outward. This indicates that energy is put out by the pump. We'll see later that the apex also points to the pressure, or output, line. The second aspect is that where liquid is being pumped, the triangle is filled in completely. If gas were the working fluid, the triangle would be open. Thus, the only difference between the symbols for a hydraulic pump and an air compressor is the closed or open triangle. This is logical when the functions of the two devices are considered, and makes even more sense when the physical construction of the two are compared. Air compressor designs are basically the same as those of hydraulic pumps; that is, the compression device is usually a piston, screw, or vane type of unit.

Returning to the earlier discussion of graphic symbols, note that the symbol represents the function but not the actual construction of the device. The pump symbol doesn't give us a clue as to the type of pump it represents. It could be any one of the types we have mentioned. It also tells us nothing about the flow rate, displacement, speed, or pressure capabilities of the pump. You need to refer to some other sources for those types of information.

The symbol shown in Figure 4.22b adds another basic graphic, the arrow.

a. Uni-directional
   Fixed displacement

b. Uni-directional
   Variable displacement

c. Uni-directional
   Variable displacement
   Pressure-compensated

d. Bi-directional
   Fixed displacement

Figure 4.22 Graphic symbols representing fluid power pumps.

A long arrow at a 45° angle across a symbol indicates that the device is adjustable or variable. (This applies not only to pumps, but to any fluid power device.) For a pump, this means *variable displacement*. Again, only the function is denoted; thus, while we know that the pump displacement can be varied, we don't know what mechanism is used to achieve the variation.

Looking back at the first pump symbol, we see that there is no such arrow. We can conclude that the pump represented by that symbol is not variable, so it must be a *fixed displacement* pump.

Yet another step in pump sophistication is denoted by Figure 4.22c. The small arrow seen in that figure is another widely used symbol that shows that the device is *pressure-compensated*, meaning that it somehow senses pressure and adjusts to perform properly within preset limits. In the case of a pump, the pressure at the outlet is sensed, and the pump geometry is automatically varied to ensure that the proper pressure is maintained by changing the pump flow. If an external mechanism is used to adjust the pump, other symbols will be attached to the pump symbol to show the device.

The two triangles in the pump symbol of Figure 4.22d mean that the flow from the pump can be reversed. In some pumps, the direction of rotation of the pump can be reversed to reverse the flow. Such pumps are variously termed *reversible, bi-directional,* or *bi-rotational*. In other pumps, reversed flow is accomplished by continuously rotating in one direction and varying the internal pumping mechanism to cause the suction and pressure strokes to occur in physically opposite positions. These variable displacement pumps are termed *over center* pumps because their variable members pass over some center or neutral position.

Summarizing Figure 4.22, then, symbols are used to describe the following pump functions:

a. Fixed displacement, uni-directional hydraulic pump.
b. Variable displacement, uni-directional hydraulic pump.
c. Variable displacement, uni-directional, pressure-compensated hydraulic pump.
d. Bi-directional, fixed displacement hydraulic pump.

4.9 Pump Efficiencies

In Chapter 2, when we introduced the concept of horsepower, we saw that if we assumed 100 percent efficiency, we could equate the input and output (hydraulic) horsepower of the pump. In reality, there is no pump that is 100 percent efficient. Because of friction—both mechanical and fluid—and other mechanical losses, and because there is usually some fluid leakage within the pumping mechanism, perfect efficiency can never be achieved. Thus, we must always put more horsepower into a pump than we can get out. How much more is a direct function of pump efficiency. There are three efficiencies associated with pumps: mechanical, volumetric, and overall efficiency.

Mechanical Efficiency

Mechanical losses due to mechanical and fluid friction in the bearing and pumping mechanisms, losses in the drive gears or coupling mechanisms, and other losses associated with simply turning the pump all diminish the *mechanical efficiency* of a pump. Such mechanical losses are difficult to measure, so mechanical efficiency is usually calculated from the other two efficiencies—volumetric and overall—both of which can be readily measured. Thus, we use the equation

$$\eta_o = \eta_v \times \eta_m$$

or   **(4.1)**

$$\eta_m = \frac{\eta_o}{\eta_v}$$

where

$$\eta_m = \text{mechanical efficiency.}$$
$$\eta_o = \text{overall efficiency.}$$
$$\eta_v = \text{volumetric efficiency.}$$

Volumetric Efficiency

Volumetric efficiency is based on the theoretical displacement of the pump. The theoretical displacement is independent of any internal leakage, and is based on the physical aspects of the unit. For instance, the theoretical displacement of a piston pump is dependent on the number and size of pistons, thus

Pump Displacement = Number of pistons × volume displaced by Each Piston

The volume theoretically delivered by a piston is the piston area (A) multiplied by its stroke (S) so we can write

$$V_p = n \times A \times S$$ **(4.2)**

where V_p is the pump displacement and n is the number of pistons.

This gives us the displacement *per revolution*. This displacement is measured in in^3/rev or cm^3/rev, although it is often incorrectly stated as just in^3 or cm^3—as, for example, "a 2.8 in^3 pump." To find the theoretical output (or flow rate) from this, we need only the rotational speed, N. Now we have

$$Q_T = V_p \times N = n \times A \times S \times N$$ **(4.3)**

where Q_T is theoretical flow rate.

Due to internal leakage in the pump (termed *slippage*) and, in some cases, fluid used to cool and lubricate pump components, the actual output (Q_A) from a pump is usually somewhat less than the theoretical output. The volumetric efficiency is the ratio of the actual to the theoretical outputs. Thus,

$$\eta_V = \frac{Q_A}{Q_T}$$ **(4.4)**

Example 4.3

An axial piston pump has nine pistons with a 0.375 in² area. The stroke is 1.5 in. The pump turns at 2200 rpm. The measured flow rate from the pump is 45 gpm. What is the volumetric efficiency of the pump?

Solution

From Equation 4.2, we see that volumetric efficiency is found by

$$\eta_v = \frac{Q_A}{Q_T}$$

Theoretically, then, the pump flow is

$$
\begin{aligned}
Q_T &= n \times A \times S \times N \\
&= 9 \times 0.375 \text{ in}^2 \times 1.5 \text{ in} \times 2200 \text{ rev/min} \\
&= 11138 \text{ in}^3/\text{min}
\end{aligned}
$$

Remember that $n \times A \times S$ actually gives in³/rev, not just in³.
Having found Q_T, we can now find efficiency as follows:

$$
\begin{aligned}
\eta_v &= \frac{Q_A}{Q_T} \\
&= \frac{(45 \text{ gal/min}) \, (231 \text{ in}^3/\text{gal})}{11138 \text{ in}^3/\text{min}} \\
&= 0.933
\end{aligned}
$$

or 93.3 percent.

Note that it was necessary to convert gallons to cubic inches in this procedure.

Overall Efficiency

The overall efficiency of a pump is the ratio of output horsepower to input horsepower:

$$\eta_o = \frac{\text{Power Out}}{\text{Power In}} = \frac{HP_{\text{OUT}}}{HP_{\text{IN}}} \tag{4.5}$$

We have seen already that the output (hydraulic) horsepower of a pump can be found using Equation 2.11, while the input horsepower comes from Equation 2.10. In its complete form, we can write the overall efficiency equation as

$$\eta_o = \frac{(p \times Q)/1714}{(T \times N)/63{,}025} \tag{4.6}$$

In SI units, pump overall efficiency is found from Equation 4.7:

$$\eta_o = \frac{(p \times Q)/60{,}000}{(T \times N)/9550} \tag{4.7}$$

The flow rate in the numerator is the actual, or measured, pump output. Proceed with caution! Remember that the use of the constants locks you into specified units for pressure, flow, torque, and speed.

Example 4.4

The pump of Example 4.3 is driven by a 75 HP electric motor. If it operates at 2500 psi, find its overall and mechanical efficiencies.

Solution

The input horsepower, *IHP*, is given as 75 HP. We can use this number directly in the denominator with no need to find torque or rpm. Thus, we have all the information necessary to find the overall efficiency. Notice that the flow rate we use here is the actual flow rate.

$$\eta_o = \frac{(p \times Q)/1714}{IHP}$$

$$= \frac{(2500 \times 45)/1714}{75}$$

$$= \frac{65.6}{75}$$

$$= 0.875 \times 100 = 87.5\%$$

Now we can find the mechanical efficiency using Equation 4.1.

$$\eta_m = \frac{\eta_o}{\eta_v}$$

$$= \frac{0.875}{0.933}$$

$$= 0.938 \times 100 = 93.8\%$$

Example 4.5

The pump of the previous example is driven by an electric motor producing 75 HP. This motor has an overall efficiency of 0.85. The cost of electricity to run the motor is \$0.085 per kilowatt hour. The system operates 8 hours per day, 220 days per year. Calculate the annual cost of the inefficiencies of the pump and the motor.

Solution

The first thing we need to do is calculate the electrical horsepower (*EHP*) required to drive the motor. We can use the definition of overall efficiency for this calculation

$$\eta_o = \frac{\text{Output Horsepower}}{\text{Input Horsepower}}$$

In the case of the electric motor, the output horsepower is 75 HP, while the input horsepower is the *EHP*. Thus,

$$\eta_o = \frac{75}{EHP}$$

Solving for *EHP*, we get

$$EHP = \frac{75}{\eta_o} = \frac{75}{0.85} = 88.2 \text{ HP}$$

This represents the total horsepower to drive the electric motor and, subsequently, to drive the pump. The difference between this number and the hydraulic horsepower output of the pump (65.6 HP as calculated in the previous example) represents the horsepower loss.

$$\text{Horsepower Loss} = 88.2 - 65.6 = 22.6 \text{ HP}$$

Now we can determine the loss in kilowatts (kW):

$$
\begin{aligned}
\text{Kilowatt Loss} &= \text{Horsepower Loss} \times 0.746 \text{ kW/HP} \\
&= 22.6 \text{ HP} \times 0.746 \text{ kW/HP} \\
&= 16.86 \text{ kW}
\end{aligned}
$$

The next step is to use the operating time to determine the kilowatt-hours (kWh) of electricity used.

$$
\begin{aligned}
\text{kWh} &= 16.86 \text{ kW} \times 8 \text{ hr/day} \times 220 \text{ day/yr} \\
&= 29{,}674 \text{ kWh/yr}
\end{aligned}
$$

The annual cost of this electricity is

$$
\begin{aligned}
\text{Cost} &= \text{kWh} \times \text{Cost/kWh} \\
&= 29{,}674 \text{ kWh/yr} \times \$0.085\text{/kWh} \\
&= \$2{,}522\text{/yr}
\end{aligned}
$$

Similar exercises can be carried out with gear and vane pumps. The displacement of a gear pump in in³/rev can be estimated by

$$V_P = 6 \times W_G \times (2\,D - L) \times \left(\frac{L - D}{2}\right) \qquad \textbf{(4.8)}$$

where

W_G = width of the gears.

D = bore diameter of the housing.

L = longest dimension of the bore chamber.

The exact displacement depends on the shape of the gear teeth. Figure 4.23 illustrates these dimensions. The theoretical output flow can be found using Equation 4.3.

The cubic inch displacement of a balanced vane pump can be estimated in a similar manner using Equation 4.9.

$$V_P = 12 \times W \times \left(\frac{L + D}{4}\right) \times \left(\frac{L - D}{2}\right) \qquad \textbf{(4.9)}$$

Figure 4.24 illustrates the applicable dimensions. Again, the theoretical flow rate can be found using Equation 4.3.

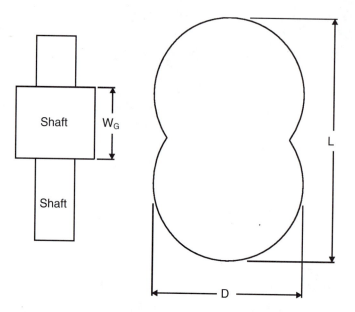

Figure 4.23 Internal gear pump displacement can be estimated based on W_G, the width of the gears; D, the bore diameter of the housing; and L, the longest dimension of the bore chamber.

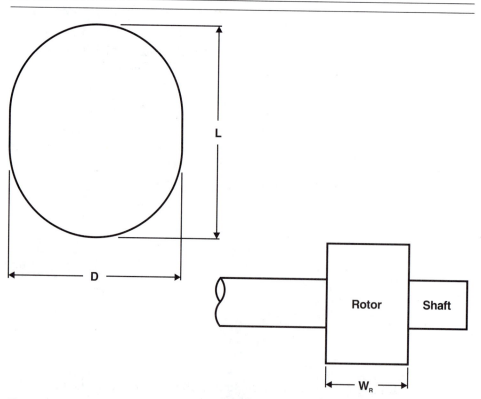

Figure 4.24 Balanced vane pump displacement can be estimated based on W_R, the width of the rotor; D, the widest dimension of the pressure ring; and L, the longest dimension of the pressure ring.

4.10 Effects of Air and Other Gases on Pump Life and Performance

The problems of air and gas originate on the suction side of the pump, where the vacuum is generated by the increase in volume in the pumping chamber. There are actually two very different phenomena that can occur. One is the evolution of air or vapor from the liquid due to a high vacuum, while the other is the presence of excessive air in the fluid from a variety of sources. We'll look at each of these problems in some detail.

High Vacuum

Saying that a high vacuum—that is, something near a perfect vacuum—in the pump suction line can be a problem seems to be contradictory. We want the pump to produce a high inlet vacuum so there will be no problem with atmospheric pressure pushing the fluid into the pumping chamber; however, we want the fluid to fill that vacuum as quickly as it is created. In other words, the *pump* should be capable of creating a high vacuum, but the *system* should be capable of filling that vacuum with fluid. Thus, a vacuum gage installed near the pump inlet would indicate a relatively low vacuum.

The problems generated by a high vacuum have to do with the volatility of the liquid and the dissolved air in the liquid. The *volatility* of a liquid is the degree to which it will vaporize under given conditions of pressure and temperature. The volatility of any liquid increases with either an increase in temperature or a decrease in pressure. Since we have a low pressure (actually, a vacuum) in the suction line, there is a chance that vaporization can occur. The higher the vacuum is, the greater the degree of vaporization. The result is that instead of having a continuous phase of liquid, we have a liquid filled with tiny vapor bubbles. This phenomenon is termed *cavitation*. The term "cavitation" is often used incorrectly to describe any and all of the air/gas problems in the pump inlet line; however, it correctly defines the vaporation process described here.

The second problem is similar to cavitation except that instead of vapor bubbles, microscopic air bubbles appear in the liquid. All hydraulic fluids contain some air dissolved into their physical structure. A crude analogy is a gallon bucket of ball bearings. Since there are spaces between the balls, the gallon bucket also contains a considerable amount of air. Although liquid molecules are not round, there are voids between them. These voids are termed *interstices*, and they are filled with air. At atmospheric pressure, petroleum oils contain about 10 percent air by volume (Figure 4.25); that is, a gallon of oil contains about a tenth of a gallon of air. This is termed the *saturation level* of the oil at atmospheric pressure. As the pressure increases or decreases, the saturation level of the fluid also increases or decreases. Figure 4.25 shows that this is essentially a linear relationship. The figure also shows the saturation level for several other typical hydraulic fluids.

The air problem on the suction side of the pump is a result of this pressure-saturation relationship. As the pressure decreases, some of the air that was in solution *evolves*, or comes out of solution. Since there is no way for the air to escape from the suction line, it simply forms bubbles in the fluid in much the same way that cavitation bubbles form. The lower the pressure (actually, the greater the vacuum), the more air will come out of solution.

While the problem of cavitation (oil vaporization) and dissolved air evolution have been presented here as separate, distinct actions, in actual operation they are both likely to occur anytime the fluid is subjected to a high vacuum in the pump inlet line. As the pump starts and the vacuum begins to develop, the air evolution will begin almost immediately, although the rate (and, consequently, the effect) of the evolution will be low initially. The fluid cavitation (vaporization) process may develop somewhat more slowly, but in a short time, the two processes will be occurring simultaneously.

Several problems can lead to a high vacuum in the suction line. Some are basic design problems such as suction lines that are too long or too small in diameter or contain too many bends and fittings. We'll discuss those problems in Chapter 12. Another design problem is placing the pump too far above the reservoir, termed *excessive lift*.

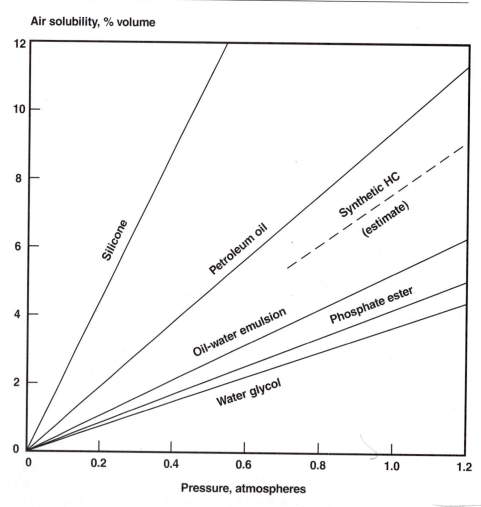

Figure 4.25 Air content of some common hydraulic fluids.
Source: Reprinted from Ref. 3, p. 5, by courtesy of Marcel Dekker, Inc.

Maintenance problems that can lead to high vacuum include clogged suction filters, collapsed hose liners, and debris (rags, paper cups, etc, that somehow get into the reservoir) in the suction line. In short, anything that restricts the flow to the pump can cause a high vacuum. The hotter the oil, the more easily it will vaporize.

Excessive Air

Excessive air, also called *entrained air*, is simply any air in the fluid in excess of the saturation level. This is air that you could easily see in a fluid sample—the fluid would be foamy or cloudy because of the air bubbles in it.

As with the high-vacuum problem, some of the causes of excessive air are design-related while others are maintenance problems. Among the design problems are return lines that are too small (leading to high fluid velocities), return lines that terminate above the fluid level, and reservoirs that are too small or have inadequate internal baffling. (We'll discuss reservoirs in Chapter 13.) Maintenance problems include air leaks in suction lines or suction filters, or gas leaks in accumulators, a low reservoir-fluid level, or installing components without filling them with fluid. A cold fluid aggravates these problems because the air cannot escape (like air bubbles in honey). Many fluids contain an antifoaming additive; but if the fluid has been abused (overheated, contaminated with water or chemicals, or has a high level of solid particles), this additive may be depleted, adding to the excessive-air problem.

Results of Air/Gas Problems

The undesirable consequences of the air/gas problem is a phenomenon termed *cavitation*, which refers to the rapid vaporization and subsequent collapse of the cavities (bubbles) in the liquid. According to Tullis, there are two types of cavitation—vaporous and gaseous.[2] *Vaporous cavitation* occurs when the cavity consists almost completely of liquid vapor. *Gaseous cavitation* occurs when there is a high proportion of air in the cavity. It is usually found only in highly aerated liquids. While both types of cavitation are potentially damaging to the pump and the fluid, vaporous cavitation is by far the more significant.

In both cases, as the fluid carrying the bubbles moves from the low-pressure to the high-pressure area of the pump, the bubbles collapse. This is the result of compression due to the high fluid pressure. As gases are compressed, heat is generated, sometimes to the point that the fluid is scorched, which damages the fluid structure and increases the oxidation rate.

Erosion damage to the pump can occur if the vapor cavities or bubbles reach the internal surfaces of the pump before they collapse. Tullis suggests two possible mechanisms. The first is a high-pressure shock wave that occurs when the cavities collapse. The pressure in these waves has been estimated at over 1,000,000 psi! This is sufficient to damage any surface. The second mechanism is a high-velocity *microjet* of fluid that occurs because of an unequal pressure distribution around the bubble due to the presence of the metal surface. When the bubble collapses, the side of the bubble opposite the pump surface reaches a higher velocity than that of the other collapsing areas. The result is a jet shooting through the center of the bubble. The jet reaches very high velocities and creates a small pit when it hits the wall (Figure 4.26).

There are numerous troublesome results of such pitting. One is that the

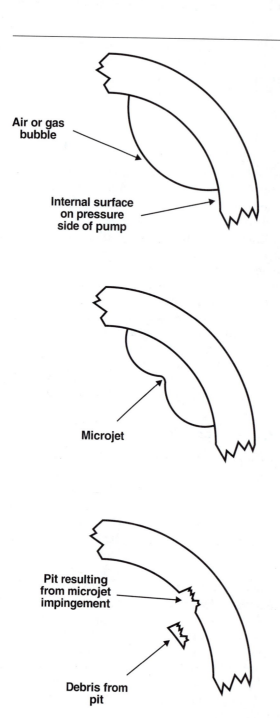

Figure 4.26 Pitting results from the collapse of air or gas bubbles on the internal surface of the pump.

resulting particles contaminate the fluid and can cause additional damage. Another is that the internal leakage of the pump increases. In vane pumps, the collapse of the cavities can allow the vanes to impact the pressure ring. This leads to additional wear and possibly vane chatter or bouncing.

The presence of either a vacuum or excessive-air problem shows itself by several symptoms. Probably the first and most noticeable of these is noise. The pump may begin to produce a loud and very annoying shriek or it may sound like it is pumping gravel. The next symptom will be a decrease in the pump flow rate. A look in the reservoir is likely to reveal a very cloudy or foamy fluid. Over the long term, the fluid may become darkened from the scorching and take on a burnt smell. This would be accompanied by an increase in acid level, sludge and varnish formation, and other indications of oxidation, as well as viscosity changes and an increased solid particulate level.

Once the symptoms begin, the first step in troubleshooting is to install a vacuum gage near the pump inlet. If the gage shows a high vacuum (most pump manufacturers recommend no more than 5 or 6 in Hg vacuum maximum), then those causes listed for a high vacuum should be investigated. If a high vacuum is not indicated, then excessive air is probably the trouble. If the gage reads atmospheric pressure (0 psig) or only a slight vacuum, the problem is very likely to be air leaking into the suction line.

In any case, action should be taken immediately to prevent serious damage to the fluid and the pump. As soon as the telltale noise is heard, corrective action should begin, because pump damage progresses very rapidly. A new pump can be destroyed in only a few minutes in severe cases.

4.11 Selection Criteria

When specifying a pump for a particular application—whether it be for a new system or a replacement for an existing system—consideration must be given to pump performance (gpm, pressure, efficiency), physical size and mounting, noise levels, and containment sensitivity. When specifying a replacement pump, the easiest thing to do is simply to choose an exact replacement—the same part number from the same manufacturer—but this might not be the best thing to do. It sometimes pays off handsomely to study the specifications of a number of other pumps, because there may be one that is better suited to your application.

The choice of a hydraulic pump should not be taken lightly—and it should not be based upon cost. Unfortunately, cost is commonly a predominant consideration as accounting and purchasing departments often make decisions they are not qualified to make. The lowest bidder may not have the best product, even though it will fit the hole and produce the pressure and flow required. A sound familiarity with pumping theory, pump mechanisms, and selection criteria is necessary to decide on the best pump for a given application.

Performance

Performance is the obvious first consideration in the selection of a pump. The single term "performance" includes the pressure and flow capabilities of the pump as well as its efficiency. When selecting a pump, you must have already determined the pressure and flow required to make your equipment perform the desired operations. Pump efficiency, while an important consideration, is usually a secondary factor in pump selection.

There are no pump performance specification standards utilized uniformly throughout the fluid power industry. The National Fluid Power Association (NFPA) has formulated a series of tests for pumps, as well as recommendations for presenting the resulting data, in NFPA Recommended Standard T3.9.17-1971. However, this standard has not been fully integrated into the industry, as seen from the fact that different manufacturers present their pump performance data in different ways, primarily utilizing tables and graphs. Figure 4.27 shows a graphical presentation incorporating pressure, flow, and volumetric and overall efficiencies on one graph. This graph—which is typical of pumps in general as far as the shapes of the curves are concerned—reveals several things that were not obvious from our previous discussions. The first is that, although Figure 4.27 is for a fixed displacement pump, the output flow does not remain constant; there is a slight, but obvious, decline in the flow rate as the pressure increases. This results from a decline in volumetric efficiency caused by increased internal leakage as the pressure rises. This change in efficiency is shown on the graph to be dependent on both pressure and rotational speed. Looking at the 99 percent efficiency curve, we see that it runs from 750 psi at 1200 rpm to 1700 psi at 1800 rpm. The other curves follow similar patterns.

When we observe the overall efficiency curves, we see some interesting

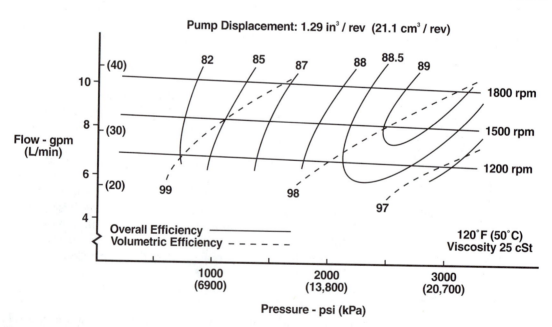

Figure 4.27 Example of in-line axial piston pump efficiency curves.
Source: Reprinted from Ref. 1, p. 36, by courtesy of Marcel Dekker, Inc.

things. First, the overall efficiency increases as pressure increases, while volumetric efficiency decreases with the same change. The second is that some of the curves occur at more than one set of values for rpm, flow and pressure. For instance, 88.5 percent efficiency occurs at both 2250 and 3250 psi for 1500 rpm. Note, however, that between those pressures, the overall efficiency increases to 89 percent.

This graph also shows that the maximum overall efficiency for this pump is slightly over 89 percent and occurs at approximately 2900 psi at 1800 rpm.

All performance graphs are not set up the same way. Some use multiple scales on the axis in order to represent efficiencies by continuous curves, others include horsepower requirements in the graphs, and some use pump displacement rather than flow rate. So be very careful in reading and interpreting performance graphs.

Physical Size and Mounting

Another major consideration when specifying a replacement pump is the physical location, including the space, the mounting facilities for the pump housing, and the coupling of the pump shaft to the prime mover.

Physical space—often referred to as the *space envelope*—is not a problem in many industrial applications, but in some situations, such as aircraft and mobile equipment, space is very limited and thus flexibility is severely reduced.

For specifying pump shafts and flange mounting, NFPA has developed NFPA Standard T3.9.2R1. This standard has been adopted by the American National Standards Institute as ANSI B93.6. A third standard, SAE J744a, also addresses pump shafts and mounting flanges, but is directed more toward automotive applications than the NFPA and ANSI standards. Figure 4.28 shows excerpts from the NFPA standard. Note that only flange mounting, in which the face of the pump is mounted flush with another surface, is addressed by these standards. There are no standards specifying foot-mounting dimensions and patterns. (Note: Foot mounts are considered by many to be unsatisfactory. The pump and shaft tend to deflect under load in many configurations. This can result in uneven bearing loading that leads to pump failure.)

Noise

There are two distinct causes of pump noise—mechanical factors and the action of air and gases in the fluid. We've already discussed the air/gas problem, but let's take a quick look at the mechanical problem.

Pumps produce three kinds of mechanical noise (Figure 4.29): airborne, structure-borne, and fluid-borne. *Airborne noise* is the physical result of the internal components of the pump responding to the forces generated within the pumping chambers as the pump rotates. The stretching and contracting of the components cause the structure of the pump to vibrate. This vibration generates pressure waves that manifest themselves as sound.

Structure-borne noise is the result of structural vibration of the pump hardware, mounts, coupling, and so forth. This vibration is in response to the same internal forces that generate airborne noise. Structure-borne noise usually adds to airborne noise.

Fluid-borne noise is the result of pressure pulsations from the pump. These

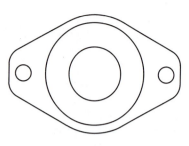

a. Two-bolt Flange

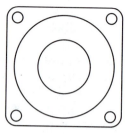

b. Four-bolt Flange

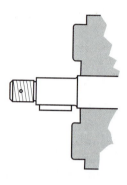

c. Straight Shaft
(Threaded and keyed)

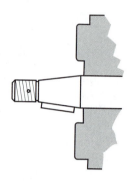

d. Tapered Shaft
(Threaded and keyed)

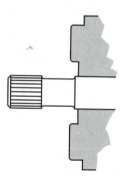

e. Splined Shaft

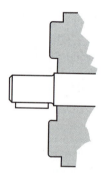

f. Straight Shaft (keyed)

Figure 4.28 Hydraulic power pump and motor mounting flange and shaft configurations (excerpt from NFPA standard.)

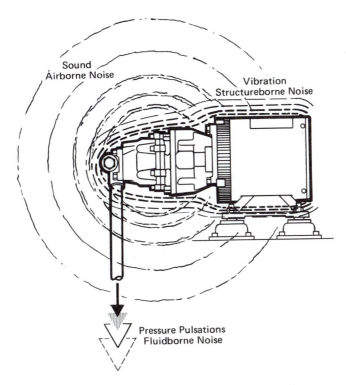

Figure 4.29 Mechanical pump noise may be airborne, structure-borne, or fluid-borne.

Source: Courtesy of Vickers, Inc.

are caused by flow variations and the resistance of the system fluid components to these pulsations. Pump design and manufacturing procedures can significantly affect the generation of fluid-borne noise. Design changes that cause a reduction in fluid-borne noise also reduce airborne noise. Because screw pumps don't produce the distinct fluid pulses of other pumps, they are inherently quieter than other designs.

While structure- and fluid-borne noise can create vibrations that can cause structural fatigue with the associated cracking and breakage, such occurrences are relatively rare because the vibrations are usually readily detected and can often be corrected by clamping, shock mounting, and so on. Thus, we are normally more concerned with airborne noise.

The airborne noise from pumps is measured in decibels. The decibel, or db, scale is a relative scale based on the threshold of human hearing. It is a measure of sound pressure level (also termed *intensity*) and not of "loudness". Intensity is a purely physical quantity, while loudness is subjective and depends on the individual listener and ambient noise.

The reference for the db scale is the level at which the average ear can detect any sound. The maximum sound intensity that can be tolerated without severe pain is approximately 10^{14} times as great as the minimum detectable level. (The threshold of pain is about 10^{12} times the minimum level.) Since this is a very wide range, a logarithmic scale provides a more convenient scaling factor.

In comparing the relative intensities of two sound levels, the logarithm of the ratio of the two levels, (I_1 and I_2) is used. This is expressed in bels as

$$\log \frac{I_2}{I_1}$$

where I_2 is greater than I_1. If I_2 is 10 times greater than I_1, then

$$\log \frac{I_2}{I_1} = \log \frac{10}{1} = 1 \text{ bel}$$

A ratio of 1000 to 1 gives 3 bels, and a ratio of 10^{14} to 1 gives 14 bels. Thus, the span of intensity measurements from initial detection to the pain threshold is only 14 bels. This small range is restrictive, so rather than using bels, we use decibels (db) and express the relative sound intensities as

$$10 \log \frac{I_2}{I_1} \tag{4.10}$$

Now we see that a 10 to 1 ratio gives 10 db, and a 10^{14} to 1 ratio gives 140 db.

If we have an increase in sound intensity of 1 db, we can calculate the relative intensities as

$$10 \log \frac{I_2}{I_1} = 1$$

or

$$\log \frac{I_2}{I_1} = 0.1$$

$$\therefore \frac{I_2}{I_1} = 1.26$$

This means that I_2 is 1.26 times as great as I_1, indicating a 26 percent increase in the sound intensity.

If the intensity is doubled, we find that the increase in decibels is

$$10 \log \frac{2}{1} = 3.01 \text{ db}$$

Usually we cheat a little and say that 3 db means that the intensity has doubled.

When considering the increase in sound intensity that results from more than one noise source, we determine the intensity of the first source, then add to it the increase in intensity that results from subsequent sources. To do this, we use the equation

$$I = I_n + 10 \log \frac{\sum_1^n I_i}{I_n} \tag{4.11}$$

Where I_n is the highest intensity level of all the sources. In this equation, $\sum_1^n I_i$ is a mathematical notation meaning that we add up the intensities of all the machines ($I_1, + I_2, + I_3 + \ldots I_n$). We then find the increase in decibels and add that increase to the original intensity.

Example 4.6

The sound intensity of a certain hydraulic power pack is 75 db. If a second, identical power pack is started, what would be the reading on a decibel meter placed exactly midway between the two machines?

Solution

The sound intensity of the first machine is 75 db. With both machines running, the total intensity would *appear* to be 150 db, but remember that the intensity scale is relative to the sound threshold intensity. Thus, we must use Equation 4.11 to get the meter reading.

$$I = I_1 + 10 \log \frac{I_1 + I_2}{I_1}$$

$$= 75 + 10 \log \frac{75 + 75}{75}$$

$$= 75 + 3 = 78$$

So our meter would read 78 db.

If we add another identical machine, a centrally placed meter would read

$$I = 75 + 10 \log \frac{75 + 75 + 75}{75} = 79.8 \text{ db}$$

Still another machine would give

$$I = 75 + 10 \log \frac{75 + 75 + 75 + 75}{75} = 81 \text{ db}$$

An *A-weighted* measuring system is used to measure pump noise. This is a standardized sound filtering system that reduces the strength of signals below 50 Hz and emphasizes the frequencies in the 500 to 10,000 Hz range in an attempt to more accurately simulate the sensitivity of the human ear. Sound levels measured in this way are expressed in *A-weighted decibels*, or (A).

Table 4.1 shows the intensities of some common noise situations. Intensities of up to 80 db(A) are readily tolerated by most people. Between 100 and 110 db(A), most people will begin to experience some discomfort. Above 120 db(A), the noise becomes intolerable and physical and sometimes mental damage can result from prolonged exposure. The Occupational Safety and Health Agency (OSHA) specifies 90 db(A) as the maximum level allowed for eight hours of continuous exposure in the workplace. Some Canadian provinces specify 85 db(A).

While it is impossible to rate the relative noise levels of the various types of pumps on an absolute scale, some generalization is possible. Screw pumps are generally acknowledged to be the quietest type. They have relatively few moving parts and a smooth, continuous through-flow, unlike other pump types that process "chunks" of fluid. Then, in loose order would come vane, piston, and gear pumps. Remember that this is a generalization, and the order could be changed by factors such as individual pump design, pressure, flow, and type of fluid. Always consult the manufacturer's data, which should be generated in accordance with NFPA Standard T3.9.12.70 or ISO 4412.

Table 4.1 Typical Noise Situations

Decibels		Typical Noise Source
160		Jet engine
150	Permanent hearing loss after short exposure	Large propellor aircraft
140		Hydraulic press
130		Riveting and chipping
120	Pain threshold	Rock concert
110	Loud power mower	
100	Discomfort threshold	Heavy city traffic
90	OSHA limit without hearing protection	Shouting
80		Busy office
70		Loud speech
60		Normal speech
50		Light vehicular traffic
40		Typical residential neighborhood
30		
20		Soft whisper
10		
0		Threshold of hearing

Contaminant Sensitivity

While a new pump may perform well under normal circumstances, if it is subjected to large quantities of solid particles in the fluid, its performance can quickly degrade to an unacceptable level. Filters are normally used in fluid power systems in an effort to prevent this degradation, which results from the wearing away of mating surfaces. Unfortunately, as we will see in Chapter 11, these efforts often accomplish little because of poor filter performance, incorrect placement of filters, extremely high contaminant ingression rates, or poor maintenance practices. Therefore, it is good insurance to utilize pumps and other components that are relatively insensitive to contamination but will still perform the required task.

The Fluid Power Research Center at Oklahoma State University has developed test procedures for determining the contaminant sensitivity for both fixed displacement and pressure-compensated pumps. These procedures—specified in NFPA T3.9.18M—involve supplying the test pump with fluid that has been heavily contaminated with AC Fine Test Dust (ACFTD) (a standard test contaminant that will be discussed in Chapter 11) in different size ranges. The flow degradation of the pump is observed with each injection of ACFTD. The test is continued for a specified period of time or until the pump flow rate has degraded to 70 percent of the original flow rate. A computer program is then used to define a figure of merit, termed the *pump Omega rating*. This rating, while providing a method for ranking the contaminant sensitivity of pumps, actually defines the fluid filtration required to ensure at least 1000 hours of life for the pump when operating in fluid where the contamination level does not exceed the test level. The Omega scale ranges from 1 to infinity, with 1 indicating maximum contaminant tolerance. In other words, Omega rating of 1 indicates that the pump could be expected to operate for 1000 hours without any filtration at all. Needless to say, very few pumps have that type of contaminant tolerance. In fact, many pumps fail long before the entire test procedure can be completed.

As with noise, contaminant sensitivity is very much a factor of pump design. Operating pressure and contaminant level play significant roles in the pump wear rate, while pump speed and flow rate influence the wear rate very little.

In general, we can say that the piston pump family is the least contaminant-tolerant as a group, while the gear pumps are the most contaminant-tolerant. As with all generalizations, there are likely to be notable exceptions to this one, so your best course of action is to select candidate pumps based on performance criteria, and so forth, then ask the manufacturer for the pump Omega ratings. Even though the test procedure has been in use for several years, many manufacturers still do not provide contaminant-sensitivity data for their pumps. As a result, you may have some difficulty in obtaining the Omega rating for all the pumps that might meet the other criteria. If you are forced to choose your pump without this information, you will have been deprived of an important decision-making tool regarding the system life and reliability of that pump.

As an alternative, most pump manufacturers provide some guidelines concerning acceptable contamination levels. Some merely refer to a filter rating (say, 15 micron absolute) which, as we will see in Chapter 11, is non-definitive. Others provide information concerning the actual fluid contamination level, using the ISO Solid Contamination Code (ISO 4406).

4.12   Troubleshooting Tips

While numerous situations can lead to pump problems, the following are most common:

Symptom	Possible Cause
Insufficient flow	• Drive unit not running, running too slowly, or turning pump in the wrong direction.
	• Excessive internal leakage due to wear.
	• Restriction or air leaks in pump suction line or highly aerated fluid. (This will usually be accompanied by characteristic pump noise. See Section 4.10.)
	• Low oil level in reservoir.
	• Cold oil or oil with high viscosity (difficult to move).
	• Hot oil or oil with low viscosity (internal leakage).
	• Compensator stuck on pressure-compensated pump.

Symptom	**Possible Cause**
System pressure too low	• System relief valve set too low. • Pressure compensator set too low. • Insufficient pump flow output.
Pump noisy	• Restriction or air leaks in suction line or highly aerated fluid. (See Section 4.10). • Low oil level in tank. • Pump running too fast. • Oil viscosity too low (wrong oil or hot oil). • Loose mount bolts. • Worn coupling or splines.
Pump wears out quickly	• Dirty fluid. • Fluid viscosity too low. • Incompatible or degraded fluid. • Pump running at unnecessarily high pressure. • Coupling misaligned. • Excessive side loading. • Cavitation due to restricted inlet line or aerated fluid.

Upon disassembling a pump, look for the following symptoms to help you to detect the causes of the pump failure.

Symptom	**Possible Cause**
Brittle, hardened, or melted O-rings	• Excessive heat. • Incompatible fluid. • Lubrication failure due to oil breakdown.
Discolored metal parts	• Excessive heat. • Incompatible fluid.
Pitted metal parts	• Cavitation
Score marks or grooves	• Contaminated oil. • Lubrication failure due to oil breakdown (discoloration usually present, also)

4.13 Summary

There are a large number of positive displacement pumps from which to choose. While their actual design and operation differ greatly, they all do the same thing—provide fluid flow to operate the system. Pressure in a system results from the load and other resistances to flow. The pump must continue to provide its rated flow at the maximum required pressure in order to move the load at the required speed.

Selecting the best pump for a particular application involves more than simply meeting the flow and pressure requirements. A good system design will not only do the job, but will do it in the most cost-effective manner. As we have already seen, considerable energy savings can be achieved in some applications by using a pressure-compensated pump instead of a fixed displacement pump. This, however, may not be the most economical choice for some applications. Operating costs, pump costs, pump life, and other factors must be considered before a qualified judgment can be made.

Pumps are generally very sensitive to contaminated fluid. To achieve optimum life and performance, care must be taken to ensure that system fluid is kept clean and cool. The pump must also receive an adequate supply of fluid from the reservoir that is free of excess air.

In subsequent chapters, we will explore the relationship of the pump with other components in the system. We'll also see how the horsepower output required of the pump is directly affected by the selection and placement of components that make up the control and output portions of the system.

4.14   Key Equations

Volumetric efficiency:

$$\eta_v = \frac{Q_A}{Q_T} \tag{4.4}$$

Overall efficiency:

$$\eta_o = \frac{HP_{OUT}}{HP_{IN}} \tag{4.5}$$

$$= \frac{\text{Power Out}}{\text{Power In}}$$

Efficiency relationships:

$$\eta_o = \eta_v \times \eta_m \tag{4.1}$$

Volumetric displacement:
   Piston pump:

$$V_p = n \times A \times S \tag{4.2}$$

   Gear pump:

$$V_p = 6 \times W_G \times (2D - L) \times \left(\frac{L - D}{2}\right) \tag{4.8}$$

   Vane pump:

$$V_p = 12 \times W_R \times \left(\frac{L + D}{4}\right) \times \left(\frac{L - D}{2}\right) \tag{4.9}$$

Theoretical flow output:

$$Q_T = V_p \times N \tag{4.3}$$

Sound Intensity Ratio (db)
($I_2 > I_1$):

$$10 \log \frac{I_2}{I_1} \tag{4.10}$$

Total Sound Intensity (db)
(I_n = highest intensity)

$$I = I_n + 10 \log \frac{\sum_1^n I_i}{I_n} \tag{4.11}$$

(handwritten note in margin: Find 4.6 & 4.7)

References

1. Raymond P. Lambeck, *Hydraulic Pumps and Motors* (New York: Marcel Dekker, Inc, 1983).
2. J. Paul Tullis, *Hydraulics of Pipelines* (New York: John Wiley, 1989).
3. Frank Yeaple, *Fluid Power Design Handbook,* 2nd Edition, Marcel Dekker, Inc., NY, 1990.

Review Problems

Note: The letter "S" after a problem number indicates SI units are used.

General

1. What is the difference between positive and non-positive displacement pumps?
2. Why are non-positive displacement pumps seldom used in fluid power systems?
3. Describe the actions that cause fluid to enter and leave a pump.
4. List the causes of high vacuum in the pump suction line.
5. List the causes of excessive air in hydraulic fluid.
6. Define cavitation.
7. What are the two mechanisms that lead to damage of the internal pump surfaces when air or gas bubbles collapse under pressure? What kind of damage occurs?
8. Draw the "family tree" of positive displacement pumps.
9. What are the important criteria to be considered when selecting a pump for a particular application?
10. What is the purpose of a pump?
11. What causes pressure in a system?

Efficiency

12. A pump produces a hydraulic horsepower output of 16 HP. It requires a 20 HP input. What is its overall efficiency?
13. A pump with an overall efficiency of 0.85 requires a 10 HP input. What is its hydraulic horsepower output?
14. A system operates at 10 gpm and 1800 psi. The pump has an overall efficiency of 0.9. What input horsepower is required to drive the pump?
15. A hydraulic pump produces a flow of 10 gpm at 2500 psi.

    a. Assuming 100% efficiency, what horsepower is required to drive the pump?

    b. The pump turns at 1800 rpm. What is the displacement in cubic inches?

    c. The pump actually requires 15 HP to drive it. What is its overall efficiency?

16. A pump has s 2.3 in³/rev displacement and is driven at 1800 rpm. Its flow output is measured at 15.3 gpm. What is its volumetric efficiency?
17. A piston pump has seven pistons. Each piston has a 0.196 in² area and a 3 in stroke. It is rotated at 1800 rpm and produces a 29.5 gpm output. What is its volumetric efficiency? If the mechanical efficiency is 0.95, what is its overall efficiency?
18. A gear pump has a maximum bore chamber length of 3 in and a bore diameter of 1.7 in. It uses gears that are 1.25 in wide. When it is turning at 1800 rpm, the measured flow from the pump is 14.5 gpm. What is its volumetric efficiency?
19. A pump has a volumetric efficiency of 0.9 and a mechanical efficiency of 0.95. What is its overall efficiency?
20. A pump has a volumetric efficiency of 0.87 and an overall efficiency of 0.85. Find its mechanical efficiency.

21. Find the approximate mechanical efficiency of the pump of Figure 4.27 at the following points:

    a. 2500 psi, 8 gpm
    b. 3000 psi, 9 gpm
    c. 1100 psi, 8 gpm

22S. A hydraulic pump has a 30 kW output. It requires a 35 kW input. What is its overall efficiency?

23S. A pump has an overall efficiency of 0.91. What power input is required for a 50 kW output?

24S. A pump produces a flow of 60 Lpm at 10,000 kPa. It has an overall efficiency of 0.86 and a volumetric efficiency of 0.90. It turns at 1800 rpm. Find the

    a. power output.
    b. pump displacement in cm^3/rev.
    c. power input required.

25S. A pump has a delivery of 100 Lpm at 1800 rpm. Its displacement is 60 cm^3/rev. What is its volumetric efficiency?

26S. A piston pump has nine pistons. Each piston has an area of 1.13 cm^2 and a 3 cm stroke. At 2200 rpm it produces 62 Lpm. It has a mechanical efficiency of 0.98. Find the

    a. volumetric efficiency.
    b. overall efficiency.

Flow Rate

27. A pump has a 1.7 in^3/rev displacement and is driven at 3200 rpm. It has a volumetric efficiency of 0.92. What is its actual flow output in gallons per minute (gpm)?

28. A pump with a 3.6 in^3/rev rotates at 1750 rpm. What is its theoretical flow output in gallons per minute (gpm)?

29. A pump with a volumetric displacement of 1.25 in^3/rev rotates at 2250 rpm. It has a volumetric efficiency of 0.83. Find its theoretical and actual flow outputs in gallons per minute (gpm).

30S. A pump with a volumetric displacement of 15.6 cm^3/rev rotates at 1750 rpm. Find its theoretical flow rate in liters per minute (Lpm).

31S. A pump has a volumetric efficiency of 0.8 and a displacement of 25 cm^3/rev. If it is rotated at 2000 rpm, find its theoretical and actual outputs in liters per minute (Lpm).

Pump Power

32. A pump with a theoretical output of 18.5 gpm actually produces only 15 gpm. The pressure at the pump outlet is 3000 psi. If the pump has a mechanical efficiency of 0.97, what input horsepower is required to drive the pump?

33. A fixed displacement pump produces 20 gpm. The relief valve is set at 3250 psi. With the actuator stalled, the system leakage rate is 2.5 gpm. How much horsepower could be saved if a pressure-compensated pump were used with the compensator set at 3250 psi?

34. A hydraulic pump normally operates at 20 gpm and 1500 psi. The system-relief valve is set at 1700 psi. When all cylinders in the system are fully extended, the internal leakage rate is 2 gpm. How much horsepower could

be saved if a variable displacement, pressure-compensated pump with a 1700 psi compensator setting was used?

Electrical Power

35. If an electric motor produces 8.7 horsepower and has an efficiency of 0.92, what is the electrical power (in both horsepower and kilowatts) required to drive it?
36. The electric motor driving the pump in Problem 14 has a 0.87 percent efficiency. Find the electrical power required in kilowatts.
37. An electric motor with an efficiency of 0.85 is producing 7.5 HP. What is the electrical power input to the motor in horsepower and kilowatts?

System Problems

38. A pump is operating at 20 gpm (75.7 Lpm) and 1800 psi (12,400 kPa). It has an overall efficiency of 0.83. It is driven by an electric motor with an efficiency of 0.87. How much power in kilowatts is the electric motor drawing?
39. The system in Problem 38 operates eight hours per day. How much electrical power is consumed in one day of operation (kilowatt hours)? Assuming that electricity costs $0.085 per kilowatt hour, how much does it cost to run the system for one day?
40. How much of the daily operating cost of the system in Problem 38 is wasted because of inefficiencies in the pump and electric motor? If the system is operated 250 days per year, what is the annual cost of these inefficiencies?
41. A fixed displacement pump has a 30 gpm (113.6 Lpm) output. The relief valve is set at 3000 psi (20,700 kPa). The system operates 16 hours a day. The system cylinder is stalled 75 percent of that time. When the cylinder is stalled, system internal leakage is 1.5 gpm (5.7 Lpm). Based on a 250-day operating year, a pump overall efficiency of 0.90, an electric motor efficiency of 0.85, and an electricity cost of $0.09 per kilowatt hour, find the annual cost of the energy wasted by the system. If the pump is replaced by a pressure-compensated pump (compensator set at 3000 psi (20,700 kPa)) that costs $4000 more that the fixed displacement pump, how long would it take the pressure-compensated pump to pay for itself through energy savings? Assume the same efficiencies for both pumps.

Hydraulic Motors

Outline

5.1 Introduction

The energy introduced into the fluid system by the pump is generally utilized in either a linear device or a rotary device. The linear device is commonly termed a *hydraulic cylinder*, but may also be termed a *linear actuator* or a *linear motor*. The subject of this chapter will be the rotary device, which is commonly called a *hydraulic motor*. A hydraulic motor is a device that converts fluid energy into rotary mechanical energy. This is exactly opposite from the pumps described in Chapter 4, which convert rotary mechanical energy to fluid energy.

Objectives

When you have completed this chapter, you will be able to

- Explain the differences between hydraulic pumps and hydraulic motors.
- Explain the differences between the common types of hydraulic motors and describe how they work.
- Discuss the flow-speed characteristics of hydraulic motors.
- Discuss the pressure-torque characteristics of hydraulic motors.
- Explain the differences between low-speed, high-torque motors and high-speed, low torque hydraulic motors.
- Calculate the various efficiencies of hydraulic motors.
- Identify the graphic symbols representing the various categories of hydraulic motors.
- Explain the operation of basic hydraulic motor circuits, including hydrostatic transmissions.

5.2   Principle of Operation

The purpose of a hydraulic motor is to convert the fluid power output from the system pump into a rotary (mechanical) output. It does this by converting fluid pressure to torque. Regardless of the motor type, the basic principle of this power conversion is the same—pressurized fluid in the power chamber exerts a force on the rotating mechanism. If the force is sufficient to overcome the resistance to rotation, the motor will turn. If there is not enough force to turn the motor, the motor will stall, but torque is still the result. In many motor designs, maximum output torque is generated at near zero rpm and decreases as speed increases.

It should be noted that this last feature is one of the major advantages of hydraulic drives. Stalling a hydraulic motor at full system pressure causes no damage to the motor, even if it is pressurized continuously. In fact, in a properly designed system, little or no fluid heating will result. If the pump has pressure compensation, it will operate in an idling condition, using only enough power to overcome mechanical losses and make up system leakage.

5.3   Types of Motors

We can categorize motors in several different ways, as we did with pumps. for instance, there are positive and non-positive displacement motors. In a positive displacement motor, all the fluid that comes in at the pressure port (theoretically) goes out through the exhaust (or tank) port. This occurs because of the sealing inherent in the design of these motors that (theoretically) does not allow fluid to transfer between the power chambers (the equivalent of pumping chambers in pumps).

In a non-positive displacement motor, such as the turbine on a hydroelectric generator, not all the fluid entering the device produces work. Much of it can bypass the turbine blades altogether. Most devices of this type are known as *hydrodynamic* (or *hydrokinetic) drives*, because they rely on the kinetic and/or potential energy in the fluid stream, which may not be "pressurized" at all. A hydrostatic device, on the other hand, derives its energy from pressurized fluid. It makes use of a positive displacement hydraulic motor, which has very little internal leakage.

Because the vast majority of hydraulic motors are of the positive displacement type, we will consider only that category. Within this large grouping, we find the subcategories of fixed displacement and variable displacement motors. Gear, vane, and piston motors make up the bulk of the fixed displacement family. Piston and vane motors may have variable displacement, although variable displacement vane motors are not commonly used. Positive displacement hydraulic motors are generally very similar in construction to hydraulic pumps. In fact, some units are designed to operate either as a pump or a motor. These pump/motors work very well in many applications, particularly where steady-state operation is required.

There are applications, however, where the demands on motors require radical departures from typical pump designs. These applications usually involve high torque at low speeds (especially low-speed startups), frequent starts and stops, frequent thermal shock due to long idle periods, and high side loading on the output shaft because of their output drive arrangements. The obvious differences in these applications and those of typical hydraulic pumps are that pumps usually operate continuously at relatively high rpm, see fairly constant fluid temperatures, are not required to reverse the direction of rotation, and any side loading is usually due to internal pressure rather than shaft loading. These differences frequently require designs peculiar to motors and not found in pumps.

Even in some cases where the motor is very similar to its pump counterpart, some design modifications are required. One instance of this is the vane motor. In a typical vane pump, centrifugal force causes the vanes to move outward in the slots to contact the pressure ring, trapping the fluid in the pumping chamber. This occurs because the rotation is a mechanical input from a prime mover. A motor, however, depends on the fluid trapped in the power chambers to impart rotation. If the vanes of a vane motor were retracted into the rotor, the fluid would simply pass through the motor and out the drain line, and the motor would not rotate. Therefore, it is necessary for the vanes of a motor to be spring-loaded outward to ensure contact with the pressure ring.

Vane Motors

Figure 5.1a demonstrates the principle of operation of a vane motor. As the pressurized fluid enters the power chamber, a force is exerted on the vane ($F = P \times A$ again). The force may be considered to act on the center of the exposed vane area, and the resulting torque for any one vane (ignoring friction) is given by Equation 5.1.

$$T = F \times d \tag{5.1}$$

where F is the resultant force and d is the distance from the center of the drive shaft to the center of the vane. Remember that the units of torque are lb·ft (or lb·in) in the English system and N·m in SI units. Observe that the torque is a function of pressure and not a function of speed. It would be assumed that the pump has sufficient flow capacity to maintain the flow to prevent pressure losses.

Balanced vane motors are much more common than unbalanced vane motors; however, because balanced vane motors, by virtue of their design, do not have internal side loads on the shaft, they do not have heavy bearings. Most have only a single antifriction bearing at the shaft output end and a small bushing at the opposite end. In applications where the shaft is subjected to heavy side loading, external bearings are usually required to absorb the side loads in order to extend motor life.

A balanced vane motor is shown in Figure 5.1b. As the vanes pass the inlet port and approach the outlet port, they are subjected to high pressure on one side and low pressure on the other. The result is a force equal to the pressure difference multiplied by the area of the vane upon which the pressure acts. The torque produced by this force depends on the distance from the centerline of the shaft to the center of the exposed area of the vane. The center of the exposed area of the vane is termed the *center of pressure*. It represents the theoretical point at which the total resultant force acts. The total torque output of the motor depends upon the number of vanes on which torque is being produced and is the sum of the individual vane torques.

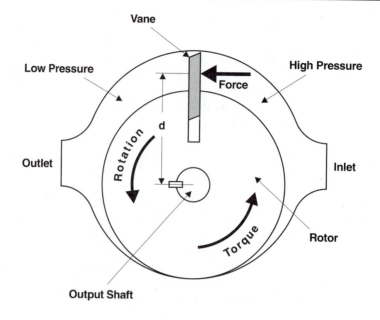

a. Torque in an unbalanced vane motor.

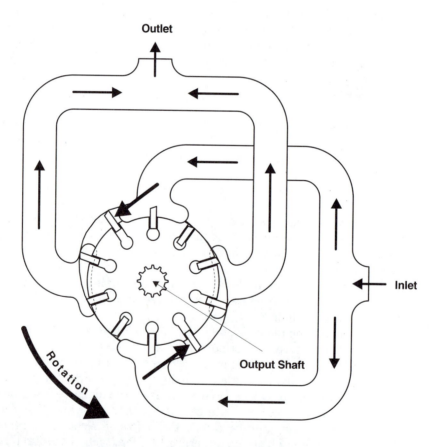

b. Torque produced on both sides of a balanced vane motor.

Figure 5.1. Torque development in an unbalanced and a balanced vane motor.

Gear Motors

Figure 5.2 shows the operation of an external gear motor. Except for the pressure-input, torque-output function, there is little difference between gear motors and gear pumps. The primary design difference is in the locations of the inlet and outlet ports inside the motor.

In addition to the external gear motor shown in Figure 5.2, there is a motor counterpart for almost every gear pump configuration. Since gear motors, by virtue of their design, have heavy side loading on their bearings because of the imbalance of internal pressures, they include heavy bearings in their construction. In applications where the motor is subjected to high external side loading, the designer may have the option to orient the motor so that the internal side loading counteracts the external side loading. Another method used to overcome heavy side loading is to provide a balanced gear motor design. In these units, high-pressure fluid is ported to the side of the bearings opposite the pressure (inlet) port of the motor. This tends to "float" the shaft and reduce rapid bearing wear. The design that provides the greatest protection for the motor bearings, though, is one that uses external bearings to handle all the side forces of the driven load.

Piston Motors

Piston motors, like pumps, include both axial piston and radial piston units. *Axial piston motors* may be either in-line or bent-axis and either fixed or

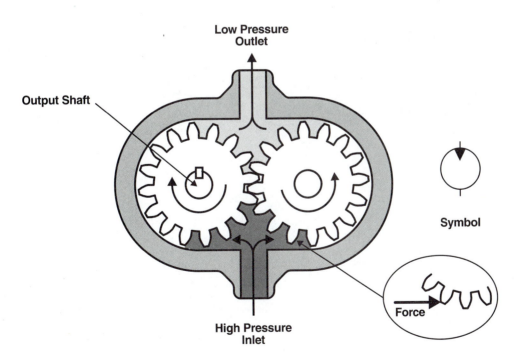

Figure 5.2. Torque development in a gear motor.

variable displacement. The operation of an in-line piston motor is shown in Figure 5.3. Again we see simply the pump action in reverse with the pressure causing the motor's pistons to move out of the cylinder block and causing a rotation as the reaction with the angled swashplate.

As is the case with pumps, the axial piston motors (both inline and bent axis) have very high efficiencies. They are generally more efficient than either gear or vane motors over the entire range of operational speeds.

The mechanisms used to change the displacement of variable displacement piston motors are the same as those used for piston pumps. As we will see later, increasing the motor displacement increases torque while decreasing

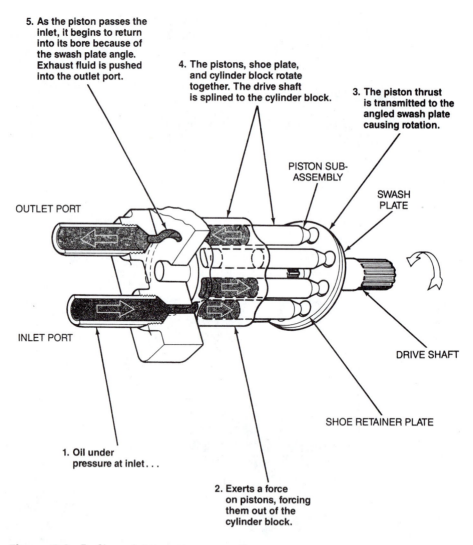

Figure 5.3. In-line piston motor operation.
Source: Courtesy of Vickers, Inc.

speed. For constant pressure and flow, this type of motor has a constant power output, but torque and speed change when displacement is changed.

Radial piston motors are basically similar to radial piston pumps. They are used primarily in medium and heavy duty applications requiring low speed and high torque. These LSHT units characteristically operate at speeds below 500 rpm with torque output from 500 lb-ft to 20,000 lb-ft or more. They are typically capable of handling overrunning loads and can provide braking to very low speeds. They are generally very quiet and have overall efficiencies in the range of 85 to 92 percent.

Like gear motors, piston motors have the same problems with side loading. The same techniques can be employed to extend motor life.

Low Speed, High Torque Motors

When the hydraulic motors we've discussed so far are used in applications that require a high torque output at low speeds, it is usually necessary to use a gearbox. In the gearbox, the gear arrangement reduces the output shaft rotating speed while increasing the output torque. It is often advantageous to use a *low-speed, high-torque (LSHT)* hydraulic motor for such applications, eliminating the space, weight, and expense of the gearbox. The design of LSHT motors effectively multiplies the volumetric displacement of the motor. Torque is directly proportional to displacement; therefore, increasing the displacement increases the torque. On the other hand, speed is inversely proportional to displacement. Consequently, increasing the displacement decreases the speed.

There are three basic LSHT motor designs—*radial piston* (mentioned in the previous section), *multi-lobed vane*, and *orbital* (or internal gear). All three types are radical variations of the piston, vane, and internal gear designs with which we are already familiar.

Figure 5.4 shows a *gerotor-type* LSHT motor. In this particular motor, the displacement multiplication is the result of the gerotor element and an internal valving system that distributes pressurized fluid to the gerotor. The result is a sixfold increase in output torque and an equivalent reduction in output speed. This provides the same mechanical advantage as a 6 to 1 gear reduction, but without requiring a gearbox.

The LSHT motor shown in Figure 5.5 uses a *geroler* mechanism, which is similar to a gerotor mechanism, but precision-machined rollers form the displacement chambers. The advantage of this design is that the rollers provide support with a rolling contact. This minimizes friction, especially at startup and low speeds, resulting in improved efficiency. The disc valve shown in the drawing rotates with the assembly and ports the fluid to the geroler elements.

Typical speeds for LSHT motors range from essentially zero to a few hundred rpm. Torques range from a few pound-inches to several thousand pound-feet. Performance data for a large geroler-type motor are shown in Figure 5.6.

Low-speed, high-torque motor applications include

- Conveyors
- Augers and screws
- Positioning and indexing
- Wheel drives for vehicle propulsion

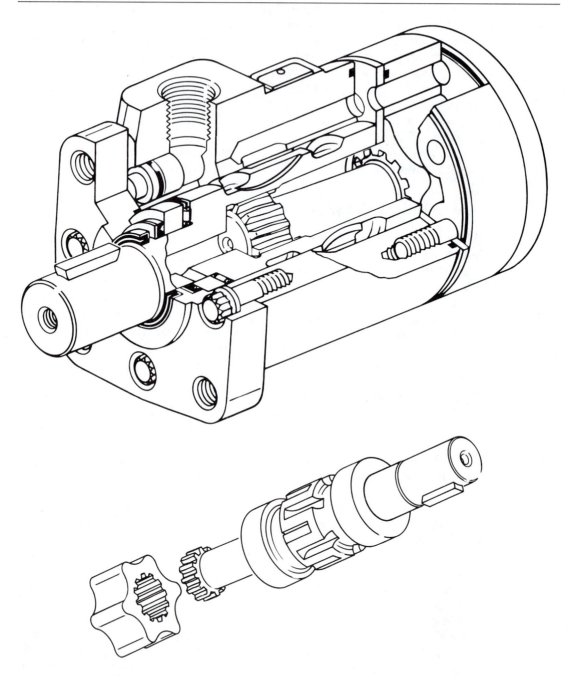

Figure 5.4. LSHT motor in which displacement multiplication is the result of the gerotor element and internal valving.

Source: Courtesy of Eaton Corporation.

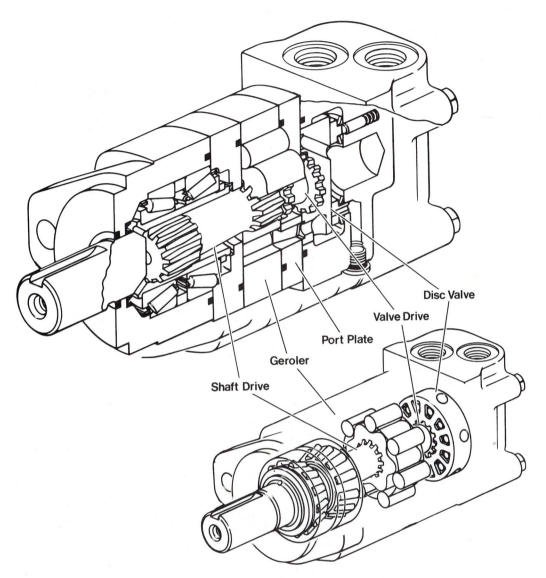

Figure 5.5. LSHT motor using geroler mechanism.
Source: Courtesy of Eaton Corporation.

- Spreaders, mowers, and grinders
- Low-speed reels and rollers
- Winch drives

5.4 Efficiency

The similarities between pumps and motors extend into the discussion of efficiency, where we are again concerned with mechanical, volumetric, and overall efficiencies.

When considering the overall efficiency of any component or system, we can safely revert to Equation 4.5. The application of this equation to hydraulic motors is a little different from its application to pumps, however. In a pump, the horsepower input came from the prime mover, while the horsepower output was in the form of hydraulic horsepower. For a hydraulic motor, the horsepower *input* is hydraulic horsepower (*HHP*) while the output is rotational mechanical horsepower (*OHP*). The overall efficiency equation for a hydraulic motor takes the form

$$\eta_o = \frac{OHP}{HHP} \tag{5.2}$$

The hydraulic horsepower available to drive the motor is found from Equation 5.3.

$$HHP = \frac{\Delta p \times Q}{1714} \tag{5.3}$$

57.4 cu. in./rev. [940 cu. cm./rev.]
Δ Pressure

Flow GPM [LPM] / Bar PSI	[15] 250	[35] 500	[50] 750	[70] 1000	[85] 1250	[105] 1500	[120] 1750	[140] 2000	[155] 2250	[170] 2500	[190] 2750	[205] 3000	[225] 3250	[240] 3500
[3,8] / 1	[235] 2080 3	[480] 4260 2	[730] 6440 1	[Nm] lb. in. RPM } Torque —Speed										
[7,5] / 2	[235] 2090 7	[480] 4270 6	[730] 6450 5	[975] 8640 5	[1220] 10820 4	[1470] 13000 3	[1715] 15190 2	[1965] 17370 .1						
[15] / 4	[235] 2080 15	[480] 4260 14	[730] 6440 13	[975] 8620 13	[1220] 10810 12	[1470] 12990 11	[1715] 15170 10	[1960] 17360 9	[2210] 19540 8	[2455] 21720 7	[2700] 23910 7	[2950] 26090 6	[3195] 28270 5	[3440] 30460 4
[30] / 8	[230] 2040 31	[475] 4220 30	[725] 6400 29	[970] 8590 28	[1215] 10770 28	[1465] 12950 27	[1710] 15140 26	[1955] 17320 25	[2200] 19500 24	[2450] 21690 23	[2695] 23870 22			
[45] / 12	[225] 1990 47	[470] 4170 46	[715] 6350 45	[965] 8540 44	[1210] 10720 43	[1460] 12900 43	[1705] 15090 42	[1950] 17270 41	[2200] 19450 40	[2445] 21640 39				
[61] / 16	[220] 1930 63	[465] 4110 62	[710] 6290 61	[960] 8480 60	[1205] 10660 59	[1450] 12840 58	[1700] 15030 58	[1945] 17210 57	[2190] 19390 56					
[76] / 20	[210] 1860 79	[455] 4040 78	[705] 6220 77	[950] 8410 76	[1195] 10590 75	[1445] 12770 74	[1690] 14960 73	[1935] 17140 72	[2185] 19320 72					
[91] / 24	[200] 1780 95	[450] 3970 94	[695] 6150 93	[940] 8330 92	[1190] 10520 91	[1435] 12700 90	[1680] 14880 89	[1930] 17070 88						
[106] / 28	[190] 1700 111	[440] 3890 110	[685] 6070 109	[930] 8250 108	[1180] 10440 107	[1425] 12620 106	[1675] 14800 105	[1920] 16990 104						
[121] / 32	[185] 1620 127	[430] 3800 126	[675] 5980 125	[920] 8160 124	[1170] 10350 123	[1415] 12530 122	[1665] 14720 121							
[136] / 36	[170] 1520 143	[420] 3710 142	[665] 5890 141	[910] 8070 140	[1160] 10260 139	[1405] 12440 138	[1650] 14620 137							
[151] / 40	[160] 1420 159	[410] 3610 158	[655] 5790 157	[900] 7970 156	[1150] 10160 155	[1395] 12340 154	[1640] 14520 153							
[170] / 45	[145] 1290 179	[395] 3480 178	[640] 5660 177	[885] 7840 176	[1130] 10020 174	[1380] 12210 174	[1625] 14400 173							
[227] / 60	[95] 860 239	[345] 3040 238	[590] 5230 236	[835] 7410 235	[1085] 9600 234	[1330] 11780 233								
[265] / 70	[60] 540 279	[305] 2720 278	[555] 4910 276	[800] 7090 275	[1045] 9270 274	[1295] 11460 273								

Full load, 100% of time

Full load intermittent loads ~50%

Figure 5.6. Performance data for a large geroler-type LSHT motor.
Source: Courtesy of Eaton Corporation.

Note that this equation is essentially the same as for the horsepower output of hydraulic pumps. The 1714 constant again constrains us to using Q in gpm and Δp in psi.

In SI units, Equation 5.3 becomes

$$P_{\text{HYD}} = \frac{\Delta p \times Q}{60,000} \qquad (5.4)$$

where Δp is in kPa and Q is in Lpm to give P in kw.

The Δp term is actually the pressure differential between the motor inlet port and its outlet (or tank) port. For our purposes, we will assume that the motor tank port is at atmospheric pressure so that all the system pressure is dissipated across the motor. In this case, the Δp is actually the system operating pressure. Be careful about applying this definition too liberally. Line losses, heat exchanges, filters, and pressurized reservoirs raise the pressure at the outlet port. In these cases, the outlet port of the motor will not be at atmospheric pressure, so the value of Δp will be less than system pressure.

If we ignore system losses, we can say that the hydraulic power input to the motor is the same as the hydraulic power output from the pump. This allows us the flexibility to utilize the system flow rate and the system pressure to calculate the hydraulic power at either end of the fluid power system.

The output of a hydraulic motor in horsepower is found by using

$$OHP = \frac{T \times N}{63,025} \qquad (5.5)$$

where OHP is the output horsepower, T (torque) is in lb·in and N is in rpm. If T is in lb·ft, use 5252 for the constant in the denominator. Once again, recall the restrictions mandated by the constant in the denominator. Output horsepower is sometimes referred to as *mechanical horsepower*.

The equivalent SI equation is

$$P_{\text{OUT}} = \frac{T \times N}{9550} \qquad (5.6)$$

where T is in N·m and N is in rpm to give P in kw.

Since there is no 100 percent efficient motor, we must account for power losses and define the motor efficiencies. *Power loss* is simply the difference between the hydraulic power input and the mechanical power output. Thus,

$$\text{Power Loss} = HHP - OHP \qquad (5.7)$$

We use the same two parameters to define the overall efficiency of the motor:

$$\eta_o = \frac{\text{Output}}{\text{Input}} = \frac{OHP}{HHP} = \frac{P_{\text{OUT}}}{P_{\text{HYD}}} \qquad (5.8)$$

Substituting from Equation 5.3 and 5.5, we can write

$$\eta_o = \frac{(T \times N)/63,025}{(\Delta p \times Q)/1714} \qquad (5.9)$$

or

$$\eta_o = \frac{(T \times N)/9550}{(\Delta p \times Q)/60,000} \qquad (5.10)$$

Some typical motor efficiency curves are shown in Figures 5.7 and 5.8.

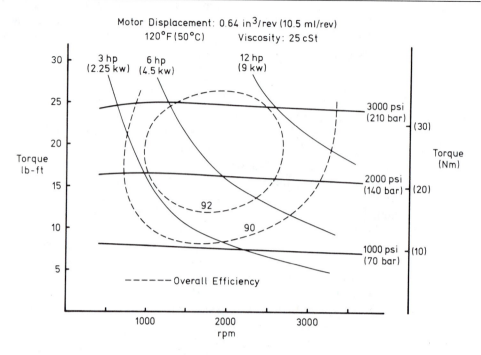

Figure 5.7. Efficiency curves for high-speed balanced vane motor.
Source: Reprinted from Ref. 1, p. 99 by courtesy of Marcel Dekker, Inc.

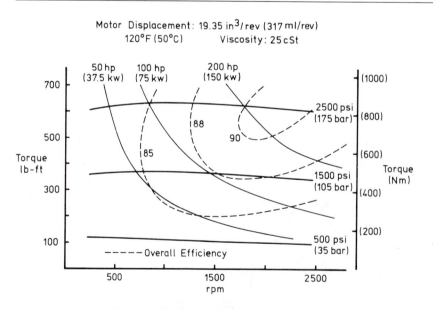

Figure 5.8. Efficiency curves for heavy duty in-line piston motor.
Source: Reprinted from Ref. 1, p. 103 by courtesy of Marcel Dekker Inc.

Example 5.1

A hydraulic motor is used in a system that has a flow rate of 18 gpm and a pressure of 1750 psi. The motor turns at 700 rpm. Assuming 100 percent efficiency, find the output torque of the motor.

Solution

Assuming 100 percent efficiency, we can set equations 5.3 and 5.5 equal. Thus,

$$HHP = OHP$$

or

$$\frac{\Delta p \times Q}{1714} = \frac{T \times N}{63{,}025}$$

Solving the T, we have

$$T = \frac{63{,}025 \times \Delta p \times Q}{1714 \times N}$$

$$= \frac{63{,}025 \times 1750 \text{ psi} \times 18 \text{ gpm}}{1714 \times 700 \text{ rpm}}$$

$$= 1655 \text{ lb·in}$$

Example 5.2

If the motor in Example 5.1 actually has a power loss of 3.38 HP, find the

a. actual torque output.
b. overall efficiency.

Solution

a. First, we must find the hydraulic horsepower from Equation 5.3.

$$HHP = \frac{\Delta p \times Q}{1714} = \frac{1750 \text{ psi} \times 18 \text{ gpm}}{1714} = 18.38 \text{ HP}$$

Now we can find the power output based on HHP and the power loss.

$$\text{Power Loss} = HHP - OHP$$

or

$$OHP = HHP - \text{Power Loss}$$

$$= 18.38 - 3.38$$

$$= 15 \text{ HP}$$

Next, we use Equation 5.5 to find the torque output.

$$OHP = \frac{T \times N}{63{,}025}$$

so

$$T = \frac{63{,}025 \times OHP}{N}$$

$$= \frac{63{,}025 \times 15}{700}$$

$$= 1350.5 \text{ lb·in}$$

b. The overall efficiency is found from Equation 5.8, where

$$\eta_o = \frac{OHP}{HHP} \times 100 = \frac{15}{18.38} \times 100 = 81.6\%$$

Two other efficiencies may also be of interest when assessing motor performance. The first of these is the volumetric efficiency, which concerns the percentage of the fluid entering the motor that produces mechanical work (in this case, torque). Fluid can fail to produce mechanical work in two ways. Some fluid may be directed through passages in the motor casing to provide cooling and lubrication for bearings and other moving surfaces or to pressurize sealing elements to reduce internal leakage. This fluid exits the motor through a case drain port. Some fluid may leak through the clearances in the motor, particularly as the motor starts to wear. This internal leakage—also termed *slippage*—is the major obstacle to maintaining a high volumetric efficiency. Because most wear, and consequently most slippage, can be attributed to particulate contamination in the fluid, the importance of maintaining low contamination levels is evident.

Equation 5.11 is used to calculate the volumetric efficiency of a hydraulic motor.

$$\text{Volumetric Efficiency} = \frac{\text{Theoretical Flow Rate}}{\text{Actual Flow Rate}}$$

$$\eta_v = \frac{Q_T}{Q_A} \tag{5.11}$$

Note that the flow rate terms in this equation are exactly reversed from Equation 4.4, which deals with hydraulic pumps. In Equation 5.11, the Q_T term defines the flow rate that would theoretically cause the motor to rotate at a given rpm, N, assuming no leakage, while Q_A is the actual flow rate required to produce that rpm. Q_A is always greater than Q_T; therefore, η_v is always less than 1. The value of Q_T, as for a pump, is based on rpm and volumetric displacement. It is expressed mathmatically as

$$\text{Theoretical Flow Rate} = \text{Volumetric Displacement} \times \text{Speed}$$
$$Q_T = V_m N \tag{5.12}$$

where V_m is the volumetric displacement (commonly in in³/rev) and N is the rotating speed (commonly in rpm).

Example 5.3

A hydraulic motor with a 2.4 in³/rev volumetric displacement is required to turn at 500 rpm. Find the theoretical flow rate needed for this speed.

Solution

Using Equation 5.12,

$$Q_T = V_m N$$

$$= 2.4 \text{ in}^3/\text{rev} \times 500 \text{ rev/min} \times \text{gal}/231 \text{ in}^3$$

$$= 5.19 \text{ gpm}$$

The actual flow rate is the amount of flow required to rotate the motor at the specified rpm, accounting for leakage. Since there are slippage and (possibly) cooling/lubricating losses in hydraulic motors, the actual flow requirement is always greater than the theoretical flow requirement.

Example 5.4

The hydraulic motor of Example 5.3 actually requires 5.5 gpm to rotate at 500 rpm. Find its volumetric efficiency.

Solution

From Equation 5.11.

$$\eta_v = \frac{Q_T}{Q_A}$$

We have already calculated the theoretical flow rate, and we are given the actual flow requirement. Therefore,

$$\eta_v = \frac{5.19 \text{ gpm}}{5.5 \text{ gpm}}$$

$$= .944 = 94.4\%$$

Another way to express volumetric efficiency is to compare the actual and theoretical motor speeds resulting from a given actual flow rate. Theoretically, a given flow rate will give a speed defined by Equation 5.13

$$\text{Theoretical Speed} = \frac{\text{Actual Flow Rate}}{\text{Volumetric Displacement}}$$

$$N_T = \frac{Q_A}{V_M} \tag{5.13}$$

Due to internal leakage, however, the actual speed will be less than the theoretical speed. The actual speed is defined by Equation 5.14.

$$\text{Actual Speed} = \frac{\text{Actual Flow Rate} \times \text{Volumetric Efficiency}}{\text{Volumetric Displacement}}$$

$$N_A = \frac{Q_A}{V_M} \eta_v \tag{5.14}$$

Thus, we can use these two equations to give us another definition of volumetric efficiency. Solving Equation 5.14 or Q_A/V_M gives

$$\frac{Q_A}{V_M} = \frac{N_A}{\eta_v}$$

Setting this equal to Equation 5.13, we get   **(5.15)**

$$\frac{N_A}{\eta_v} = N_T$$

or

$$\text{Volumetric Efficiency} = \frac{\text{Actual Speed}}{\text{Theoretical Speed}}$$

$$\eta_v = \frac{N_A}{N_T}$$

(5.16)

Again, since the actual speed is always less than the theoretical speed, the value of volumetric efficiency is always less than 1.

Example 5.5

A hydraulic motor has a volumetric displacement of 3 in³/rev. Find its theoretical speed for a flow rate of 10 gpm. The actual speed is 700 rpm. What is its volumetric efficiency?

Solution

The theoretical speed is found from

$$Q = V_M N$$

or

$$N_T = \frac{Q}{V_M} = \frac{10 \text{ gal/min} \times 231 \text{ in}^3\text{/gal}}{3 \text{ in}^3\text{/rev}} = 770 \text{ rpm}$$

Now, using Equation 5.16, we get

$$\eta_v = \frac{700 \text{ rpm}}{770 \text{ rpm}} = 0.91 = 91\%$$

Another efficiency that is of interest is termed *mechanical efficiency* and involves mechanical losses due to friction, inertia, distortion, and so on. As with hydraulic pumps, mechanical efficiency can be calculated using the relationship between it and the overall and volumetric efficiencies as shown in Equation 5.17 and 5.18:

$$\eta_o = \eta_m \times \eta_v$$

(5.17)

or

$$\eta_m = \frac{\eta_o}{\eta_v}$$

(5.18)

Mechanical efficiency is defined as the ratio of the theoretical torque output (T_T) of the motor and its actual torque output (T_A). This is shown in Equation 5.19

$$\text{Mechanical Efficiency} = \frac{\text{Actual Torque}}{\text{Theoretical Torque}}$$

$$\eta_M = \frac{T_A}{T_T}$$

(5.19)

5.5 Rotary Actuators

There are some applications in which continuous, unlimited rotation is not needed, but where the power and torque capabilities and component cost make a hydraulic motor attractive. One such case is a stationary industrial mechanism performing pick-and-place operations and in heavy-duty, large-payload robots. In such mechanisms, rotation is usually limited to 360 degrees or less because of hydraulic connections, structural obstacles, and safety.

The hydraulic motors used in these applications are termed *oscillating motors, limited-rotation motors,* or *rotary actuators.* They are commonly vane-type motors with a small number of vanes, usually one or two. The single-vane unit is usually limited to around 280 degrees of rotation. Dual-vane actuators can rotate only 200 degrees or so. They are bi-directional, with the direction of rotation determined by a valve that directs the fluid into first one port, than the other. These motors are usually designed to produce high torque (up to 500,000 lb·in (56,500 N·m)), but relatively low speeds. Single-vane and double-vane rotary actuators are shown in Figure 5.9.

Figure 5.10 shows a rotary actuator that uses a helical spline to generate torque. These units usually have 90, 180, 270, or 360 degrees of rotation, depending on the length and pitch of the helix. Torques of up to 1,000,000 lb·in (113,000 N·m) are possible.

Another type of rotary actuator, shown in Figure 5.11, is the *skotch yoke actuator.* These devices use short-stroke cylinders attached to the yoke, which translates the linear motion of the cylinders into rotary motion of the shaft. The range of rotation is limited to 90 degrees or less, but torques up to 45,000,000 lb·in (5,080,000 N·m) are possible.[1]

The *rack-and-pinion actuators* shown in Figure 5.12 also convert the linear motion of a hydraulic cylinder into rotary motion of the pinion, or shaft. The stroke of the cylinder can be controlled to determine the angle of shaft rotation, which is limited only by the maximum cylinder stroke. Two cylinders can be used to provide additional torque, which can be as much as 52,000,000 lb·in (5,876,000 N·m).[2]

5.6 Symbols

The graphic symbols for hydraulic motors are similar to those for hydraulic pumps. As seen in Figure 5.13, the basic elements are circles, equilateral triangles, and arrows. The difference lies only in the orientation of the triangles. In *pump* symbols, the apex of the triangle points outward. This can be thought of as indicating that the fluid power flows outward from a pump. In *motor*

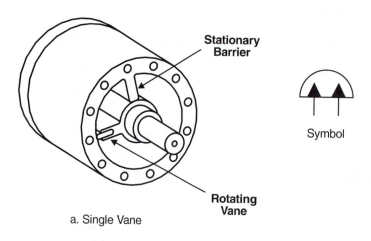

a. Single Vane

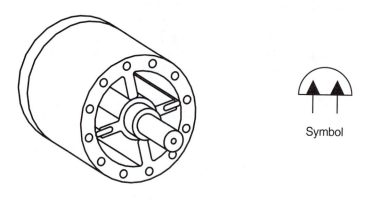

b. Double Vane

Figure 5.9. Single-vane rotary actuators are usually limited to about 280 degrees of rotation and double-vane to about 200 degrees.

symbols, the apex is pointing inward, indicating that fluid power flows into a motor. Since symbols indicate function only, the type of motor (gear, vane, piston) cannot be determined from the symbol.

The symbol for the oscillating (limited-rotation), or rotary, actuator utilizes a semicircle rather than a full circle (Figure 5.14). The bi-directional nature of these units is shown by the two inward-pointing triangles used in the symbol.

5.7 Selection Criteria

As might be expected, the selection criteria for hydraulic motors are basically the same as for pumps. They include performance, physical size and mounting, noise levels, and contaminant sensitivity.

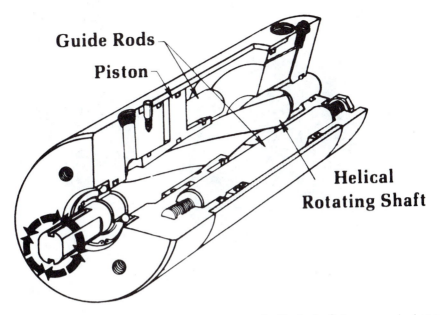

Figure 5.10. This rotary actuator uses a helical shaft to generate torque.
Source: Courtesy of Parker Hannifin Corporation.

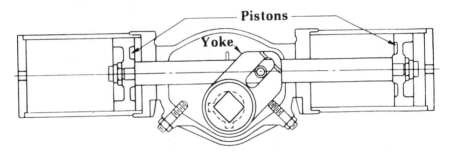

Figure 5.11. This skotch yoke rotary actuator uses short-stroke cylinders attached to a yoke, which translates the linear motion of the cylinders into rotary motion of the shaft.
Source: Courtesy of Parker Hannifin Corporation.

The performance of a hydraulic motor should be determined by the procedures contained in SAE (Society of Automotive Engineers) Recommended Practice J746. These procedures cover the following performance parameters:

1. SAE volumetric rating.
2. SAE running torque characteristics.
3. SAE stall torque characteristics.
4. Power output.

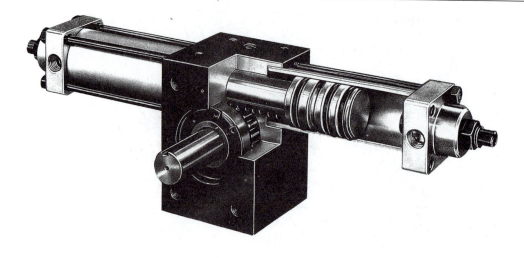

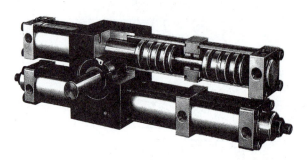

Figure 5.12. Rack-and-pinion rotary actuators convert the linear motion of a hydraulic cylinder into rotary motion of the shaft.

Source: Courtesy of PHD, Inc., Ft. Wayne, IN.

| Uni-directional, fixed displacement | Bi-directional (reversible), fixed displacement | Uni-directional, variable displacement | Bi-directional (reversible) variable displacement |

Figure 5.13. Hydraulic motor symbols.

Figure 5.14. Oscillating motor (rotary actuator) symbol.

5. Power loss.
6. Torque efficiency.
7. Overall efficiency.

The same document recommends a format for presenting the test results graphically. This is shown in Figure 5.15. Many manufacturers do not adhere exactly to this report format, so care must be exercised in interpreting manufacturers' data.

In the design of a system or the specification of a replacement motor, the required torque output is usually known. Based on this torque requirement and the operating pressure of the system the "size" of the motor (which actually refers to the displacement of the motor) can be calculated using the following equation:

$$\text{Volumetric Displacement} = \frac{2\pi \times \text{Torque}}{\text{Pressure Differential} \times \text{Mechanical Efficiency}}$$

$$V_m = \frac{2\pi T}{\Delta p \; \eta_m} \tag{5.20}$$

where

V_m = volumetric displacement of the motor (in³/rev or cm³/rev).
T = actual torque output (lb · in or N · m).
Δp = pressure differential across the motor (psi or kPa).
η_m = mechanical efficiency.
2π = conversion from radians to revolutions.

As we've discussed already, we can often consider the pressure in an open-loop system to be atmospheric (0 psig or 0 kPag) at the outlet of the motor, so Δp is actually the system operating pressure for most applications.

Once the motor displacement has been established, the actual volumetric flow rate required to provide any desired speed can be calculated from

$$\text{Actual Flow Rate} = \frac{\text{Volumetric Displacement} \times \text{Speed}}{\text{Volumetric Efficiency}}$$

$$Q_A = \frac{V_m N}{\eta_v} \tag{5.21}$$

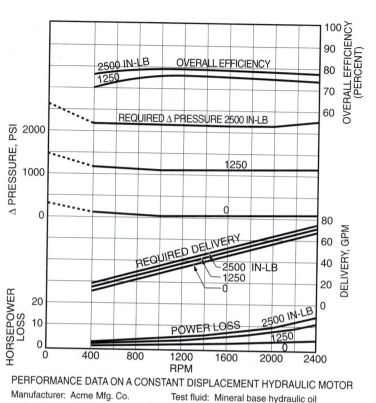

PERFORMANCE DATA ON A CONSTANT DISPLACEMENT HYDRAULIC MOTOR

Manufacturer: Acme Mfg. Co.   Test fluid: Mineral base hydraulic oil
Series or type: ZYX   Fluid viscosity @ test temperature 100SUS
Model: 8Z24   Temperature of test fluid: 120 F
Motor rotation: Double
SAE volumetric rating: 35.5 GPM/1000 RPM

Figure 5.15. Motor performance data recommended by SAE J746.
Source: Reprinted with permission from SAE J746, © 1992, Society of Automotive
Engineers, Inc.

where

$$V_m = \text{motor displacement (in}^3\text{/rev or cm}^3\text{/rev).}$$
$$N = \text{actual speed (rpm).}$$
$$\eta_v = \text{volumetric efficiency.}$$

Example 5.6

A hydraulic motor is required to produce 100 lb·in (11.3 N·m) of torque at 3000
rpm. The system pressure is 3000 psi (20,685 kPa). Find the size (displacement)
of motor required and the system flow rate needed if the mechanical and
volumetric efficiencies are 0.90 and 0.95, respectively.

Solution

We find the motor displacement from Equation 5.20.

$$V_m = \frac{2\pi T}{\Delta p \eta_m}$$

If we assume that the outlet port is at atmospheric pressure, then Δp is equal to the system pressure, so we get

$$V_m = \frac{2\pi \times 100 \text{ lb·in}}{3000 \text{ lb/in}^2 \times 0.90}$$

$$= 0.233 \text{ in}^3/\text{rev}$$

We can now use Eq. 5.21 to find the flow rate to turn the motor at 3000 rpm:

$$Q_A = \frac{V_m N}{\eta_v}$$

$$= \frac{0.233 \text{ in}^3/\text{rev} \times 3000 \text{ rev/min}}{0.95}$$

$$= 734.8 \text{ in}^3/\text{min} \times \text{gal}/231 \text{ in}^3$$

$$= 3.18 \text{ gpm}$$

Solving this problem in SI units gives

$$V_m = \frac{2\pi \times 11.3 \text{ N·m}}{(20.685 \text{ kPa})(0.90)}$$

$$= 3.8 \text{ cm}^3/\text{rev}$$

Using Equation 5.21, we find the required flow rate to be

$$Q_A = \frac{V_m N}{\eta_v}$$

$$= \frac{3.8 \text{ cm}^3/\text{rev} \times 3000 \text{ rev/min}}{0.95}$$

$$= 12,000 \text{ cm}^3/\text{min} = 12 \text{ Lpm}$$

If we solve Eq. 5.20 for torque, we find that actual torque can be calculated as

$$T = \frac{V_m \Delta p \eta_m}{2\pi} \qquad \qquad \textbf{(5.22)}$$

From this equation, we see that the torque output from a motor is a function of both its displacement and its operating pressure. Therefore, for a fixed displacement motor, the torque is dependent on the system pressure.

Solving Eq. 5.21 for speed, we get

$$N = \frac{Q_A \eta_v}{V_m} \qquad \qquad \textbf{(5.23)}$$

This shows us that the speed of any given motor is a function of the flow rate of fluid through the motor. There is no pressure term in this equation. We have already seen that one advantage of a hydraulic motor is that it produces full torque even at near-zero rpm and can remain stalled indefinitely at full torque with no damage to it or the system as long as proper pressure relief is provided. Varying the flow rate changes the speed. This flow rate change can be accomplished by several means, including changing pump speed, changing pump displacement, or using valves as we'll see in later chapters.

Applying Eq. 5.22 and 5.23 to *variable* displacement motors, we see that if pressure and output flow are held constant, reducing the displacement results in a decrease in torque and an increase in speed. Theoretically, taking the displacement to zero would give zero torque and infinite speed.

Flange mounts and shaft designs for hydraulic motors for use on construction equipment and industrial machinery have been standardized in SAE J744. The noise output of a motor can be measured by methods detailed in NFPA T.3.9.14M. The contaminant sensitivity (Omega rating) of hydraulic motors can be established using NFPA T.3.9.25. This test is similar to the contaminant sensitivity test for pumps, except that speed degradation is the test parameter for motors, while flow degradation is the test parameter for pumps. It is commonly referred to as the Omega test.

5.8   System Analysis

We now have enough information to try our hand at a simplistic system design problem. To do this, we'll combine our knowledge of pumps from Chapter 4 with the new information on motors to size the components in a hypothetical hydraulic system.

Example 5.7

A hydraulic motor is required to produce approximately 15 lb·ft (20.3 N·m) of torque at 3000 rpm. The volumetric efficiency of the motor is 92 percent. The system operating pressure is 2000 psi (13,800 kPa). The electric motor driving the pump runs at 1800 rpm. Ignoring all losses between the pump and the hydraulic motor, and using the performance curves in Figures 5.16 and 5.17, determine the horsepower input from the electric motor and find the overall efficiency of the system.

Solution

We'll start at the output end of the system and size the motor first. We find the motor displacement from Equation 5.20.

$$V_m = \frac{2\pi T}{\Delta p \eta_m}$$

We know the torque and the pressure drop from the problem statement. Looking at the performance curves for the motor (Figure 5.17), we see that the overall efficiency is about 0.90. The volumetric efficiency is given in

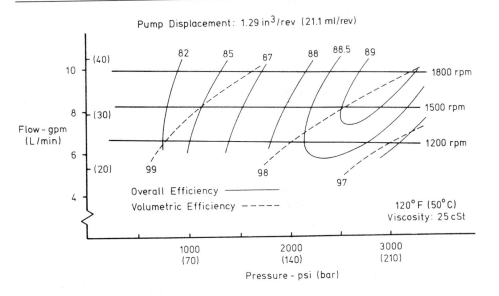

Figure 5.16. Performance curves for in-line piston pump.
Source: Reprinted from Ref. 3, p. 31, by courtesy of Marcel Dekker, Inc.

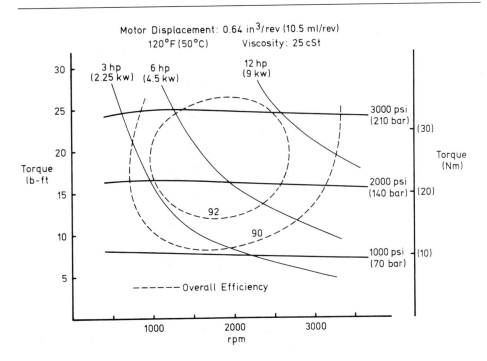

Figure 5.17. Performance curves for in-line piston motor.
Source: Reprinted from Ref. 3, p. 48, by courtesy of Marcel Dekker, Inc.

the problem statement as 0.92; however, the equation calls for *mechanical* efficiency. Using Equation 5.18, we find

$$\eta_m = \frac{\eta_o}{\eta_v} = \frac{0.90}{0.92} = 0.98$$

Thus, the required motor displacement is

$$V_m = \frac{2\pi \times 15 \text{ lb·ft} \times 12 \text{ in/ft}}{0.98 \times 2000 \text{ lb/in}^2}$$

$$= 0.58 \text{ in}^3/\text{rev}$$

Now we can find the flow rate required to operate the motor at 3000 rpm from Equation 5.21:

$$Q = \frac{V_m N}{\eta_v}$$

$$= \frac{0.58 \text{ in}^3/\text{rev} \times 3000 \text{ rev/min}}{0.92}$$

$$= 1891 \text{ in}^3/\text{min} \times \text{gal}/231 \text{ in}^3$$

$$= 8.19 \text{ gpm}$$

Now that we know the required system pressure and flow rate, we can find the hydraulic horsepower needed to drive the motor from Equation 5.3.

$$HHP = \frac{\Delta p \times Q}{1714}$$

$$= \frac{2000 \text{ psi} \times 8.19 \text{ gpm}}{1714}$$

$$= 9.56 \text{ HP}$$

Since we are ignoring system losses, we can say that this also represents the full output of the pump.

We now look at Figure 5.16, where we see that the pump has an overall efficiency of around 0.88 at 2000 psi and 8.19 gpm. Using this information, we can find the horsepower required of the electric motor from our basic definition of efficiency (output divided by input). Thus,

$$\eta_o = \frac{HHP}{IHP}$$

or
$$IHP = \frac{HHP}{\eta_o}$$

$$= \frac{9.56}{0.88}$$

$$= 10.86 \text{ HP}$$

Finally, to find the system efficiency, we again use the definition of efficiency. In this case, the output is the horsepower output of the hydraulic motor, which we calculate from Equation 5.5, (for T in pound-inches), $$OHP = \frac{T \times N}{63{,}025}$$

or $$OHP = \frac{T \times N}{5252}$$

(for T in pound-feet), which gives us

$$= \frac{15 \text{ lb} \cdot \text{ft} \times 3000 \text{ rpm}}{5252}$$

$$= 8.57 \text{ HP}$$

Thus, we get a system efficiency of

$$\eta_{\text{SYS}} = \frac{8.57}{10.86} = 0.79 = 79\%$$

Solving this problem using SI units, we get

$$V_m = \frac{2\pi T}{\Delta p \eta_m} = \frac{2\pi \times 20.3 \text{ N} \cdot \text{m}}{13{,}800 \text{ kPa} \times 0.98}$$

$$= 9.4 \text{ cm}^3/\text{rev}$$

$$Q = \frac{V_m N}{\eta_v} = \frac{9.4 \text{ cm}^3/\text{rev} \times 3000 \text{ rev/min}}{0.92}$$

$$= 30{,}652 \text{ cm}^3/\text{rev} = 30.652 \text{ Lpm}$$

$$P_{\text{HYD}} = \frac{\Delta p \times Q}{60{,}000} = \frac{13{,}800 \text{ kPa} \times 30.652 \text{ Lpm}}{60{,}000}$$

$$= 7.05 \text{ kw}$$

$$P_{\text{IN}} = \frac{P_{\text{HYD}}}{\eta_o} = \frac{7.05 \text{ kw}}{0.88}$$

$$= 8.01 \text{ kw}$$

$$P_{\text{OUT}} = \frac{T \times N}{9550} = \frac{20.3 \times 3000}{9550}$$

$$= 6.4 \text{ kw}$$

$$\eta_{\text{SYS}} = \frac{P_{\text{OUT}}}{P_{\text{IN}}} = \frac{6.4 \text{ kw}}{8.01 \text{ kw}}$$

$$= 0.799 = 79.9\%$$

The slight difference in the two answers is due to rounding during conversions. This is actually a fairly high system efficiency. As we will see in Chapter 12, losses through pipes, fittings, and valves can be significant.

We can actually take this system calculation one step further and include the electric motor efficiency in our calculations. That will allow us to calculate the electrical horsepower (*EHP*) required. We define the overall efficiency of the electric motor as

$$\eta_o = \frac{HP_{\text{OUT}}}{HP_{\text{IN}}}$$

Here HP_{OUT} is the horsepower provided to drive the pump, which we have already designated IHP in Chapter 4. The HP_{IN} term is the electrical horsepower, which we'll label EHP. Thus,

$$\eta_o = \frac{IHP}{EHP} \tag{5.24}$$

or

$$EHP = \frac{IHP}{\eta_o} \tag{5.24a}$$

Note that in these equations, we use the overall efficiency (η_o) of the electric motor.

Example 5.8

The electric motor driving the pump in Example 5.7 has an overall efficiency of 0.87. Find the electric horsepower required.

Solution

Using Equation 5.24a, we get

$$EHP = \frac{IHP}{\eta_o}$$

$$= \frac{10.86}{.87}$$

$$= 12.48 \text{ HP}$$

We can now find a total system efficiency from our basic definition:

$$\eta_o = \frac{HP_{\text{OUT}}}{HP_{\text{IN}}}$$

Here, HP_{OUT} is the output from the hydraulic motor (OHP) and HP_{IN} is the electric horsepower required to drive the electric motor. Therefore,

$$\eta_o = \frac{8.57}{12.48} = .69 = 69\%$$

In working these system problems, it is absolutely essential that you keep track of efficiencies, flows, and so on, according to the individual component for which they are specified.

The overall efficiency of any system can also be found as the product of all the system components. In this case, we could write

$$\eta_{o(\text{SYSTEM})} = \eta_{o(\text{ELECT. MOTOR})}\,\eta_{o(\text{PUMP})}\,\eta_{o(\text{HYD. MOTOR})} \tag{5.25}$$

$$= 0.87 \times 0.88 \times 0.90 = 0.69$$

The two methods for finding the system efficiency are equivalent.

5.9 Hydrostatic Transmissions

A hydrostatic transmission in its simplest form is nothing more than a closed-loop system consisting of a hydraulic pump and a hydraulic motor. The term "closed loop" used here means that the fluid that leaves the hydraulic motor returns directly to the pump inlet rather than going to the system reservoir. In an open-loop system, all fluid returns to the reservoir.

There are five possible pump-motor combinations, shown in Figures 5.18 and 5.19. The characteristics of each will be discussed. The block designated "Valve Package" normally contains a relief valve, check valves, and so forth. The three circuits in Figure 5.18 provide only single-direction operation. Forward and reverse operation is possible with the circuits in Figure 5.19.

Figure 5.18a shows a fixed displacement pump and a fixed displacement motor. In this transmission, the torque and horsepower required to drive the pump are functions of the load (and, therefore, the system pressure), but the pump speed and flow rate are usually constant. Below the relief valve setting, the motor torque and horsepower output vary with load, while the motor speed is constant. When the pressure reaches the relief valve setting, motor torque remains constant, but motor speed will depend on the amount of flow going through the relief valve. The resulting performance is variable torque at constant speed below the relief valve setting and constant torque and variable speed at the relief valve setting.

In Figure 5.18b we have a fixed displacement pump with a variable displacement motor. The pump in this system operates like the one just described; however, the motor performance varies as the motor displacement is adjusted. The motor speed is a function of both the motor displacement and the flow over the relief valve, while the torque varies inversely with speed. This transmission produces a constant power output.

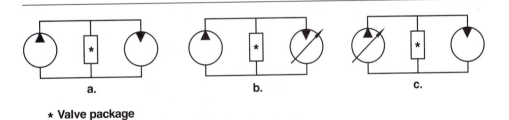

a.   b.   c.

* Valve package

Figure 5.18. Single-direction hydrostatic tranmissions.

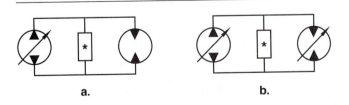

a.   b.

* Valve package

Figure 5.19. Reversible hydrostatic transmissions.

A variable displacement pump and a fixed displacement motor are shown in Figure 5.18c. In this unit, the pump pressure and speed are constant, but the torque and horsepower inputs to the pump vary. The motor inlet pressure and torque output are constant, but the speed and horsepower output vary according to the pump displacement.

A common variation of this transmission uses an over-center (reversible) pump and a bi-directional motor as shown in Figure 5.19a. This unit provides forward and reverse directions from the motor with infinite speed variations, from full-forward to zero to full-reverse, with no gearing or directional valve operation.

The circuit in Figure 5.19b utilizes a variable displacement pump and a variable displacement motor. The motor torque and speed outputs are functions of the displacements of both the pump and the motor. The maximum torque output will occur when the pump is at minimum displacement and the motor is at maximum displacement. Maximum motor speed will occur at maximum pump displacement and minimum motor displacement.

5.10 Troubleshooting Tips

Following are common malfunctions that might be experienced when using hydraulic motors:

Symptom	Possible Cause

Motor will not turn	1. Insufficient pressure
	  a. Pump not running
	  b. Relief valve setting too low
	  c. Relief valve stuck open
	2. Adjustment set incorrectly on variable displacement motor
	3. Excessive internal wear allowing internal leakage
	4. Internal damage causing the mechanism to jam
	5. External mechanism jammed or stuck
	6. Too much load
	7. Fluid viscosity too low (exessive internal leakage)

Motor turns too slowly	1. Insufficient flow
	  a. Pump not running
	  b. Pump output insufficient
	  c. Flow control valve set incorrectly
	  d. Excessive flow over relief valve
	2. Excessive internal wear allowing internal leakage
	3. Adjustment set incorrectly on variable displacement motor
	4. Too much load
	5. Fluid viscosity too low

Motor turns in wrong direction	1. Directional control valve or plumbing connected incorrectly. 2. Over-center mechanism set incorrectly on reversible motor

Motor wears out quickly   1. Dirty fluid
                          2. Fluid viscosity too low
                          3. Coupling misaligned
                          4. High side loading

5.11   Summary

Hydraulic motors are rotary output devices that convert a fluid power input into a rotary mechanical power output. The general term "hydraulic motor" usually refers only to those motors that have a continuous rotation capability. Devices that have a limited rotation capability—from a few degrees to a few hundred degrees of rotation—are termed rotary actuators, oscillating motors or limited rotation motors. The construction of most hydraulic motors is similar to their counterparts in the pump family. The exception to this is the low-speed, high-torque motor.

The outputs from rotary devices are speed and torque. The rotating speed is a function of volumetric displacement and flow rate. The output torque is a function of volumetric displacement and the pressure drop across the motor.

5.12   Key Equations

Output power:

U.S. customary

$$OHP = \frac{T \times N}{63,025} \tag{5.5}$$

SI

$$P_{\text{OUT}} = \frac{T \times N}{9550} \tag{5.6}$$

Overall efficiency:

U.S. customary

$$\eta_o = \frac{OHP}{HHP} \tag{5.8}$$

SI

$$\eta_o = \frac{P_{\text{OUT}}}{P_{\text{HYD}}}$$

Volumetric efficiency:

$$\eta_v = \frac{Q_T}{Q_A} \tag{5.11}$$

$$\eta_v = \frac{N_A}{N_T} \tag{5.16}$$

Mechanical efficiency:

$$\eta_m = \frac{T_A}{T_T} \tag{5.19}$$

Efficiency relationship: $\eta_o = \eta_v \times \eta_m$ (5.17)

Theoretical flow required for speed N: $Q_T = V_m N$ (5.12)

Actual flow required for speed N: $Q_A = \dfrac{V_m N}{\eta_v}$ (5.21)

Torque output:
$$T = \frac{V_m \Delta p \eta_m}{2\pi}$$
(5.22)

Electrical horsepower:
$$EHP = \frac{IHP}{\eta_o}$$
(5.24a)

References

1. *Industrial Hydraulic Technology,* Bulletin 0221-B1, Parker Hannifin Corporation, Cleveland, OH, 1986.
2. Ibid.
3. Raymond P. Lambeck, *Hydraulic Pumps and Motors,* Marcel Dekker, Inc. NY, 1983.

Review Problems

Note: The letter 'S" after a problem indicates SI units are used.

General

1. Explain the functional difference between a hydraulic motor and a hydraulic pump?
2. Explain the operational difference between a high-speed, low-torque motor, and a low-speed, high-torque motor.
3. What determines the torque output of a hydraulic motor?
4. What determines the speed output of a hydraulic motor?
5. Explain the difference between an open-loop system and a closed-loop system. Use ISO symbols to draw an example of each.
6. Explain what is meant by the Omega rating of a hydraulic motor.
7. Explain how a hydrostatic transmission works.
8. Make a chart listing each of the types of hydrostatic transmissions, the speed, torque, and horsepower characteristics of each, and a circuit diagram of each.
9. List three ways to control the flow rate going to a hydraulic motor.
10. Discuss the effects of changing the displacement on a variable displacement motor.

Power

11. A hydraulic motor receives flow at 12 gpm. The system pressure is 2000 psi, and the fluid returns to the reservoir at atmospheric pressure. Find the hydraulic horsepower available to the motor.
12. If the reservoir in Review Problem 11 is pressurized to 200 psi, what is the horsepower available?
13. A hydraulic motor produces 700 lb · in of torque at 1000 rpm. What it its output horsepower?
14. A hydraulic motor produces 100 lb · ft of torque at 5000 rpm. What is its output horsepower?
15. A hydraulic motor has a 25 horsepower output at 400 rpm. What is its torque output?
16S. A hydraulic motor has a pressure drop of 12,000 kPa. It receives a flow rate of 40 Lpm. What is the hydraulic power available to the motor?

17S. A hydraulic motor produces 150 N · m of torque at 1000 rpm. Find its output power.

18S. A hydraulic motor has a 20 kw output at 500 rpm. What is its torque output?

Efficiency

19. A hydraulic motor produces an output horsepower of 15 from a hydraulic horsepower of 18. What is its overall efficiency?

20. A hydraulic motor requires a 20 HP input to produce a 15 HP output. What is its overall efficiency?

21. A hydraulic motor produces 175 lb·ft of torque at 600 rpm. If its overall efficiency is 0.85, what hydraulic horsepower input is required?

22. A hydraulic motor with a flow rate of 10 gpm at 1500 psi has an overall efficiency of 0.82. What is its output horsepower?

23. The system pressure is 3000 psi and the flow rate is 30 gpm to operate a hydraulic motor with an output of 200 lb · ft at 1200 rpm. Find the motor's overall efficiency.

24. If the actual flow rate required by a motor is 18 gpm and the theoretical flow rate is 15 gpm, what is its volumetric efficiency?

25. A hydraulic motor with a volumetric efficiency of 0.88 requires 10 gpm to turn at 500 rpm. What is the theoretical flow rate required?

26. A flow rate of 15 gpm produces a speed of 600 rpm. The theoretical speed is 700 rpm. What is the volumetric efficiency?

27. A given flow will theoretically produce 1000 rpm from a hydraulic motor with a volumetric efficiency of 0.87. What is the actual speed?

28. The theoretical torque output is 500 lb · in, while the actual torque output is 400 lb · in. What is the mechanical efficiency?

29. A hydraulic motor has a theoretical torque output of 65 lb · ft and a mechanical efficiency of 0.93. What is the actual torque required?

30S. A hydraulic motor has a hydraulic power input of 12 kw and an output of 9.5 kw. What is its overall efficiency?

31S. A hydraulic motor produces 200 N · m of torque at 500 rpm. What hydraulic power input is required if the overall efficiency of the motor is 0.87?

32S. At the inlet to a hydraulic motor, the pressure has been measured at 20 MPa and the flow rate at 90 Lpm. The motor has a torque output of 1300 N · m at 1200 rpm. Find its overall efficiency.

33S. A hydraulic motor turns at 600 rpm when the actual flow rate is 40 Lpm. Its volumetric efficiency is 0.85. What is the theoretical flow required?

34S. A flow rate of 50 Lpm turns a hydraulic motor at 500 rpm. Its theoretical speed at that flow rate is 575 rpm. What is its volumetric efficiency?

Motor speed

35. A hydraulic motor has a displacement of 2.8 in³/rev. What is its theoretical speed for a flow rate of 10 gpm?

36. A hydraulic motor has a displacement of 5 in³/rev. What is its theoretical speed for a flow of 30 gpm?

37. A hydraulic motor has a displacement of 6.5 in³/rev and a volumetric efficiency of 0.91. Find its actual and theoretical speeds for a flow rate of 25 gpm.

38. What flow rate (in gpm) is required to produce 900 rpm from a 3.5 in³ motor with a volumetric efficiency of 0.85?

39. A motor has a 2.5 in³ displacement. It turns at 1200 rpm. What flow rate is required if its volumetric efficiency is 0.83?

40. A hydraulic motor with a mechanical efficiency of 95 percent and a volumetric efficiency of 87 percent produces 25 lb · ft of torque. The pressure drop through the motor is 1700 psi. It is required to turn a gear box at 700 rpm. What flow rate in gpm is required?

41S. A hydraulic motor has a displacement of 50 cm³/rev. What is its theoretical speed for a flow rate of 100 Lpm.

42S. A motor has a 30 cm³/rev displacement and a volumetric efficiency of 0.91. It turns at 1500 rpm. What flow rate is required?

43S. A gear box is driven by a hydraulic motor that turns at 700 rpm. The pressure drop through the motor is 12,000 kPa. The gear box requires a 40 N · m torque input. The motor has mechanical and volumetric efficiencies of 0.95 and 0.85, respectively. What flow rate is required to turn the motor at the required speed?

Motor torque

44. A 5 in³ motor has a mechanical efficiency of 0.95. Find the torque output of the motor for a pressure differential of 2500 psi.

45. How much torque can be produced by a 4 in³ motor with a pressure differential of 2000 psi? It has a mechanical efficiency of 0.95.

46. A 5 in³ hydraulic motor has a mechanical efficiency of 0.75. What pressure is required to produce a 500 lb · in output?

47. A hydraulic motor has a 20 in³ displacement and a volumetric efficiency of 0.75. How much torque can it produce if the system pressure is 3000 psi?

48. A conveyor system requires 200 lb · in of torque from a hydraulic motor that has a 1.2 in³/rev displacement. If the mechanical efficiency is 96 percent, what is the minimum pressure required to operate the motor? Under what circumstances would the pressure requirement be higher, assuming the torque requirement remains the same?

49. A hydraulic motor must produce 5000 lb · in of torque at 1500 rpm. The system flow rate is 8 gpm. If the overall efficiency of the motor is 90 percent, what system pressure will be required?

50S. An 80 cm³ motor has a mechanical efficiency of 0.96. The pressure drop through the motor is 18,000 kPa. What is the torque output of the motor?

51S. Find the pressure required to produce an output of 800 N · m from a 100 cm³ motor that has a mechanical efficiency of 0.92.

52S. At 1500 rpm, a hydraulic motor has a torque output of 60 N · m. The motor has a 0.85 overall efficiency. What pressure is required if the flow rate is 35 Lpm?

Motor sizing

53. A hydraulic system operates at 2000 psi and 20 gpm. It operates a hydraulic motor at 1250 rpm. The motor has a mechanical efficiency of 0.92 and an overall efficiency of 0.88. What is the volumetric displacement of the motor? What is its output torque?

54. What size hydraulic motor is required to produce 31,000 lb · in of torque at 2000 psi? Assume a mechanical efficiency of 0.87.

55S. What size hydraulic motor is required to produce 3500 N · m of torque at 7000 kPa if the mechanical efficiency is 0.96?

System problem

56. A 5 in³ hydraulic motor has a mechanical efficiency of 0.92 and a volumetric efficiency of 0.87. It produces 800 lb · in of torque at 1300 rpm. System flow is provided by a piston pump that has a theoretical flow rate of 35 gpm and a mechanical efficiency of 0.9. The pump is driven by an electric motor with a 0.85 efficiency. Find the electric horsepower and kilowatts required to operate the system. What is the overall efficiency of the system?

57S. A hydraulic motor produces a torque of 100 N · m at 1500 rpm. The motor has a volumetric displacement of 80 cm³, a mechanical efficiency of 0.95 and a volumetric efficiency of 0.89. The system pump has a theoretical flow rate of 125 Lpm and a mechanical efficiency of 0.93. The electric motor that drives the pump has an efficiency of 0.87. Find the

    a. power required to operate the system.
    b. overall efficiency of the system.

Hydraulic Cylinders

Outline

6.1 Introduction

In addition to hydraulic motors, the power output from a fluid power system can also be a hydraulic cylinder. Cylinders are sometimes called *linear actuators* or *linear motors,* because they convert fluid power into linear mechanical force and motion.

We have already looked at cylinders in our discussion of the basic fluid power concepts, so we'll not be on totally unfamiliar ground.

Objectives

When you have completed this chapter, you will be able to

- Discuss the various types of cylinders and the terminology used to describe them.
- Calculate the force and speed capabilities of cylinders.
- Discuss the flow-speed and pressure-force relationships of cylinders.
- Calculate inertial and friction loads and understand how they affect pressure requirements.
- Explain the concept of regeneration in a cylinder circuit.
- Discuss cylinder application and mounting options.

6.2 Cylinder Terminology

Hydraulic cylinders can be described in several different ways. The most common descriptions refer either to their functioning or to their construction.

Types of Cylinders

Functionally, cylinders can be designed as either single-acting or double-acting, depending on the number of working-fluid ports. Figure 6.1 illustrates these two types. The *single-acting* cylinder has been discussed earlier to demonstrate some basic concepts. Note that in the cylinder in Figure 6.1a, there is only one fluid port, so it can provide force only to extend the cylinder. The cylinder moves inward (retracts) due either to gravitational force (often assisted by the load) or a spring force. In order for it to retract, a flow path must be opened so that the fluid in the cylinder can escape. This type of cylinder is commonly used for jacks and lifts.

There is no fluid port on the rod end of the cylinder in Figure 6.1a. However, there is often a small amount of leakage past the piston seal into the rod end, so a screen-covered drain hole is usually provided to prevent the accumulation of fluid in that chamber. This hole also allows the upper part of the cylinder to "breathe" to prevent problems with either compression of air or creating a vacuum. If fluid were to accumulate on the rod end with no way to escape, the cylinder would not be able to extend completely because of a situation termed *hydraulic lock,* which is not peculiar to cylinders, but can occur any time fluid is trapped in a chamber and prevents the movement of a component or element. The relative incompressibility of hydraulic fluids makes this problem more serious in hydraulic systems than in pneumatic systems.

The single-acting cylinder of Figure 6.1b is configured so that it is *retracted* by fluid power. Since there is no pressure port on the blind end of the cylinder, it must be extended by the load (as in this illustration), or by a spring in the

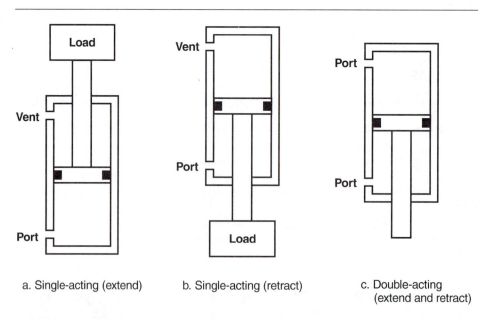

a. Single-acting (extend)   b. Single-acting (retract)   c. Double-acting
                                                                        (extend and retract)

Figure 6.1 Cylinders can be either single-acting or double-acting, depending on the number of working-fluid parts.

blind end. A variation of the single-acting cylinder design is the *hydraulic ram* shown in Figure 6.2. In this unit, there is no piston as such; rather, the rod (ram) fills the entire bore, leaving just enough clearance to allow the ram to move in and out.

The *double-acting* cylinders in Figure 6.3 have two fluid ports, both of which can be either a pressure port or a return port. As a result, the double-acting cylinder can be extended and retracted hydraulically.

Another way of describing cylinders is by their construction features, such as whether they are single-ended or double-ended. The cylinders in Figures 6.1 and 6.2 have only one rod extending from the housing and thus are termed a *single-ended* (or *differential*) cylinders. Actually, since most cylinders are single-ended, the designation is rarely used. Figure 6.3b shows a *double-ended* cylinder. Note the rod extending from each end of the cylinder. This type is also termed *double rod* or *double rod end.* (For comparison, a single-ended double-acting cylinder is shown in Figure 6.3a). A double-ended cylinder is usually also double-acting. Therefore, a double-ended, double-acting cylinder is described simply as "double-ended."

In order to provide a large force with a small-diameter cylinder, a *tandem* cylinder such as the one shown in Figure 6.4 is sometimes used. The pressure acts on both pistons. Therefore, the total force capability is almost twice that of a standard cylinder of the same size. The most common applications for this type of cylinder are in mobile equipment, where limited space is often a consideration.

Dual (or *duplex*) cylinders are shown in Figure 6.5. When pressurized fluid enters the center port of the cylinder in Figure 6.5a, both pistons will extend, provided there is sufficient pressure to move both loads. If the loads are differ-

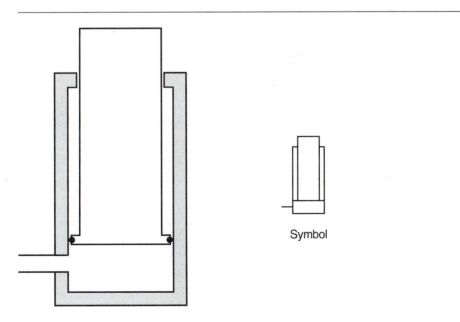

Symbol

Figure 6.2 The hydraulic ram is a variation of single-acting cylinder.

ent, the one requiring the least pressure will extend first. The configuration in Figure 6.5b allows the pistons to be operated individually or at the same time, depending on how the fluid is directed to the ports. Duplex cylinders are commonly employed to operate double doors such as the landing gear doors on some aircraft.

The cylinder shown in Figure 6.6 is called a *telescoping* cylinder for obvious reasons. Telescoping cylinders may have two or more moving sections (called *stages*), but seldom more than four or five. They are used in applications where a long stroke is required but relatively little installation space is available. A completely retracted cylinder may be as little as one-third to one-fourth of its fully extended length. Telescoping cylinders can be either single-acting or double-acting.

The cylinder in Figure 6.6 is a three-stage, single-acting cylinder. The largest

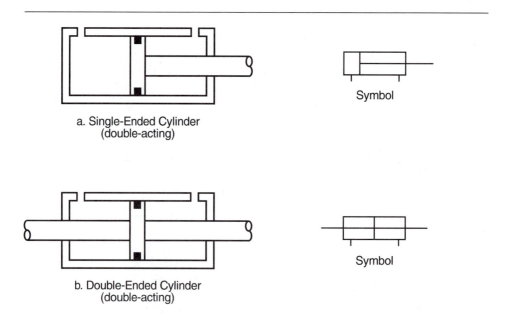

Symbol

a. Single-Ended Cylinder
(double-acting)

Symbol

b. Double-Ended Cylinder
(double-acting)

Figure 6.3 Cylinder styles include single-ended and double-ended.

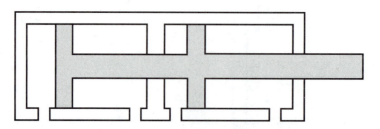

Figure 6.4 In the tandem cylinder pressure acts on both pistons, providing almost twice the force capability of a standard cylinder of the same size.

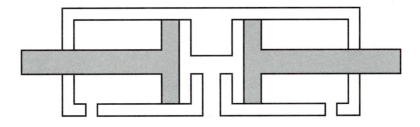

a. Both pistons are pressurized at the same time.

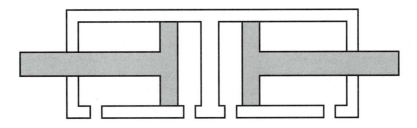

b. Each piston can be pressurized independently

Figure 6.5 Dual cylinders are commonly employed to open double doors.

tube section is the main cylinder. It is not counted as a stage. The smallest section is commonly termed the *plunger*.

Other Terminology

Figure 6.7 shows the various parts that make up a typical double-acting cylinder. The end of the cylinder where the rod extends is called the *rod end, head end,* or *gland end.* The opposite end (on a single-ended cylinder, at least) is called the *cap end, blind end,* or *blank end.*

A cylinder usually contains three seals or sets of seals. The seals on the piston prevent the leakage of fluid between the two chambers of the cylinder. There are commonly two seals for the rod. One is the *rod pressure seal,* which prevents fluid leakage from the rod end. The second is called the *rod wiper seal* or *scraper.* Its function is to prevent the ingress of dirt, water, and other contaminants as the rod is retracted. A wiper seal has virtually no pressure-holding capability.

The inside diameter (ID) of the cylinder is termed the *bore.* When we talk about a "four-inch cylinder," for instance, we are referring to a cylinder that has a nominal four-inch bore or ID. (In fact, the bore of a standard four-inch cylinder is 4.004 to 4.005 in.) The bore is slightly larger than the piston, and

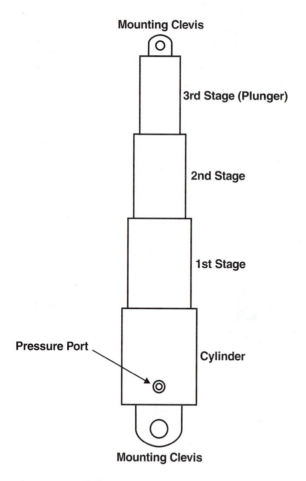

Figure 6.6 Telescoping cylinders have two or more moving sections called *stages*.

piston seals are used to seal the resulting clearance and prevent fluid leakage from one side of the piston to the other. These seals move with the piston and actually form part of the area upon which the pressure acts. For this reason, the actual area involved in calculating cylinder force is the piston area plus the exposed seal area. This total area equals the cross-sectional area of the cylinder bore. In this text, the terms *area* and *piston area* will be used interchangeably to mean the total area upon which the pressure in the cap end of the cylinder is exerted. Likewise, the terms *bore* and *piston diameter* will be used interchangeably. Finally, the dimension used to describe the cylinder will be used literally. For example, consider a four-inch (4 in) cylinder to have a four-inch internal diameter (and a four-inch piston) even though that is not exactly the case.

The distance through which the piston with its attached rod can travel is known as the *stroke* of the cylinder.

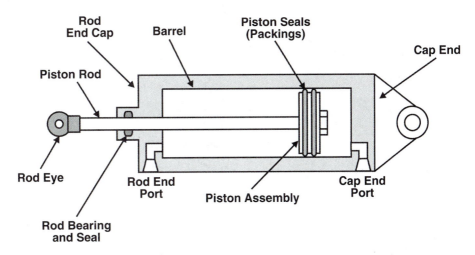

Figure 6.7 Cross section of a typical double-acting cylinder. This cylinder has a pivot mounting on both the rod and the cap end.

6.3   Cylinder Calculations

We have already seen several types of calculations that can be used to determine system pressure, cylinder diameter (bore), and force output. We have for instance, calculated the force output of a cylinder using Equation 6.1 (which is the same as Equation 2.1).

$$F = p \times A \qquad (6.1)$$

Remember that we are actually referring to the force output *capability* in this equation. It would probably be better to use this equation in the following form:

$$p = \frac{F}{A} \qquad (6.2)$$

This equation tells us that the system pressure depends on the force required to move the load. This is a more realistic concept, because we already know that pressure results from a resistance to flow. In the case of a cylinder, the load constitutes the resistance to flow.

In a single-ended, double-acting cylinder, there are two different force capabilities (one for extension and one for retraction) at any given pressure because there are different areas on the two sides of the piston. On extension, the pressure acts in the entire face of the piston (Figure 6.8a). Thus, we can write Equation 6.1 as

$$F_{\text{ext}} = p \times A_{\text{piston}} \qquad (6.3)$$

On retraction, however, the surface area on which the pressure acts is

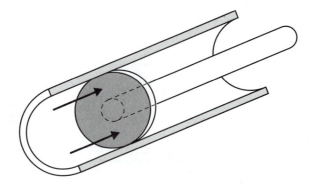

a. On extension, the full piston area (A$_P$) is exposed to fluid pressure

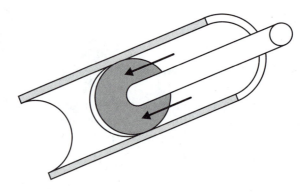

b. On retraction, only the annular area around the rod (A$_P$-A$_R$) is exposed to fluid pressure

Figure 6.8 The total effective area in a single-ended, double-acting cylinder depends on the direction of movement.

reduced by the area of the rod (Figure 6.8b). In this case, we write Equation 6.1 as

$$F_{\text{ret}} = p \times (A_{\text{piston}} - A_{\text{rod}}) \qquad (6.4)$$

The term $(A_{\text{piston}} - A_{\text{rod}})$ actually refers to the annular area around the rod. This is the area upon which the pressure is exerted. Because there is less area, the force capability on retraction is reduced.

Example 6.1

Calculate the force output of a 4 in cylinder that has a 2 in rod if the system pressure is 2000 psi.

Solution

For the extension stroke, we use the whole piston face area. The designation "4 in cylinder" actually means that the cylinder has a nominal 4 in bore, so the diameter of the piston is taken as four inches. Thus,

Solution

$$A_{\text{piston}} = \frac{\pi D^2}{4}$$

$$= 12.56 \text{ in}^2$$

The extension force available is therefore

$$F_{\text{ext}} = p \times A_{\text{piston}}$$

$$= 2000 \text{ lb/in}^2 \times 12.56 \text{ in}^2$$

$$= 25{,}120 \text{ lb}$$

The area of the rod is

$$A_{\text{rod}} = \frac{\pi D^2}{4}$$

$$= 3.14 \text{ in}^2$$

so that the retraction force available is

$$F_{\text{ret}} = p \times (A_{\text{piston}} - A_{\text{rod}})$$

$$= 2000 \text{ lb/in}^2 \times (12.56 \text{ in}^2 - 3.14 \text{ in}^2)$$

$$= 18{,}840 \text{ lb}$$

or 25 percent less than the force available on extension.

If we have a double-ended, double-acting cylinder on which both rods are the same diameter, we have the same force capability for both extension and retraction. If the rods are not the same diameter—which almost never happens—we have different areas upon which the pressure can act, resulting in different forces.

An interesting phenomenon in hydraulic cylinders is a pressure increase across the piston, termed *pressure intensification*. It occurs when the flow from the rod end of the cylinder is somehow restricted. This phenomenon is best illustrated by an example.

Example 6.2

A hydraulic cylinder has a 15 cm bore and a 5 cm rod. The pressure in the cap end of the cylinder is 21 MPa. If the fluid flow from the rod end of the cylinder is completely blocked, find the pressure in the rod end.

Solution

The pressure in the cap end produces a force of

$$F = p_{\text{sys}} \times A_{\text{piston}}$$

$$= 21 \text{ MPa} \times 177 \text{ cm}^2$$

$$= 372 \text{ kN}$$

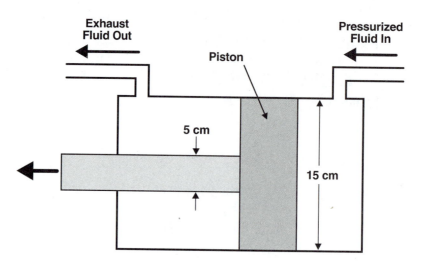

Piston Rod Moving Out

This force is transmitted mechanically through the piston structure and is exerted on the fluid at the rod end. The surface area on the rod side is less than in the piston side because of the rod. Therefore, the pressure on the rod side is found to be

$$p = \frac{F}{A_{\text{piston}} - A_{\text{rod}}}$$

$$= \frac{372 \text{ kN}}{177 \text{ cm}^2 - 19.6 \text{ cm}^2}$$

$$= 23.6 \text{ MPa}$$

As the rod diameter increases relative to the piston diameter, the degree of intensification increases. This increased pressure would be felt in the cylinder and throughout the circuit to the point of the blockage. Thus, the piping, seals, and all components in this area—as well as the cylinder itself—must be able to withstand pressures that may be considerably higher than the maximum *system* pressure. The magnitude of this intensified pressure depends upon the relationship between the piston and rod diameters.

The velocity with which a cylinder extends and retracts is a function of the system flow rate and can be calculated from Equation 6.5.

$$v = \frac{Q}{A} \qquad\qquad \textbf{(6.5)}$$

where

v = cylinder velocity.

Q = flow rate.

A = piston area subject to the fluid flow.

This equation should look familiar to you. It is, in fact, Equation 2.13, which we used to calculate the velocity of fluid flowing through a pipe. Do you see how it also applies to the velocity of movement of a hydraulic cylinder? The cylinder is actually nothing more than a pipe with a piston in it. If we removed the piston (and plugged the hole where the rod goes through the end cap), we would immediately recognize the cylinder as simply a larger-diameter pipe

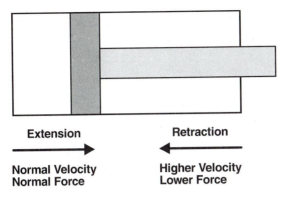

Extension

→

**Normal Velocity
Normal Force**

Retraction

←

**Higher Velocity
Lower Force**

Figure 6.9 Comparison of extension and retraction velocity and force for a constant flow rate.

section. We would apply the continuity equation ($A_1v_1 = A_2v_2$) to it without a second thought.

Although there is no flow *through* the cylinder with the piston installed, we can still apply the continuity equation as long as there is piston movement (extension or retraction), because the piston will be pushed ahead of the fluid at the fluid velocity. **NOTE:** Equation 6.5 does not include a pressure term. This implies that increasing system pressure will not increase the speed of the cylinder as long as there is sufficient pressure to provide the force required to move the load in the first place. This is, in fact, the case.

On the other hand, Equation 6.5 does contain a flow rate term. The only way to vary the speed of any given cylinder is to vary the fluid flow rate. We will see how that is done in a later chapter.

The area term in the denominator of Equation 6.5 should lead us to see an important concept concerning cylinder speed in a single-ended, double-acting cylinder. On the extension stroke, the area of concern is the total area of the piston (including the moving seal), because that is the area upon which the fluid pressure is acting. On the retraction stroke, however, the rod area is not subjected to pressure, so the area in the equation becomes the piston area minus the rod area. Since the area is reduced, the velocity will be higher on retraction than on extension if the flow rate remains the same for both directions (Figure 6.9).

Example 6.3

A hydraulic cylinder has a piston area of 10 in^2 and a rod area of 4 in^2. Find the extension and retraction speeds for a 3 gpm flow rate.

Solution

From Equation 6.5, we can write for the extension stroke that

$$v_{ext} = \frac{Q}{A_p}$$

$$= \frac{3 \text{ gal/min} \times 231 \text{ in}^3/\text{gal}}{10 \text{ in}^2} = 69.3 \text{ in/min}$$

Since it is usually preferrable to express velocities in ft/sec, we can convert this answer to get

$$v_{ext} = \frac{69.3 \text{ in}}{\text{min}} \times \frac{\text{min}}{60 \text{ sec}} \times \frac{\text{ft}}{12 \text{ in}} = 0.096 \text{ ft/sec}$$

For the retraction stroke, we have

$$v_{ret} = \frac{Q}{A_{piston} - A_{rod}}$$

$$= \frac{3 \text{ gal/min} \times 231 \text{ in}^3/\text{gal}}{10 \text{ in}^2 - 4 \text{ in}^2}$$

$$= 115.5 \text{ in/min}$$

$$= 0.16 \text{ ft/sec}$$

Example 6.4

Find the velocities and forces produced in both extension and retraction by a 10 cm cylinder with a 5 cm rod operating in a 15 MPa, 15 Lpm system.

Solution

We use Equation 6.1 to find the forces. Solving for the areas, we have

$$A_{piston} = 78.5 \text{ cm}^2$$

$$A_{rod} = 19.6 \text{ cm}^2$$

Therefore,

$$F_{ext} = p \times A_{piston}$$

$$= 15 \text{ MPa} \times 78.5 \text{ cm}^2$$

$$= 111.8 \text{ kN}$$

And

$$F_{ret} = p \times (A_{piston} - A_{rod})$$

$$= 15 \text{ MPa} \times (78.5 \text{ cm}^2 - 19.6 \text{ cm}^2)$$

$$= 88.4 \text{ kN}$$

The velocities are found from Equation 6.5

$$v_{ext} = \frac{Q}{A_{piston}}$$

$$= \frac{15 \text{ L/min} \times \text{m}^3/1000 \text{ L} \times \text{min}/60 \text{ s}}{78.5 \text{ cm}^2 \times \text{m}^2/10000 \text{ cm}^2}$$

$$= 0.032 \text{ m/s}$$

And

$$v_{ret} = \frac{Q}{A_{piston} - A_{rod}}$$

$$= \frac{15 \text{ L/min} \times \text{m}^3/1000 \text{ L} \times \text{min}/60 \text{ s}}{(78.5 \text{ cm}^2 - 19.6 \text{ cm}^2) \times \text{m}^2/10,000 \text{ cm}^2}$$

$$= 0.042 \text{m/s}$$

Remember, for a single-ended, double-acting cylinder, force capability is greater on extension than on retraction for a given pressure, but velocity is greater on retraction than on extension for a given flow rate.

Another phenomenon that results from the differential piston areas is called *flow differential*. It simply means that the flow leaving a cylinder port is different from the flow entering the opposite port. The relationship can be shown by using Equation 6.5. In this equation, the velocity in extension is

$$v = \frac{Q_{IN}}{A_{piston}}$$

where Q_{IN} is the flow entering the blind end. Now we can modify that equation slightly to consider the flow leaving the cylinder:

$$v = \frac{Q_{OUT}}{A_{piston} - A_{rod}}$$

where Q_{OUT} is the flow leaving the rod end. Since both of these equations define the same velocity, we can equate them to get

$$\frac{Q_{IN}}{A_{piston}} = \frac{Q_{OUT}}{A_{piston} - A_{rod}}$$

Solving this equation for Q_{OUT}, we get

$$Q_{OUT} = \frac{Q_{IN}\,(A_{piston} - A_{rod})}{A_{piston}} \tag{6.6}$$

This tells us that, for *extension*, the flow rate coming out of the cylinder is less than the flow rate going in. The ratio of the areas of the two sides of the piston determines how much the flow is reduced.

We could do the same thing for the flow rates on retraction. This would show that, on retraction, the flow rate leaving the blind end of the cylinder would be *greater* than the flow rate entering the blind end. It would be *increased* by the ratio of the total piston area to the annular area. This can be important when sizing system components—valves and filters especially—based on flow rate requirements.

Example 6.5

A hydraulic cylinder has a 3 in bore and a 2 in rod. Determine the flow out of the cylinder when it is extended by a flow rate of 10 gpm.

Solution

Using Equation 6.6, we get

$$Q_{OUT} = \frac{Q_{IN}\,(A_{piston} - A_{rod})}{A_{piston}}$$

$$A_{piston} = 7.07 \text{ in}^2$$

$$A_{rod} = 3.14 \text{ in}^2$$

Therefore,

$$Q_{OUT} = \frac{10 \text{ gpm} \times (7.07 \text{in}^2 - 3.14 \text{ in}^2)}{7.07 \text{ in}^2} = 5.56 \text{ gpm}$$

The horsepower output of a cylinder is calculated from the basic definition of horsepower given in Equation 2.8.

$$HP = \frac{F \times d}{550t}$$

(2.8)

where

$$d = \text{distance moved.}$$

$$t = \text{time required to move through that distance.}$$

Remember that the 550 carries the units of ft·lb/sec. Often, the velocity of the cylinder is known. Since

$$v = \frac{d}{t}$$

we therefore can also express the output as

$$HP = \frac{F \times v}{550}$$

(6.7)

In the SI system, Equation 6.7 would be

$$P = F \times v$$

(6.7a)

where force in newtons and velocity in meters per second gives power in watts.

Example 6.6

A hydraulic cylinder moves a 4000 lb load two feet in eight seconds. What is the horsepower output of the cylinder?

Solution

Using Equation 2.8, we get

$$HP = \frac{F \times d}{550t}$$

$$= \frac{4000 \text{ lb} \times 2 \text{ ft}}{550 \text{ ft·lb/sec} \times 8 \text{ sec}}$$

$$= 1.8 \text{ HP}$$

Example 6.7

A hydraulic cylinder has a 25 kN output at a velocity of 3 m/s. What is the power output?

Solution

From Equation 6.7a,

$$P = F \times v$$

$$= 25 \text{ kN} \times 3 \text{ m/s}$$

$$= 75 \text{ kW}$$

6.4 Regeneration

The term *regeneration* can be applied to any system that diverts all or part of its output and adds it to its input to enhance some aspect of its performance. In a hydraulic cylinder, *regeneration* means that the fluid flowing out of the rod end of the cylinder during extension is returned to the inlet side of the cylinder in order to increase the extension velocity. This concept allows the extension speed to be increased significantly without increasing the pump flow output.

Figure 6.10 illustrates, in a grossly simplified way, the regeneration principle. As the cylinder extends, the fluid in the rod end is expelled. In a normal circuit, this fluid would return to the system reservoir at low pressure. In a regenerative circuit, the fluid is injected into the pressure line and is added to the total flow rate going into the cap end. From Equation 6.5, we see that increasing the flow rate into the cap end will increase the extension velocity.

Let's investigate this concept more thoroughly. Consider an instant in time when you have just shifted a valve to allow the system of Figure 6.10 to be pressurized. Now "freeze the action" and see what you have. You must agree that, at that instant, before there is any movement of the piston, the pressure is the same in both ends of the cylinder. If the pressure is equal, what about the forces? The force on the cap end is greater than the force on the rod end.

Now start the motion. What happens? Because of the imbalance of forces, the piston starts to extend. When that happens, the fluid from the rod end is forced out of the cylinder. Where does it go? It can only go through the line

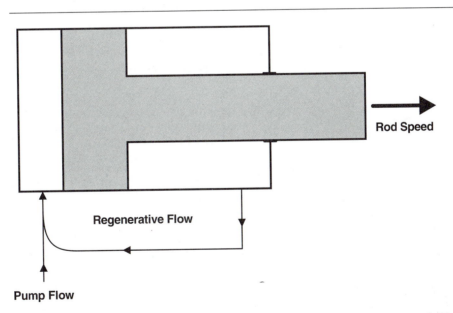

Figure 6.10 An illustration of the regeneration concept. The circuit shown here will only extend. Additional piping and valves are needed to provide a practical circuit.

into the cap end. The result is an increase in cylinder speed. (Keep in mind that Figure 6.10 shows only the concept, not how it is actually done. Some valving is required to complete the system, but we're not ready for that yet.)

To calculate the extension speed that is possible in a given regenerative circuit, we again use Equation 6.5. In this case, the flow rate to the cap end of the cylinder is the sum of the pump output and the output flow from the rod end of the cylinder, or

$$Q_{total} = Q_{pump} + Q_{rod} \tag{6.8}$$

The flow rate from the rod end can be found by solving Equation 6.5 for flow rate to get

$$Q = A \times v \tag{6.9}$$

Here the area of concern is the piston area minus the rod area, so we have

$$Q_{rod} = (A_{piston} - A_{rod}) \times v \tag{6.10}$$

Substituting into Equation 6.5 gives

$$v = \frac{Q_{total}}{A_{piston}}$$

$$= \frac{Q_{pump} + Q_{rod}}{A_{piston}}$$

$$= \frac{Q_{pump} + (A_{piston} - A_{rod}) \times v}{A_{piston}}$$

Multiplying both sides by the area of the piston gives

$$A_{piston} \times v = Q_{pump} + (A_{piston} \times v) - (A_{rod} \times v)$$

or

$$A_{rod} \times v = Q_{pump}$$

Solving this for velocity gives us

$$v_{regen} = \frac{Q_{pump}}{A_{rod}} \tag{6.11}$$

Notice that we have added the subscript "regen" to indicate that this equation is used to calculate the extension velocity in a regenerative system.

Example 6.8

A hydraulic cylinder has a 4 in bore and a 2 in rod. The system flow rate is 10 gpm. Find the extension velocity of the cylinder with and without regeneration.

Solution

Without regeneration, the extension velocity of the cylinder is a function of pump flow only. We again use Equation 6.5:

$$v_{ext} = \frac{Q_{pump}}{A_{piston}}$$

For a 4 in bore, we have

$$A_{piston} = 12.56 \text{ in}^2$$

Therefore,

$$v = \frac{10 \text{ gal/min} \times 231 \text{ in}^3/\text{gal} \times \text{min}/60 \text{ sec}}{12.56 \text{ in}^2}$$

$$= 3.06 \text{ in/sec}$$

$$= 0.26 \text{ ft/sec}$$

With regeneration, the velocity is a function of both the pump flow and the flow from the rod end of the cylinder. We account for this by using Equation 6.11. The area of the 2 in rod is 3.14 in^2, so

$$v_{\text{regen}} = \frac{Q_{\text{pump}}}{A_{\text{rod}}}$$

$$= \frac{10 \text{ gal/min} \times 231 \text{ in}^3/\text{gal} \times \text{min}/60 \text{ sec}}{3.14 \text{ in}^2}$$

$$= 12.26 \text{ in/sec}$$

$$= 1.02 \text{ ft/sec}$$

From this example, we can see the tremendous speed increase that can result from regeneration. In this case, we get almost four times the speed without having to resort to a larger pump. Unfortunately, this increased speed capability, like most other improvements in performance, is not without cost. In this case, the price is paid in a loss of force capability.

Looking again at our circuit in Figure 6.10, we realize that, in the instant before any movement begins, the pressure on both sides of the piston is the same. At that point, we have a situation such as that depicted in Figure 6.11.

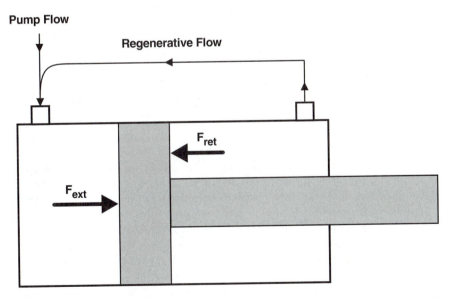

Figure 6.11 In the regenerative mode, opposing forces are working on both faces of the piston.

A force is generated on each side of the piston that is proportional to the area over which the pressure is exerted. Thus, a portion of the force on the cap side of the piston is canceled by the force on the rod side.

The force on the cap side, which is the force trying to extend the piston, is

$$F_{ext} = p \times A_{piston} \tag{6.12}$$

The force exerted on the rod end is attempting to retract the piston, and is found from

$$F_{ret} = p \times (A_{piston} - A_{rod}) \tag{6.13}$$

The net extension force available is the difference between these two, so

$$F_{net} = F_{ext} - F_{ret}$$

$$= p \times A_{piston} - p \times (A_{piston} - A_{rod})$$

or

$$F_{regen} = p \times A_{rod} \tag{6.14}$$

Note again the use of the subscript "regen" to emphasize that this is the force available when extending a regenerative system.

Example 6.9

The system in Example 6.8 operates at 2000 psi. Find the extension force available with and without regeneration.

Solution

Without regeneration, the force available is

$$F_{ext} = p \times A_{piston}$$

$$= 2000 \text{ lb/in}^2 \times 12.56 \text{ in}^2$$

$$= 25,120 \text{ lb}$$

For regeneration, we employ Equation 6.14 to find the extension force.

$$F_{regen} = p \times A_{rod}$$

$$= 2000 \text{ lb/in}^2 \times 3.14 \text{ in}^2$$

$$= 6,280 \text{ lb}$$

This represents a significant loss in capability.

There are numerous instances in which this loss in extension force capability is acceptable. For example, in the case of a feed mechanism where a high speed is desired to move a workpiece into a workplace, a high force would be required to perform some other function. This could be accomplished by the appropriate use of a directional control valve that provides regeneration during the feed cycle, then is shifted to isolate the rod end from the blind end to provide the higher force required during the work portion of the stroke.

The achievement of the higher extension speed from a small pump can be a significant energy saving, especially if a fixed displacement pump is used. This can be shown by comparing the horsepower requirements for obtaining the same cylinder speed with and without regeneration.

Example 6.10

Determine the horsepower required to obtain the higher cylinder extension speed for the circuit of Examples 6.8 and 6.9, both with and without regeneration.

Solution

With regeneration, the horsepower required is

$$HP = \frac{p \times Q}{1714}$$

The Q used here is simply the pump output. The flow from the rod end of the cylinder is free as far as the pump horsepower requirement is concerned; therefore, with regeneration, we have

$$HP = \frac{p \times Q}{1714}$$

$$= \frac{2000 \text{ psi} \times 10 \text{ gpm}}{1714}$$

$$= 11.67 \text{ HP}$$

If we did not use regeneration, we would have to increase the size of the pump to produce the total flow required to obtain the higher speed. The total flow required is found from

$$Q = A_{\text{piston}} \times v_{\text{regen}}$$

We have to use the regenerative velocity here, because this is the velocity we must achieve. Thus,

$$Q = 12.56 \text{ in}^2 \times 12.26 \text{ in/sec}$$

$$= 154 \text{ in}^3/\text{sec} \times \text{gal}/231 \text{ in}^3 \times 60 \text{ sec/min}$$

$$= 40 \text{ gpm}$$

This tells us that without regeneration our pump would have to put out 40 gpm to give us the required extension speed. The horsepower requirement now becomes

$$HP = \frac{p \times Q}{1714}$$

$$= \frac{2000 \text{ psi} \times 40 \text{ gpm}}{1714}$$

$$= 46.67 \text{ HP}$$

Thus, by using regeneration, we reduce our energy consumption by a factor of four.

While the savings just illustrated could be significant, you will probably never find a regenerative system that uses a 4 to 1 area ratio as was used in this example. Much more common is a 2 to 1 area ratio, meaning that the piston area is twice the rod area. This ratio normally provides the best trade-off between speed increase and force decrease. There are also some other interesting advantages of using this ratio, as you will discover when you work one of the review problems at the end of the chapter.

Regeneration can be used to increase the cylinder extension speed, but not the retraction speed. There are two reasons for this. First, if both the cap and rod end ports were connected to the pressure line, the piston would always try to extend because of the larger piston area on the piston side. Second, the volume of fluid in the cap end is greater than in the rod end, so there would be no place to put the fluid exhausted from the blind end. (**Note:** Regeneration is one hydraulic concept that doesn't work in pneumatic systems.)

6.5   Cylinder Cushioning

In many applications, it is desirable to provide a means of slowing the piston prior to full extension or retraction. This is done to prevent impact loading at the ends of the strokes. The process is termed *cushioning* and is commonly accomplished by restricting the flow out of the cylinder as the piston nears the end of its stroke.

Recall that we have already discussed the concept of cylinder speed and have seen that the only way to control the speed of a hydraulic cylinder is to control the rate at which we pump fluid into the cylinder. In a double-acting cylinder, as we pump fluid into one end so that the piston moves toward the other end, the fluid in that (the second) end must be expelled. If we somehow close off the exhaust port so that no fluid can flow out, the piston will not move; therefore, no fluid can flow into the cylinder, either. If we then open the exhaust port slightly, so that there is a very slow outflow, the piston will move slowly. Consequently, there can be only a low incoming flow rate. In essence, then, we are controlling the incoming flow rate (hence, cylinder speed) by controlling the outgoing flow rate.

In a cushioned cylinder, the flow rate leaving the cylinder is controlled, but instead of using a separate flow control valve, we are using the internal design of the cylinder itself to restrict the outgoing flow. This internal design normally includes a protrusion (termed a *spear*) at the end of the piston and a sleeve on the rod, a port or cavity into which the spear or sleeve is inserted at each end of the stroke, and a built-in needle valve to allow the cushion effect to be adjusted. Most cylinders also include a check valve that allows fluid to bypass the cushion mechanism when the direction of travel is reversed, so that fluid pressure acts on the full piston area. Figure 6.12 shows a typical cushioned cylinder.

There are many different spear shapes employed in cushioning. Some of them are shown in Figure 6.13. These shapes are used in an effort to provide a controlled deceleration during the cushion stroke and to avoid shock-loading in the cylinder when the cushion zone is reached.

When cushioning is used, the flow restriction that allows the cushioning forces to develop also encourages the pressure intensification discussed earlier. The cylinder designer must be certain that both the cylinder structure and the seals used can withstand the increased pressures.

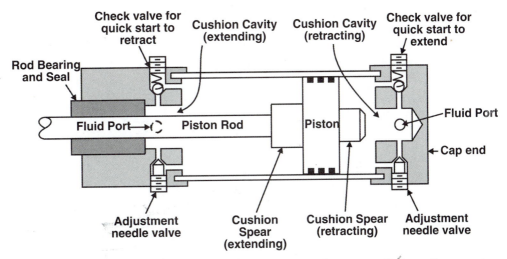

Figure 6.12 In the typical cushioned mechanism, the spears fit into the cylindrical cavities.

6.6 Cylinder Load Calculations

The force of output required from the cylinder is determined by the load the cylinder must move. Let's look at how we actually determine what is meant by "load." There is little question about its meaning when we are considering a vertical cylinder; in that case, the "load" is simply the weight of the object that we are attempting to move, ignoring friction and inertia. The weight constitutes the resistance to upward motion.

As we move away from the vertical, the "load" due to that weight begins to decrease, because only the portion of the weight that is directed along the axis of the cylinder resists the motion of the cylinder. Finally, when the direction of motion is horizontal, the load *theoretically* presents no resistance to motion, because none of the weight is directed along the axis of the cylinder, so *theoretically,* no force would be required to move the load. In reality, there would still be friction and inertia resisting the motion.

If the load is sliding across a horizontal surface, a force is required that is proportional to the weight of the load, *wt,* and the friction factor (or coefficient of friction), *f,* between the surface and the load. This force is given by

$$F = wt \times f \tag{6.15}$$

Example 6.11

Determine the cylinder size needed to push a 10,000 lb weight across a horizontal surface at a constant velocity with a coefficient of friction of 0.15. The maximum pressure available is 2000 psi.

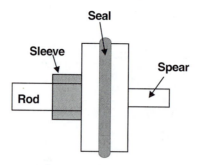

a. Straight spear

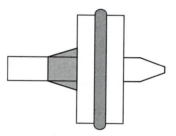

b. Tapered spear

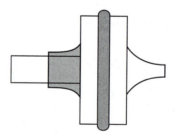

c. Inverted parabola spear

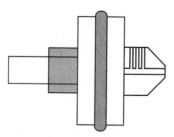

d. Piccolo spear with sharp-edged
   orifices connecting into an axial
   passageway.

Figure 6.13 Typical spear shapes

Solution

In this case, the force required depends on the coefficient of friction, so

$$F = wt \times f$$

$$= 10{,}000 \text{ lb} \times 0.15$$

$$= 1500 \text{ lb}$$

To find the diameter, we use

$$A = \frac{F}{p}$$

$$= \frac{1500 \text{ lb}}{2000 \text{ lb/in}^2}$$

$$= 0.75 \text{ in}^2$$

and

$$D = 2\sqrt{\frac{A}{\pi}}$$

$$= 2\sqrt{\frac{0.75 \text{ in}^2}{\pi}}$$

$$= 0.98 \text{ in}$$

If the cylinder is not either vertical or horizontal, the proportion of the weight of the load that must be considered in calculating the cylinder load depends on the angle (θ) between the horizontal and the centerline of the cylinder (Figure 6.14). Since we're concerned with the cylinder force requirements, we can express this as

$$F_{\text{wt}} = wt \times \sin\theta \tag{6.16}$$

Example 6.12

Find the force required to extend the load shown here at a constant velocity.

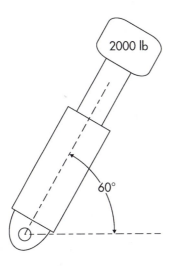

Solution

From Equation 6.16,

$$F = wt \times \sin \theta$$
$$= 2000 \text{ lb} \times \sin 60°$$
$$= 1732 \text{ lb}$$

For a quick reference, we can use the *load factor* values listed in Table 6.1 instead of the sine of the angle. To use the load factor (f_L), we modify Equation 6.16 to

$$F_{\text{wt}} = wt \times f_L \tag{6.17}$$

Now, to solve the problem of Example 6.12, we simply find the load factor, f_L, for 60 degrees and plug it into Equation 6.17. From the table, we see that when θ is 60 degrees, f_L is 0.866. Therefore,

$$F_{\text{wt}} = wt \times f_L$$
$$= 2000 \text{ lb} \times 0.866$$
$$= 1732 \text{ lb}$$

Now consider the case shown in Figure 6.15 in which the load is being pushed up an inclined surface at a constant speed. Here we must consider the force required to overcome the surface friction. We can't use Equation 6.15 directly, however, because only the fraction of the weight of the load that acts

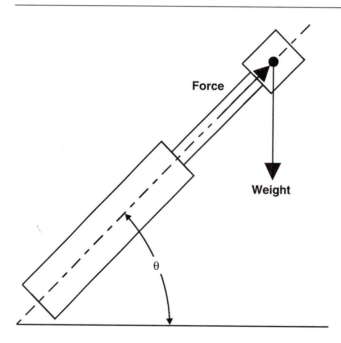

Figure 6.14 In an angled cylinder, the proportion of the weight of the load that must be considered in calculating the cylinder load depends on the angle (θ).

Table 6.1 Load Factors, f_L

Angle, θ (in degrees)	Load Factor, f_L ($\sin\theta$)
90	1.000
85	0.996
80	0.985
75	0.966
70	0.940
65	0.906
60	0.866
55	0.819
50	0.766
45	0.707
40	0.643
35	0.574
30	0.500
25	0.423
20	0.342
15	0.259
10	0.174
5	0.087

perpendicular to the surface contributes to the force. The fraction of the weight that we are concerned with depends on the angle, θ. We express the friction force as

$$F_f = wt \times \cos\theta \times f \qquad (6.18)$$

We can also express this in terms of the *weight factor, f_w*, shown in Table 6.2. In that case, we have

$$F_f = wt \times f_w \times f \qquad (6.19)$$

This force must be added to the load factor force of Equation 6.17 to get the total force required. Thus, for a load being pushed up an incline with surface friction, the total force required is

$$F_{tot} = (wt \times f_L) + (wt \times f_w \times f)$$

or

$$F_{tot} = wt \times [f_L + (f_w \times f)] \qquad (6.20)$$

The forces we have found in these and previous examples have been the

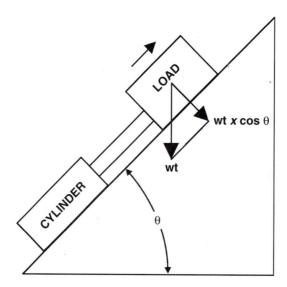

Figure 6.15 If a load is being pushed up an inclined surface at a constant speed, we must consider the force required to overcome the surface friction.

forces required to move the load at a constant velocity, which we term the *steady state* condition. Due to inertia, additional force is required to accelerate the load from zero to its constant (maximum) velocity. The magnitude of this additional force depends on the distance allowed for acceleration. We can use the graph shown in Figure 6.16 to help calculate this force. From the graph—based on the maximum cylinder speed and the acceleration distance—we find a value called the "acceleration force factor" (f_a). We use it in a modification of Equation 6.17 to calculate the force required for acceleration. This gives us

$$F_a = wt \times (f_a + f_L) \tag{6.21}$$

where F_a is the force required for acceleration.

Example 6.13

Find the force required to accelerate the load in Example 6.12 from zero to 20 ft/min in 0.25 inches.

Solution

On the graph in Figure 6.16, we first locate the maximum velocity, 20 ft/min, on the vertical axis. We then follow that line horizontally until it intersects the slope for the acceleration distance, 0.25 in, or ¼ in. We find the acceleration force factor, f_a, by going straight down to the horizontal axis. We see that this gives us a value of 0.084.

Next, we find the load factor, f_L, from Table 6.1. For 60 degrees, this is 0.866. Using these values in Equation 6.21 gives

$$F_a = wt \times (f_a + f_L)$$
$$= 2000 \text{ lb} \times (0.084 + 0.866)$$
$$= 1900 \text{ lb}$$

Table 6.2 Weight Factors, f_w

Angle, θ (in degrees)	Weight Factor, f_w (cosθ)
0	1.000
5	0.996
10	0.985
15	0.966
20	0.940
25	0.906
30	0.866
35	0.819
40	0.766
45	0.707
50	0.643
55	0.574
60	0.500
65	0.423
70	0.342
75	0.259
80	0.174
85	0.087
90	0.000

If friction is present, the friction factor (f) must also be included in the acceleration force equation. Thus, we modify Equation 6.21 to get

$$F_a = wt \times [f_a + f_L + (f_w \times f)]$$

Example 6.14

The load in Example 6.13 has a friction factor of 0.2. Find the acceleration force.

Solution

$$F_a = wt \times [f_a + f_L + (f_w \times f)]$$
$$= 2000\,\text{lb} \times (0.084 + 0.866 + 0.500 \times 0.2)$$
$$= 2000\,\text{lb} \times 1.05 = 2100\,\text{lb}$$

To this point, we have looked only at upward acceleration. In addition, we could have downward acceleration, as well as upward and downward *deceleration*. The equation for upward acceleration is

$$F_a = wt \times [f_a + f_L + (f_w \times f)] \tag{6.23}$$

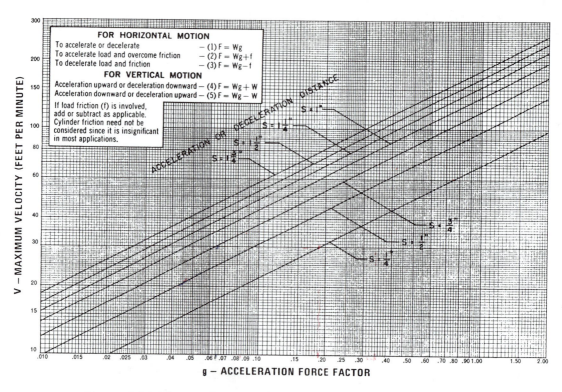

Figure 6.16 The acceleration force factor (f_a) can be calculated from the maximum cylinder speed and the acceleration distance.
Source: Courtesy of Parker Hannifin Corporation.

The equation for downward acceleration is

$$F_a = wt \times [f_a - f_L + (f_w \times f)] \tag{6.24}$$

The equation for upward deceleration is

$$F_a = wt \times [f_a - f_L - (f_w \times f)] \tag{6.25}$$

The equation for downward deceleration is

$$F_a = wt \times [f_a + f_L - (f_w \times f)] \tag{6.26}$$

The significance of these force calculations is, of course, that the pressure requirements depend on the force requirements. Therefore, for any particular situation, we can find the pressure required to accelerate or decelerate the load as well as to move it at a constant velocity.

Example 6.15

Find the pressure required to accelerate the load of Example 6.14 and the pressure required to move the load at a constant speed. The cylinder has a 2 in² piston.

Solution

The pressure to accelerate the load required is found from Equation 6.2. In this case, the value of F is the acceleration force we found in Example 6.14. Thus,

$$p = \frac{F_a}{A} = \frac{2100 \text{ lb}}{2 \text{ in}^2} = 1050 \text{ psi}$$

The pressure required to move it at a constant speed is based on the load factor. In Example 6.13, we find this to be 1900 lb. Therefore,

$$p = \frac{F}{A} = \frac{1900 \text{ lb}}{2 \text{ in}^2} = 950 \text{ psi}$$

You'll have the opportunity to explore some of the other situations in working the review problems.

To this point, we have discussed acceleration using the acceleration distance to determine the forces required. We can also base the calculation of these forces on the time required to change from one speed to another. This is done using Newton's Second Law of Motion, which states that the time rate of change (acceleration) of linear momentum of a body is proportional to the force acting upon it and occurs in the direction in which the force acts. We express this concept mathematically as

$$\text{Force} = \text{Mass} \times \text{Acceleration}$$
$$F = m \times a$$

(6.27)

or

$$F = \frac{wt}{g} \times a$$

(6.28)

where

$$a = \frac{v_2 - v_1}{t}$$

$v_1 = $ initial velocity.

$v_2 = $ final velocity.

$g = $ acceleration of gravity (32.2 ft/sec^2 or 9.81 m/s^2).

$t = $ time.

As with the acceleration force factor discussed earlier, these equations give us only the force involved with the change in velocity (acceleration). To find the total force required for the motion, we must include the force required to produce the steady-state motion. This force is equal to the weight of the load in vertical upward motion. The total force requirement during acceleration is given by

$$F_a = wt + \left(\frac{wt}{g} \times a \right) = wt \times \left(1 + \frac{a}{g} \right)$$

(6.29)

Example 6.16

A 2 in hydraulic cylinder is required to move a 1000 lb load vertically upward. Determine

a. The steady state force required to move the load at 10 ft/sec.
b. The total force required to accelerate the load from a stationary position to 10 ft/sec in 0.5 seconds.
c. The pressures required for steady-state motion and acceleration.

Solution

a. Ignoring friction losses, the steady-state force (F_{ss}) required to move the load vertically at 10 ft/sec (or *any* constant speed) is equal to the weight of the load; therefore,

$$F_{ss} = 1000 \text{ lb}$$

b. The force required to accelerate the load is found from Equation 6.29

$$F_{tot} = wt + \left(\frac{wt}{g} \times a \right)$$

$$= 1000 \text{ lb} + \left(\frac{1000 \text{ lb}}{32.2 \text{ ft/sec}^2} \times \frac{10 \text{ ft/sec} - 0 \text{ ft/sec}}{0.5 \text{ sec}} \right)$$

$$= 1621 \text{ lb}$$

c. We find the pressures from Equation 6.2

$$p_{ss} = \frac{F_{ss}}{A} = \frac{1000 \text{ lb}}{3.14 \text{ in}^2} = 318 \text{ psi}$$

$$p_a = \frac{F_a}{A} = \frac{1621 \text{ lb}}{3.14 \text{ in}^2} = 516 \text{ psi}$$

Notice that Equation 6.29 is similar to Equation 6.21. In Equation 6.29, we have assumed vertical motion, so $f_L = 1$. Also, in place of f_a, which is based on acceleration *distance*, we have used a/g, which is based on acceleration *time*. Thus, in acceleration problems where the acceleration parameter is time instead of distance, simply substitute a/g for f_a in Equations 6.21 through 6.26.

We have considered only cases where the load has been in direct line with the axis of the cylinder; however, while cylinders always act linearly, they don't always move the load directly. In many cases, they are used to actuate mechanisms that in turn move the load. Figure 6.17 shows some typical arrangements. We will not explore all of these configurations, but we will look at the first-, second-, and third-class levers.

A first-class lever is shown in Figure 6.18. This device is used to raise and lower a load that is attached to a lever that is mounted on a clevis mount (or pivot) between the cylinder and the load. The cylinder must be clevis-mounted on both ends to allow it to move through the arc of the lever.

The force required to move the load is a function of the resistance presented by the load, the relative distances (d_1 and d_2) of the lever arms from the pivot

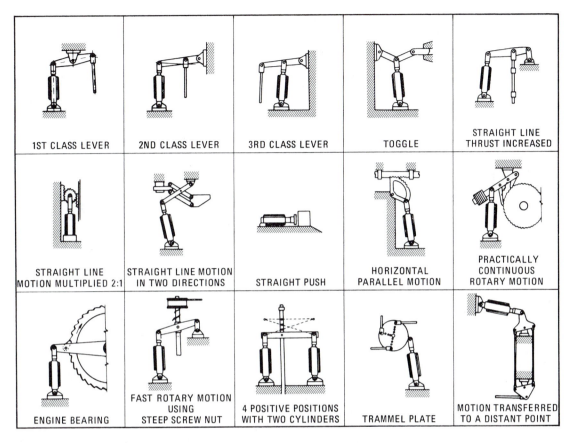

1ST CLASS LEVER	2ND CLASS LEVER	3RD CLASS LEVER	TOGGLE	STRAIGHT LINE THRUST INCREASED
STRAIGHT LINE MOTION MULTIPLIED 2:1	STRAIGHT LINE MOTION IN TWO DIRECTIONS	STRAIGHT PUSH	HORIZONTAL PARALLEL MOTION	PRACTICALLY CONTINUOUS ROTARY MOTION
ENGINE BEARING	FAST ROTARY MOTION USING STEEP SCREW NUT	4 POSITIVE POSITIONS WITH TWO CYLINDERS	TRAMMEL PLATE	MOTION TRANSFERRED TO A DISTANT POINT

Figure 6.17 Applications of cylinders for providing a variety of fundamental mechanical motions.
Source: Courtesy of Parker Hannifin Corporation.

point, and the angle of the centerline of the cylinder from the vertical. Basically, we have a "see-saw" problem in this type of lever, and we are looking for a moment balance. In the special case when the arm is horizontal, this will result when the cylinder force (F) multiplied by its lever arm (d_1) equals the load resistance (L) multiplied by its lever arm (d_2). For a constant velocity, then, we can write

$$F \times d_1 = L \times d_2$$

where ϕ is the angle between the cylinder axis and the vertical. To find the force required, we solve for F to get

$$F = \frac{L \times d_2}{d_1} \qquad\qquad \textbf{(6.30)}$$

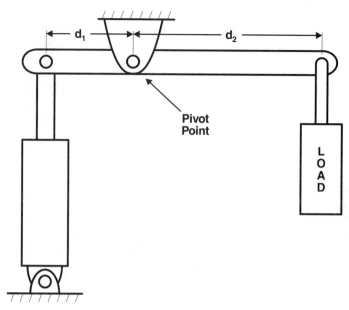

Figure 6.18 This first class lever is used to raise and lower a load that is attached to a lever pivoted on a clevis mount (or pivot) between the cylinder and the load.

In the general case where the arm is not horizontal, the angle of the arm from the horizontal as well as the angle of the cylinder from the vertical must be used.

Example 6.17

A first-class lever is used to move a load resistance of 2000 lb. The distance from the pivot to the cylinder is 6 in and from the pivot to the load is 12 in. Find the force required to move the load.

Solution
From Equation 6.30,

$$F = \frac{L \times d_2}{d_1}$$

$$F = \frac{2000 \text{ lb} \times 12 \text{ in}}{6 \text{ in}} = 4000 \text{ lb}$$

In a second-class lever (Figure 6.19), the load is attached between the cylinder and the pivot point. Again, we are looking for a moment balance based on

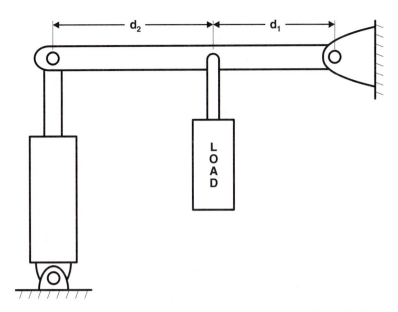

Figure 6.19 In this second-class lever, the load is between the cylinder and the pivot point.

lever-arm length and the cylinder angle. In the horizontal case, the force output from the cylinder is found from Equation 6.31.

$$F = \frac{L \times d_1}{(d_1 + d_2)} \tag{6.31}$$

The third-class lever (Figure 6.20) requires a similar equation to find the cylinder force. In the horizontal case, however, the cylinder is between the pivot point and the load, so the equation is modified to give Equation 6.32.

$$F = \frac{L \times (d_1 + d_2)}{d_1} \tag{6.32}$$

Another common application for hydraulic cylinders is shown in Figure 6.21. Here we see the cylinder pulling a cable to raise a load. In the direct-pull situation shown in this figure, no load advantage is realized. The only advantage is the convenience and practicality of the arrangement for some specific applications. To raise the 2000 lb load 12 in, the cylinder must produce a force of 2000 lb and have a stroke of 12 in. This is a 1 to 1 load ratio.

The pulley arrangement in Figure 6.22 does provide some load advantages. By attaching a pulley to the load and running the cable through the pulley and anchoring it above the load as shown, we can distribute the load between the cylinder and the anchor point. In this arrangement, the load is split evenly between the two, essentially using two cables to support the load. This gives us a 2 to 1 load ratio. As a result, the cylinder force required to lift the 2000 lb load is only 1000 lb. There is a penalty, however, and that is that a cylinder

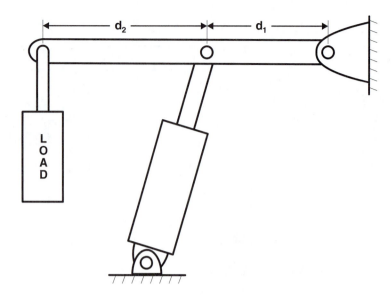

Figure 6.20 In this third-class lever, the cylinder is between the pivot point and the load.

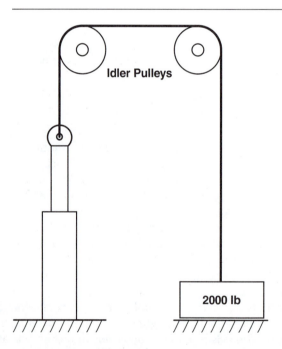

Figure 6.21 Direct pull puts the full load on the cylinder.

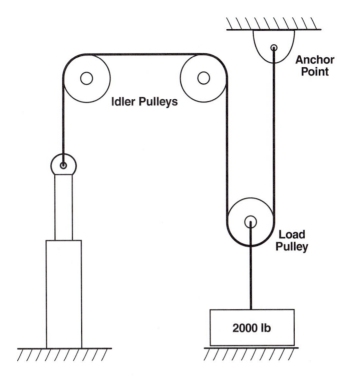

Figure 6.22 The load pulley arrangement divides the load evenly between the cylinder and the anchor point but doubles the cylinder stroke.

stroke of 24 in is required to raise the load 12 in. In other words, a 2 to 1 load ratio results in a 1 to 2 stroke ratio.

By adding a load-supporting idler pulley, as shown in Figure 6.23, we are effectively adding another support for the load. This distributes the load among four cables. The result now is a 4 to 1 load ratio. Thus, a cylinder force of 500 lb will raise the load. The stroke ratio is the inverse of the load ratio, 1 to 4. Therefore, a 48 in stroke will be required to raise the load 12 in.

6.7 Selection Criteria

Many factors must be considered in selecting a cylinder for a particular application. These include diameter, stroke, bore, rod size, mounting, and contaminant sensitivity.

Size

The size (bore and stroke) of the cylinder is of primary importance. The stroke must be sufficient to provide the required reach capability. The bore must be

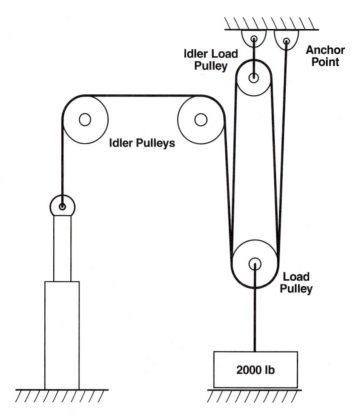

Figure 6.23 Adding the idler load pulley further distributes the load, giving a 4 to 1 load ratio, but a 1 to 4 stroke ratio.

large enough to ensure that the necessary force will be generated to move the design load. Cylinders are usually oversized slightly to allow for static friction (stiction), seal friction, and inertia. Excessive oversizing must be avoided, however, because of the increased weight and cost of large cylinders. Increasing the cylinder bore also decreases the speed capability for a given system flow rate.

As we have already seen, the cylinder bore is directly related to the force output required and the pressure available. To calculate the minimum diameter that will satisfy a given condition for extending a load, we solve Equation 2.1 for area:

$$A = \frac{F}{p} \qquad\qquad (6.33)$$

After finding the area, we use the area equation to find diameter as follows:

$$A = \frac{\pi D^2}{4}$$

Therefore,

$$D^2 = \frac{4A}{\pi}$$

so that

$$D = 2\sqrt{\frac{A}{\pi}} \tag{6.34}$$

When finding the area required for retracting the cylinder, we must account for the rod area that does not contribute to the retraction force. In this case, we again use Equation 6.33 to find the area required to generate the necessary force; however, this is only the *active* area upon which the pressure acts, not the *total* piston area needed to find the cylinder bore. To get the total area, we must add the rod area to the active area.

Example 6.18

Determine the cylinder bore needed to retract a cylinder against a 5000 lb load. The cylinder has a 2 in rod. The pressure available is 1200 psi.

Solution

Using Equation 6.33, we find the active area

$$A_{\text{active}} = \frac{F}{p} = \frac{5000\,\text{lb}}{1200\,\text{lb/in}^2} = 4.17\,\text{in}^2$$

Next we add the rod area to find the total piston area

$$A_{\text{total}} = A_{\text{active}} + A_{\text{rod}}$$
$$= 4.17\,\text{in}^2 + 3.14\,\text{in}^2 = 7.31\,\text{in}^2$$

Now we can use Equation 6.34 to find the cylinder bore.

$$D = 2\sqrt{\frac{A}{\pi}} = 2\sqrt{\frac{7.31\,\text{in}^2}{\pi}} = 3.05\,\text{in}$$

The bore and rod sizes for fluid power cylinders have been standardized to a large extent. National Fluid Power Association (NFPA) Recommended Standard T3.6.11, ANSI Standards B93.3 and B93.8, and ISO 2091 list the standard dimensions for both the cylinder bore and piston rod diameters. Therefore, once a minimum diameter has been calculated, as in Example 6.18, you should consult one of these standards or a manufacturer's catalog and specify the next larger cylinder size.

Mounting

Because hydraulic cylinders are used in so many different applications, numerous mounting configurations are necessary. The NFPA in their recommended standards T3.6.39.7, T3.6.7-SI, and T3.6.11 shows the mounting types and dimensions for industrial cylinders. This type of information is also contained in ANSI B93.15. These standards apply primarily to the so called *square head* (also called *tie rod*) cylinders used in industrial applications, and are intended to facilitate dimensional interchangeability among general-purpose cylinders from different manufacturers.

Figure 6.24 illustrates several standard mounting configurations that are designated according to the NFPA mounting codes. The NFPA divides cylinders into two categories. Air and light-duty hydraulic cylinders constitute one group. The other group is the heavy-duty hydraulic cylinders. The same mounting codes are used for both groups, but the two groups are not dimensionally interchangeable.

Piston Rod Diameter

In addition to specifying the piston diameter that will ensure sufficient force for the intended application, the diameter of the piston rod must be specified to ensure that it has sufficient structural strength to prevent buckling or other load-related failures.

On a rod that operates primarily in tension, load-related failures are rare. Cylinders that operate in compression (which is the more common mode) are subject to columnar buckling if the load exceeds the structural capabilities of the rod. The specification of the rod diameter depends on several factors including the

- Load on the cylinder.
- Mounting method used.
- Length of the rod and amount of extension.
- Type of support for the rod (if any).

Figure 6.25 shows two different cylinders undergoing columnar buckling. The effect of the mounting method is illustrated. Rigid mounting of both the cylinder body and the rod end significantly reduce the likelihood of buckling when compared with a cylinder and rod with clevis mounts. The use of stop tubes inside the cylinder, as shown in Figure 6.26 significantly add to the rod support area and reduce the likelihood of buckling.

Contaminant Sensitivity

The life and reliability of a hydraulic cylinder depends to a large extent on how resistant the cylinder is to solid contaminants in the fluid. The wear of a hydraulic cylinder usually shows up as internal leakage around the piston seal or external leakage around the rod seal. In either case, such leakage will cause the cylinder to drift and fail to hold its position, operate sluggishly, and fail to position accurately.

Particles may damage a cylinder in two ways. One is damage to the seal by channels cut through it. The second is wear on the cylinder bore that results in either leakage paths in the form of tiny grooves cut into the metal of the bore, or a general wearing of the diameter of the bore in a localized area. This type of wear is sometimes termed *barreling* because the bore takes on the shape of an old wooden stave barrel (Figure 6.27).

While there is currently no widely accepted standard for evaluating the contaminant sensitivity of hydraulic cylinders, the Fluid Power Research Cen-

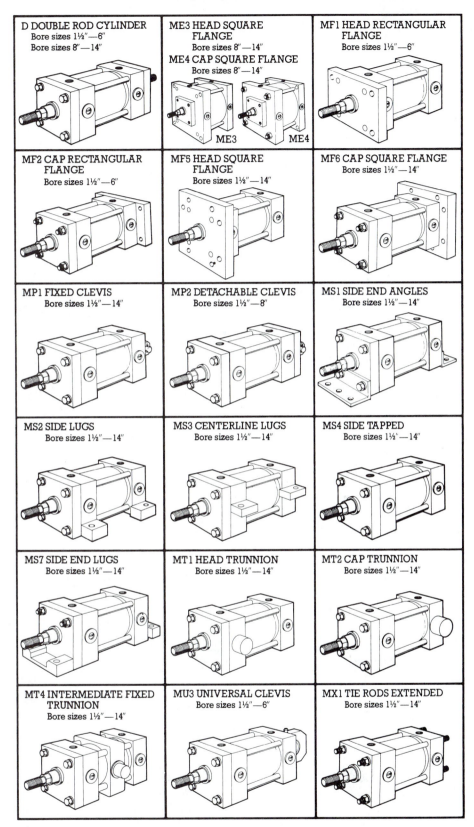

Figure 6.24 NFPA cylinder mounting configurations.
Source: Courtesy of Parker Hannifin Corporation.

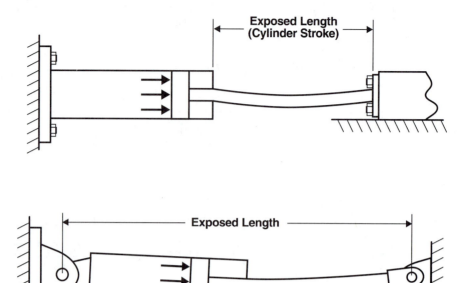

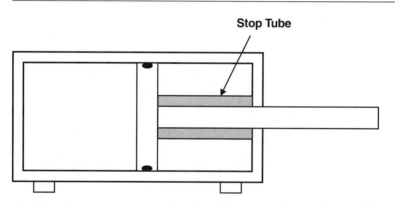

Figure 6.25 Buckling in hydraulic cylinders.

Figure 6.26 A stop tube can be used to provide additional support for the rod and thus reduce the likelihood of buckling.

ter at Oklahoma State University has developed a technique that has proved successful. The cylinder is subjected to a predetermined operating period using hydraulic fluid contaminated to a known level with AC Fine Test Dust (ACFTD). The rate of creep under a constant, static load is then measured to evaluate the rate at which fluid is leaking past the seal.

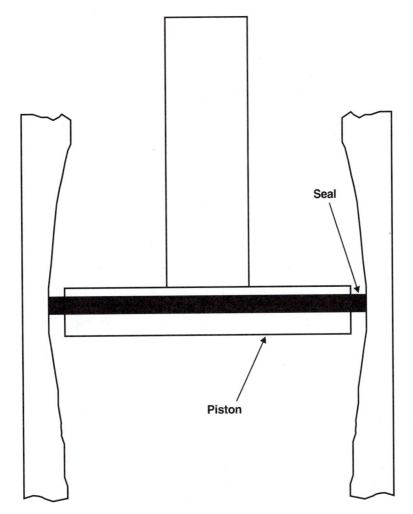

Figure 6.27 Grossly exaggerated example of cylinder wear due to fluid contaminants.

The ingression of particulate contamination into hydraulic systems through cylinder rod wiper seals can be a major problem for both the cylinder and the rest of the system. This ingression occurs when atmospheric dirt collects in the film of oil on the extended rod. The function of the wiper seal is to scrape the dirt off the rod as it retracts into the cylinder. The ability of different wiper seals to effectively accomplish this very important task varies greatly.

The effectiveness of rod wiper seals can be evaluated by the method described in SAE J1195. This method measures the amount of ACFTD that enters a fluid chamber as the rod extends into a dust box and then retracts through the wiper seal.

6.8   Troubleshooting

When troubleshooting problems in cylinder circuits, it is necessary to understand the basic concepts of cylinder operation as expressed in Equations 6.1 and 6.5. From these equations, we see that force is a function of pressure, while speed is a function of flow. With that in mind, we can use the following list to assist in the troubleshooting process.

Symptom	Possible Cause
Cylinder will not move the load	1. Pressure too low 2. Cylinder too small 3. Piston seal leaking
Load moves too slowly	1. Pump not producing sufficient flow 2. Flow control valves set too low 3. Piston seal leaking
Cylinder drifts	1. Piston seal leaking 2. Rod seal leaking 3. Leaking through directional control valve spool
Rod seal leaking	1. Damaged seal (torn or worn) 2. Deterioration due to fluid incompatibility or excessive heat 3. Excessive pressure
Piston seal leaking	1. Wear due to contaminated fluid 2. Deterioration due to fluid incompatibility or excessive heat 3. Excessive pressure
Chatters during movement	1. Misaligned load 2. High seal static friction 3. Cylinder too small 4. Pilot-operated check valve

6.9   Summary

Hydraulic cylinders may be single- or double-acting, as well as single- or double-ended. Variations on the common cylinder configurations include telescoping, tandem, and dual cylinders, all of which are used in applications where space for mounting a cylinder is limited.

The output force capability of a cylinder is a function of the pressure available to the cylinder and the area exposed to that pressure. The actual force output will be only the force required to overcome the resistances (load, friction, and inertia) required to move the load. The speed at which the cylinder can move the load is a function of the fluid flow rate available to the cylinder and the area exposed to the flow. In single-ended, double-acting cylinders—which are the most commonly used—for any given pressure and flow rate the extension force is higher than the retraction force, while the retraction speed is higher than the extension speed.

In a regenerative circuit, the exhaust flow from the rod end of a single-ended cylinder is ported to its blind end. This increase in inlet flow causes the cylinder to extend at a higher speed than would be possible using only the pump flow. The increase in speed is at the expense of force capability, which is reduced by the same percentage as the speed is increased. The most commonly used cylinders for regenerative circuits have a 2 to 1 area ratio.

When a complete cylinder performance analysis is needed, the forces required to overcome the inertia of the load must be considered. These forces depend upon the direction—upward, downward, horizontally, or at an angle—in which the load is being moved. Both acceleration and deceleration must be considered. When loads are being moved in a direction other than vertically upward, the angle from the horizontal must also be considered. Only the portion of the weight which is parallel to the centerline of the cylinder determines the force requirement from the cylinder.

Cylinders are often used to generate forces to move loads that are not parallel to the cylinder's centerline. In these cases, the affects of angles, lever arms, cables, pulleys, and linkages must be calculated to determine the force and speed that is required from the cylinder to produce the desired motion from the load.

Operating cylinders with dirty fluid can lead to wear, seal damage, and premature cylinder failure due to excessive internal leakage. Cylinders with internal leakage may be unable to move the load at all or may move it too slowly. Depending on the orientation of the load and the system design, a leaking cylinder may not be capable of holding the load in place when flow to the cylinder has been stopped.

6.9   Key Equations

Cylinder force:

$$F = p \times A \tag{6.1}$$

Cylinder velocity:

$$v = \frac{Q}{A} \tag{6.5}$$

Cylinder output power
U.S. Customary

$$HP = \frac{F \times d}{550t} = \frac{F \times v}{550} \tag{6.7}$$

Cylinder output power
SI

$$P_{OUT} = \frac{F \times d}{t}$$
$$= F \times v \tag{6.7a}$$

Regenerative velocity:

$$v_{regen} = \frac{Q_{pump}}{A_{rod}} \tag{6.11}$$

Regenerative force:

$$F_{regen} = p \times A_{rod} \tag{6.14}$$

Acceleration forces: **Note:** For time-based accelerations and decelerations, substitute a/g for f_a in the following equations:

Upward:

$$F_a = wt[f_a + f_L + (f_w \times f)] \tag{6.23}$$

Downward:

$$F_a = wt [f_a - f_L + (f_w \times f)] \tag{6.24}$$

Deceleration forces:

Upward:

$$F_a = wt[f_a - f_L - (f_w \times f)] \tag{6.25}$$

Downward:

$$F_a = wt[f_a + f_L - (f_w \times f)] \tag{6.26}$$

First-class lever:

$$F = \frac{L \times d_2}{d_1} \tag{6.30}$$

Second-class lever:

$$F = \frac{L \times d_1}{(d_1 + d_2)} \tag{6.31}$$

Third-class lever:

$$F = \frac{L \times (d_1 + d_2)}{d_1} \tag{6.32}$$

Review Problems

Note: The letter "S" after a problem number indicates SI units are used.

General

1. Explain the difference between a single-acting and a double-acting cylinder.
2. Explain the difference between a single-ended and a double-ended cylinder.
3. In a telescoping cylinder, what determines which stage of the cylinder will move first?
4. In a telescoping cylinder, what determines the maximum load that can be moved to the maximum extension with a given pressure.
5. List the factors that determine the force output from a hydraulic cylinder.
6. List the factors that determine the speed of operation of a hydraulic cylinder.
7. Define regenerative circuit.
8. Explain when a regenerative circuit could be used.
9. What is the major advantage of a regenerative circuit?
10. What is the major disadvantage of a regenerative circuit?
11. What is the optimum area ratio for a regenerative cylinder?
12. Explain pressure intensification.

13. Explain flow differential.
14. Explain the purpose of a cushion in a hydraulic cylinder.
15. If a cylinder moves the load too slowly, what are the possible causes?

Cylinder Force

16. A cylinder with a 3 in² piston area has a pressure of 1200 psi. What is its force output?
17. A cylinder with a 4 in² piston area has a pressure of 1500 psi. What is its force output?
18. A cylinder with a 3 in bore operates at 1000 psi. What is its extension force output?
19. A cylinder with a 3 in bore and a 2 in rod operates at 2000 psi. What are its extension and retraction force capabilities?
20. A tandem cylinder with a 3 in bore and a 1 in rod operates with a pressure of 3000 psi. Find the

   a. output force.
   b. diameter of a normal cylinder required to produce the same output force at the same pressure.

21S. A cylinder with a 20 cm² piston area is operating at 7000 kPa. What is its force output?
22S. A cylinder has a 9 cm bore and a 5 cm rod. The maximum pressure available to it is 14 MPa. What are its maximum forces in extension and retraction?

Pressure (Ignore friction and inertia in the following problems)

23. What pressure is required to produce a 4000 lb force in a 3 in cylinder?
24. A 3 in cylinder is lifting a 10,000 lb load. What pressure is required?
25. A cylinder with a 4 in bore and a 2 in rod extends and retracts a 12,000 lb load resistance. How much pressure is required for extension? For retraction?
26. A cylinder with a 3.5 in bore and a 1.5 in rod extends and retracts a load. The load resistance on extension is 8000 lb. On retraction, the load resistance is 4000 lb. What pressures are required for extension and retraction?
27. A cylinder with a 5 in bore and a 3 in rod has a pressure of 2000 psi on the blind end. If the outlet port in the rod end is blocked off, what is the pressure in the rod end?
28S. A cylinder with a 12 cm² piston must produce a 4 kN output force. What pressure is required?
29S. A cylinder operates against a 20 kN force on extension and a 10 kN force on retraction. The piston diameter is 10 cm. The rod diameter is 4 cm. What pressures are required for extension and retraction?
30S. A cylinder has a 12 cm bore and a 5 cm rod. A pressure of 5 MPa is applied to the blind end while the rod-end port is blocked. What is the pressure in the rod end?

Cylinder Speed

31. A cylinder with a 2 in bore has a flow rate of 8 gpm. What is its extension speed in ft/sec?
32. A cylinder with a 3 in bore has a flow rate of 14 gpm. What is its extension speed in ft/sec?

33. A 4 in cylinder has a 2 in rod. If the flow rate to the cylinder is 20 gpm, what is its retraction speed in ft/sec?
34. Find the extension and retraction speeds in ft/sec for a 3 in cylinder with a 2 in rod if the system flow rate is 10 gpm.
35. Find the flow rate in gpm required to extend a 3 in cylinder at 1 ft/sec.
36. Find the flow rate in gpm required to extend a 2 in cylinder at 1.5 ft/sec.
37. Find the flow rate in gpm required to retract a 3 in cylinder with a 2 in rod at 1 ft/sec.
38. Find the flow rate in gpm required to retract a 4 in cylinder with a 1 in rod at 1.5 ft/sec.
39. What is the maximum flow rate that can be found *anywhere* in a system operating at 10 gpm for a cylinder with a 3 in bore and a 1 in rod?
40S. Calculate the extension speed of a 12 cm cylinder for a flow rate of 30 Lpm.
41S. Find the extension and retraction speeds for a 10 cm cylinder with a 5 cm rod if the flow available is 40 Lpm.
42S. What flow rate (Lpm) is required to extend a 15 cm cylinder 80 cm in 8 seconds?
43S. What flow rate (Lpm) is required to extend a 20 cm cylinder in 0.3 m/s?
44S. What flow rate (Lpm) is required to retract a cylinder with a 9 cm bore and a 4 cm rod in 0.25 m/s?

Cylinder Performance

45. A system operates at 20 gpm and 1500 psi. Find the extension and retraction speeds and forces for a cylinder with a 4 in bore and a 1.5 in rod.
46. A system operates at 30 gpm and 3000 psi. Find the extension and retraction speeds and forces for a 5 in cylinder with a 3 in rod.
47. A hydraulic cylinder with a 4 in bore lifts a 12,000 lb load 18 in in 3 seconds. Ignoring inertia, find the

   a. pressure required.
   b. flow rate required.
   c. horsepower output of the cylinder.

48S. A double-acting cylinder has a 4.5 cm bore and a 2.5 cm rod. If the system pressure is 7000 kPa and the flow rate is 12 Lpm, find the

   a. extension speed.
   b. retraction speed.
   c. maximum extension force capability.
   d. maximum retraction force capability.

49S. Make the same calculation as Review Problem 48S for a 10 cm cylinder with a 5 cm rod operating at 20 MPa and 60 Lpm.

Regeneration

50. A regenerative system uses a cylinder with a 3 in bore and a 1 in rod. The pump output is 6 gpm at 1000 psi. Find:

   a. extension speed without regeneration.
   b. extension speed with regeneration.
   c. extension force capability without regeneration.
   d. extension force capability with regeneration.

51. For the system in Review Problem 50, calculate the horsepower output of the pump using regeneration. If regeneration is not used, how much

horsepower would be required to move the cylinder at the same speed as the regenerative circuit?

52S. A regenerative system uses a cylinder with a 10 cm bore and a 6 cm rod. The system pump operates at 25 Lpm and 7500 kPa. Find the

    a. extension speed without regeneration.
    b. extension speed with regeneration.
    c. retraction speed.
    d. extension force without regeneration.
    e. extension force with regeneration.
    f. retraction force.
    g. power output of the pump using regeneration.
    h. power output required from the pump to provide the high extension speed if regeneration is not used.

53. At what ratio of piston to rod areas would the extension and retraction speeds be the same for a cylinder in a regenerative circuit? Demonstrate your answer using a cylinder with a 4 in (10 cm) bore.

Acceleration

54. A 3 in cylinder accelerates a 5000 lb load vertically from 0 to 1.2 ft/sec in 0.3 seconds. Determine the pressures required for acceleration and steady-state operation.

55. A 900 lb load is being raised vertically by a 4 in cylinder at 30 ft/min. On startup, it reaches its maximum speed in 0.75 inches. Determine the pressures required for acceleration and steady-state operation.

56. A 2.5 in cylinder with a 1 in rod pushes a 10,000 lb load across a surface with a friction factor of 0.4. The total travel is 22 in. The maximum velocity is 2 ft/sec. The load is accelerated and decelerated in 0.2 sec. Plot the pressure versus time profile for the system.

System Problems

57. A cylinder with a 3 in bore raises a 10,000 lb load vertically 18 inches in 2 seconds. The system is operated by a pump that has an overall efficiency of 0.91. The pump is driven by an electric motor with an efficiency of 0.87. Ignoring acceleration and seal friction, find the

    a. cylinder horsepower output.
    b. system flow rate (gpm) and pressure (psi).
    c. hydraulic horsepower of the pump.
    d. input horsepower to the pump.
    e. electric horsepower input to the motor.
    f. overall efficiency of the entire system.

58S. A hydraulic system uses a 9 cm cylinder to extend a 25 kg load 0.7 m in 3 seconds. The system pump has an overall efficiency of 0.87. It is driven by an electric motor with an efficiency of 0.85. Find the

    a. cylinder power output
    b. system flow rate (Lpm) and pressure (kPa).
    c. power output of the pump.
    d. power output of the electric motor.
    e. electrical power input to the electric motor.
    f. overall efficiency of the entire system.

Angled Cylinders

59. The cylinder shown here has a 4 in bore and a 2 in rod. Find the pressure required to extend the load at a constant speed.

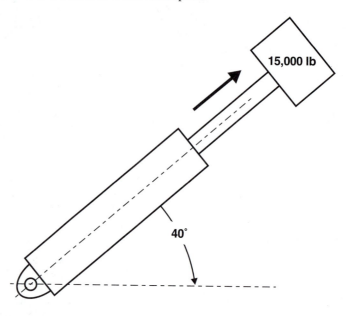

Review Problem 6.59

60. A cylinder with a 2 in bore is required to accelerate a 4000 lb load vertically upward to 30 ft/min in 0.25 inches. Find the

    a. acceleration pressure.
    b. steady-state pressure.
    c. flow rate.

61. When being lowered, the load shown here must decelerate from 40 ft/min to a stop in 0.75 inches. What will be the pressure in the rod end of the cylinder during the deceleration?

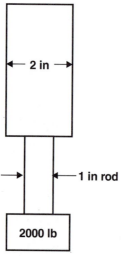

Review Problem 6.61

62. A 400 lb load is pushed across a horizontal table by a 1 in cylinder with a 0.5 in rod. It accelerates and decelerates in 0.25 inches. The maximum speed is 50 ft/min. The table has a coefficient of friction of 0.15. Find the

   a. acceleration pressure.
   b. deceleration pressure.
   c. steady-state pressure.
   d. flow rate.

Chapter 7

Directional Control Valves

Outline

7.1 Introduction

Between the power input device (pump) and the power output device (hydraulic motor or cylinder), we usually find one or more valves that provide various control functions for the hydraulic system. These valves fall into three groups—directional control, pressure control, and flow control. In this chapter, we'll discuss directional control valves and their applications. We'll cover pressure control in Chapter 8, and flow controls in Chapter 9.

Objectives

When you have completed this chapter, you will be able to

- Explain directional control valve (DCV) terminology.
- Recognize and be able to draw circuits using the standard ISO symbols for DCV control circuits.
- Explain the functions of two-way, three-way, and four-way DCVs as well as check valves and deceleration valves and discuss the applications of each.
- Explain the different center positions used in three-position DCVs.
- Explain the various types of actuators that can be used with DCVs and be able to recognize and draw the ISO symbols for these actuators.
- Understand the difference between open center and closed center systems.
- Explain how pilot-operated valves work and how they are applied.
- Discuss the advantages, disadvantages, and applications of cartridge valves used for directional control.
- Explain valve subplates, their designations, and how they are used.

Directional control valves (DCVs) are defined in ANSI B93.2 as valves whose basic function is to direct or prevent flow through selected passages. In other words, the DCVs in fluid power systems serve the same function as the traffic light at a busy intersection—they start and stop the flow in various parts of the system and direct flow to and from the power output devices in the system. They can range in complexity from a very simple ball check valve to a bank of pilot-operated valves that control a dozen or more functions from a single valve set.

The definition of a DCV can be misleading. The *primary* function is to control direction, but remember that nothing in a fluid power system acts independently of the other components in the system. A DCV can be expected to have some effect on pressure, although in a properly sized valve, this effect will be minimal. A DCV may affect flow, also. In fact, it is common practice in mobile applications to use a DCV to control flow to an actuator on a crane, winch, dozer blade, and so on. In Chapter 10, we'll see that one advantage of some electrohydraulic valves is their ability to control direction and flow simultaneously.

This section is not intended to make you an instant valve expert, but rather to expose you to the basic DCV functions and familiarize you with common valve terminology. We'll deal with the terminology first to simplify the communications problem.

7.2   Directional Control Valve (DCV) Terminology

To give a complete yet concise description of a DCV, we need to describe several attributes of the valve, including

- The porting of the valve.
- The number of discrete valve settings or positions.
- The methods of actuation (including any latching devices, springs, detents, etc).
- Whether the valve is open or closed in its unactuated position (for some two-position valves) or the type of center position (for three-position valves).

The number of connecting ports on a DCV are expressed as the "ways" of the valve. Technically, we are referring to the number of flow paths through the valve in the working position. It is easier, however, to refer to the number of flow ports by which the valve is connected to the system. Thus, a valve that has three different flow ports is a *three-way* valve, even if one port is blocked. In some literature, *port* is substituted for *way*.

We usually discuss DCVs as having discrete functional positions—open or closed, forward or reverse, and so on—even though there are many exceptions to this, such as infinitely positionable valves. A valve that has two discrete positions is described as a *two-position* valve.

When we describe a directional control valve as being *normally open,* we are saying that, in its unactuated position, there is an open flow path from the pressure port to a working port and/or the tank port so that fluid from the pump is free to flow through the valve. A *normally closed* valve has the pressure port blocked in its unactuated position. Flow from the pressure port does not occur until the valve is shifted. For those who are familiar with electrical switching, this terminology is exactly opposite, because with switches, the normally closed switch allows current flow.

If we are describing a three-position valve, it is necessary to specify the *center* of the three positions. This is sometimes called the *neutral position*. When we reach the section in which these three-position valves are discussed, we will see some of the many different center-position configurations and how they affect valve and system operation.

7.3   Check Valves

Having said all that, we will begin the discussion of specific valve types with the *check valve*—a valve that corresponds very little to our description of typical DCVs. A check valve is a simple one-directional valve that allows fluid to flow through it in only one direction and prevents flow in the other direction. This directional control is achieved by using a simple ball, poppet, spool, or flapper that is pushed off the valve seat when fluid flow is in the forward direction and seated firmly when fluid tries to flow in the reverse direction. Figure 7.1 shows a simple check valve. In this particular valve, the control mechanism is a ball. The spring shown is very "light" and is used only to bias the ball toward the closed position. Normally, a pressure less than 5 psi is required to open the valve against the spring.

While a check valve is technically a two-way, two-position DCV, the graphic symbol that is almost universally accepted is not what one would expect for such a valve (see Figure 7.1). The "V" represents the seat, while the circle represents the control mechanism, regardless of the actual device used. The symbol is drawn so that the forward flow can be imagined to push the mechanism off the seat while reversed flow seats the mechanism. Although the valve includes a small bias spring to keep the valve seated when the system is not operating, the spring is commonly omitted from the symbol. This type of valve can be compared to a funnel with a steel ball in it. The ball prevents the flow of fluid from the spout end of the funnel.

The simple check valve shown in Figure 7.1 has many applications, some

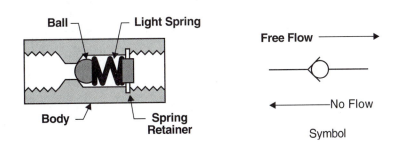

Figure 7.1 An in-line check valve may use a ball, as shown here, or a poppet.

of which are shown in Figure 7.2. It can be used simply to facilitate maintenance, such as when it is used in a pressure line from a pump to prevent fluid
loss from the system while the pump is being replaced. Or it can be used to
bypass other components (usually pressure or flow control valves)—that is, to
allow flow in the reverse direction to go around these valves. A third application
is to act as a safety bypass—that is, to provide an alternative flow path around
filters or heat exchangers in case the pressure differential across these devices
exceeds the acceptable level. The symbols for these safety bypass valves normally include a spring to indicate that they require more pressure to open
than a normal check valve. The opening pressure can be specified. It may be
as low as 10 to 15 psi, but it seldom exceeds 100 psi.

Check valves can also be used to provide a positive lock for holding suspended
loads in place. The simple check valve of Figure 7.1 could not be used for this
application, however, because it allows flow in one direction only. This means
that the load could be raised, but it could never be lowered. To overcome this
problem, a pilot-operated check valve (Figure 7.3) must be used to allow reverse
flow through the valve. This valve actually consists of two sections—a simple
check valve (shown here as a poppet type) and a plunger similar to a small,
single-acting hydraulic cylinder. In the free flow direction, the poppet is pushed
off its seat and allows fluid to pass through the valve. Reverse flow seats the
poppet as in a normal check valve to stop the flow. The difference in function
occurs when pressure is applied at the pilot port. This causes the plunger to
extend and push the poppet off its seat. The unseated poppet then allows fluid
to flow "backward" through the valve so that the load can be lowered. The
dashed line in the symbol represents the pilot line.

Figure 7.4 shows a typical application of a pilot-operated check valve. In
this circuit, the cylinder is extended normally. When the cylinder stops, the
load "sits" on the check valve. When the cylinder is to be retracted, the pressure
signal to the blind end of the cylinder is sensed through the pilot line. The
check valve is pushed open, and fluid is allowed to flow from the blind end of
the cylinder so the load can be lowered.

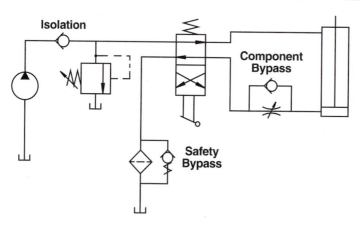

Figure 7.2 A simple check valve has many applications.

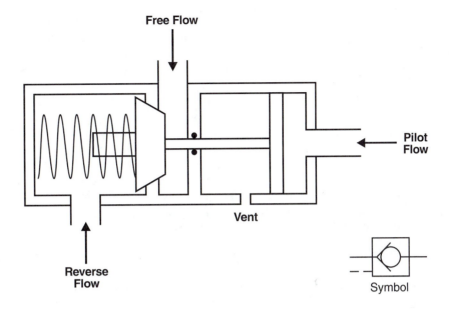

Figure 7.3 A pilot-operated check valve allows reverse flow. This valve is called a pilot-to-open valve because the pilot signal is used to open the valve.

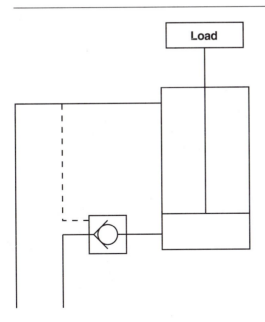

Figure 7.4 In a typical application of a pilot-operated check valve.

The valve shown in Figure 7.3 is called a *pilot-to-open* check valve because the pilot signal is used to open the valve. There is also a *pilot-to-close* check valve that uses the pilot signal to close the valve and prevent flow through it in what is normally the free flow direction. This type valve, shown in Figure 7.5, is used in a very few special applications.

Another variation on the basic check valve is the *shuttle* valve shown in Figure 7.6. This device is basically two check valves with a single ball. There are two inlet ports and a single outlet port. The ball is positioned by the relative pressures at the two inlet ports. The higher pressure will cause the ball to close off the opposite port. This allows flow from the higher pressure port to pass through the valve and out the outlet port. Shuttle valves are commonly found in emergency circuits in aircraft hydraulic systems. If pressure is lost in the primary system so that the emergency system must be used, the pressure in the emergency system shifts the shuttle valve to close off the primary system port.

7.4 Two-Way Valves

Two-way directional valves usually serve as shutoff valves, either allowing no flow or full flow, depending upon the position of the valve. As the designation suggests, these valves have two ports. Two-way valves can be termed either *normally open* or *normally closed*. The term applied to the valve indicates whether it allows flow (normally open) or does not allow flow (normally closed) when it is in the unactuated position. Figures 7.7 and 7.8 use conceptual drawings to distinguish between the two.

In Figure 7.7a, the normally closed valve is shown in its normal, or unactuated, position, where the spring is holding the spool in a position that blocks the flow. The spool land is separating the pressure (or inlet) port, P, from the

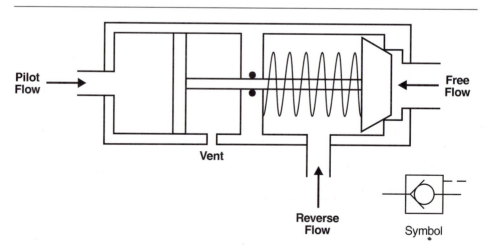

Figure 7.5 A pilot-to-close check valve is used in a very few special applications.

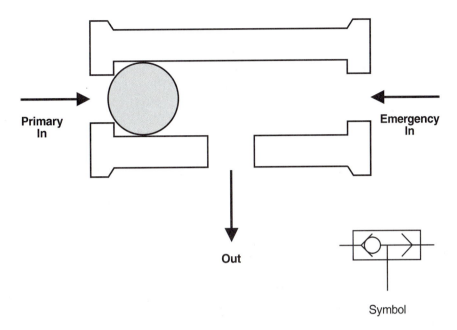

Primary
In

Emergency
In

Out

Symbol

Figure 7.6 The shuttle valve is basically two check valves with a single ball. It is commonly used in emergency circuits in aircraft hydraulic systems.

working (or outlet) port, A. In this "normal" position, there is no flow through the valve. Symbolically, this position is shown by a single valve position envelope containing two blocked ports.

The actuated position of the valve is shown in Figure 7.7b. In this figure, the spool has been pushed to the right so that the spool land is past the A port, providing a flow path from the P port to the A port. This is shown symbolically by a single valve position envelope with an arrow connecting the P port to the A port, indicating the open flow path. When the lever is released, the spring returns the spool to its original position.

To get the full valve symbol, we put together the two envelopes that represent the two spool positions and tack on the return spring symbol and the actuator symbol as shown in Figure 7.7c. Note that the conduit attachments and port designators are associated with the block representing the normal, or unactuated, spool position.

A normally open valve is shown in Figure 7.8a. Here we see that the inlet port (P) and the outlet port (A) are connected when the valve is in its unactuated position. The graphic symbol for this position is exactly the same as for the open position in the normally closed valve, with the arrow indicating the open flow path.

If we push the lever, the spool land moves to block the opening between ports P and A, closing the valve (Figure 7.8b). This closed valve position is symbolically the same as the closed position for the normally closed valve. When we release the lever, the spring opens the valve.

The complete symbol for the valve is shown in Figure 7.8c, which incorporates the two valve position symbols, the return spring and the actuator. Again,

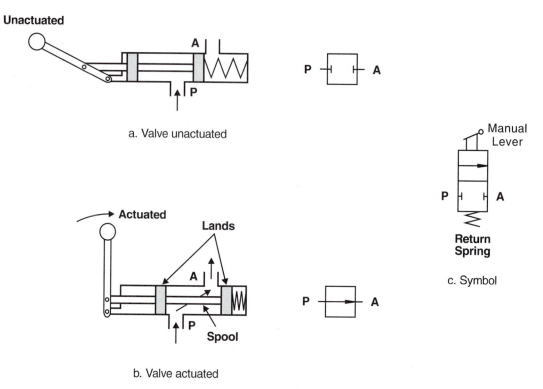

a. Valve unactuated

b. Valve actuated

c. Symbol

Figure 7.7 Compare this two-way normally closed valve with the normally open valve in Figure 7.8.

note the location of the connecting fluid lines and port designators. These are important in verbally describing the valve.

Speaking of describing valves, you should, by now, be catching on to the idea of valve designations based on the number of ports (excluding gage and drain ports), positions, and so forth. For instance, the full description of the valve in Figure 7.7 is "two-way, two-position, normally closed, manually actuated, spring-returned, directional control valve."

How would you describe the valve of Figure 7.8? As with the previous valve, it is a directional control valve with two connecting ports and two distinct positions. It also has a return spring and a manual actuator. The only difference is that it is open in its unactuated position. Therefore, we would term it a "two-way, two-position, normally open, manually operated, spring-returned, directional control valve."

Figure 7.9 shows a normally closed, two-way valve being used to start and stop a hydraulic motor. The valve spool is held in its closed position by the spring (Figure 7.9a). A push of the lever moves the spool to a position that allows flow from the P port to the A port, and hence to the motor (Figure 7.9b). The motor will operate as long as the spool is held in this position. As soon as the lever is released, the spring will push the spool back to its closed position.

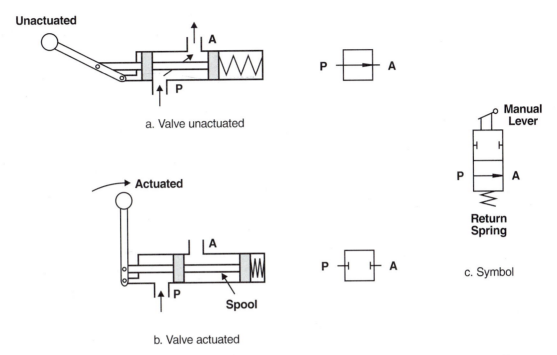

Figure 7.8 Compare this two-way normally open valve with the normally closed valve in Figure 7.7.

This stops the flow to the motor. Remember from Chapter 1 that valves are always shown in their unactuated position, so the drawing in figure 7.9a is the correct way to draw the circuit. Figure 7.9b is drawn only to illustrate the functional concept. (Note, by the way, that this is not a good application for two-way valves, but it serves to illustrate the valve operation.)

7.5 Three-Way Valves

Three-way valves have three connecting ports. The third port provides a return flow path to the reservoir or tank. As with the two-way valves, they can be normally open or normally closed.

In Figure 7.10a, we see a normally closed three-way valve. In the unactuated position, pressure port P is separated from the tank and working ports (T and A, respectively) by the spool land. While there is a flow path (A to T) there is no path for fluid from the pressure line, hence the designation "normally closed." The symbol for this position shows port P blocked and port A connected to port T by the arrow.

When the valve is actuated, a flow path is provided between the pressure port and the working port, while the spool land separates those two ports from the tank port. The symbol for this position shows the tank port blocked, while port P is connected to port A by an arrow (Figure 7.10b).

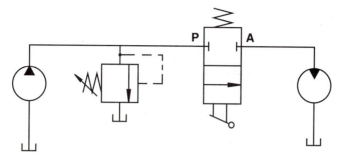

a. Valve unactuated.

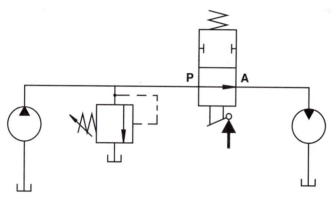

b. Valve actuated.

Figure 7.9 This two-way, two-position valve is shown being used to operate a hydraulic motor for illustrative purposes only, since it is not an ideal application.

The complete graphic symbol for a three-way, two-position, normally closed, manually actuated, spring-returned, directional control valve is shown in figure 7.10c. The symbol for a normally open valve is shown in Figure 7.11.

Figure 7.12 illustrates a circuit using a three-way, two-position DCV to control the operation of a single-acting, spring-returned cylinder. In the unactuated position, the spring holds the spool in the position shown. This allows fluid flow from the P port to the A port to cause the cylinder to extend. When the valve is actuated, the spool shifts (to the left in this drawing). This connects the A port to the T (tank) port. The flow path allows the fluid from the blind end of the cylinder to be pushed out of the cylinder and back to the tank as the spring retracts the cylinder. Releasing the handle allows the spring to push the spool back to the unactuated position so that the cylinder can extend again.

Three-way valves are not widely used in hydraulic applications, but they are used frequently in pneumatic systems. Few hydraulic valve manufacturers

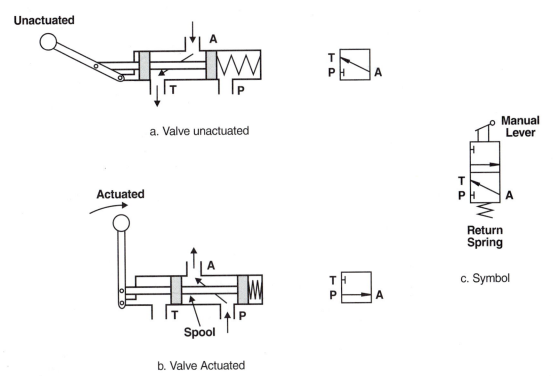

a. Valve unactuated

b. Valve Actuated

c. Symbol

Figure 7.10 A three-way valve has three connecting ports. This is a normally closed type.

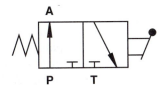

Figure 7.11 A three-way, two-position, manually operated, spring-return, normally open DCV.

produce three-way valves; therefore, if a three-way valve is needed, the common practice is to block one port (either A or B) on a four-way valve, discussed next.

7.6   Four-Way Valves

The transition from three-way to *four-way valves* involves adding another port to the three with which we are already familiar. This allows both working and

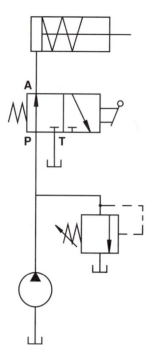

Figure 7.12 Circuit using a two-position, three-way DCV.

return flow through the same valve position. Figure 7.13 shows a cutaway drawing of a four-way, two-position directional control valve and the graphic symbols used to represent it.

Figure 7.13a shows the valve with the spool shifted to the left-handed position. Here we see that the pressure (P) port and the A port are connected by a flow path. Likewise, the B port and the tank (T) port are connected. When the valve is shifted to the right (as in figure 7.13b), the pressure port is connected to the B port, while the A port is connected to the tank port.

Figure 7.14 shows a typical application for this valve. The spring pushes the spool to its unactuated position, in which P is connected to A, and B is connected to T. In this position, the pressure flow to the blind end of the cylinder causes the piston to extend. The fluid from the rod end is exhausted through port B to port T and back to the reservoir.

Actuating the valve (sliding the spool to the left) opens a flow path from P to B that directs the flow to the rod end of the cylinder. This causes the piston to retract. The fluid that is exhausted from the blind end returns to the tank through ports A and T. When the lever is released, the spring returns the spool to its unactuated position and causes the piston to extend once again.

In describing the valve in Figure 7.14, we again must include all actuators and springs. Thus, we describe this valve as a four-way, two-position, lever-operated, spring-returned DCV.

Our analysis of the circuit in Figure 7.14 shows that this two-position valve provides no ability to stop the cylinder in any intermediate position. While we

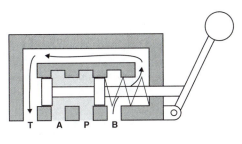

a. Valve unactuated

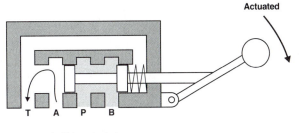

b. Valve actuated

Figure 7.13 Cutaway of a four-way, two-position directional control valve showing flow through the valve in each spool position.

can reverse its direction, we can stop it only at the end positions. The only way to stop the cylinder in any other position is to use a three-position valve that incorporates a center position that blocks off the A and B ports. Figure 7.15 shows one such valve. The center position of this valve is termed "closed" because in that position, all four ports are blocked and there is no flow through the valve. We would describe this as a four-way, three-position, closed-center, lever-operated, spring-centered (note the centering spring on each end) directional control valve.

The unactuated (or normal) position of the valve in figure 7.15 is the center position. (This is usually the case with three-position valves.) The centering springs always push the spool to the center position when the lever is released. Thus, the valve symbol is drawn to show the fluid lines connected to the center envelope.

In the circuit shown in Figure 7.16, shifting the valve to the right allows flow from the P port to the A port to the blind end of the cylinder. As the cylinder extends, fluid from the rod end passes through the B port to the T

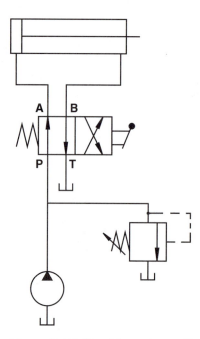

Figure 7.14 Double-acting cylinder with a four-way, two-position, lever-operated, spring-returned DCV.

port and back to the tank. Releasing the lever at any time allows the springs to center the spool. The blocked ports stop all flow through the valve. Consequently, the cylinder stops and is held in position.

When the spool is shifted to the left, fluid will flow from the P port to the B port and out to the rod end of the cylinder, causing it to retract. Flow from the blind end of the cylinder passes through the A port to the T port and back to the reservoir.

The configuration of the center position in a three-position valve is selected based on the system design and performance requirements. Figure 7.17 shows several of the more popular center configurations. The *closed center,* which we have already discussed, locks the actuator in position while also blocking the pressure and tank ports. The blocked pressure port means the system pressure is available right up to that valve port.

Another advantage of the closed-center valve is its ability to be used in a *parallel circuit,* meaning one in which the flow is (or can be) split between two or more branches. Figure 7.18 shows a parallel circuit with three branches. With closed-center valves, any one branch can be operated without actuating the other two valves. Actually, more than one branch can be operated at a time, but pressure and flow considerations must be taken into account. An application example is a backhoe in which the bucket tilt, lift, and swing can be operated either individually or simultaneously.

A drawback to this concept is that the pump is constantly operating against full system pressure (as set by the relief valve or pump compensator). (This is

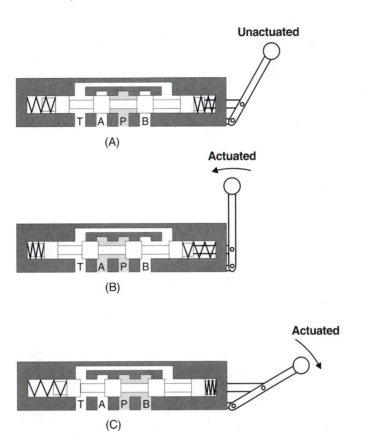

Figure 7.15 Cutaway of a four-way, three-position directional control valve showing the flow paths in each spool position.

often compared to sitting at a stoplight with one foot on the accelerator and one foot on the brake.) If pump flow is being dumped across the relief valve at a high pressure drop, considerable energy is wasted, and this wasted energy will be converted to heat in the fluid. The energy loss can be alleviated by connecting the pressure port to the tank port using a *tandem center* or an *open center* (see Figure 7.17 again).

A cutaway drawing of a tandem center valve is shown in Figure 7.19. The spool configuration for this valve is such that the A and B ports are blocked by the spool lands in the center position, as in the closed-center valve. This locks the cylinder in place. The P and T ports, however, are connected through internal porting, which allows the pump flow to pass through the valve at low pressure (usually less than 100 psi) so that the pump is unloaded when the valve is not actuated. Shifting the spool connects the ports to cause the cylinder to extend or retract.

When a tandem center is used, as in Figure 7.20, the actuator is still locked, but now the pump flow passes through the valve and returns to tank at low

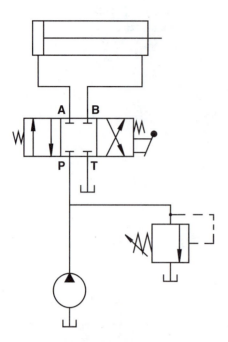

Figure 7.16 Double-acting cylinder circuit with a four-way, three-position, lever-operated, spring-centered, closed-center DCV.

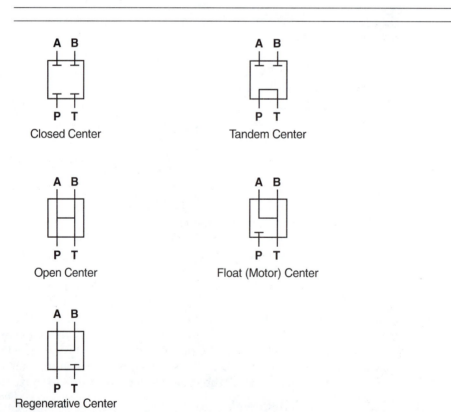

A B

Closed Center

A B

Tandem Center

A B

Open Center

A B

Float (Motor) Center

A B

Regenerative Center

Figure 7.17 Directional control valve center configurations.

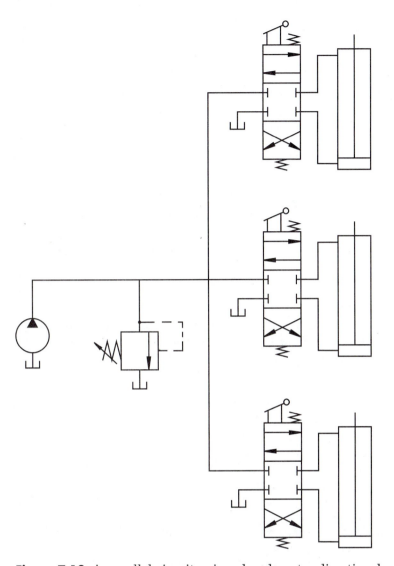

Figure 7.18 A parallel circuit using closed-center directional control valves.

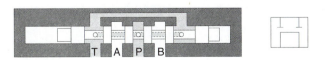

Figure 7.19 Cutaway drawing of a tandem-center directional control valve.

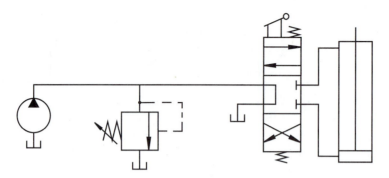

Figure 7.20 A cylinder circuit using a tandem-center directional control valve.

pressure. This allows the system pressure to be maintained at a relatively low level (usually around 100 psi). The resulting energy savings can be significant as we'll see in Example 7.1. One disadvantage of this type of center, though, is that there is a brief time delay before pressure is available to the actuator when the valve is shifted.

Another, perhaps more significant disadvantage of tandem center valves is that they cannot be used in parallel circuits. A quick analysis of Figure 7.21 will show why this is so. Let's say that the spool of valve A is shifted to direct fluid to the blind end of cylinder A and open a flow path to the reservoir for the exhaust flow from the rod end. Will the cylinder move? Probably not, especially if there is a load on the cylinder. Why? Remember that the maximum pressure depends on the minimum resistance to flow. Since valve B is still centered, there is a very low resistance to flow because the pump flow can return to the reservoir through the tandem center of valve B. Therefore, the entire circuit is at low pressure.

Tandem center valves can be used in *series circuits,* however; that is, those in which all flow passes through all the valves before it returns to tank. In the series circuit in Figure 7.22 either valve A or valve B can be actuated to operate its cylinder. Other than the small pressure drop as the fluid passes through the center position of the unactuated valve, there are no problems with this operation. If both valves are actuated at the same time, some interesting things happen. When both valves are shifted to the right, cylinder B will be extended by the flow from cylinder A. The distance that cylinder A can extend, as well as its extension velocity, will depend on the flow leaving the blind end of cylinder A. The pressure required to extend cylinder A will depend on the loads on both cylinders, because the pressure in cylinder A must overcome not only its own load, but also the back-pressure that results from the load on cylinder B.

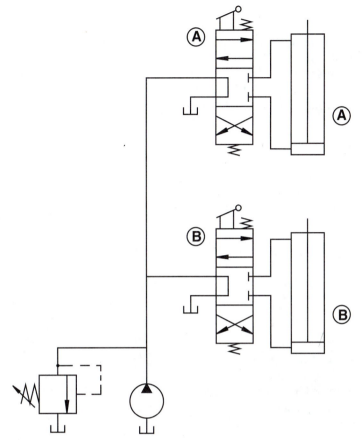

Figure 7.21 The use of tandem-center valves in a parallel circuit is not practical.

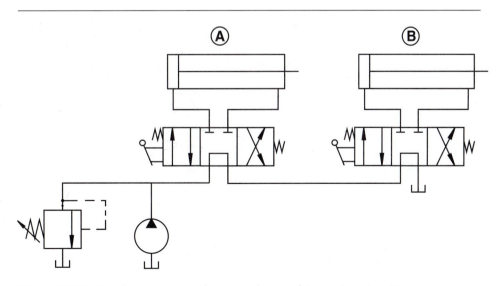

Figure 7.22 Tandem-center valves can be used in series circuits.

Example 7.1

Compare the horsepower requirements of a closed-center and a tandem-center directional control valve in their center positions. The circuit uses a pressure relief valve set at 1500 psi (10,300 kPa) and a fixed displacement pump that produces 10 gpm (37.85 Lpm). The pressure drop through the tandem center is 100 psi (690 kPa) at 10 gpm (37.85 Lpm). Assume no losses between the pump and the valve.

Solution

With the closed-center valve in its center position, the full flow of the pump is dumped across the relief valve at 1500 psi (10.3 MPa). From Chapter 4, we know that the horsepower loss across the relief valve is based on the pressure drop (Δp) across the valve and the flow rate through the valve; therefore, we get

$$HP_{\text{LOST}} = \frac{\Delta p_{\text{RV}} \times Q_{\text{RV}}}{1714}$$

$$= \frac{1500 \times 10}{1714}$$

$$= 8.75 \text{ HP}$$

When a tandem-center valve is used, there is still a horsepower loss. This time, however, the tandem center provides a low-pressure flow path back to the tank for the entire pump flow. We still find the horsepower loss from the same equation, but now we use the pressure drop and flow through the tandem center of the DCV instead of the relief valve. This gives us

$$HP_{\text{LOST}} = \frac{\Delta p_{\text{DCV}} \times Q_{\text{DCV}}}{1714}$$

$$= \frac{100 \times 10}{1714}$$

$$= 0.58 \text{ HP}$$

Solving this problem in SI units, we find the power loss for the closed-center valve to be

$$P_{\text{LOST}} = \frac{\Delta p_{\text{RV}} \times Q_{\text{RV}}}{60,000}$$

$$= \frac{10,300 \text{ kPa} \times 37.85 \text{ L/min}}{60,000}$$

$$= 6.5 \; kw$$

The loss through the tandem center is

$$P_{\text{LOST}} = \frac{\Delta p_{\text{DCV}} \times Q_{\text{DCV}}}{60,000}$$

$$= \frac{690 \text{ kPa} \times 37.85 \text{ Lpm}}{60,000}$$

$$= 0.44 \text{ kW}$$

In addition to the energy and cost savings that result from reduced horse-power loss in the system, there is a long-term benefit associated with the fluid itself. Recall from Chapter 3 that the oxidation rate of the oil is aggravated by high fluid temperature. Also recall that inefficiencies (losses) in the system result in fluid heating.

One of the major causes of fluid heating is pressure drops across the valves. Relief valves are usually the primary source of heating because of the high pressure drop as the fluid goes from system (relief valve) pressure setting to tank pressure. This generates heat in direct proportion to the pressure drop. We can calculate the heat generation rate (HGR) across any pressure drop where no work is done using Equation 7.1.

$$\text{Heat Generation Rate} = \text{Horsepower Loss} \times 42.4 \text{ Btu/min} \qquad \textbf{(7.1)}$$

$$HGR = HP_{\text{LOST}} \times 42.4 \text{ Btu/min}$$

The 42.4 Btu/min is the heat equivalent of horsepower.

In the SI system, the heat generation rate is expressed in watts or kilowatts and is exactly the same as the hydraulic power loss across the valve.

Example 7.2

Find the heat generation rate for the closed-center and tandem-center arrangement of Example 7.1.

Solution

We use Equation 7.1 for both valves. For the closed-center valve,

$$HGR = HP_{\text{LOST}} \times 42.4 \text{ Btu/min}$$

$$= 8.75 \times 42.4 \text{ Btu/min} = 371 \text{ Btu/min}$$

For the tandem-center valve, center of the DCV:

$$HGR = 0.58 \times 42.4 \text{ Btu/min} = 24.6 \text{ Btu/min}$$

In SI units, the heat generation rate equals the power loss.

We can also calculate the temperature rise (ΔT) in the two arrangements. To do this, we use

$$\text{Temperature Rise} = \frac{\text{Heat Generation Rate}}{\text{Specific Heat} \times \text{Weight Flow Rate}}$$

$$\Delta T = \frac{HGR}{c_p \times W} \qquad \textbf{(7.2)}$$

Here, c_p is the specific heat of the oil. This defines the energy absorbing capability of the fluid per unit mass per unit of temperature change. For hydraulic oils, c_p is usually between 0.45 and 0.5 Btu/(lb°F), or 1.9 to 2 kJ/(kg°C).

The weight flow rate (W) defines the weight of fluid passing a certain point in a given time. It is calculated by multiplying the volume flow rate (Q) by the specific weight (γ), or

$$\text{Weight Flow Rate} = \text{Specific Weight} \times \text{Volume Flow Rate}$$

$$W = \gamma Q \qquad \textbf{(7.3)}$$

Example 7.3

Determine the fluid temperature rise resulting from the two valve arrangements in Example 7.1. The specific heat of the oil is 0.45 Btu/(lb°F), or 1.9 kJ/(kg°C), and the specific weight is 58.7 lb/ft³, or 939.6 kg/m³.

Solution

For the closed-center valve, the flow across the relief valve must be converted to a weight flow rate using Equation 7.3.

$$W = \gamma Q$$

$$= 58.7 \text{ lb/ft}^3 \times 10 \text{ gal/min} \times \text{ft}^3/7.48 \text{ gal}$$

$$= 78.5 \text{ lb/min}$$

Using this value in Equation 7.2, we get

$$\Delta T_{RV} = \frac{HGR}{c_p \times W}$$

$$= \frac{371 \text{ Btu/min}}{0.45 \text{ Btu/lb°F} \times 78.5 \text{ lb/min}}$$

$$= 10.5°F$$

Using SI units for these calculations gives us

$$W = \gamma Q$$

$$= 939.6 \text{ kg/m}^3 \times 37.85 \text{ L/min} \times \text{m}^3/1000 \text{ L}$$

$$= 35.6 \text{ kg/min}$$

So that

$$\Delta T_{RV} = \frac{HGR}{c_p \times W}$$

$$= \frac{6.5 \text{ kW}}{(1.9 \text{ kJ/(kg°C)}) \times 349 \text{ N/min} \times \text{min/60 sec}}$$

$$= \frac{6.5 \text{ kW} \times \text{kJ/kW} \cdot \text{sec}}{(1.9 \text{ kJ/(kg°C)}) \times 35.6 \text{ kg/min} \times \text{min/60 sec}}$$

$$= 5.8°C$$

This means that the temperature of every drop of oil that passes through the relief valve is increased by 10.5°F, or 5.8°C, in the time required to go through the valve mechanism.

The fluid going through the tandem-center valve also experiences a temperature rise. The weight flow rate is the same as that through the relief valve, so Equation 7.2 gives

$$\Delta T_{DCV} = \frac{HGR}{c_p \times W}$$

$$= \frac{24.6 \text{ Btu/min}}{0.45 \text{ Btu/lb°F} \times 78.5 \text{ lb/min}}$$

$$= 0.7°F$$

or 0.39°C.

The reduced heating rate of the tandem center has many benefits—longer fluid life due to a reduced oxidation rate; better fluid viscosity, resulting in reduced pump wear; and reduced additive depletion, to name a few. The fact that the pump is operating at a lower pressure also means that pump wear is reduced.

A primary rule in fluid power system life and reliability concerns the fluid— keep it clean and keep it cool. In Chapter 11, we discuss keeping the fluid clean. In Chapter 13, we discuss the use of large reservoirs and heat exchangers to keep it cool. Bear in mind, however, that it is far better not to generate the heat in the first place than to use devices to remove the heat after it has been generated. As we've seen in these examples, simply choosing a tandem-center DCV instead of a closed-center DCV might prevent heat generation, assuming all other system requirements can be met.

As mentioned earlier, an open-center valve such as the one shown in Figure 7.23 can also be used to unload the pump when the system is idling; however, as the figure shows, the A and B ports are connected to the tank port, as is the pressure port. Thus, while the open center will unload the pump, it will not lock the cylinder in place. As a result, if the cylinder is not horizontal, any load will very likely drive the cylinder to its fully extended or fully retracted position (depending on the cylinder orientation) because of the low-pressure flow path back to the tank. If there is no load on the cylinder, the piston may drift to its fully extended position if the system pressure is high enough to overcome the friction of the piston and rod seals. Figure 7.24 is the circuit diagram for a cylinder circuit using an open-center valve. Again, because of the spring centering, all lines are shown connected to the center position.

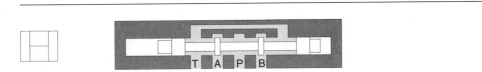

Figure 7.23 Cutaway drawing of an open-center directional control valve.

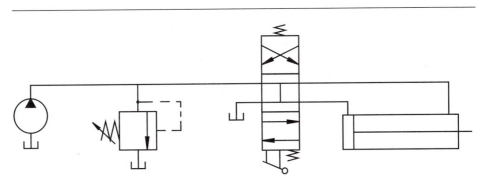

Figure 7.24 Circuit diagram for an open-center directional control valve.

Figure 7.25 shows a float center. In the center position of the *float-center* valve in Figure 7.17, note that the A and B ports are connected internally to the T port, as in the open-center valve. Here, however, the P port is blocked. As in any case where the P port is blocked, all pump flow goes back to the tank through the relief valve, so the pump is not unloaded when the valve is centered. As with the open-center valve, a cylinder cannot be locked in place because of the low-pressure flow path to the tank.

This valve configuration is frequently termed *motor center* because it is commonly used to control a hydraulic motor (Figure 7.26). When the spool is shifted to the left, the motor will rotate in one direction. Shifting the spool across center to the opposite position will cause the motor to reverse and rotate in the opposite direction. When the spool is centered, both of the motor ports are connected to the tank line. This means that the motor will not be stopped, but can "freewheel" and coast to a stop. This configuration results in much less shock loading on the motor side of the valve than there would be with closed-and tandem-center valves, but it provides no capability to stop and hold a load at a precise point.

In Chapter 6 we discussed the concept of regeneration with hydraulic cylinders. Because we hadn't yet talked about valves at that time, we just did a little hand waving and said we needed some valve to give us the capability to retract the cylinder. A valve with the *regeneration center* shown in Figure 7.17 gives us that capability. The cutaway drawing of this valve in Figure 7.27 shows how, in the center position, the pressure port is connected to both A and B ports while the tank port is blocked. As shown in Figure 7.28, this provides

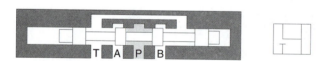

Figure 7.25 Cutaway drawing of a float-center directional control valve.

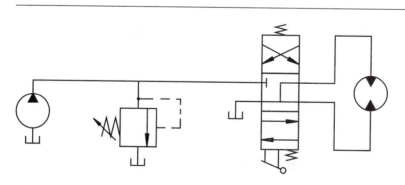

Figure 7.26 A float-center valve used in a hydraulic motor circuit.

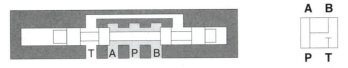

Figure 7.27 Cutaway drawing of a regenerative-center directional control valve.

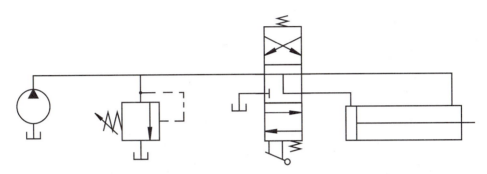

Figure 7.28 A regenerative-center directional control valve circuit.

the regenerative part of the circuit by connecting the pump pressure line and the return line from the rod end of the cylinder and running both into the blind end of the cylinder. If we shift the spool to the right, we get a P–A, B–T combination causing the cylinder to extend in the normal manner (providing lower speed, but higher force). To retract the cylinder, we shift the spool to the left for the P–B, A–T combination.

In order to achieve regeneration with this valve, we have had to sacrifice the ability to stop (or at least depressurize) the cylinder in mid-stroke. An alternative circuit that allows us to use regeneration while retaining the ability to stop and hold the load is shown in Figure 7.29. Here, a closed-center valve is used. The B port is blocked externally, and a tee is used in the pressure line to tie in the return from the rod end of the cylinder.

With the valve in the center position, pressure is applied to the annular piston face on the rod end, but, because the A port is blocked by a spool land, the fluid cannot be exhausted from the blind end. Therefore, a hydraulic lock holds the cylinder in place. Shifting the spool to the right completes the P–A flow path and causes the cylinder to extend. Flow from the rod end is forced into the pressure line and provides the regenerative speed.

To retract the cylinder, shift the spool to the left. This connects the P–B and A–T flow paths, but the B port is blocked externally. This P–B flow path is not needed to retract the cylinder, however, because flow is already going to the rod end from the tee in the pressure line. The A–T flow path allows the blind end fluid to be exhausted, so the cylinder will retract.

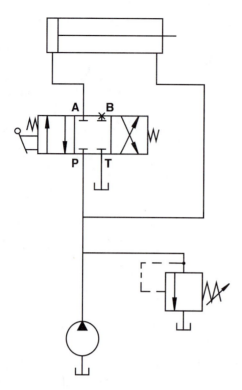

Figure 7.29 In this alternative regenerative circuit that allows the cylinder to be stopped and held in mid-stroke, port B is blocked.

While the circuit of Figure 7.29 allows us to stop the cylinder in mid-stroke, it eliminates the option to operate the cylinder at either the normal or regenerative speed as we could with the regenerative-center spool. The circuit in Figure 7.30 gives us both capabilities—mid-stroke stop and hold, and two-speed extension. With the three-way, two-position valve in its unactuated position, the cylinder extends and retracts normally, because the rod end is connected to the B port of the closed-center DCV. Shifting the three-way valve to the left blocks the line from the B port and connects the rod end of the cylinder to the tee in the pressure line to give regenerative extension speed as in the circuit of Figure 7.28. Normal retraction occurs with the three-way valve shifted.

A word of CAUTION is appropriate at this point. We've talked about the ability of closed- and tandem-center valves to stop and hold a load. While they will theoretically "hold" the load, in fact they are unlikely to do so for any heavy load unless the cylinder is nearly horizontal. A closed-center spool will hold as long as the pump is operating, but once the pump is stopped, the load will probably creep downward. This is not a fault of the valve, but an inevitable physical fact. For the valve spool to be moved within the bore of the valve body, there must be some radial clearance, however small. This is shown exaggerated in Figure 7.31. Due to the load on the cylinder, there is a high pressure in the blind end of the cylinder which, from Pascal's Law, we know will also exist in

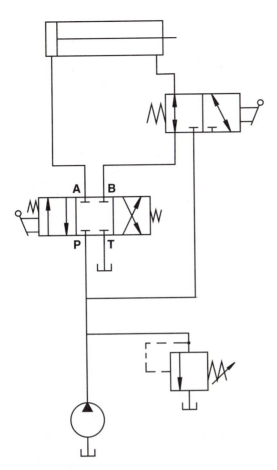

Figure 7.30 The addition of a three-way, two-position DCV in the rod-end line provides both regenerative and normal extension speed capability, as well as the ability to stop in mid-stroke.

the line all the way to port A. The tank port, however, is at low pressure. The result is leakage from port A to port T and a subsequent lowering of the load.

The rate at which the load will come down depends on the magnitude of the pressure and the size of the clearance. The size of the clearance depends on the quality of the valve and on the amount of wear that has occurred due to contaminated oil.

Remember, any time there is a clearance and a pressure differential, there will be leakage, and this leakage could represent a safety hazard. This is where the pilot-operated check valve comes in. See Figure 7.32, where it provides a safety lock for a suspended load. With the DCV in the center position, the load "sits" on the check valve. When the DCV spool is shifted upward, pump flow from P to B raises the load, while the flow from the blind end is exhausted through A–T. When the spool is shifted downward, pump flow goes to the blind end through P–A, but because the check valve is seated, the load cannot go

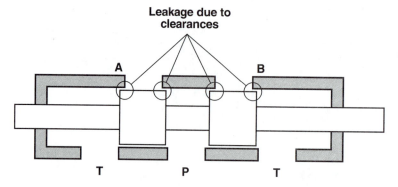

Figure 7.31 Spool clearances allow flow from A and B ports to T. This allows the load to drift.

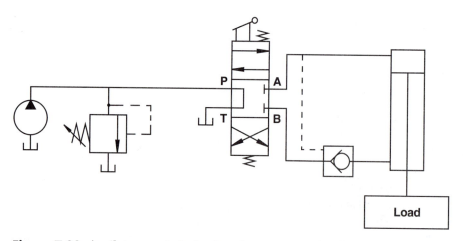

Figure 7.32 A pilot-operated check valve prevents load creep due to leakage through a closed-center DCV.

down. This results in a buildup of pressure in the line from port A to the cylinder. The increased pressure is sensed through the sensing line (the dashed line in the diagram) and fed into the plunger in the check valve. The extended plunger pushes the check valve open and allows the load to be lowered.

Another point that we need to discuss here is the concept of open- and closed-center *systems*—not to be confused with open- and closed-center *valves* or open and closed *loop* systems. Simply stated, an *open-center system* is any system where the pump is unloaded when the system is idle (as with open- and tandem-center valves). A *closed-center system* is any system where the pump is loaded—that is, operating at high pressure—when the system is idle (as with closed- and float-center valves, as well as any valve that connects the pressure port to either the A or B port, or both, when the system is idle). This can include two-position as well as three-position valves.

An *open loop system* is one in which the fluid exhausted from the actuators is returned to a reservoir before it is recirculated by the pump. A *closed loop system* may have a reservoir, but the exhaust fluid from the actuator (usually a hydraulic motor) is the inlet flow to the pump—as in a hydrostatic transmission. Note that the terms open loop and closed loop are also frequently used in reference to electronic feedback systems.

7.7   Deceleration Valve

A *deceleration valve*, shown in Figure 7.33, is used to provide automatic speed control for a cylinder (which leads many people to classify it as a flow control valve rather than a directional control valve). The plunger-like spool in the valve is shaped so that it changes the size of the flow path through the valve as the spool moves in the bore. This has the same effect on flow through the valve as turning the handle on the faucet in the sink. Increasing the size of the flow path increases the flow, while decreasing the size reduces the flow.

The valve is positioned so that a cam rides over the mechanical actuator which is an extension of the spool. As the cam pushes the spool into the valve, flow to the cylinder is decreased, and the cylinder slows. the follower can be shaped to allow the cylinder speed to be varied throughout its stroke.

Because of the special nature of this valve, a variety of symbols can be used to represent it. The two shown in Figure 7.33 are the most common. One depicts the valve as a two-way, two-position, normally open DCV. The lines parallel to the symbol indicate that the valve can be positioned anywhere between the fully open and the fully closed positions. The dashed line from the valve body through the spring to the reservoir symbol represents a drain line. It is necessary to drain the spring cavity of any leakage that gets past the spool. If the fluid was allowed to accumulate in the spring cavity, the motion of the spool could be restricted. The end result would probably be damaged equipment because the cylinder would try to continue to move the follower over the unyielding valve stem.

The second symbol shows only a single envelope in which the flow path is normally open. The symbol shows that the mechanism is held open by the spring and closes when the mechanical actuator is activated. The spring cavity drain is also shown.

Figure 7.34 shows a deceleration valve placed to vary the extension speed of the cylinder. As the cylinder extends, the cam contacts the valve stem and pushes it down, restricting flow to the blind end and slowing the cylinder. The check valve allows the cylinder to retract at full speed, even though the deceleration valve may be partially, or even fully, closed.

7.8   Valve Actuator Symbols

Figure 7.35 shows the symbols for the most common *valve actuators*. We've seen some of these symbols already, especially the hand lever. All of the top three symbols represent manual actuators. The symbol to the left of these is

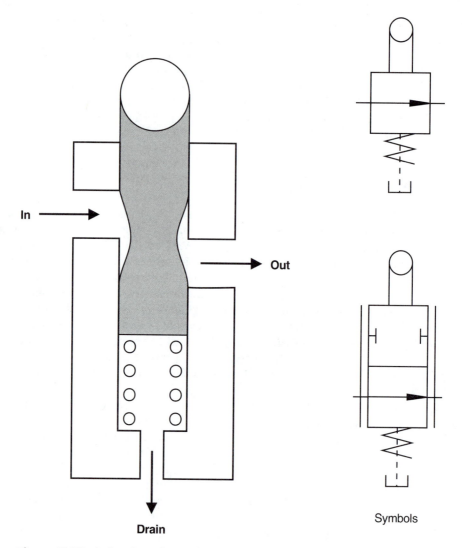

In

Out

Drain

Symbols

Figure 7.33 A deceleration valve can be represented by a variety of symbols. Two of the most common are shown here.

the generic symbol for a manual actuator. The pilot actuators use either hydraulic or pneumatic pressure to shift the valve spool. Omitting the pilot symbol is a commonly used simplification when there is no ambiguity that could cause the person reading the circuit to misinterpret the symbol.

The spring symbol and the detent symbol represent exactly opposite actions. The spring returns the spool (or other mechanism) to its unactuated position. The detent, on the other hand, holds the valve in position even when the actuator is released. This is usually accomplished by using a spring-loaded ball that drops into a notch on the spool stem. Remember that for the purpose of drawing symbols, springs always push.

Follower or cam

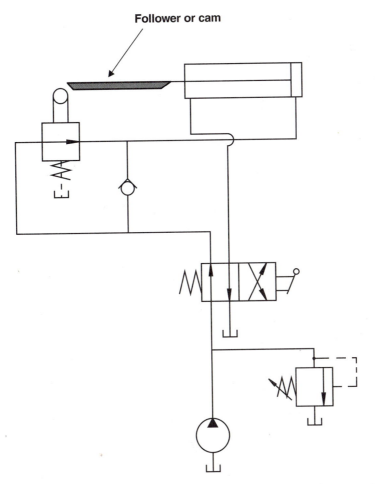

Figure 7.34 A deceleration valve can be used to change cylinder speed automatically.

The solenoid, proportional solenoid, and servo actuator are all electrically operated. These will be covered in detail in Chapter 10. For now, the primary thing to keep in mind about these electrically operated actuators is that they almost always push the valve spool when they are actuated.

7.9   Pilot-Operated Valves

Pilot-operated valves employ a pressure source to move the valve mechanism in response to either a pressure buildup in some other portion of the system or the shifting of a small valve that sends a pilot signal to a larger valve. The pilot-operated check valve is an example of response to a pressure buildup.

The circuit with the pilot-operated valve shown in Figure 7.36 is a modifica-

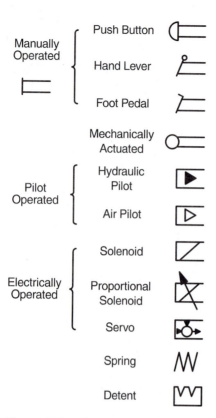

Figure 7.35 Actuator symbols for common valve actuators.

tion of the regenerative circuit shown in Fig. 7.29. The use of the pilot-operated DCV provides automatic switching between the normal and regenerative modes in response to changes in the pressure in the line leading to the blind end of the cylinder. The spring holds the valve in the regenerative position as long as the pilot pressure is low. When the pilot pressure becomes high enough to move the spool against the spring force, the spool shifts to the normal mode.

This type of operation might be applied to a log splitter. When soft wood is being split, relatively little force is required, so regeneration can be used to run the logs through quickly. If a hardwood log happens to get mixed in, or if a knot is encountered, more force will be required. The higher resistance causes an increase in pressure, which is sensed through the pilot line. This shifts the valve into the normal mode, with its slower speed but higher force capability.

Another use for pilot operation is to overcome the high resistances involved in shifting large directional valves. This is usually accomplished by mounting a small valve on the larger valve. The operator actuates the small valve, which in turn uses hydraulic pressure to move the large spool. Pilot operators are commonly used where the fluid port size on the main valve is ⅜ inch or larger.

Before we consider how a pilot-operated system works in a valve, let's look at a simple partial fluid power system consisting of a small four-way, two-

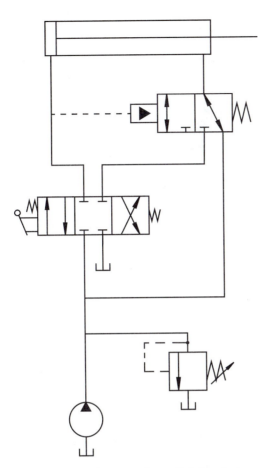

Figure 7.36 Automatic switching between the regenerative and normal modes is accomplished by using a pilot-operated directional control valve.

position, manually operated directional control valve and a double-acting, double-ended cylinder, as shown in Figure 7.37. Assume that the valve spool is pushed to the right. This causes fluid to be ported into the left end of the cylinder. The pressure creates a force on the piston and pushes the piston to the right. When the spool is shifted to the left, the cylinder moves to the left.

Now let's hook up the lines coming out of the small directional control valve into the ends of the spool chamber of a large, three-position directional control valve (Figure 7.38) rather than into the cylinder. If the spool of the small valve is again shifted to the right, fluid will be ported into the left end of the large valve, causing its spool to shift to the right just as the cylinder did earlier. This connects the main valve pressure port to port B and port A to the tank port. Shifting the small valve to the left will shift the large spool in the opposite direction. The output from this large valve can be connected to any actuator.

In this example, the small valve would be termed the pilot valve, while the larger valve would be called the main valve. The pilot valve is usually mounted

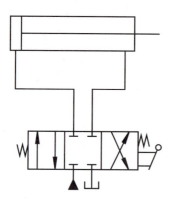

Figure 7.37 Small four-way, three-position, manually operated directional control valve and double-acting, double-ended cylinder.

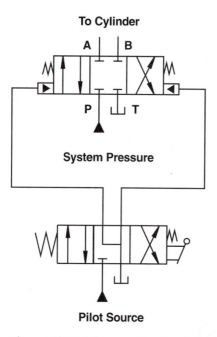

Figure 7.38 The small valve here is the pilot valve, and the large one is the main valve.

on the main valve, and fluid porting is internal to the valve body (Figure 7.39). The valve is then said to be a manually actuated, pilot-operated directional control valve. The concept can be applied to any type of directional control valve with any porting and center configuration.

Figure 7.40a shows the complete graphic symbol for a pilot-operated solenoid

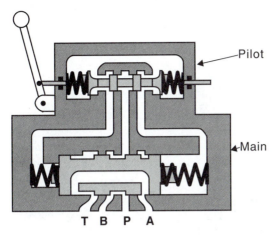

Figure 7.39 Cutaway drawing of a pilot-operated directional control valve.

valve. In the symbol, each stage is shown just as if it were a completely separate valve, which, in reality, it is. The dashed box surrounding the two symbols is the clue that they are integrated into the same housing. Because of the complexity of this composite symbol, it is seldom used. Recall the basic concept of graphic symbology that says that the symbol should indicate what the component does, but not how it does it. In keeping with this concept, the simplified symbol of Figure 7.40b would normally be used to represent the valve of Figure 7.40a. This simplified representation is really nothing more than the symbol for the main spool with the pilot input symbol added. Both the pilot spool and the main spool can be of any configuration, but only the main spool is detailed in the symbol.

The pressure that is used by the pilot valve to operate the main valve is termed the *pilot pressure*. There are two possible sources for the pilot pressure. Normally, it is channeled from the pressure port of the main valve to the pressure port on the pilot valve through an internal flow passage cast into the main valve body. This is shown symbolically in Figure 7.40a.

In certain cases, it may be desirable to use an external pilot source. One such case occurs when the system pressure fluctuates over a wide range, providing an unstable pilot source. Another case occurs when the system pressure is 1500 psi or higher, which could cause damage to the main spool because of the high force resulting from the high pressure. (External pilot pressures are usually 300 to 400 psi.) Figure 7.41 shows the symbols for a valve with an external pilot source.

Normally, the tank port of the pilot valve is connected through an internal flow passage to the tank port of the main valve, as shown in Figure 7.40. In cases where a back pressure exists in the tank line that might interfere with the valve operation, an external drain can be used (Figure 7.42).

Another variation to the basic pilot valve is the use of a *choke control*, which provides the ability to control the rate at which the main spool shifts. This technique is used for three purposes: (1) to control the acceleration or decelera-

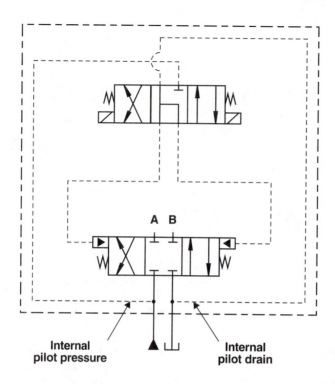

Internal
pilot pressure

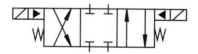

Internal
pilot drain

a. Complete graphic symbol.

b. Simplified symbol.

Figure 7.40 Pilot-operated directional control valve symbols.

tion rate of the actuator; (2) to prevent damage to the main valve that could result from the impact of the rapidly moving spool; or (3) to reduce system shock (termed *water hammer*) that can occur when a large spool shifts rapidly from one position to another.

The choke control is actually a separate valve body that is inserted between the pilot valve and the main valve as shown in Figure 7.43. (It is sometimes

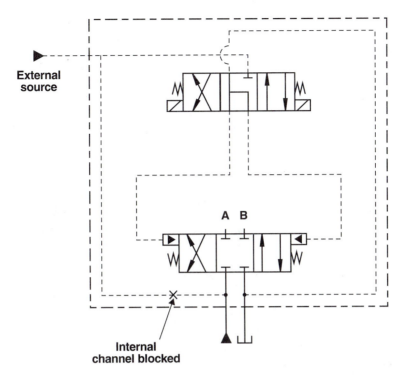

External source

Internal channel blocked

a. Complete graphic symbol.

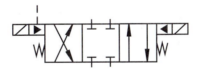

b. Simplified symbol.

Figure 7.41 Pilot-operated directional control valve with external pilot source.

referred to as a *sandwich valve*.) This valve body contains two flow control valves along with bypass check valves. Adjusting the flow control valves provides control of the flow rate for the main spool. From the earlier analogy, we know that the main spool is similar to a cylinder. Thus, as the pilot flow rate changes, the main spool shifting speed changes.

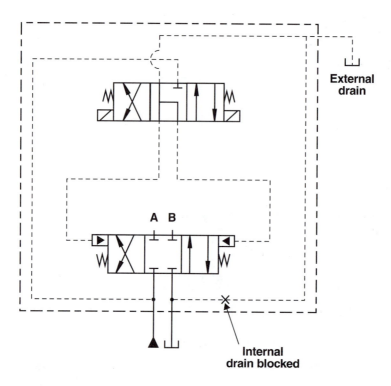

a. Complete graphic symbol.

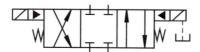

b. Simplified symbol.

Figure 7.42 Pilot-operated directional control valve with external drain.

7.10 Valve Mounting

There are three options for mounting directional control valves—line, subplate, or cartridge. A *line-mounted valve* is one that has threaded ports into which pipes or fittings are attached. Anytime this type valve is changed, all associated plumbing must be unscrewed, then screwed into the new valve. This practice

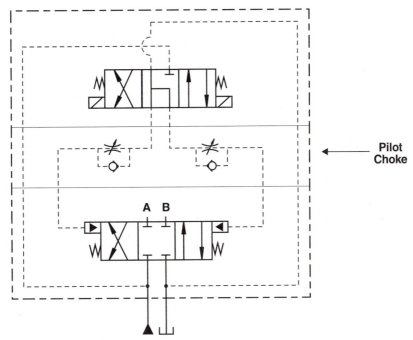

Figure 7.43 Pilot chokes provide control of the main spool shifting speed.

often leads to leaks because of thread wear, and it invariably generates contamination that is likely to find its way into the system.

In the *subplate mount*, the valve has no threaded ports. The valve ports are all on the bottom of the valve and are arranged in a standard pattern, depending on the valve size and type. The port pattern matches the pattern on the subplate surface. The subplate ports are drilled through the subplate and brought out through threaded ports on either the side or the bottom of the plate. In this arrangement, no plumbing is disturbed when the valve is changed. Instead, to remove the valve, the mounting bolts are removed and the valve is simply lifted off the subplate. O-rings are fitted into the recesses on the ports of the replacement valve to ensure proper sealing. When installing subplate mounted valves, it is essential that the bolts be tightened in the correct order and to the manufacturer's recommended torque.

Subplates are standardized for most types of valves. The design—port pattern, bolt-hole pattern and spacing, surface flatness and finish—are all specified in NFPA Standards T3.5.1 and T3.5.9M, ANSI B93.7M and B93.40M, and ISO 4401. The standard subplate patterns for directional control valves are shown in Figure 7.44. The figure includes the designations for specific patterns and sizes. The standardization of the subplate patterns allows universal interchangeability among different manufacturers' valves and subplates.

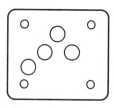

a. ISO-4401-05/NFPA-D05
(Replaces NFPA-02)

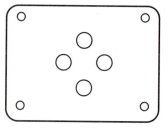

b. ISO-4401-03/NFPA-D03
(Replaces NFPA-D01)

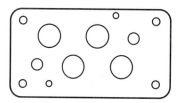

c. ISO-4401-08/NFPA-D08
(D06 has same pattern,
but different port size.)

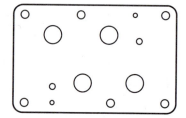

d. ISO-4401-10/NFPA-D10

Figure 7.44 Subplates for mounting directional control valves.

Cartridge Valves

Cartridge valves are not a type of valve in themselves; rather, they perform the same functions as other valves, but they are installed in a cavity in a valve block. Figure 7.45 is a photograph of some cartridges and valve blocks. One of the advantages of cartridge valves is the flexibility they provide—one can machine a manifold to accept several cartridges. Consequently, an entire system control section can be incorporated into one small manifold rather than in several separate valves located around the system. This provides for a very compact system that is easily maintained. In addition, the use of manifolds simplifies the physical circuit by eliminating piping and fittings. This reduces the likelihood of system failure due to line breaks or fitting leaks. Some pumps, motors, and cylinders are designed with cartridge valves cavities machined directly into them so that control functions can be accomplished directly on the device.[1]

All control functions—directional, pressure, and flow—can be provided by cartridge valves. Thus, the complete control function for an entire, complex circuit can be contained in a single manifold.

There are two basic types or configurations of cartridge valves—slip-in and screw-in. *Slip-in cartridges* are usually poppet-type devices that are normally controlled by a conventional valve to provide a complete hydraulic function. *Screw-in cartridges* may be either the spool or poppet type. They usually provide

Figure 7.45 Cartridge valves (top) are installed in a cavity in a valve block (bottom).

Source: Courtesy of Sun Hydraulics Corporation.

a complete hydraulic function—directional, pressure, or flow control—without the need for another valve.[2]

The slip-in type cartridges are very much like simple poppet-type check valves both in appearance and in function. They allow flow in one direction and block flow in the opposite direction. Figure 7.46 shows the construction of a typical slip-in cartridge valve. The insert—which consists of a sleeve, poppet, and spring—is inserted into the properly machined manifold block. Sealing is provided by O-ring seals. The insert is held in the block by the cover assembly, which is bolted to the manifold. The passageway connecting the X port to the spring chamber provides a pilot signal to the spring side of the poppet, while the main flow through the poppet is through the A and B ports.

A typical application for slip-in cartridge valves is shown in Figure 7.47. The opening and closing of the poppets is controlled by the solenoid-operated pilot valve. In the center position, the pilot valve ports pressure to the spring cavities of the four poppets to hold them closed. This stops flow to and from the cylinder and holds it in position. When solenoid S-1 is energized, the spool is pushed to the right. This pressurizes the spring cavities of poppets 1 and 3, keeping them closed. The cavities of poppets 2 and 4 are opened to tank, so they are free to open. System pressure pushes poppet 2 open and allows flow to the blind end of the cylinder. The exhaust flow from the rod end of the cylinder pushes poppet 4 open and flows through the poppet to tank. Thus, the cylinder extends.

If solenoid S-2 is energized, the pilot spool shifts to the left. The result is that poppets 2 and 4 are held closed, while poppets 1 and 3 are allowed to open. Flow through poppet 3 goes to the rod end of the cylinder. Exhaust flow

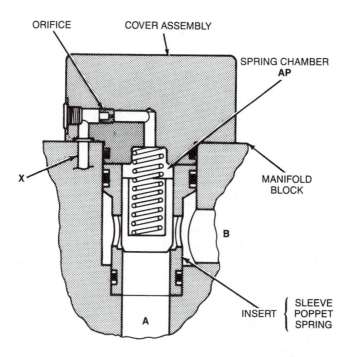

Figure 7.46 Slip-in cartridge valves are much like simple poppet-type check valves.

Source: Courtesy of Vickers, Inc.

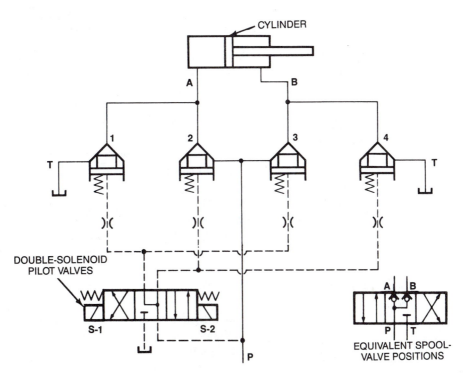

Figure 7.47 This cylinder circuit is a typical application for slip-in cartridge valves.

Source: Courtesy of Vickers, Inc.

from the blind end of the cylinder passes through poppet 1 to tank. In this case, the cylinder retracts.

At first glance, this appears to be a very complex way to operate a cylinder. The advantages of the circuit—reduced size and cost of the directional control valve—are not readily obvious. Recall our discussion in Chapter 6 of the relative flow rates from the blind end and the rod end of double-acting cylinders during extension and retraction. We saw that the flow from the blind end during retraction could be much higher than the pump flow. If conventional valves are used in this type of circuit, the valve spool and ports must be sized to handle that high flow. Slip-in cartridge valves, on the other hand, can be individually sized to handle the flow required through their individual ports. Since the pilot valve simply directs pressure to the poppet spring cavities, it can be the smallest valve available. The result is a considerable saving in cost and space. Standard subplate mounts can be used as an interface between the pilot valve and the manifold block.

Screw-in cartridge valves get their name from the external threads used to secure them in the cavities of the valve block or manifold. While there are numerous functions available in cartridge valves, there are only four basic cavity designs in a variety of sizes. (The cavity designs vary among manufacturers, so Company X's cartridge will probably not work in Company Y's cavity.)

The cavity designs are termed two-way, three-way, three-way short, and four-way. The designators refer to the number of ports in the cavity. The three-way short cavity has three ports, but one is used as a pilot port. These four cavities accommodate all of the individual cartridges, regardless of function. For example, Figure 7.48 shows six different cartridges that fit into the same two-way cavity.

If a single control function is desired, it is usually convenient to purchase both the cartridge and the valve body with the appropriate cavity from the valve manufacturer. However, if an entire control circuit is desired, it is usually necessary to machine the manifold specially for the circuit. The tools to machine the cavities can be purchased from the cartridge manufacturer. The design and machining of manifolds is an individualized effort, and is more of an art than a science. The physical design and layout of the manifold is relatively unimportant as long as the required functions can be accomplished.

7.11   Valve Selection

When selecting a directional control valve for a specific application, the choice may or may not be obvious, depending on the requirements of the system. In some cases, the system requirements can be met with two or more valve types. We saw this in Example 7.1 where either a closed-center or a tandem-center DCV could satisfy the primary requirement of stopping and holding the cylinder mid-stroke. A secondary selection parameter in that example was the energy consumption of the system, for which the tandem center was the obvious choice. Other selection criteria include the actuator, mounting, direct or pilot operation, flow rate requirements, physical characteristics such as size and the location of the ports, and, of course, price. Putting too much emphasis on the price of the valve, however, can be a costly mistake, as the following example shows.

Example 7.4

The cylinder portrayed in this circuit diagram has a 4 in bore and a 2 in rod and moves an 11,000 lb load. The cylinder extends at 6.5 in/sec. The system operates 16 hr/day, 250 days/yr on a 50 percent duty cycle (meaning 50 percent of the time it is extending under load, and for the remainder of the cycle, it is either retracting under no load or resting.) Cylinder speed is controlled by the flow control valves. The pilot-operated check valves lock the load in place when the directional control valve is centered.

Compare the cost of the system components and the first year of operation based on using a certain D05 size valve versus a certain D06 size valve. The D05 valve costs \$750, while the D06 valve costs \$1318.[3]

Solution

The first step is to determine the pressures and flow rates involved. Based on the cylinder size and velocity, we can find the inlet flow required,

$$Q = A_P \times v = 12.56 \text{ in}^2 \times 6.5 \text{ in/sec} \times 60 \text{ sec/min} \times \text{gal}/231 \text{ in}^3$$

$$= 21.2 \text{ gpm}$$

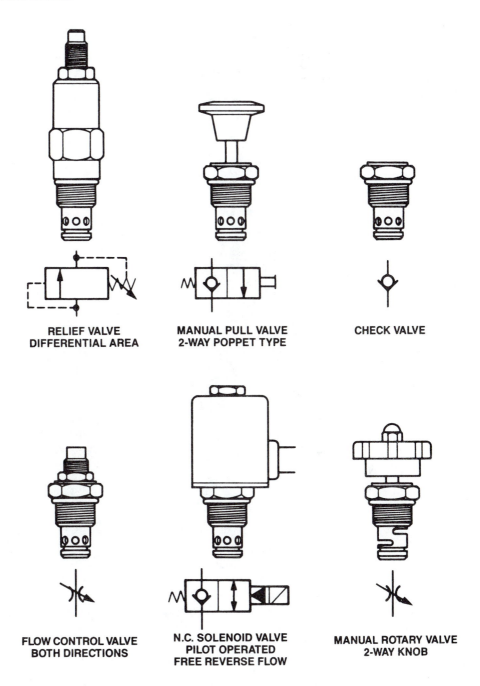

Figure 7.48 All of these screw in cartridge valves fit into the same cavity.
Source: Courtesy of Vickers, Inc.

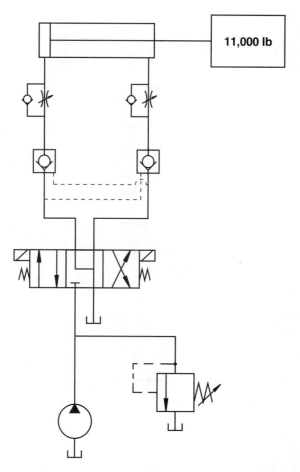

Example 7.4

and the resulting outlet flow,

$$Q = (A_P - A_R) \times v$$

$$= (12.56 \text{ in}^2 - 3.14 \text{ in}^2) \times 6.5 \text{ in/sec} \times 60 \text{ sec/min} \times \text{gal}/231 \text{ in}^3$$

$$= 15.9 \text{ gpm}$$

At 15.9 gpm, the D05 valve set (including all three valves shown in the circuit drawing) has a total pressure drop of 195 psi, while the D06 valve set has a pressure drop of only 37 psi. To find the effect of these back-pressures on the pressure required to extend the cylinder, we must convert them to forces that add to the 11,000 lb load to resist the cylinder extension. We use Equation 6.1 for this, remembering that the area required is the piston area minus the rod area. Thus, for the D05 set, the force due to back-pressure is

$$F_{BP} = p \times (A_P - A_R) = 195 \text{ lb/in}^2 \times (12.56 \text{ in}^2 - 3.14 \text{ in}^2)$$

$$= 1837 \text{ lb}$$

For the D06 set, the force due to back-pressure is only

$$F_{BP} = p \times (A_P - A_R) = 37 \text{ lb/in}^2 \times (12.56 \text{ in}^2 - 3.14 \text{ in}^2)$$

$$= 349 \text{ lb}$$

We add these to the load to get the total resistive force.

$$\text{D05:} \qquad F_{TOTAL} = 11{,}000 \text{ lb} + 1837 \text{ lb}$$

$$= 12{,}837 \text{ lb}$$

$$\text{D06:} \qquad F_{TOTAL} = 11{,}000 \text{ lb} + 349 \text{ lb}$$

$$= 11{,}349 \text{ lb}$$

From these force calculations, we now find the pressure required to move the load:

$$\text{D05:} \qquad p = \frac{F}{A_P} = \frac{12{,}837 \text{ lb}}{12.56 \text{ in}^2} = 1022 \text{ psi}$$

$$\text{D06:} \qquad p = \frac{F}{A_P} = \frac{11{,}349 \text{ lb}}{12.56 \text{ in}^2} = 909 \text{ psi}$$

Next we must find the pressure at the pump outlet, using the pressures we just calculated plus the pressure drop through the valving going into the cylinder. For the valves in question, the pressure drops at 21.2 gpm are

$$\text{D05:} \qquad 410 \text{ psi}$$

$$\text{D06:} \qquad 52 \text{ psi}$$

This means that the total pump outlet pressures are

$$\text{D05:} \qquad 1432 \text{ psi}$$

$$\text{D06:} \qquad 961 \text{ psi}$$

Now we find the hydraulic horsepower of each system

$$\text{D05:} \qquad HHP = \frac{p \times Q}{1714} = \frac{1432 \times 21.2}{1714} = 17.7 \text{ HP}$$

$$\text{D06:} \qquad HHP = \frac{p \times Q}{1714} = \frac{961 \times 21.2}{1714} = 11.9 \text{ HP}$$

Assume that the pump used has an overall efficiency of 0.83. This gives us an input horsepower requirement of

$$\text{D05:} \qquad IHP = \frac{HHP}{\eta_o} = \frac{17.7}{0.83} = 21.3$$

$$\text{D06:} \qquad IHP = \frac{HHP}{\eta_o} = \frac{11.9}{0.83} = 14.3$$

Based on the horsepower requirements, we would need a 25 HP motor costing $2548 for the D05 valve versus a 15 HP motor costing $1857 for the D06 valve. Assuming an 87 percent efficiency for both motors, we now calculate the kilowatts of electricity to run the system.

$$\text{D05:} \qquad EHP = \frac{IHP}{\eta} = \frac{21.3}{0.87} = 24.5 \text{ HP}$$

$$24.5 \text{ HP} \times 0.746 \frac{\text{kW}}{\text{HP}} = 18.3 \text{ kw}$$

$$\text{D06:} \quad EHP = \frac{IHP}{\eta} = \frac{14.3}{0.87} = 16.4 \text{ HP}$$

$$16.44 \text{ HP} \times 0.746 \frac{\text{kW}}{\text{HP}} = 12.3 \text{ kw}$$

Now we can find the killowatt-hours to determine the annual operating cost based on a cost of $0.085/kwh for electricity.

$$\frac{\text{kWh}}{\text{yr}} = \text{kW} \times \text{hr/day} \times \text{days/yr}$$

$$\text{D05:} \qquad \frac{\text{kWh}}{\text{yr}} = 18.3 \text{ kW} \times 16 \text{ hr/day} \times 0.5 \text{ (duty cycle)} \times 250 \text{ days/yr}$$

$$= 36{,}600 \text{ kWh/yr}$$

$$\text{Cost} = \frac{\text{kWh}}{\text{yr}} \times \frac{\$}{\text{kWh}}$$

$$= 36{,}600 \text{ kWh/yr} \times \$0.085/\text{kWh}$$

$$= \$3{,}111/\text{yr}$$

$$\text{D06:} \qquad \frac{\text{kWh}}{\text{yr}} = 12.3 \text{ kW} \times 16 \text{ hr/day} \times 0.5 \text{ (duty cycle)} \times 250 \text{ days/yr}$$

$$= 24{,}600 \text{ kWh/yr}$$

$$\text{Cost} = 24{,}600 \text{ kWh/yr} \times \$0.085/\text{kWh}$$

$$= \$2{,}091/\text{yr}$$

The following table summarizes the total costs:

	D05	D06
Valve set	750	1318
Electric motor	2548	1857
Electricity cost	3111	2091
Total first-year cost	$6409	$5266

This shows that by using the larger valve, we see a saving in the first year of $1143, even though the valve itself is more expensive. In subsequent years, we save $1020 per year in electricity costs. Other, less visible, savings will come from increased pump life due to the lower pressure, and increased fluid life due to cooler fluid.

7.12   Troubleshooting Tips

The directional control valves discussed in this chapter are, functionally, relatively simple devices. Because of this, they are much less prone to failure than more complex components such as pumps. The vast majority of directional control valve problems are due to contaminated fluid.

Symptom	Possible Causes
Cylinder or motor operates in wrong direction	Lines between valve and cylinder or motor connected incorrectly
Load will not stay in place when valve centered	1. Wrong center configuration 2. Excessive leakage through DCV due to contaminant wear 3. Spool jammed open slightly by contamination 4. Spool lands eroded by aerated fluid or cavitation 5. Pilot-operated check valve leaking 6. Cylinder piston seals leaking
Cannot shift valve	1. Spool jammed by solid particles 2. Spool jammed by residue from deteriorated fluid (due to heat or chemical reactions) 3. Drain line blocked causing hydraulic lock
Check valve leaks	1. Seat or mechanism worn due to contamination 2. Mechanism held off seat by particle contamination
Fluid heats excessively	1. Leakage through or around worn or improperly positioned spool 2. Check valve not properly seated

7.13   Summary

Directional control valves direct the system fluid along the proper flow paths. While this is their primary function, they can also affect flow and pressure. Therefore, the selection of both the appropriate configuration and size (flow capacity) is important to both the function and efficiency of the system.

The valve actuators we have discussed here are fairly commonly used; however, industrial systems use a much higher proportion of electrically actuated valves. We will discuss those valves in Chapter 10.

In a real system, the directional control valve function is interrelated to the system's pressure control valves. We'll look at pressure controls in the next chapter.

7.14 Key Equations

The equations used in this chapter were introduced in earlier chapters with the exception of the equation for calculating the power loss across a valve, heat generation rate, and temperature rise.

Power loss across valves

U.S. Customary: $HP_{LOST} = \dfrac{\Delta p \times Q}{1714}$

SI: $P_{LOST} = \dfrac{\Delta p \times Q}{60,000}$

U.S. Customary: $HGR = HP_{LOSS} \times 42.4 \text{ Btu/min}$ **(7.1)**

SI: $HGR = P_{LOST}$

Temperature rise: $\Delta T = \dfrac{HGR}{c_p \times W}$ **(7.2)**

References

1. Pippenger, John J., *Hydraulic Cartridge Valve Technology* (Jenks, OK: Amalgam Publishing Company, 1989).
2. *Industrial Hydraulics Manual*, 2d Ed. (Rochester Hills, MI: Vickers, Inc., 1989.)
3. This example is based on Brian J. Roberts, "Don't Neglect Pressure Drop" *Hydraulics and Pneumatics Magazine*, November 1987.

Review Problems

Note: The letter "S" after a problem number indicate SI units are used.

General

1. What is meant by the word *position* in describing directional control valves?
2. What is the difference between a normally open and a normally closed directional control valve?
3. What is the difference between a closed-center valve and an open-center valve?
4. What is the difference between a closed-center system and an open-center system? Discuss the advantages and disadvantages of each.
5. What is the difference between an open-loop system and a closed-loop system? Draw a circuit diagram representing each type.
6. What is the purpose of a deceleration valve? How does it work? Why is a spring cavity drain needed?
7. Why are pilot-operated directional control valves used?
8. In the circuit shown for this review problem, where does the pump flow go when the directional control valve is centered? List some advantages and disadvantages of this arrangement.

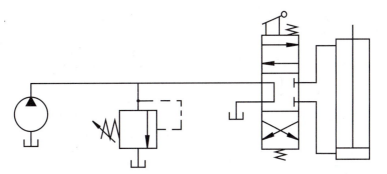

Review Problem 7-8

9. In the partial circuit shown for this review problem, the cylinder drifts downward, even with the pilot-operated check valve working properly. What is causing the problem?

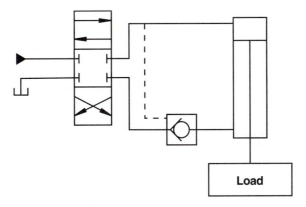

Load

Review Problem 7-9

10. In the partial circuit shown for this review problem, will the load drift downward if the piston seal leaks? Fully support your answer.

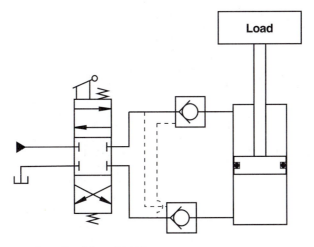

Load

Review Problem 7-10

Power Loss

11. Fluid flows at 40 gpm through a device that causes a 60 psi pressure drop. Calculate the horsepower lost through the device.

12. The graph for this review problem shows the flow-pressure drop curves for a directional control valve. Construct a set of curves showing horsepower loss versus flow rate for the valve.

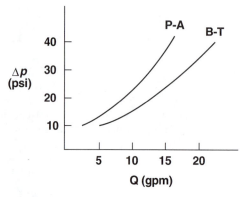

Review Problem 7–12

13S. Find the power loss through a valve in which a 100 Lpm flow rate has a 250 kPa pressure drop.

14S. Construct a set of curves showing the power loss versus flow rate for a valve that has the pressure drop-flow characteristics shown in the graph for this review problem.

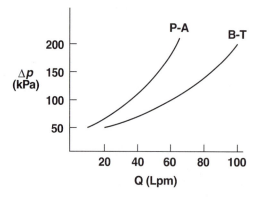

Review Problem 7–14S.

15. Find the heat generation rate and the temperature rise caused by the loss of 30 HP through a system. The fluid has a specific heat of 0.48 Btu/(lb°F) and a specific gravity of 0.86. The flow rate is 40 gpm.

16. Fluid flowing at 60 gpm experiences a 500 psi pressure drop. The fluid has a specific gravity of 0.9 and a specific heat of 0.45 Btu/(lb°F). Find the heat generation rate and the temperature rise.

17S. A fluid with a specific heat of 1.85 kJ/(kg°C) and a specific gravity of 0.87 is flowing through a valve at 125 Lpm. The pressure drop across the valve is 200 kPa. Find the heat generation rate and the temperature rise through the valve.

Circuit Diagrams

18. Explain the operation of the circuit shown for this review problem.

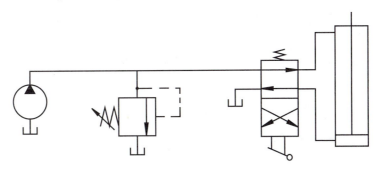

Review Problem 7–18

19. Explain the operation of the circuit shown for this review problem.

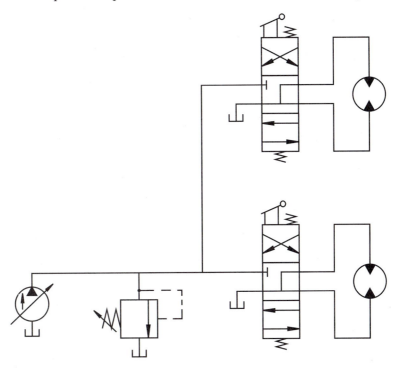

Review Problem 7–19

Use Standard ISO Symbols in the Following Circuit Diagrams.

20. Draw a circuit for operating a bi-directional hydraulic motor. The motor must be operated in either direction, but cannot be stopped without stopping the pump.
21. Draw a circuit for operating a bi-directional hydraulic motor. Use a float-center directional control valve.
22. Draw an open-center *circuit* for operating a double-acting cylinder.
23. Draw a closed-center *circuit* with a pressure-compensated pump for operating a double-acting cylinder.

24. A circuit is required for operating two double-acting hydraulic cylinders in parallel. The cylinders are not operated simultaneously. Both can be stopped in mid-stroke. The system uses a pressure-compensated pump.

25. In a die casting operation, a hydraulic cylinder (called the *clamp cylinder*) is used to close and clamp the mold. A second cylinder (called the *shot cylinder*) then pushes a plunger to inject molten aluminum into the mold. The directional control valves are lever-operated. When the clamp cylinder closes the mold, the lever on the DCV is released, and the cylinder is stopped and held. When the shot cylinder completes its stroke, the lever on its DCV is released, and the cylinder retracts. A pressure-compensated pump is used to reduce the energy losses when the system is idle. Draw the circuit diagram for this system.

26. In a machining process, a clamping cylinder holds a workpiece while a grinding operation is performed. After the workpiece is secured, the operator starts a single-direction hydraulic motor using a three-way, two-position valve operated by a foot pedal. The motor turns the grinding wheel. A hydraulic cylinder then lowers the grinding wheel to the workpiece. When the operator releases the DCV for this cylinder, it retracts automatically. Releasing the foot pedal stops the motor. The clamping cylinder is then retracted. Draw a hydraulic circuit for the operation.

Chapter 8

Pressure Control Valves

Outline

8.1 Introduction

The term *pressure control* is applied to a group of valves whose action is determined by system pressure. The family includes the following valve types:

- Pressure relief valves
- Pressure reducing valves
- Unloading valves
- Counterbalance valves
- Brake valves
- Sequence valves

"Pressure-control" is something of a misnomer for these valves because only the pressure-reducing valves actually *control* pressure. The rest *sense* pressure and automatically adjust their internal mechanisms in response to the sensed pressure. Look for this feature as we discuss each valve type. Also, observe the secondary effect of the valve operation on the system. In most cases, flow, direction, or both are affected as the valve responds to the pressure.

Objectives
When you have completed this chapter, you will be able to

- List the different types of pressure control valves.
- Explain the operation and applications for the different types of pressure control valves.
- Recognize and draw the ISO graphic symbols representing the pressure control valves.
- Analyze and draw circuits using pressure control valves.
- Calculate the energy losses through pressure control valves.
- Explain the difference between direct-acting and pilot-operated pressure control valves.

8.2   Pressure Relief Valves

Pressure relief valves are responsible for limiting the maximum pressure reached in the fluid power system. They accomplish this by diverting all or part of the output of the pump back to the reservoir rather than allowing it to continue to flow into the system. Remember that pressure is the result of resistance to flow. As the pump attempts to "force" more fluid into the system, the pressure will continue to rise. We can limit the pressure increase if we can stop the flow of fluid into the system (by using a pressure-compensated pump, for instance), or we can provide an alternate flow path to get the fluid back to the reservoir. The latter is the function of the relief valve.

As a practical example, consider that we are operating a hydraulic cylinder with a fixed displacement pump. As long as the cylinder is free to move, the maximum pressure in the system will be governed by the load we are moving. When the cylinder reaches the end of its stroke and stops (bottoms out), what happens to the system pressure? Since we have a fixed displacement pump, it will continue to try to pump more fluid into a system that can accept no more fluid. Therefore, the pressure will increase, and will continue to do so until either the prime mover stalls or something gives. That "something" might be the pump housing, the pump shaft, a fitting, a line, or some other piece of hardware that will fail under the high internal pressure.

In most systems, a pressure relief valve is installed and adjusted to open and divert the excess flow back to the reservoir before that failure occurs. It won't automatically divert the entire pump output, however. Instead, it will dump just enough of the flow to maintain the system pressure at the selected level. Depending on system leakage (both internal and external) and other flow requirements, it may divert anywhere from a small fraction of the pump output all the way to the entire flow in its efforts to prevent that selected pressure from being exceeded.

Relief valves can be used to accomplish two different things in fluid power systems. In the example we just used, the relief valve is used as a safety device to prevent overpressurizing the system. The second use is as a pressure limiter to maintain the system pressure at some predetermined level. In the process, the relief valve limits the maximum force or torque output capability of the system actuator.

Recall from our discussion of pressure-compensated pumps that those devices could also accomplish this second task. In fact, they can do the job at a considerable energy saving when compared to the relef valve. Remember that this is possible because a fixed displacement pump must produce its full output flow against the pressure at which the relief valve is set (usually about 150 to 200 psi above the normal system operating pressure). However, when a pressure-compensated pump is used, the pump flow output is reduced to the minimum required to make up the system leakage and maintain the pressure at the selected level (usually just enough to hold the load).

Please do not interpret this discussion as implying that a relief valve is not necessary when a pressure-compensated pump is used. Compensators do fail, especially in highly contaminated systems. When they do, the relief valve must perform its role as a safety device. Otherwise, disaster is likely to occur in the form of a major system failure.

Relief valves (and all pressure control valves, for that matter) can be mounted in the same ways as directional-control valves—in-line, subplate, and cartridge. In-line mounting is somewhat more common than the other two methods. As with directional-control valves, the subplate mounts for pressure control valves are standardized, with relatively few patterns and sizes available. Cartridges provide the same advantages here as in directional valves, especially in the ability to use a single manifold to contain all of the valving necessary for the system.

Direct-Acting Relief Valves

A *direct-acting relief valve* (Figure 8.1) consists of a valve body with two ports (referred to as pressure and tank, inlet and outlet, or primary and secondary), a shutoff mechanism, a spring, and, usually, an adjustment device. It uses the heavy spring to force a ball or poppet onto a seat or to hold a spool in a position to block the pressure or tank port. This seals the valve and prevents flow through it as long as the force exerted on the mechanism by the system pressure is below the force exerted by the spring. The term "direct acting" means that the pressure acts directly on the valve mechanism. When the pressure exceeds the preset level as determined by the spring force, the resultant force becomes greater than the spring force and begins to push the mechanism off its seat. This allows a portion of the flow to pass through the valve and back to the tank. If the pressure continues to increase, the valve will open more in response to the increased force until the flow it is passing is sufficient to stop the pressure increase and maintain the pressure at the desired level. At some time during the system operation, the valve will probably be required to pass the full pump output. Therefore, relief valves must always be sized so that they can dump the entire pump flow if necessary.

The valve shown in Figure 8.1a has a nonadjustable configuration. The sealing mechanism is a poppet. The spring force is fixed and cannot be changed. The valve depicted in Figure 8.1b is adjustable. The compressive force on the spring can be varied by running the adjustment screw in and out. This changes the spring force and, consequently, the pressure required to cause the valve to open. Most relief valves are adjustable.

The symbols shown in the figure describe the functions of the valves. They are very similar to the symbols used for other pressure control valves, so you must look very carefully to distinguish among the symbols you find in circuit diagrams. The symbol begins with a single square to which are connected the inlet (pressure) and outlet (tank) lines. Within the square is a flow arrow that is offset to one side, indicating that there is normally no flow through the valve (that is, it is normally closed). The spring attached to the symbol on the side opposite the flow arrow indicates that the mechanism is pushed closed by the spring, while the dashed line on the same side as the flow arrow shows that the valve is opened (actuated) by pressure. Since this pressure-sensing line is attached to the inlet line, we see that actuation results from the upstream (inlet) pressure. If the valve is adjustable (as is the one in Figure 8.1b), this will be indicated by a diagonal arrow across the spring symbol. Relief valves will always be shown in parallel with the main flow line, while the outlet line will always be shown going directly to the tank.

The sealing mechanism (ball, poppet, or spool) is held closed by spring force. The force exerted by the spring depends on the strength of the spring and the

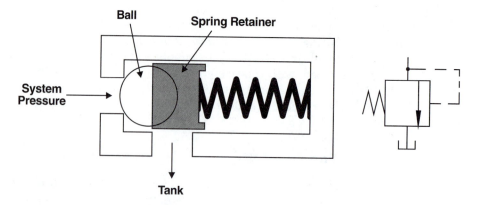

a. A direct acting relief valve with no adjustment is preset for a single pressure by the spring.

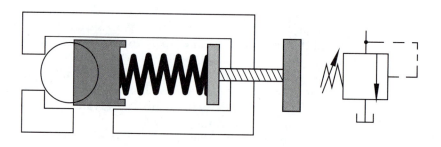

b. The setting of an adjustable direct acting relief valve can be changed by changing the compression of the spring.

Figure 8.1 In a direct-acting relief valve, the pressure acts directly on the valve mechanism.

---

distance it is compressed. The relationship between the force and compression distance is defined by the spring constant, k. The units of k are lb/in (N/cm). If we say that a spring has a spring constant of 50 lb/in (87.5 N/cm), we would mean that each inch that we compress the spring would require a force of 50 lb (222.5 N). Therefore, 150 lb (667.5 N) would be required to compress the spring 3 in (7.6 cm). The equation defining this is

$$\text{Force Required} = \text{Spring Constant} \times \text{Distance Compressed}$$

$$F = k \times d \tag{8.1}$$

where F is the force required and d is the distance the spring has been compressed.

Example 8.1

A spring has a spring constant of 35 lb/in (61 N/cm). How much force is required to compress the spring 2 in (5 cm)? How much to compress it 3 in (7.6 cm)?

Solution

Use Equation 8.1:

$$F = k \times d$$

To compress the spring 2 in (5 cm) requires a force of

$$F = 35 \text{ lb/in} \times 2 \text{ in}$$
$$= 70 \text{ lb}$$

or

$$F = 61 \text{ N/cm} \times 5 \text{ cm}$$
$$= 305 \text{ N}$$

For a 3 in (7.6 cm) compression, the force required is

$$F = 35 \text{ lb/in} \times 3 \text{ in}$$
$$= 105 \text{ lb}$$

or

$$F = 61 \text{ N/cm} \times 7.6 \text{ cm}$$
$$= 464 \text{ N}$$

If we apply this concept to a relief valve mechanism, it becomes clear that the valve does not hold one constant pressure throughout its entire movement. Rather, it *begins* to open at some pressure and opens more as the pressure increases until it is fully open. The pressure at which a relief valve first opens and begins to divert flow back to the tank is termed the *cracking pressure* (Figure 8.2). This pressure is usually considerably below the desired maximum pressure setting. The maximum opening of the valve occurs when the pressure reaches a point termed the *full flow pressure*. At this level, the entire pump output is being diverted back to the tank. The difference between the cracking and full flow pressure is termed the *pressure override*. A third pressure of interest is the *reseat pressure*. That is the pressure at which the valve once again seals off all flow back to the reservoir. Because of flow forces (*drag*) from the flowing fluid that act to hold the valve open, the reseat pressure may be considerably below the cracking pressure.

Let's look at an example of what might happen with a typical, direct-acting relief valve.

Example 8.2

A relief valve uses a poppet with a 0.5 in² (3.23 cm²) area on which the system pressure acts. The poppet is held by a spring that has a spring constant of 3000 lb/in (5.25 kN/cm). When the valve is assembled, the spring is initially compressed 0.25 in (0.64 cm). The mechanism is then adjusted so that the maximum compression of the spring is 0.35 in (0.89 cm), giving the poppet only 0.10 in (0.25 cm) of travel. Find the cracking and full flow pressures for the valve.

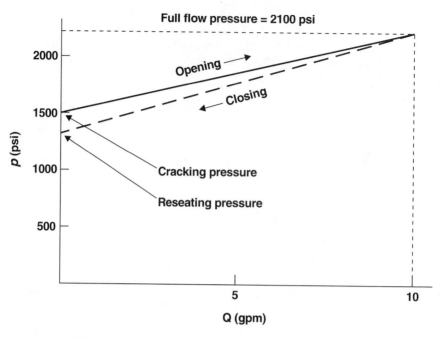

Figure 8.2 Typical relief valve pressure flow (P–Q) characteristics include the cracking pressure, where the valve first opens and begins to divert flow back to the tank. (The numbers refer to Example 8.3.)

Solution

Since the spring is initially compressed a quarter of an inch, the force required to compress it was

$$F = k \times d$$
$$= 3000 \text{ lb/in} \times 0.25 \text{ in}$$
$$= 750 \text{ lb}$$

or

$$F = 5.25 \text{ kN/cm} \times 0.64 \text{ cm}$$
$$= 3.36 \text{ kN}$$

This also means that the spring is pushing the poppet into its seat with a force of 750 lb (3.36 kN). Until the fluid pressure rises to the point where the resultant force on the face of the poppet exceeds the spring force, the valve will not open. That force is based on the pressure and the area over which it acts (the poppet face, in this case). That pressure is the *cracking pressure* for which we are looking:

$$p = \frac{F}{A}$$
$$= \frac{750 \text{ lb}}{0.5 \text{ in}^2}$$
$$= 1500 \text{ psi}$$

or

$$p = \frac{3.36 \text{ kN}}{323 \text{ mm}^2}$$
$$= 10.4 \text{ MPa}$$

Thus, when the system pressure exceeds 1500 psi (10.4 MPa), the poppet will move off the seat slightly, and fluid will begin to flow through the valve.

The *full flow pressure* occurs when the spring is completely compressed; therefore, we first need the force required to compress it the full 0.35 in (0.89 cm).

$$F = k \times d$$
$$= 3000 \text{ lb/in} \times 0.35 \text{ in}$$
$$= 1050 \text{ lb}$$

or

$$F = 5.25 \text{ kN/cm} \times 0.89 \text{ cm}$$
$$= 4.67 \text{ kN}$$

The pressure required, then, is

$$p = \frac{F}{A}$$
$$= \frac{1050 \text{ lb}}{0.5 \text{ in}^2}$$
$$= 2100 \text{ psi}$$

or

$$p = \frac{4.68 \text{ kN}}{3.23 \text{ cm}^2}$$
$$= 14.5 \text{ MPa}$$

Thus, at 2100 psi (14.5 MPa), the full pump output will dump over the relief valve. The pressure override in this case is (2100 − 1500 = 600 psi), or (14.5 − 10.4 = 4.1 MPa).

The effect of pressure override in system performance can be seen in the following example.

The effect of pressure override in system performance can be seen in the following example.

Example 8.3

A 4 in (10.2 cm) cylinder is operated by a fixed displacement pump that has an output of 10 gpm (37.85 Lpm) and a maximum operating pressure of 3000 psi (20.7 MPa). In order to control the maximum force output of the cylinder to prevent damage to a fixture, a relief valve with the characteristics shown in Figure 8.2 is used with the full flow pressure set at 2100 psi (14.5 MPa). The desired cylinder extension speed is 0.18 fps (0.055 mps). This speed can be achieved with a 21,000 lb (93.4 kN) load. However, a modification is made to the product that increases the load to 22,000 lb (97.9 kN). Can the required speed still be achieved?

Solution

With a 21,000 lb (93.5 kN) load, we see that the required pressure is

$$p = \frac{F}{A}$$

$$= \frac{21{,}000 \text{ lb}}{12.56 \text{ in}^2}$$

$$= 1672 \text{ psi}$$

or

$$p = \frac{93.4 \text{ kN}}{8103 \text{ mm}^2}$$

$$= 11.5 \text{ MPa}$$

From Example 8.3, we know that at that pressure, the relief valve has already partially opened and is dumping approximately 2.8 gpm (10.6 Lpm) back to the tank, leaving 7.2 gpm (27.25 Lpm) to go to the cylinder. At that point, the velocity at which the cylinder extends is

$$v = \frac{Q}{A}$$

$$= \frac{7.2 \text{ gal/min} \times 231 \text{ in}^3/\text{gal} \times \text{ft}/12 \text{ in} \times \text{min}/60 \text{ sec}}{12.56 \text{ in}^2}$$

$$= 0.18 \text{ fps}$$

or

$$v = \frac{27.25 \text{ L/min} \times \text{m}^3/1000 \text{ L} \times \text{min}/60 \text{ sec}}{81.03 \text{ cm}^2 \times \text{m}^2/10{,}000 \text{ cm}^2}$$

$$= 0.56 \text{ m/s}$$

This is the desired speed.

When the load is increased to 22,000 lb (97.9 kN), the system pressure becomes

$$p = \frac{F}{A}$$

$$= \frac{22{,}000 \text{ lb}}{12.56 \text{ in}^2}$$

$$= 1752 \text{ psi}$$

or

$$p = \frac{97.9 \text{ kN}}{81.03 \text{ cm}^2}$$

$$= 12.1 \text{ MPa}$$

At that pressure, the relief valve has opened somewhat further and is now dumping approximately 4 gpm (15.1 Lpm) back to the tank. Thus, the flow rate to the cylinder is about 6 gpm (22.75 Lpm). The maximum cylinder velocity now is

$$v = \frac{Q}{A}$$

$$= \frac{6 \text{ gal/min} \times 231 \text{ in}^3/\text{gal} \times \text{ft/12 in} \times \text{min/60 sec}}{12.56 \text{ in}^2}$$

$$= 0.15 \text{ fps}$$

or about 0.046 m/s.

This does not satisfy the minimum velocity requirement of the process.

What could be done about the situation in the example to reach the required velocity? Obviously, we must somehow increase the flow rate to the cylinder, because we have already seen that the cylinder velocity is a function of the flow rate.

One way to increase the flow rate would be to adjust the pressure relief valve to a higher full flow pressure. This would increase the valve's cracking pressure (and, in fact, shift the entire curve upward), thereby reducing the amount of fluid diverted by the valve at a given pressure. **Caution!** Before this step is taken, you must be certain that the system can tolerate the increased pressure.

There are those who would interpret the action just suggested as meaning that the cylinder velocity is dependent on system pressure. *That is absolutely not the case!* The reason we get increased cylinder velocity from adjusting the relief valve is that we have also adjusted the secondary function of the valve— the fluid flow through it. Attempting to increase cylinder or hydraulic motor speed by adjusting the relief valve may result in damage to the system or injury to personnel due to exceeding the pressure limits of the system. Using the relief valve in this manner is also highly inefficient and leads to energy loss and heat generation. Although we have not yet discussed flow control valves, we will note here that if a flow control valve is used in the circuit, increasing the relief valve setting will increase the flow through the flow control valve. In both cases, the cylinder pressure will remain constant because it is determined by the load, not by the relief valve.

Pilot-Operated Relief Valves

A pilot-operated relief valve is actually two valves in one. One design of this type of valve is shown in Figure 8.3. The top, or pilot, portion of the valve is actually a direct-acting relief valve that operates exactly like the direct-acting valves described in the previous section. Its purpose is to control the pressure in the pilot chamber above the disk on the main spool.

The lower, or main, valve consists of a valve body with pressure and tank ports, a spool that includes a balanced disk and a poppet in a single piece, a seat for the poppet, and a light bias spring. The disk on the spool is termed "balanced" because the areas on either side are precisely the same. The result is that equal pressures top and bottom will produce equal and opposite forces, and the poppet will remain closed.

Notice that the symbol used to represent a pilot-operated relief valve is exactly the same as the symbol for a direct-acting relief valve. Unless the vent port is being used (we'll discuss this later), there is no way to tell from the symbol that the valve is pilot-operated.

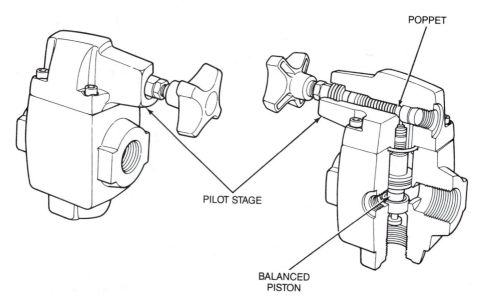

POPPET

PILOT STAGE

BALANCED
PISTON

Figure 8.3 A pilot-operated relief valve is actually two valves in one.
Compare the pilot section, which is a direct-acting valve, with the lower,
main valve body.
Source: Courtesy of Vickers, Inc.

Figure 8.4 illustrates the operation of a pilot-operated valve. The system
pressure is exerted on the lower face of the balanced disk. An orifice in the
disk allows this same pressure to be exerted on the top of the disk. Because
of a light biasing spring, the poppet remains seated and prevents flow through
the tank port.

This system pressure is also sensed on the face of the pilot poppet (usually
referred to as the *dart* because of its small size and sharp point), which is
backed by a heavy spring adjusted for the desired pressure. When the pressure
on the pilot poppet reaches the preset level, the resultant force offseats the
poppet. This allows the fluid in the upper chamber to flow down through the
pilot drain and out the tank port. The result is a reduction in pressure in the
upper chamber. The unbalanced forces that result from the pressure differences
cause the spool to move upward and the poppet to lift off its seat. Part of the
pump output can then be diverted back to the reservoir through the tank port.
Note that very little flow passes through the pilot section; the majority of the
valve flow goes past the main valve poppet to the tank port.

Pilot-operated relief valves are generally preferred over the direct-acting
valves for several reasons. Chief among these are that they operate more
smoothly and that their cracking pressure is usually around 90 to 95 percent
of their full flow pressure. In other words, the pressure override is usually less
than 100 psi. This means that more flow is available to the system throughout
the operating range of these valves than is possible with a direct-acting valve
(Figure 8.5). As a result, if a pilot-operated valve were used in the system of

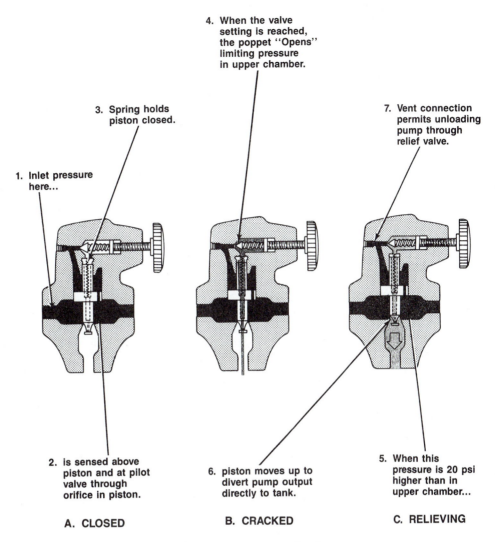

4. When the valve setting is reached, the poppet "Opens" limiting pressure in upper chamber.

3. Spring holds piston closed.

7. Vent connection permits unloading pump through relief valve.

1. Inlet pressure here...

2. is sensed above piston and at pilot valve through orifice in piston.

6. piston moves up to divert pump output directly to tank.

5. When this pressure is 20 psi higher than in upper chamber...

A. CLOSED

B. CRACKED

C. RELIEVING

Figure 8.4 Operation of a pilot-operated relief valve.
Source: Courtesy of Vickers, Inc.

Example 8.3, there would have been sufficient flow to operate at the required speed with the heavier load while maintaining the 2100 psi pressure limit.

In addition to their smooth operation and improved pressure-flow characteristics, pilot-operated relief valves have another advantage over direct-acting valves in that they can be operated remotely. To understand how this works, let's have another look at Figure 8.3. What actually determines the system pressure at which the main spool opens? The answer is, the pressure in the upper chamber, which is in turn determined by the compression of the spring on the pilot valve.

Consider what would happen if we removed the plug from the vent port of

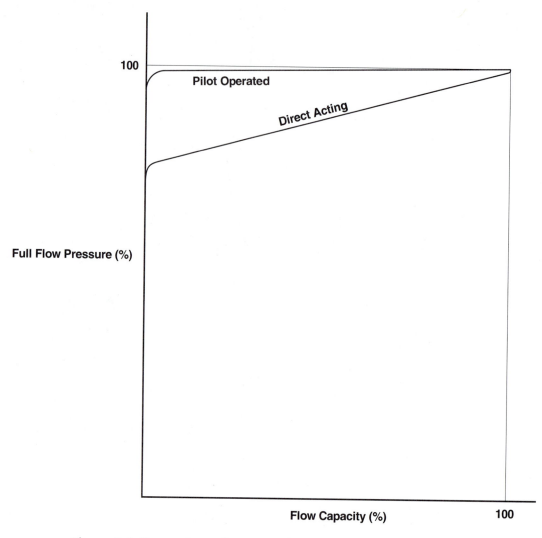

Figure 8.5 Comparison of pressure-flow characteristics for typical direct-acting and pilot-operated relief valves shows why pilot-operated valves are generally preferred.

the valve in Figure 8.3 and connected a remote pilot valve as shown in Figure 8.6. Now what determines the pressure at which the main spool opens? The answer is still the pressure in the pilot chamber; but what determines the pressure in the pilot chamber? Since the maximum pressure is determined by the minimum resistance to flow, the direct-acting valve with the lowest pressure setting will control the pilot chamber pressure, and, consequently, the system pressure. Normally, the remote pilot relief valve is set at the lower pressure.

A remote control such as this could be applied to any system that requires pressure adjustments but has its main relief valve located in a hazardous position. Rather than exposing the operator to the hazard, the pilot valve is

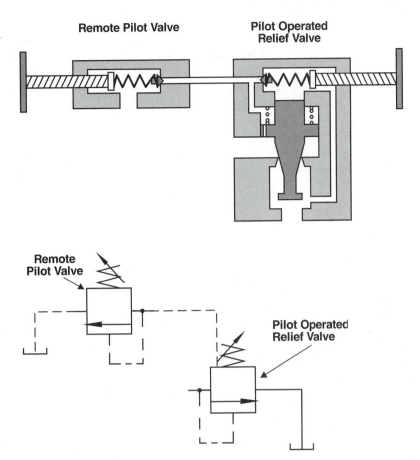

Figure 8.6 Here the plug has been removed from the vent port of the valve in Figure 8.3 and a remote pilot has been connected.

set at the maximum pressure *ever* required, and the operator controls the actual system pressure by using the remote control valve mounted on the control console.

The symbol in Figure 8.6 represents this application. The dashed line attached to the valve body beside the spring symbol represents the piping attached to the vent port on the pilot operated valve. It provides the input to the remote pilot valve. This is actually a pressure-sensing line, although there will be a flow rate of a few cubic inches per minute through it when the remote pilot valve is open.

Another advantage of the pilot-operated relief valve concerns the use of the vent port. If we simply removed the plug from the vent port, what would happen to the system pressure? Pulling the plug would drop the pilot chamber pressure to zero, so any system pressure high enough to overcome the weak biasing spring (usually around 25 psi) would push the main spool wide open and dump the entire pump flow at that pressure. While we would never simply remove the plug, we could pipe a shutoff, or vent, valve of some sort into the vent port as shown in Figure 8.7. When this vent valve is closed, the small pilot valve controls the pressure. If the vent valve is opened, system pressure drops to essentially zero. This system can be used to provide a low-pressure emergency shutdown capability.

The vent port also provides a troubleshooting feature. One of the more common failure modes of valves designed similarly to the one shown in Figure 8.3 is that the orifice in the disk becomes plugged with contaminants in the fluid—usually Teflon tape or other soft, stringy material. When this happens, the valve will fail to reseat after it has opened; thus, system pressure will be lower than required because of the constant flow through the relief valve. There may also be an unusual amount of heat generation.

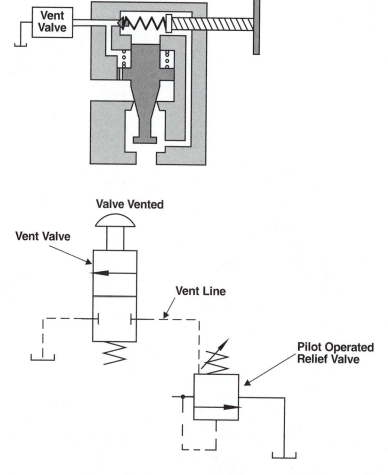

Figure 8.7 A vented pilot-operated relief valve can be used to provide low-pressure emergency shutdown capability.

One technique to determine if the orifice is plugged is to shut down the system and remove the vent plug from the pilot section. When the system is restarted, there should be a small but continuous flow of fluid from the port. If there is no flow, the orifice is probably blocked.

Power Loss across Relief Valves

Any time we have a pressure drop across a valve, power is lost. This loss is a function of the pressure drop across the valve and the flow rate through the valve. We can calculate the power loss from Equation 8.2 as

$$\text{Horsepower Loss} = \frac{\text{Pressure Drop} \times \text{Flow Rate}}{1714}$$

$$HP_{\text{LOSS}} = \frac{\Delta p \times Q}{1714} \tag{8.2}$$

or

$$P_{\text{LOSS}} = \frac{\Delta p \times Q}{60,000} \tag{8.2a}$$

where Δp is the pressure drop across the relief valve and Q is the flow rate through the Relief valve. These are the same equations we used to find the power loss across directional-control valves. Assuming that the exhaust port of the valve is essentially at tank pressure, and that tank pressure is usually atmospheric pressure in industrial applications, Δp is approximately equal to the system pressure, while Q depends on the opening of the valve. At full flow pressure, the total delivery of the pump is being dumped across the valve at full system pressure. Thus, the total power input to the pump is being wasted through pump inefficiencies and relief-valve losses. This constitutes a tremendous waste of energy. Any serious system designer or energy-cost-conscious user would want to take a long, hard look at this method of controlling system pressure. The use of a pressure-compensated pump could prove a worthwhile alternative. (Refer to Chapter 4.) Remember that any such energy loss is converted to heat that must be dissipated or it will overheat the oil and bring about viscosity loss, chemical damage, and probably material damage to the system.

Example 8.4

In a cylinder circuit the flow from a fixed displacement pump operating at 10 gpm (37.85 Lpm) is dumped across a relief valve at 2000 psi (13.8 MPa) when the cylinder reaches the end of its stroke. The duty cycle of the system is such that the cylinder is held at one end or the other for 45 seconds out of every minute, eight hours per day. The system is in operation 200 days per year. Determine the total annual cost of wasted energy across the relief valve if electricity to run the prime mover costs $0.08 per kilowatt-hour. Also, determine the temperature rise and the heat generation rate across the relief valve. Assume a specific heat of 0.5 Btu/(lb°F) (2.1 kJ/(kg°C)) and a specific gravity of 0.90.

Solution

The first step in this type of problem is to find the horsepower lost across the valve. Using Equation 8.2, we get

$$HP_{\text{LOSS}} = \frac{\Delta p \times Q}{1714}$$

$$= \frac{2000 \times 10}{1714}$$

$$= 11.67 \text{ HP}$$

Now convert this to kilowatts:

$$kW_{\text{LOSS}} = 0.746 \text{ kW/HP} \times 11.67 \text{ HP}$$
$$= 8.7 \text{ kW}$$

We now need to get this to kilowatt-hours. We do that by determining the number of hours per year during which energy is used at that rate, then multiplying by our kilowatt figure. This is based on the duty cycle in which the relief valve is open for 45 seconds per minute of operation.

$$\text{Hours} = \frac{45 \text{ sec}}{\text{min}} \times \frac{60 \text{ min}}{\text{hr}} \times \frac{8 \text{ hr}}{\text{day}} \times \frac{200 \text{ day}}{\text{yr}} \times \frac{\text{hr}}{3600 \text{ sec}}$$

$$= 1200 \text{ hr/yr}$$

Thus, the annual kilowatt-hour (kWh) loss is

$$\text{kWh} = 8.7 \text{ kW} \times 1200 \text{ hr/yr}$$
$$= 10{,}440 \text{ kWh/yr}$$

From this, we see that our annual cost for wasting energy is

$$\text{Cost} = \text{kWh/yr} \times \$/\text{kWh}$$
$$= 10{,}440 \text{ kWh/yr} \times \$0.08/\text{kWh}$$
$$= \$835.20/\text{yr}$$

Note that this cost includes only the electricity. It does not include additional maintenance, pump wear, and so forth, due to the continuous high-pressure operation.

We can calculate the heat generation rate (HGR) from Equation 7.1 as

$$HGR = HP \times \frac{42.4 \text{ Btu/min}}{HP}$$

$$= 11.67 \text{ HP} \times \frac{42.4 \text{ Btu/min}}{HP}$$

$$= 494.8 \text{ Btu/min}$$

To calculate the temperature rise (T), we use Equation 7.2.

$$\Delta T = \frac{HGR}{c_P \times W}$$

The weight flow rate, W, is

$$W = \gamma Q = Sg \times \gamma_{\text{WATER}} Q$$
$$= 0.90 \times 62.4 \text{ lb/ft}^3 \times 10 \text{ gal/min} \times \text{ft}^3/7.48 \text{ gal}$$
$$= 75.1 \text{ lb/min}$$

Therefore, the temperature rise is

$$\Delta T = \frac{494.8 \text{ Btu/min}}{0.5 \text{ Btu/lb°F} \times 75.1 \text{ lb/min}}$$

$$= 13.2°\text{F}$$

The solution of this problem is somewhat easier in SI units, because the power loss calculation gives us the loss in kilowatts; thus,

$$P_{\text{LOSS}} = \frac{\Delta p \times Q}{60{,}000}$$

$$= \frac{13{,}800 \text{ kPa} \times 37.85 \text{ L/min}}{60{,}000}$$

$$= 8.7 \text{ kw}$$

This leads us to the same cost per year for wasted energy. The heat generation rate is the same as the power loss. The temperature rise is calculated as

$$\Delta T = \frac{HGR}{c_P \times W}$$

where the weight flow rate, W, is

$$W = \gamma Q = Sg \times \gamma_{\text{WATER}} \times Q$$
$$= 0.90 \times 9.81 \text{ kN/m}^3 \times 37.85 \text{ L/min} \times \text{m}^3/1000 \text{ L}$$
$$= 0.334 \text{ kN/min}$$

Therefore,

$$\Delta T = \frac{8.7 \text{ kW} \times \text{kJ/kW} \cdot \text{sec}}{2.1 \text{ kJ/(kg°C)} \times 0.334 \text{ kN/min} \times 102 \text{ kg/kN} \times \text{min/60 sec}}$$

$$= 7.3°\text{C}$$

If the heat generated cannot be dissipated by the existing system, some modifications must be made to the system. These might include adding a larger reservoir, providing additional airflow across the existing reservoir, and so forth. A far better approach, however, would be to eliminate the heat generation rather than trying to dissipate the heat.

The use of a pressure-compensated pump could significantly reduce these annual costs as well as the heat generation. Although such a pump would initially be more expensive than the fixed displacement pump of the example, it would pay for itself in a relatively short time by reducing energy losses. Other alternatives, as we will see later, might be to use an accumulator and a smaller pump, or to use a tandem-center directional control valve or an unloading valve to allow the pump to operate at low pressure during the stationary time of the cylinder.

8.3 Pressure Reducing Valves

Pressure reducing valves are used when we need to operate one or more branches of a fluid power circuit at pressures lower than the maximum system pressure. Figure 8.8 illustrates the operation of a direct-acting pressure reducing valve. The valve in the figure is normally fully open to allow unrestricted fluid flow from the main system to the branch circuit. As in an adjustable direct-acting relief valve, the pressure setting is varied by adjusting the compression on the control spring, which varies the force required to move the spool. Unlike the relief valve, however, the pressure *downstream* of the valve is sensed. It is this pressure to which the valve responds. The valve symbol reflects this normally open, downstream-sensing function.

The dashed line attached to the spring symbol in Figure 8.8 shows that an external drain line must be provided to drain fluid from the spring cavity. There will always be a small amount of leakage past the moving mechanism into the cavity in which the spring is installed. If a drain line were not provided, the cavity would eventually fill with fluid and the valve could no longer open—a situation referred to as *hydraulic lock*.

In relief valves, the secondary port is the tank port. Since the tank port is unpressurized, the spring cavity drains through the tank port; therefore, no external drain line is needed. However, if the secondary port is pressurized—as it is in pressure-reducing valves as well as sequence valves—an external, unpressurized line is needed.

As long as there is a large flow demand downstream of the valve, the valve will remain open and allow virtually full system flow through it. As the down-

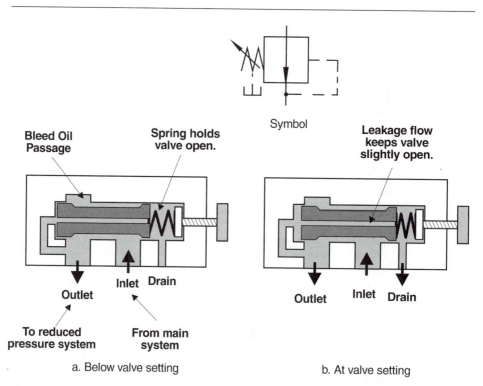

Figure 8.8 Operation of a direct-acting pressure reducing valve.
Source: Courtesy of Vickers, Inc.

stream flow requirements decrease to below full system flow, the downstream pressure begins to build up. This pressure results in a force on the end of the spool opposite the spring. When this force exceeds the spring force, the spool will slide toward the spring end, partially restricting the valve's outlet port. This causes a corresponding pressure drop through the valve. The spool will adjust as necessary to ensure that the desired downstream pressure is not exeeded. The drain port in the valve allows a small amount of flow back to tank to ensure that the valve never closes completely. This is necessary to protect the downstream branch from overpressurization due to leakage that may occur in a fully closed valve, as well as to prevent the valve from locking closed in certain circumstances. Figure 8.9 shows a circuit using a pressure reducing valve.

A pilot-operated pressure reducing valve (Figure 8.10) operates very much like a pilot-operated relief valve. In fact, the pilot stage is nothing more than a direct-acting relief valve that opens and closes in response to the pressure on the poppet. Its operation allows fluid to flow from the valve cavity, upsetting the balance of forces that have held the spool in equilibrium. Spool movement

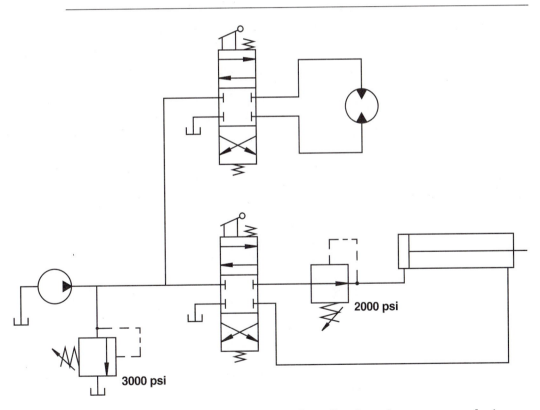

Figure 8.9 This circuit shows a typical application of a pressure-reducing valve.

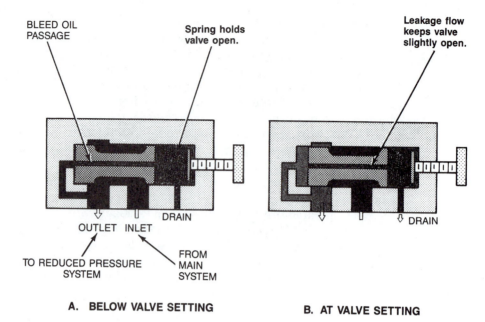

Figure 8.10 Operation of pilot-operated pressure-reducing valve.
Source: Courtesy of Vickers, Inc.

continues until a new equilibrium point is found that again balances the forces to provide the required downstream pressure. The flow from the pilot chamber through the pilot valve must be drained directly to tank.

While there are similarities between pressure relief valves and pressure reducing valves, there are several major differences. These are listed in Table 8.1.

Because a pressure reducing valve can represent a major pressure drop, it can also be a major source of heat generation. Thus, they are usually not

Table 8.1 Comparison of Pressure Relief Valve and Pressure Reducing Valve

Pressure Relief Valve	Pressure Reducing Valve
Senses pressure upstream of the valve.	Senses pressure downstream of the valve.
Outlet port returns to tank.	Outlet port attached to branch circuit.
Outlet flow at tank pressure (usually near atmospheric).	Outlet flow at branch circuit pressure.
Valve normally closed.	Valve normally open.

recommended for applications where they would see long-term or continuous flow.

8.4   Unloading Valves

An *unloading valve* serves as an open, low-pressure flow path for the entire pump output to the reservoir. As such, it reduces the pressure in its branch of the system to near zero psi. Note the important difference here between a pressure relief valve and an unloading valve. While a relief valve provides a flow path to the reservoir, it restricts the amount of fluid passed to that necessary to maintain the system pressure at the prescribed limit. Consequently, there is a very large pressure drop across the relief valve, with the associated temperature increase and so on.

A typical use of an unloading valve would be in a system such as that shown in Figure 8.11 which requires a high flow rate at relatively low pressure during a portion of the cycle, then a low flow rate at high pressure for the remainder of the cycle. This is often accomplished by using two pumps to provide the initial flow requirement, then unloading one pump when the high flow is no longer required. This is commonly referred to as a *high-low system*.

In the circuit in Figure 8.11, both pumps initially provide flow for the system. When system pressure reaches 1000 psi, the pressure is sensed by the unloading valve. It opens to provide a low-pressure (100 psi or less) flow path to dump the 30 gpm pump output back to tank. The resulting low pressure in that part of the circuit causes the check valve to seat. This isolates the 10 gpm pump

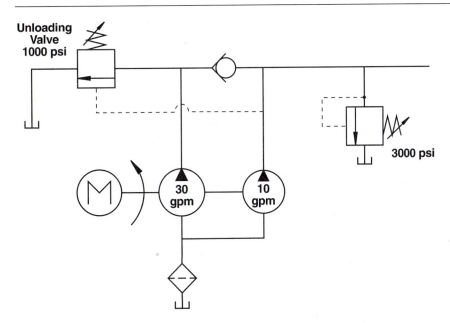

Figure 8.11 A typical application for an unloading valve is a high-low system such as this.

from the 30 gpm pump so that only the 10 gpm goes to the system. The major benefit of this circuit design is the energy savings that result from unloading the high-volume pump when it isn't needed.

8.5   Counterbalance Valves

Counterbalance valves are often used in circuits with vertically operating cylinders supporting heavy loads such as that shown in Figure 8.12. The purpose is to provide a device to prevent the load from dropping too rapidly when the directional control valve controlling the cylinder is shifted to lower the load. The counterbalance valve remains closed, holding the load in place until a pressure higher than the valve setting overcomes the spring force and causes the valve to open. The small size of the flow path through the open valve represents a major restriction that results in a back-pressure that supports, or counterbalances, the load. The speed at which the load is lowered can then be precisely controlled by a flow control valve placed in the inlet line to the top of the cylinder (the blind end in this case). If the counterbalance valve is set properly, minor load changes, fluid temperature changes, and so on, will have virtually no effect on the rate at which the load descends. If it is set too low, the load will descend too rapidly, because there will be insufficient back-pressure to counterbalance the load. If it is set too high, energy will be wasted, but the load will still come down smoothly as long as there is sufficient pressure to open the valve.

Pressure-sensing for these valves can be either at the valve inlet, as in

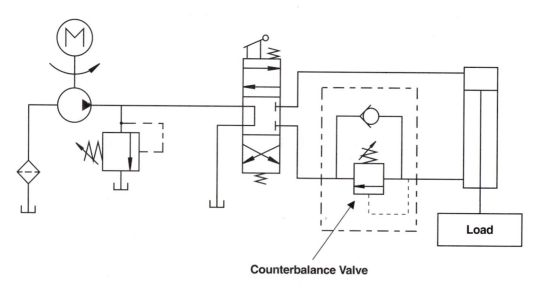

Counterbalance Valve

Figure 8.12 Counterbalance valves are typically used in circuits such as this with vertically operating cylinders supporting heavy loads.

Figure 8.12, or remotely, as in the partial circuit of Figure 8.13. Inlet-sensing provides the capability to lower the load in case of a power failure by reducing the valve pressure setting; however, this design is sensitive to load changes. Remote sensing eliminates the load sensitivity, but it also eliminates the ability to get the load down without system pressure. Because of their placement in the system and the back-pressure they create, counterbalance valves cause high pressure drops with the associated heat generation and temperature rise.

As the valve symbols in Figure 8.12 and 8.13 indicate, counterbalance valves are normally closed and require a pressure on their inlet side (either immediate or remote) to actuate them. This means that a check valve must be provided to bypass the counterbalance valve in order to raise the load. The check valve may be internal or external. In Figure 8.12, the enclosure symbol around the counterbalance valve and check valve indicates that the check valve is an integral part of the valve assembly. The absence of the box in Figure 8.13 indicates that the check valve is piped in separately.

8.6   Brake Valves

A *brake valve* is a variation of the counterbalance valve. It is intended for use with hydraulic motors in applications such as winch circuits with overrunning loads (Figure 8.14). Here the valve receives pressure signals from the motor inlet and outlet lines. These signals are used to position a control spool in the valve to open or restrict the flow path from the motor outlet as necessary. In normal operation, the pressure at the motor inlet is higher than at the outlet, so the valve is held open to a preset point. If the load begins to drive the motor, the motor actually becomes a pump. The result is that the pressure in the inlet line goes down, while the outlet pressure goes up. This causes the brake valve to move toward its closed position and thus restrict the flow coming out of the

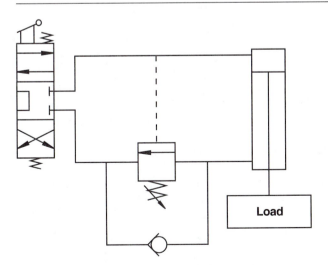

Figure 8.13 Here the counterbalance valve has remote sensing.

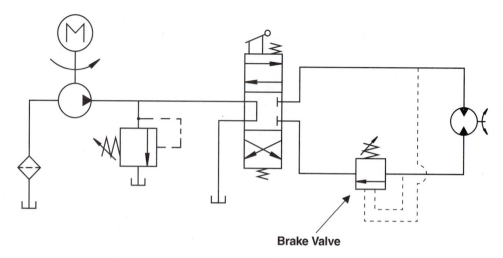

Brake Valve

Figure 8.14 Hydraulic motor control is provided by a brake valve.

motor. The effect is to provide a meter-out flow control that slows the motor and holds it at its preset speed.

8.7   Sequence Valves

The purpose of a *sequence valve* is to control the order of operation of two branches of a system. The sequence valve is normally closed during the operation of the primary branch in which hydraulic cylinders are normally the output devices. When the cylinder(s) have completed the operation at the end of the stroke, the pressure in that part of the circuit will begin to rise. At a pressure below the setting of the relief valve or the pump compensator, the sequence valve will open, porting fluid to the secondary circuit. Figure 8.15 is a diagram of a sequence valve circuit. Let's do a little bit of circuit analysis here to determine exactly how the sequencing circuit operates. With the directional control valve in the center position, both cylinders are held in place because of the closed center. All of the pump output is dumping back to tank at the relief valve pressure. If the spool is shifted downward, ports P and A are connected, as are ports B and T. Flow goes to cylinder B and to the sequence valve, but because the sequence valve is closed, no flow goes to cylinder A. As cylinder B extends, the fluid in the rod end is pushed out and goes through port B to T and back to tank. When cylinder B reaches the end of its stroke, the pressure increases. When it reaches the pressure setting of the sequence valve, the pressure overcomes the internal spring and forces the sequence valve open. This allows flow to cylinder A. As A extends, the rod end fluid is pushed through port B and T of the directional valve and back to tank. When cylinder A reaches the end of its stroke, all pump flow is again dumped over the relief valve to tank.

To retract the cylinders, the directional valve is shifted to the opposite end

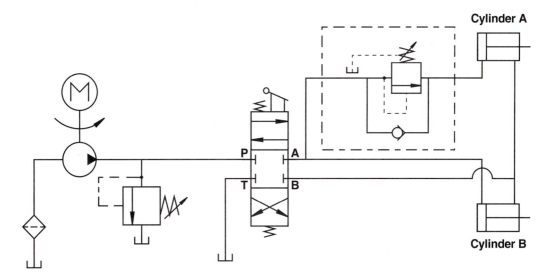

Figure 8.15 In this circuit, the sequencing valve controls the order of operation of two branches of a system.

to tie the pressure port to B port, and the A port to the tank port. Fluid flow goes to the rod end of both cylinders. As cylinder B retracts, the fluid from the blind end goes through the directional control valve (A to T) and back to tank. The fluid from the blind end of cylinder A cannot go through the sequence valve, because it is closed (no pressure is available to open it). Instead, it flows through the check valve, then through the directional control valve and back to tank.

Which cylinder will retract first? From this circuit, we really can't tell. The safe answer is that the cylinder that requires the least pressure will *always* retract first. If it is important that the cylinders retract in a particular order, an additional sequence valve would be necessary to delay the retraction of the second cylinder until the first one had completed its retraction.

Because sequence valves operate in response to the pressure in the inlet line, any change in the system (such as the addition of a flow control for the first cylinder) can result in loss of sequence. In most cases, a pilot-operated sequence valve as shown in Figure 8.16, can be used to correct the problem by sensing the pressure near the inlet of the first cylinder rather than using system pressure.

8.8   Troubleshooting Tips

When troubleshooting problems associated with pressure control valves, it is important to keep in mind the purpose and function of each of the valves as it operates within the context of the system. Remember that most of these valves simply sense pressure and respond to it. Remember, also, that the fluid flow will always take the path of least resistance. Knowing these principles,

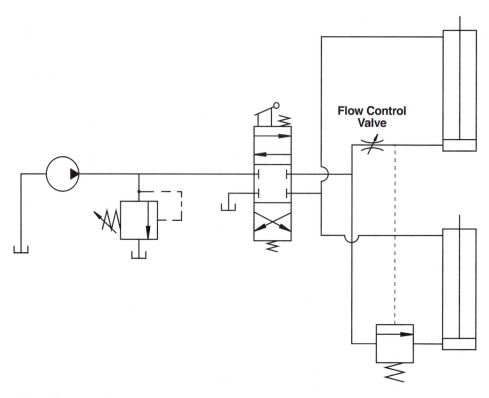

Figure 8.16 This circuit uses a pilot-operated sequence valve and a flow control valve.

combined with the following possible causes, should help you to troubleshoot the problem systematically and quickly determine the cause of the problem. Intelligent use of pressure gages will help, also.

Symptom	Possible Cause
No pressure or pressure too low	1. Pump not running 2. Low-pressure path to reservoir somewhere in circuit 3. Valve stuck open 4. Orifice plugged in pilot operated relief valve 5. Broken spring 6. Remote pilot valve set too low on pilot-operated relief valve 7. Pump compensator set too low 8. Gage reading incorrect
Pressure too high	1. Valve stuck closed 2. Internal drain line plugged 3. Gage reading incorrect
Can't adjust pressure	1. Valve stuck

Symptom	Possible Cause
	2. Lower pressure path to reservoir somewhere in circuit
	3. Another relief valve in circuit set to lower pressure
Valve opens too slowly	1. Incorrect valve type
Valve opens too fast, gives harsh action	1. Using direct-acting instead of pilot-operated valve
Valve noisy or erratic	1. Wrong size valve
	2. Incorrect valve type
	3. Valve dirty
	4. Air in system
	5. Weak spring
	6. Reseating pressure too low
Oil temperature too high	1. Relief valve set higher than necessary
	2. Valve chattering
	3. Relief valve too small
Improper reduced pressure (pressure reducing valve)	1. Improper setting
	2. Valve stuck
	3. Gage reading incorrect
Reduced pressure branch locked up.	1. Internal drain plugged in reducing valve
System will not sequence properly.	1. Drain port blocked
	2. Sequence valve pressure setting too low
	3. System modification has caused increased pressure in line upstream of sequence valve
	4. Sequence valve pressure setting higher than relief valve setting

8.9   Summary

Pressure control valves are used for many purposes in fluid power systems—safety, pressure limitation, and sequencing. They may be operated directly by system pressure or pilot-operated by pressures sensed at various points throughout the system.

The ISO symbols for pressure control valves—summarized in Figure 8.17—are all very similar. This is because their functions, and, in fact, their physical construction, are so similar. (Some manufacturers produce a single valve that can be used to perform several different functions.) When analyzing circuits, you usually must consider the context in which the symbol is used rather than the symbol itself to determine whether the valve is a relief valve, an unloading valve, a counterbalance valve, or a sequence valve. The key is usually where the line from the secondary port goes.

Relief valves used in conjunction with fixed displacement pumps can be major sources of energy loss and heat generation in closed-center or deadheaded circuits. If this represents a significant problem, alternative designs (pressure-compensated pumps or open-center systems) should be considered.

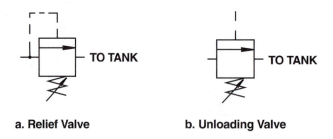

a. Relief Valve   b. Unloading Valve

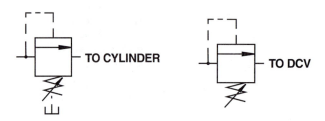

c. Sequencing Valve   d. Counterbalance Valve

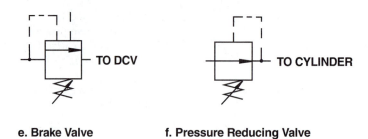

e. Brake Valve   f. Pressure Reducing Valve

Figure 8.17 ISO symbols for pressure control valves are all very similiar.

Although relief valves can effect system flow, they should never be used to control system flow. The flow control valves that we will discuss in Chapter 9 are specifically designed for that purpose.

8.10 Key Equations

Spring force: $F = k \times d$ **(8.1)**

Review Problems

Note: The letter "S" after a problem number indicates SI units are used.

General

1. List the six different types of pressure control valves.
2. Which pressure control valves can be vented or pilot-operated?
3. Which pressure control valves require external drains for the spring cavities? Why can't they be drained internally?
4. What happens to a sequence valve when the drain port is plugged?
5. When does a pressure control valve need a check valve in parallel with it?
6. Explain the difference between direct-acting and pilot-operated pressure control valves.
7. How is a pressure reducing valve different from all other pressure control valves?
8. What is the purpose of a pressure relief valve?
9. What is the purpose of an unloading valve?
10. What is the purpose of a sequencing valve?
11. What is the purpose of a brake valve?
13. What is the purpose of a pressure reducing valve?

Power Loss

14. Fluid is flowing through a relief valve at 1500 psi and 10 gpm. What is the horsepower loss through the valve?
15. The full flow pressure of a relief valve is 2000 psi. At that setting, the flow through the valve is 20 gpm. What is the horsepower loss through the valve?
16. A relief valve has the pressure-flow characteristics shown in Figure 8.2 on page 280. What is the horsepower loss across the valve when the flow through the valve is 8 gpm?
17S. An open relief valve is passing fluid at 40 Lpm. The pressure drop across the valve is 2000 kPa. Find the power loss through the valve.
18S. How much power is lost through a relief valve when the flow rate through it is 100 Lpm and the pressure drop is 8 MPa?

Spring Force

19. A spring has a spring constant of 20 lb/in. How much force is required to compress it 2.5 inches?
20. How much force is required to compress a spring 0.5 inches if its spring constant is 750 lb/in?
21. A relief valve has a poppet with a surface area of 0.8 in² exposed to the fluid pressure. The valve spring has a spring constant of 2000 lb/in. If the spring is compressed 0.35 inches when the valve is closed, what is the cracking pressure of the valve?
22. The valve in Review Problem 21 must be opened 0.3 inches to allow full flow. What is the full flow pressure of the valve?
23S. A spring has a spring constant of 35 N/cm. How much force is required to compress it 1.5 cm?
24S. A nonadjustable relief valve has a poppet with a surface area of 5 cm². the spring has a spring constant of 4000 N/cm. When the valve was assembled, the spring was compressed 0.95 cm. What pressure is required to move the poppet off its seat? (In other words, what is the cracking pressure of the valve?)
25S. The poppet of the valve in Review Problem 24S must be opened 0.8 cm to provide full flow. What is the full flow pressure of the valve?

Heat Generation

26. What is the heat generation rate through a relief valve if 15 HP is lost across the valve?

27. If 75 HP is being dumped across a relief valve, what is the heat generation rate?

Temperature Rise

28. Fluid is flowing through a relief valve at 30 gpm. The pressure is 300 psi. The fluid has a specific weight of 58.3 lb/ft^3 and a specific heat of 0.48 Btu/lb°F. Find the temperature rise across the valve.

29. How much will the fluid temperature rise if fluid with a specific heat of 0.5 Btu/lb°F and a specific gravity of 0.90 is flowing through a relief valve at 20 gpm and 1600 psi?

30S. Fluid is flowing through a relief valve at 120 Lpm and 17.5 MPa. The fluid has a specific weight of 9.2 kN/m^3 and a specific heat of 1.9 kJ/(kg°C). Find the temperature rise across the valve.

31S. How much will the fluid temperature rise for a fluid with a specific heat of 1.93 kJ/(kg°C) and a specific gravity of 0.87 when it flows through a valve at 75 Lpm with a pressure drop of 600 kPa?

32. A system contains 85 gal of fluid that has a specific gravity of 0.9 and a specific heat of 0.49 Btu/(lb°F). When the system is deadheaded, fluid flows through the relief valve at 25 gpm with a pressure drop of 2300 psi. Assuming that there is no heat dissipation from the system, how much will the system temperature increase if the system is deadheaded for 30 minutes? (**Hint:**) The temperature rise calculation must be based on the weight of fluid in the system rather than the weight flow rate through the valve.)

Circuit Practice

33. Using standard ISO symbols, draw a complete circuit diagram for a sequencing circuit to provide the following operating sequence:
    Extend 1
    Extend 2
    Retract 1
    Retract 2

34. A hydraulic broaching machine uses a fixed displacement pump, two hydraulic cylinders to clamp a part in place, and a third cylinder to broach the part. The clamp cylinders both have 2 in bores. The broaching cylinder has a 5 in bore and produces a force output of 40,000 lb. It extends at 1.5 ft/sec. The operator uses a four-way, three-position, solenoid-operated, spring-returned, closed-center directional control valve to initiate the operation. Once started, the extension sequence progresses automatically. Clamp cylinder 1 extends first, then clamp cylinder 2, then the broaching cylinder. When the operator shifts the directional control valve to retract the cylinders, there is no control over the retraction sequence. The relief valve is set so that its cracking pressure is 100 psi below the maximum pressure required to operate the system. The cracking pressure is 80 percent of its full flow pressure. The fluid has a specific gravity of 0.90 and a specific heat of 0.48 Btu/(lb°F).

    Using standard ISO symbols, draw the complete circuit diagram for the machine. Calculate the pump flow rate required, the cracking pressure of the relief valve, the heat generation rate, and the temperature rise across the relief valve when the directional control valve is centered. Suggest ways in which the design could be improved to reduce the energy loss.

Flow Control Valves

Outline

9.1 Introduction

As we discussed in the chapters on hydraulic cylinders and motors, the only way to control the speed of these devices is to control the rate at which the fluid flows into or out of the actuator. This control can be accomplished by varying the speed of a fixed displacement pump, by varying the output of a variable displacement pump, or by using a *flow control valve* to meter the fluid flow at the required rate.

Flow control valves control flow by two methods, depending on the design. One type of valve restricts the flow rate into or out of the component. The second type controls the flow by diverting fluid flow away from the component.

Objectives

When you have completed this chapter, you will be able to

- Explain the difference between flow restricting and flow diverting types of flow control valves.
- Explain the difference between compensated and noncompensated types of flow control valves and describe the applications for each.
- Explain the effects of pressure and temperature changes on flow through flow control valves.
- Calculate the flow through flow control valves.
- Discuss the operation of priority valves and flow divider valves.
- Explain the difference between meter-in and meter-out circuits.
- Calculate the pressures in various system locations that result from the use of meter-in and meter-out flow control.
- Recognize the symbols used for flow control valves.
- Draw and analyze circuit diagrams involving flow control valve symbols.

9.2   Flow Restricting Valves

A *flow restricting valve* is one that is placed in series with the actuator and constitutes a major restriction in the flow path. It allows a portion of the fluid flow to pass through it and blocks the remaining, or excess, flow, which is usually returned to the reservoir through the relief valve.

Needle Valves

The simplest and most common of the flow restricting valves is a simple *needle valve* (Figure 9.1). This type of valve consists of a valve body that contains an orifice and a tapered stem (or needle) which can be screwed in and out to vary the orifice opening. The stem may have a flat or a groove to provide for better control. The symbol shown in Figure 9.1 represents all of the simple restrictor-type valves, regardless of the actual restrictor mechanism.

There are other types of flow control valves, such as globe valves, gate valves, and ball valves. These valves provide less sensitive flow control than needle valves and are seldom used in fluid power applications.

A simple needle valve restricts flow in both directions through the valve. There is, however, a preferred direction of flow, normally indicated by an arrow on the valve body. When the flow is in this direction, the pressure loss across the

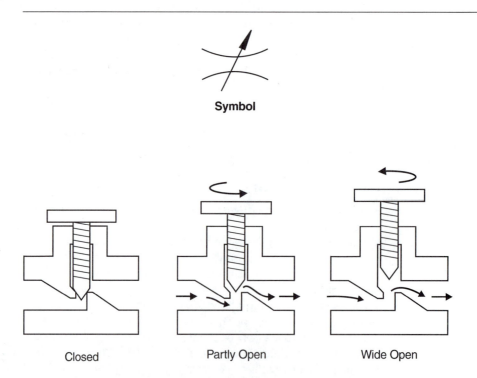

Symbol

Closed   Partly Open   Wide Open

Figure 9.1 The needle valve is the most common of the flow restricting valves.

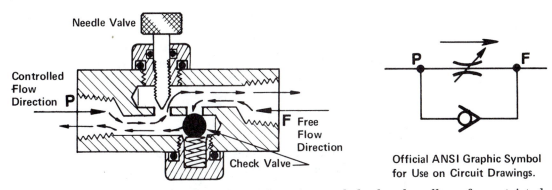

Figure 9.2 A needle valve with an internal check valve allows for restricted flow in one direction and free flow in the other direction.
Source: Industrial Fluid Power, Vol. 1, Published by Womack Educational Publications.

valve is slightly less and the valve is self-cleaning and not likely to accumulate contaminant particles that might build up and restrict flow.

The addition of an internal check valve allows for restricted flow in one direction and free flow in the other direction (Figure 9.2). Flow entering the P port of the valve is obstructed by the check valve and directed through the metering orifice. When flow is reversed, the fluid enters the F port. Flow in that direction opens the check valve and flows freely through the valve. No control is exerted in the reverse direction, so the valve must be sized to accommodate the maximum flow that can be expected in that direction.

Before we can discuss the performance characteristics of any flow control valve, we need to establish an understanding of the device from which such valves were derived, the simple orifice. An *orifice* is a device (usually a plate or a plug) that has a calibrated hole drilled in it (Figure 9.3). When placed in a flow path, by whatever means, the orifice restricts the amount of flow that can pass through it. For any given orifice design there are three factors that

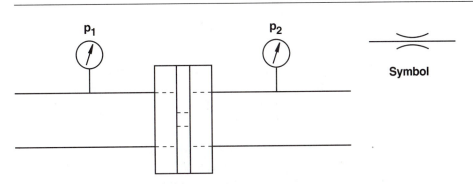

Figure 9.3 An orifice is a device with a calibrated hole drilled in it.

affect the flow rate through the orifice—the size (area) of the orifice (A), the pressure drop across the orifice ($\Delta p = p_1 - p_2$ in Figure 9.3), and the viscosity of the fluid, which is a function of the fluid temperature.

In a simplified equation (often referred to as the *orifice equation*) we bring these factors together to define the flow rate through the orifice as

$$Q = C \times A \sqrt{\frac{2g\Delta p}{\gamma}} \tag{9.1}$$

where C is a constant that is a characteristic of the particular orifice design, A is the area of the orifice, g is the acceleration of gravity, and γ is the specific weight of the fluid (which is a function of temperature). In order to change the flow rate through the orifice, one or more of those parameters must be changed. For any given orifice, C and A are fixed, g remains constant, and the specific weight is relatively constant for an operating fluid. That leaves only the differential pressure as a variable parameter.

This presents a major disadvantage in trying to use a fixed orifice as a control device in common fluid power systems. Any time the pressure differential across the orifice changes, the flow rate through the orifice changes. As we have seen, the speed of any fluid power actuator (motor or cylinder) depends on fluid flow rate. Thus, if the upstream pressure varies for any reason or if the downstream pressure varies (as will happen if the load changes), then the speed of the actuator changes.

To help overcome this problem, devices such as needle valves are used. Needle valves are nothing more than variable orifices that use the stem to change the size of the orifice. This gives us the ability to compensate for changes in the differential pressure in order to maintain a constant flow rate through the valve and, consequently, a constant actuator speed. It also allow us to establish a desired flow rate without removing a fixed orifice and replacing it with one of a different size.

The ability to vary the orifice size in a needle valve also causes the orifice constant, C, to change. Consequently, we usually modify the orifice equation to the form shown in Equation 9.2.

$$Q = C_v\sqrt{\Delta p/Sg} \tag{9.2}$$

where

Q = volume flow rate.
C_v = flow coefficient.
Δp = pressure differential across the valve (psi).
Sg = specific gravity of the liquid.

The flow coefficient, C_v, is defined as the volume flow rate of water at 60°F that will flow through the valve in one minute at a pressure drop of 1 psi, and has the rather strange units of gpm/$\sqrt{\text{psi}}$ (Lpm/$\sqrt{\text{kPa}}$). The value is peculiar to each valve type and design, and it must be determined experimentally for each valve design by testing the valve in accordance with NFPA Standard T3.5.7. The C_v value is commonly presented in manufacturers' data along with the flow rates and pressure drops through their valves. The value seen in manufacturers' data is typically the value for a fully open valve and is termed the "rated" C_v.

Example 9.1

Determine the flow rate through a flow control valve that has a flow coefficient of 12 (17.3) and a pressure drop of 15 psi (103 kPa). The fluid is a hydraulic oil with a specific gravity of 0.86.

Solution

Using Equation 9.2, we have

$$Q = C_v \sqrt{\Delta p / Sg}$$

$$= 12\sqrt{15 \text{ psi}/0.86}$$

$$= 50 \text{ gpm}$$

That is,

$$Q = 17.3\sqrt{103 \text{ kPa}/0.86}$$

$$= 189.3 \text{ Lpm}$$

Example 9.2

A valve has a flow coefficient of 20 (28.8). Find the pressure drop across the valve if an oil with a specific gravity of 0.90 is flowing through it at 45 gpm (170 Lpm).

Solution

For this problem, we must solve Equation 9.2 for Δp.

$$Q = C_v \sqrt{\Delta p / Sg}$$

$$\Delta p = Q^2 Sg / C_v^2$$

$$= 45^2 \times 0.90/20^2$$

$$= 4.56 \text{ psi}$$

In SI units,

$$\Delta p = 170^2 \times 0.90/28.8^2$$

$$= 31.4 \text{ kPa}$$

A drawback of simple needle valves is their response to changes in pressure differential. Figure 9.4 shows the pressure-flow characteristics of a typical needle valve. It is obvious from the plot that this valve provides no real measure of control unless the pressure differential remains constant. While a constant pressure differential might occur in many fluid transfer and process operations, it is not typical of most working fluid power systems. The normal working system may require more sophisticated flow control than can be provided by a simple needle valve. A pressure-compensated flow control valve, discussed next, is often necessary.

Pressure-Compensated Flow Control Valves

An adjustable, *pressure-compensated valve* is similar to a simple needle valve in that a pointed stem is moved in and out of a sharp-edged control orifice to

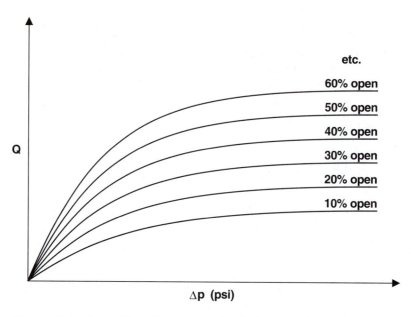

Figure 9.4 Flow (Q) and pressure (Δp) characteristics for a typical needle valve.

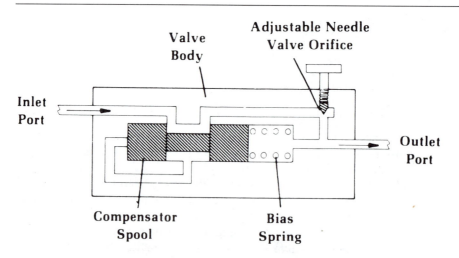

**Pressure Compensated Flow
Control Valve - Restrictor Type**

Figure 9.5 A pressure-compensated valve provides more stable flow control than a simple needle valve.

Source: Courtesy of Parker Hannifin Corporation.

change the orifice size. The difference is that there is a sliding spool on the inlet side of the pressure-compensated valve (Figure 9.5). As the upstream pressure increases, a force is exerted on the face of the compensating spool that causes it to slide back into the control cavity and partially restrict the control orifice. An increase in the downstream pressure pushes the spool the other way. The purpose of the compensator is to create a pressure drop that keeps the pressure differential across the adjustable needle valve constant— usually about 100 psi (689.5 kPa). This allows the valve to maintain a constant downstream flow rate even though the pressure drop across the entire valve may be changing frequently and over a wide range. In Figure 9.6, each line in the family of curves represents a different valve setting. Notice that there is an apparent increase in the flow rate at the lower pressure settings. This occurs because of the positive force exerted in the compensator spool by the spring. Until the inlet pressure is high enough to generate sufficient force to begin to compress the spring, the valve acts as a simple needle valve. The valve also acts as a simple needle valve if the flow is backward through it.

It is obvious that a pressure-compensated valve would be much more suitable than a noncompensated valve in any system where a steady actuator speed is

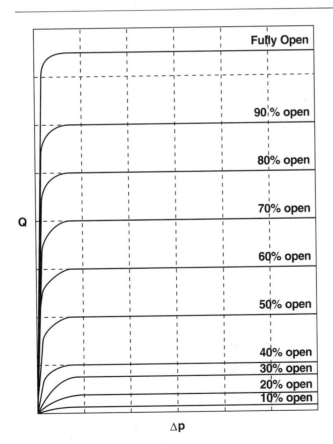

Figure 9.6 Typical pressure-compensated flow control valve performance.

required. These valves will typically maintain output flow within ± 5 percent of the selected setting over a wide pressure range. This is sufficiently accurate for the majority of industrial and mobile equipment applications. Temperature compensation can also be added to these valves to provide an additional measure of control.

Pressure-compensated valves may also be equipped with internal check valves. As in needle valves, these check valves allow fluid to flow freely through the valve in the reverse direction. The controlled flow direction will be clearly indicated on the valve by either an arrow on the valve body or by marking the inlet and outlet ports.

All types of flow control valves can be mounted in the same ways as other valves—in-line, subplate, or cartridge manifolds.

9.3   Flow Diverting Valves

Instead of providing a flow-controlling restriction, some flow control valves divert a portion of the fluid flow back to the reservoir. These *flow diverting valves* are alternatively termed *bypass flow control valves* and *three-port flow control valves*. Besides the inlet and outlet ports of the restricting-type valves, the three-port valves have a third port, connected to a low-pressure fluid line, usually one that returns directly to the reservoir or to a return manifold. They are somewhat similar to the pressure-compensated restrictor-type valves in that a spring-loaded sliding spool containing a fixed orifice is employed as the flow control mechanism; however, instead of partially blocking the control orifice to provide the adjustments in flow rate, the three-port valve bypasses the excess fluid back to the reservoir. The spool responds to the pressure differential across the valve to maintain the proper orifice opening (Figure 9.7). These valves are often temperature-compensated as well. As we will see in

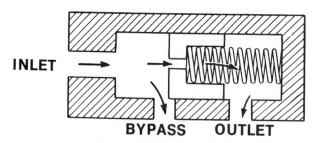

Full Flow To Bypass, Outlet Restricted

Figure 9.7 A bypass (or three-port) flow control valve performs its function by diverting part of the pump flow to the tank at low pressure, thus saving energy cost and heat generation.

Source: Reproduced by permission of Deere & Company, © 1987 Deere and Company. All rights reserved.

Section 9.6, advantages of this valve over the restrictor-type valves are reduced heating of the fluid and a significant reduction in the horsepower input required during portions of the operating cycle.

9.4 Other Flow Control Valves

Two other types of flow control valves are sometimes used in fluid power circuits. They are often referred to as *flow divider valves* because they divide the flow between two or more circuits.

Priority Valves

Priority flow divider valves are usually used to ensure that certain essential system functions are performed at the expense of nonessential functions in case part of the flow capability of the system is lost for some reason.

Figure 9.8 illustrates the operation of a priority flow divider valve. As long as the pump flow exceeds the flow requirements of the primary system, the priority outlet is partially restricted by the spool. This occurs because the annulus surrounding the fixed orifice is acted upon by the fluid entering the valve. In this position, the secondary port is open to allow fluid to flow to the secondary portion of the system.

If the pump output is reduced, the control spring causes the spool to slide toward the inlet port. The spool then begins to restrict the secondary outlet,

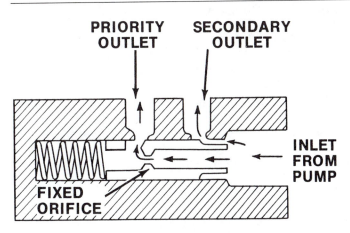

Priority Flow Divider

Figure 9.8 A priority flow divider valve ensures that certain essential system functions are performed at the expense of nonessential functions if necessary.

Source: Reproduced by permission of Deere & Company, © 1987 Deere and Company. All rights reserved.

OUTLET NO. 2 OUTLET NO. 1

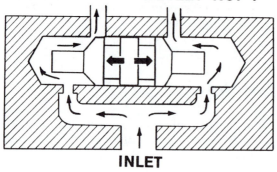

INLET

Proportional Flow Divider

Figure 9.9 A proportional flow divider divides whatever volume flow rate is available between two circuits.

Source: Reproduced by permission of Deere and Company, © 1987 Deere and Company. All rights reserved.

---

thereby reducing the flow to the secondary system. As the pump flow continues to decrease, the secondary port finally becomes completely blocked, and all available flow is routed through the priority port.

Proportional Flow Divider

A *proportional flow divider valve* (Figure 9.9) receives whatever volume flow rate is available and divides it between two circuits. This type of valve uses a free-floating spool that moves in response to the pressures on each end. The lands on the spool alternately cover and uncover the outlet ports to the circuits it serves, in essence varying the port sizes. This allows a constant flow rate to each of the circuits. The flow can be divided equally between the two circuits, or divided proportionally between them in ratios up to 90 percent to 10 percent.

These valves find application in many split-flow systems, particularly in differentially actuated power steering systems.

9.5 Flow Control Valve Applications

As stated earlier, flow control valves regulate the speed of a hydraulic actuator by controlling the volume flow rate to the actuator. This can be accomplished in three ways:

- Controlling the flow rate into the actuator (termed *meter-in*).
- Controlling the flow rate out of the actuator (termed *meter-out*).

- Diverting a portion of the fluid before it reaches the actuator (termed *bleed-off*).

In some instances, a combined meter-in and meter-out circuit may be used.

Flow Control Circuits

In a *meter-in circuit,* a flow control valve is placed in the line to the inlet port of the actuator. The actuator may be either a cylinder, rotary actuator, or a hydraulic motor. A meter-in circuit of a cylinder is shown in Figure 9.10. Adjusting the valve setting varies the flow rate into the cylinder. Meter-in circuits are normally used only in those arrangements where the load is always resisting the force of the actuator. Where the force is not resistive, as in the case of a suspended load, it is possible for the load to drop uncontrolled, even if the meter-in valve is fully closed.

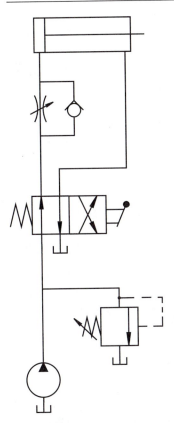

Figure 9.10 Meter-in speed control (extending) works by means of a flow control valve placed in the inlet line of the actuator.

Example 9.3

In the meter-in system shown here, determine the pressure on each gage as the cylinder is extending. The relief valve is set at 2000 psi (13.8 MPa).

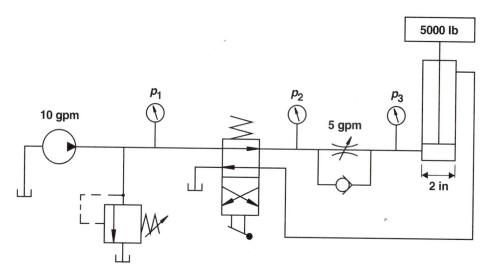

Example 9.3

Solution

We know that the load determines the pressure in the cylinder, so we can readily calculate p_3, downstream from the flow control valve.

$$p_3 = \frac{F}{A} = \frac{5000 \text{ lb}}{3.14 \text{ in}^2} = 1592 \text{ psi}$$

or 10.98 MPa.

We can now estimate the pressure at p_2. (The actual pressure depends on the characteristics of the relief valve, as we saw in Chapter 8.) The flow control valve is set to pass 5 gpm (18.9 Lpm), but the fixed displacement pump has a 10 gpm (37.8 Lpm) flow rate. Where is the other 5 gpm (18.9 Lpm) going?

The only place it *can* go is through the relief valve and back to the reservoir. For it to do that, there must be sufficient pressure to open the relief valve. The relief valve is set at 2000 psi (13.8 MPa), so we can assume that value for both p_1 and p_2. (The actual pressure would be somewhat less than that because of relief valve characteristics and other losses in the system.)

A *meter-out circuit* is shown in Figure 9.11. The valve placed in the outlet line controls the rate at which the fluid leaves the actuator. (Keep in mind that the terms *inlet* and *outlet* depend on the direction of travel of the cylinder.) The meter-out technique is usually preferred over the meter-in technique be-

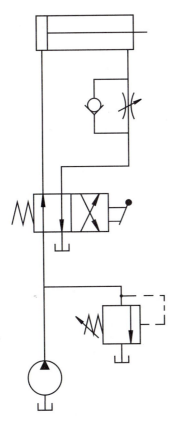

Figure 9.11 Meter-out speed control (extending) work-by means of a flow control valve placed in the outlet line of the actuator.

cause it adds stability to the circuit and eliminates the possibility of a suspended load running away.

The meter-out circuit presents the likelihood of pressure intensification when the cylinder is being extended. Because the flow control valve restricts the outlet flow, a back-pressure results on the rod end of the cylinder. The degree of intensification depends on the load, the flow restriction, and the size of the rod relative to the piston.

Example 9.4

For this meter-out system used with a suspended load, determine the pressure on each gage during extension if there is no load and for a load of 5000 lb (22,250 N). Assume a constant speed.

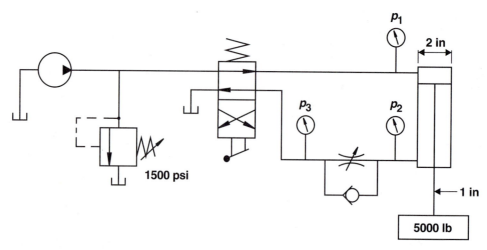

Example 9.4

Solution

The assumption of a constant speed means that the forces in each direction are equal—if they weren't, we would have either acceleration or deceleration. This equal-force situation is the key to the solution of the problem.

We will assume, also, that the flow control valve is causing a restriction not only on the flow leaving the cylinder, but also on the flow entering the cylinder. (If you question this assumption, you can calculate the incoming flow based on cylinder extension velocity and piston area; $Q = A \times v$.) Based on this assumption, we can estimate the pressure on the cap end as before, taking it to be the relief valve pressure of 1500 psi (10.3 MPa). This means that there is a pressure-generated force pushing the piston downward of

$$F = p \times A = 1500 \text{ lb/in}^2 \times 3.14 \text{ in}^2 = 4710 \text{ lb}$$

or 20,960 N.

Our constant-speed stipulation means that there is an equal force pushing upward, but where does that force come from? It must come from pressure in the rod end, which is found from

$$p_2 = \frac{F}{A_P - A_R} = \frac{4710 \text{ lb}}{(3.14 - .785)\text{in}^2} = 2000 \text{ psi}$$

or 13.8 MPa.

You have probably already recognized this as simply a pressure-intensification problem as we discussed in Chapter 6. Remember, though, that this is a no-load situation. If there is a load, things are a little different.

When we add the 5000 lb (22,250 N) load, we still have the same pressures upstream, so we still have the 1500 psi (10.3 MPa) on the cap end with the resultant 4710 lb (20,960 N) downward force; however, the consequence of adding the 5000 lb (22,250 N) load is an additional 5000 lb (22,250 N) downward force. Still assuming constant speed, we must now have a (5000 + 4710) = 9710 lb (43,210 N) upward force opposing the downward force.

Again, the upward force results from the pressure in the rod end, which now becomes

$$p_2 = \frac{F}{A_P - A_R} = \frac{9710 \text{ lb}}{(3.14 - .785)\text{in}^2} = 4123 \text{ psi}$$

or 28.4 MPa.

In this circuit, what will be the pressure p_3? In all such situations, where we have no information about pressure drops through the directional control valve, lines, and so forth, we assume that the return flow is at (or near) atmospheric pressure. Therefore, we assume $p_3 = 0$ psi (0 kPa).

You must be very careful in how you use meter-out flow controls for two reasons—very high pressures may result due to pressure intensification in the cylinder and the accompanying pressure drop across the valve may cause a high rate of heat generation (see Section 9.6).

Figure 9.12 shows a circuit in which a three-port flow control valve is used. This type of valve can be used only on the inlet side of the actuator. A major

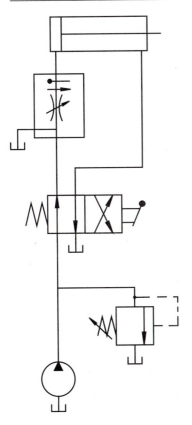

Figure 9.12 This three-port flow control valve can be used only on the inlet side of the actuator.

advantage of the three-port valve is that the system pressure is determined entirely by the load and never exceeds that required to move the load. In circuits utilizing the two-port flow control valves, the pressure upstream of the valve will continue to rise until it reaches the setting of the system relief valve. A major disadvantage is that three-port valves do not provide great accuracy in speed control. This type of speed control is unsuitable for parallel circuits where there is more than one actuator.

A variation on the use of a port flow control valve is shown in Figure 9.13. In this circuit, termed a *bleed-off* circuit, a restrictor type flow control valve is used, but it is connected in parallel with the line leading to the cylinder. The outlet port of the valve dumps directly to the reservoir. This provides a flow path by which the excess flow (not needed to operate the cylinder) can return to the reservoir without going through the relief valve. The advantage of this bleed-off circuit is that, as with the three-port valve, the load rather than the relief valve determines the system operating pressure.

Pressure Drop

There is a pressure drop through any flow control valve that results in horsepower loss and heat generation. In a meter-in circuit, for example, the pressure drop across the flow control valve depends on the force required by the actuator. If there is no load on the cylinder in Figure 9.14, the pressure at the control valve outlet is essentially zero. The result is a high pressure drop that results in a high horsepower loss. As the load increases, the force required to move the load increases, so the pressure in the portion of the circuit between the valve and the cylinder increases. The pressure drop decrease means that there is less horsepower lost and less heat generated. What happened to the horse-

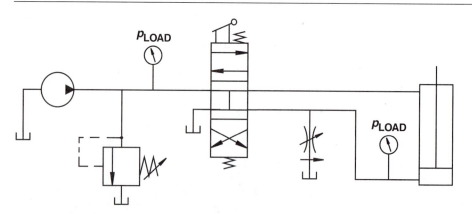

Figure 9.13 In a bleed-off flow control circuit, a restrictor type valve is connected in parallel with the line leading to the cylinder.

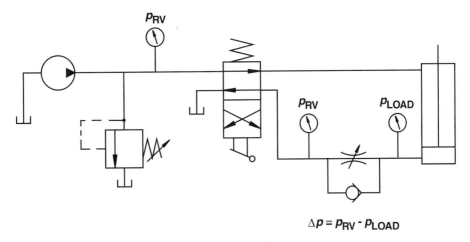

$$\Delta p = p_{RV} - p_{LOAD}$$

Figure 9.14 The pressure drop across the flow control valve in a meter-in circuit depends on the force required by the actuator.

power difference between the two cases? It was used to do useful work in lifting the load.

We calculate the power loss across a flow control valve in the same way that we calculate the power loss across directional and pressure control valves:

$$HP_{\text{LOSS}} = \frac{\Delta p \times Q}{1714}$$

$$P_{\text{LOSS}} = \frac{\Delta p \times Q}{60,000}$$

In this case, the Δp term is the pressure differential across the flow control valve, while Q is the flow rate through the flow control valve. The pump outlet flow is of no concern here; we are interested only in the flow through the flow control valve.

Example 9.5

In the circuit shown here, the pressure upstream of the flow control valve is 2000 psi (13.8 MPa or 13,800 kPa), and the flow rate through the valve is 10 gpm (37.85 Lpm). The cylinder bore is 4 in (10.16 cm). Assuming there is no load on the cylinder, find the horsepower lost across the valve. Also, find the horsepower loss if the load is 10,000 lb (44,500 N).

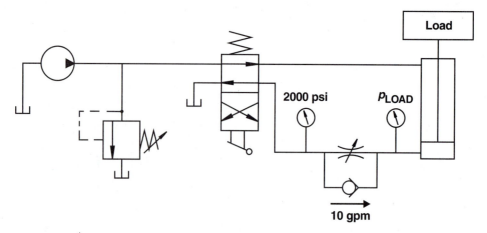

Example 9.5

Solution
With no load, the pressure at the outlet of the valve is zero, while the inlet pressure remains at 2000 psi (13.85 MPa). Thus, we can find the power loss by

$$HP_{\text{LOSS}} = \frac{\Delta p \times Q}{1714}$$

$$= \frac{2000 \text{ lb/in}^2 \times 10 \text{ gal/min}}{1714}$$

$$= 11.7 \text{ HP}$$

or

$$P_{\text{LOSS}} = \frac{13,800 \text{ kPa} \times 37.85 \text{ Lpm}}{60,000}$$

$$= 8.7 \text{ kw}$$

Now add the load. The pressure required to move the 10,000 lb (44,500 N) load is

$$p = \frac{F}{A}$$

$$= \frac{10,000 \text{ lb}}{12.56 \text{ in}^2}$$

$$= 796 \text{ psi}$$

or 5488 kPa.

Thus, the pressure drop across the valve is

$$2000 - 796 = 1204 \text{ psi}$$

or 8302 kPa.

The power loss across the valve, then, is

$$HP_{\text{LOSS}} = \frac{\Delta p \times Q}{1714}$$

$$= \frac{1204 \text{ lb/in}^2 \times 10 \text{ gal/min}}{1714}$$

$$= 7 \text{ HP}$$

or

$$P_{\text{LOSS}} = \frac{8302 \text{ kPa} \times 37.85 \text{ Lpm}}{60,000}$$

$$= 5.24 \text{ kw}$$

The significance of this power loss is that all power produced by the pump but not used to do useful work constitutes wasted energy. In fluid power systems, this wasted energy is converted to heat. We can calculate the amount of heat generated by converting the horsepower loss to Btu/min using Equation 7.1.

$$HGR = 42.4 \text{ Btu/min} \times HP$$

To find the temperature rise of the fluid flowing through the valve, we divide the heat generated by the specific heat (c_p) of the oil and the weight flow rate (W) of the oil. This is expressed in Equation 7.2.

$$\Delta T = \frac{HGR}{c_p \times W}$$

Example 9.6

Calculate the temperature rise across the valve of Example 9.5 in the case where the pressure drop across the valve is 2000 psi (13.8 MPa). The oil used has a specific gravity of 0.86 and a specific heat of 0.45 Btu/lb°F.

Solution

In Example 9.5, we calculated the horsepower loss across the valve to be 11.7. Using Equation 7.1, we find that the heat generated is

$$HGR = 42.4 \text{ Btu/min} \times HP$$

$$= 42.4 \text{ Btu/min} \times 11.7$$

$$= 496 \text{ Btu/min}$$

In SI units, the heat generation rate equals the power loss, 8.7 kw in this case.

We can find the weight flow rate from

$$W = \gamma Q = Sg(\gamma_{\text{water}})Q$$

$$= .86 \times 62.4 \text{ lb/ft}^3 \times 10 \text{ gal/min} \times (\text{ft}^3/7.48 \text{ gal})$$

$$= 71.7 \text{ lb/min}$$

or 32.5 kg/min.

Now we can find the temperature rise from Equation 7.2:

$$\Delta T = \frac{HGR}{C_p \times W}$$

$$= \frac{496 \text{ Btu/min}}{0.45 \text{ Btu/lb°F} \times 71.7 \text{ lb/min}}$$

$$= 15.4°F$$

or 8.5°C.

This means that every drop of oil that flows through the valve has its temperature increased by 15.4°F (8.5°C) as it goes through. If this temperature increase is not dissipated by some means, the bulk temperature of the oil will very quickly exceed the recommended operating limits of the oil.

When a three-port flow control valve is used, the system pressure is governed by the force required to move the load. Consequently, the pressure drop across the valve will be much less than that across a meter-in valve. Thus, if the somewhat less precise control of a bleed-off type flow control can be tolerated, significant energy savings can be realized. As the previous example shows, this decreased pressure drop will also reduce the heat buildup in the system.

9.6   Combined Effects of System Pressure Drops

In this chapter, we have looked at the pressure drops and heat generation associated with flow control valves. In Chapter 8, we looked at those same things for pressure control valves. In the real world, though, these are not separate, unrelated events. Instead, the pressure drops across both types of valves, and, indeed, every other device in the circuit combine to give the total heat generation rate for the system. Let's look at an example.

Example 9.7

The circuit shown here uses a meter-out flow control to limit the motor speed in one direction. The pressure drops and flow rates are as shown in the figure. Calculate the total heat generation rate for the circuit.

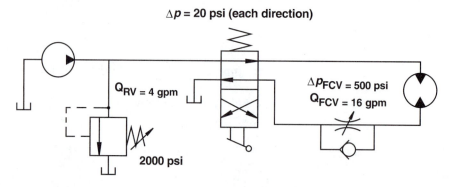

Example 9.7

Solution

We first use the pressure drop and flow rate for each component to find the power loss through each component.

For the relief valve,

$$HP_{RV} = \frac{\Delta p_{RV} \times Q_{RV}}{1714}$$

$$= \frac{2000 \times 4}{1714} = 4.67 \text{ HP}$$

or 3.48kW.

The flow through the directional control valve (both directions), the hydraulic motor, and the flow control valve is 16 gpm. Therefore, for the directional control, the horsepower loss in one direction is

$$HP_{DCV} = \frac{\Delta p_{DCV} \times Q_{DCV}}{1714}$$

$$= \frac{20 \times 16}{1714} = 0.19 \text{ HP}$$

or 0.14kW.

This gives us a total of 0.38 HP (0.28kw) lost through that valve.

The flow control valve loss is

$$HP_{FCV} = \frac{\Delta p_{FCV} \times Q_{FCV}}{1714}$$

$$= \frac{500 \times 16}{1714} = 4.67 \text{ HP}$$

or 3.48kW.

(We don't need to calculate the hydraulic motor horsepower, because that represents work rather than a loss.) The total loss for the system, then, is

$$HP_{SYS} = HP_{RV} + HP_{DCV} + HP_{FCV}$$

$$= 4.67 + 0.38 + 4.67$$

$$= 9.72 \text{ HP}$$

or 7.25kW.

The resulting heat generation rate for the system is

$$HGR = 42.4 \text{ Btu/min} \times 9.72$$

$$= 412 \text{ Btu/min}$$

or 7.25kW.

In a normal system, much of this heat will be dissipated through the components, reservoir, and system piping, but it may be necessary to add an oil cooler (heat exchanger) if there is not sufficient natural dissipation. (We'll discuss heat dissipation rates of reservoirs and oil coolers in Chapter 13.)

The effect of this heat generation rate on the temperature of the fluid in the reservoir (or the total system for that matter) can be calculated using Equation 9.3:

$$\text{Temperature Rise Rate} = \frac{\text{Heat Generation Rate}}{\text{Specific Heat} \times \text{Weight of Oil}}$$

or   **(9.3)**

$$\Delta T = \frac{HGR}{C_p \times wt}$$

This equation in almost identical to Equation 7.2. The difference is that here we must use the actual weight of the oil rather than the weight flow rate as we did before. We calculate this value from Equation 9.4:

$$\text{Oil Weight} = \text{Specific Weight} \times \text{Volume}$$

or   **(9.4)**

$$wt = \gamma V$$

Example 9.8

For the system of Example 9.7, calculate the rate at which the temperature of the oil in the reservoir will increase, assuming no natural heat dissipation. The reservoir contains 100 gal (378.5 L) of oil that has a specific weight of 57 lb/ft^3 9.06 kN/m^3 and a specific heat of 0.48 Btu/lb°F (1.97 kJ/kg°C).

Solution

Before we can use Equation 9.3, we must determine the weight of the oil in the reservoir. Converting 100 gal to cubic feet gives

$$V = 100 \text{ gal} \times (\text{ft}^3/7.48 \text{ gal}) = 13.4 \text{ ft}^3$$

or 0.375m^3.

so that the weight of the oil is

$$wt = \gamma V = 57 \text{ lb/ft}^3 \times 13.4 \text{ ft}^3 = 762 \text{ lb}$$

or 342 kg.

From Equation 9.3, we find that the temperature rises at

$$\Delta T = \frac{HGR}{C_p \times wt}$$

$$= \frac{412 \text{ Btu/min}}{0.48 \text{ Btu/lb°F} \times 762 \text{ lb}}$$

$$= 1.13°\text{F/min}$$

or 0.62°C/min.

This shows that in, say, 30 minutes, the oil temperature would increase by 33.9°F (18.8°C). While this doesn't seem excessive, remember that the increase is continuous unless some heat dissipation occurs. At this rate, recommended maximum operating temperatures could be exceeded rather quickly.

9.7   Troubleshooting Tips

When troubleshooting problems in systems using flow control devices, keep in mind both the purpose of flow control valves and their effects on other system parameters, especially pressure. Remember that the speed of operation of a cylinder or motor is flow related; therefore, if the load does not move at the correct speed, look for a flow rate problem. Flow meters are very useful tools for troubleshooting flow-related malfunctions.

Symptom	Possible Cause
Load will not move	1. Pump not running 2. Flow control valve closed 3. Insufficient pressure
Load moves too fast or too slowly	1. Flow control valve set incorrectly 2. Insufficient pump output (too slow) 3. Incorrect fluid viscosity
Pressure-compensated valve not compensating	1. Valve installed backward 2. Valve too small 3. Compensator jammed by contaminants 4. Compensator spring broken 5. Contaminants blocking internal passages
Cannot control load speed with needle valve	1. Valve with integral check valve installed backward 2. Needle jammed by contaminants 3. Seat worn 4. Using meter-in instead of meter-out on descending load 5. Need a pressure-compensated valve 6. Fluid too hot or too cold
System runs hot	1. Excessive pressure drop across flow control valve 2. Excessive flow across relief valve (pump too big)

9.8   Summary

Flow control valves provide a means to control the speed of operation of hydraulic motors and cylinders. They do this by controlling the rate at which fluid enters (meter-in or bypassing) or leaves (meter-out) the cylinder or motor. Flow restricting arrangements (meter-in and meter-out) provide stable control, but also cause high pressure drops that waste energy and cause high heat generation rates. Only the meter-out arrangement provides speed control for descending loads.

The symbols for flow control valves (summarized in Figure 9.15) are all based on the orifice symbol. Adjustments, compensators, and other special functions are indicated in the same manner as with other valves.

a. Fixed orifice.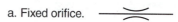

b. Variable orifice (flow control valve).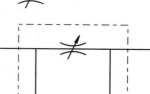

c. Flow control valve with internal check valve.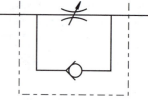

d. Pressure-compensated flow control valve.

e. Temperature-compensated flow control valve.

Figure 9.15 Flow control valve symbols.

9.9 Key Equations

Orifice equation:

$$Q = C \times A \sqrt{\frac{2g\Delta p}{\gamma}} \quad \textbf{(9.1)}$$

Needle valve flow:

$$Q = C_v \sqrt{\frac{\Delta p}{Sg}} \quad \textbf{(9.2)}$$

System temperature rise rate (no dissipation):

$$\Delta T = \frac{HGR}{c_P \times wt} \quad \textbf{(9.3)}$$

Review Problems

Note: The letter "S" after a problem number indicates SI units are used.

General

1. What function is accomplished by flow control valves?
2. What are some ways other than flow control valves to control system flow rate?
3. What is the purpose of a pressure-compensated flow control valve?
4. What is the operating concept of a pressure-compensated flow control valve?
5. List the three ways to use flow control valves to control actuator speed. What are the advantages and disadvantages of each location?
6. What are the three factors that determine the flow rate through an orifice?
7. What is meant by the "rated C_v" of a flow control valve?
8. What flow control valve design is most commonly used in fluid power systems? Why?
9. What is the purpose of a priority valve?

10. Explain the difference between a priority flow divider and a proportional flow divider.

Flow

11. A flow control valve has a $C_v = 0.92$. The pressure drop across the valve is 100 psi. The specific gravity of the fluid is 0.88. Find the flow rate through the valve.
12. The pressure upstream of a flow control valve is 600 psi. The downstream pressure is 100 psi. The specific weight of the fluid is 58.7 lb/ft³. The C_v of the valve is 1.8. Find the flow rate through the valve.
13. Fluid with a specific gravity of 0.86 is flowing through a flow control valve at 30 gpm. The valve C_v is 3.8. What is the pressure drop across the valve?
14. Find the flow rate through a flow control valve with a C_v of 7 and a pressure drop of 1500 psi. The specific gravity of the fluid is 0.88.
15. Fluid with a specific weight of 57.2 lb/ft³ is flowing through a flow control valve at 20 gpm. The C_v is 10. Find the pressure drop across the valve.
16. As a flow control valve is being opened, each position of the valve has an equivalent C_v valve. Assuming values of 1, 2, 3, 4, and 5, plot a family of curves showing the pressure-flow characteristics of the valve using differential pressures of 500, 1000, 1500, 2000, 2500, and 3000 psi.
17S. The pressure drop across a flow control valve is 700 kPa. The specific gravity of the fluid is 0.90. The valve has a C_v of 1.32. Find the flow rate through the valve in Lpm.
18S. The pressures upstream and downstream of a flow control valve are 4500 kPa and 700 kPa, respectively. The specific weight of the fluid is 9.4 kN/m³. The valve C_v is 2.6. Find the flow rate through the valve in Lpm.
19S. A valve with a C_v of 5.3 is flowing at a rate of 120 Lpm. The specific gravity of the fluid is 0.85. Find the pressure drop across the valve.
20S. The pressure drop across a flow control valve is 10,000 kPa. The C_v of the valve is 11. The specific gravity of the fluid is 0.87. Find the flow rate through the valve.
21S. A fluid with a specific weight of 9.5 kN/m³ is flowing through a flow control valve at a rate of 80 Lpm. The C_v is 10. Find the pressure drop across the valve.

System

22. A fixed displacement pump produces 20 gpm. The system relief valve is set at 1000 psi. Flow rate through the flow control valve is 15 gpm to operate a 3 in cylinder that is raising a 5000 lb load. The specific gravity of the fluid is 0.86. Find

    a. The value of C_v.
    b. The horsepower lost across the relief valve.
    c. The horsepower lost across the flow control valve.
    d. The temperature rise across the relief valve.
    e. The temperature rise across the flow control valve.

23. For the circuit shown for this review problem (see illustration on page 330), the flow control valve has a C_v of 2.4. The specific weight of the fluid is 55.86 lb/ft³. The system relief valve is set at 1000 psi. Find the extension speed of the cylinder.
24. In the system shown for this review problem, what are the pressures on each of the pressure gages while the cylinder is extending?
25. In the system shown for this review problem, what are the pressures on each gage while the cylinder is extending if there is no load on the cylinder? What are the pressures if there is a 3000 lb load on the cylinder?

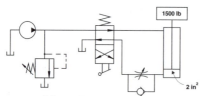

Review Problem 9-23

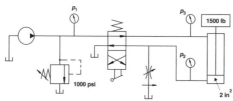

Review Problem 9-24

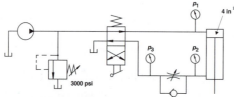

Review Problem 9-25

26. Using the assumption that the specific heat for hydraulic oils is generally around 0.45 Btu/(lb°F) and the specific gravity is around 0.9, derive a value that can be used as a rule of thumb for estimating the temperature rise in °F for each 100 psi pressure drop across any device.

27S. Using the assumption that the specific heat for hydraulic oils is generally around 1.9 kJ/kg°C and the specific gravity is around 0.9, derive a value that can be used as a rule of thumb for estimating the temperature rise in °C for each 1000 kPa pressure drop across any device.

Circuit Practice

28. Write a complete description of the operation of the circuit shown in figure for Review Problem 23. Include both the extension and retraction cycles. Explain the flow paths taken by all fluid flow, including return flow from the cylinder.

29. Repeat the exercise of Review Problem 28 for the figure in Review Problem 24.

30. A machining process requires that a clamp cylinder be used to hold a part in place while a second cylinder lowers a drill to drill a hole through the part. The operator uses a four-way, two-position, lever-operated, spring-returned directional control valve to control the process. When the operation starts, the clamp cylinder extends. When the part is locked in place, the drill cylinder extends automatically. When the drilling is completed, the operator shifts the directional valve to retract the cylinders. The drill cylinder retracts first, then the clamp cylinder retracts automatically. The pressure on the clamp cylinder must be less than system pressure to prevent damage to the part. The lowering speed of the drill cylinder must be controlled to prevent damage to the drill. Using standard ISO symbols, draw the complete hydraulic circuit diagram for the machine.

Chapter 10

Electrohydraulics

Outline

10.1 Introduction

In modern industrial hydraulic and pneumatic equipment, sensing and control is usually provided by electrical or electronic means. The most basic machines will probably utilize simple solenoid valve circuits, while highly sophisticated servo systems will be found in applications requiring very high degrees of accuracy or rapid response to input commands or changes in system parameters.

Electrohydraulic circuits are often found on mobile as well as industrial equipment. In fact, the vast majority of the early work on electrohydraulic servovalves was for aircraft flight control system applications—primarily ailerons, elevators, and rudders. Today some form of electronic control can be found on virtually everything from automotive transmissions to spacecraft.

Objectives
When you have completed this chapter, you will be able to

- Explain the differences between the various electrohydraulic valves.
- Explain how each of the valves operates.
- Analyze and draw circuits using the ISO symbols for electrohydraulic valves.
- Read and draw relay logic diagrams.
- Explain the functions of commonly used electrical sensors.
- Discuss and calculate "gain" as applied to electrohydraulic systems.

10.2 Solenoid Valves

A *solenoid valve* is any basic valve—directional, flow, or pressure control—that is operated by a solenoid rather than manually or mechanically. (We will exclude proportional valves and servovalves from this category, also.) The directional control valves are by far the most common solenoid valves.

The Solenoid

A *solenoid* is, in effect, a linear electric motor. It consists of three major components: the coil, the C-stack, and the plunger—as shown in Figure 10.1a. The coil is insulated copper wire wound around a plastic piece called a *bobbin*. This coil is installed in a frame of ferromagnetic material (iron) shaped roughly like the letter C—thus, it is usually called the *C-frame* or the *C-stack*. The movable member slides through the center of the coil. Since it is shaped like the letter T, it is often referred to as the *T-plunger,* or simply the *plunger.* The distance between the base of the plunger and the base of the C-stack is termed the *airgap.*

When the solenoid is energized, the current passing through the coil creates a magnetic field that pulls the plunger into the coil until it contacts the C-frame and is stopped mechanically. It will stay in that position as long as the coil is energized. When the coil is de-energized, a spring in the valve spool will return the plunger to its original position.

A relatively new variation on the standard T-plunger type solenoid is the design shown in Figure 10.1b. In this design, the plunger is cylindrical instead of the T-shaped, rectangular cross section. It operates inside a cylindrical core that is sealed on one end but open to the valve spool cavity on the other. This core is always full of oil, which is why it is called a *wet armature solenoid.* Slots in the plunger allow circulation of the oil and prevent hydraulic lock. There are two major advantages of this design over the T-plunger design. First, because of the sealed core and the static seal, wet armature solenoids almost never leak. The T-plunger design has the pushpin moving through a dynamic seal, so eventually some leakage is likely to occur. Since the solenoid is not sealed, this leakage will show up externally sooner or later.

The second advantage is that the wet armature solenoids tend to stay cooler than the T-plunger solenoids. The movement of the plunger causes the oil in the core to circulate and transfer heat away from the coil, which surrounds the core. Most modern solenoid valves use wet armature solenoids.

Some solenoids are designed to operate on AC, while others operate on DC power. The voltage and AC frequency is different for different designs, so you must be careful to select the solenoid that matches the power supply that is being used. In general, AC solenoids are more powerful and operate faster. They have the drawback, however, of an AC phenomenon called *in-rush current.* When the solenoid is energized, the initial current flow is very high. In fact, it may reach four to ten times the value of the *holding current* (the current that flows when the plunger is fully seated). This high current generates heat very rapidly. The heat, in turn, causes the insulation on the coil to deteriorate, especially if the plunger fails to close for any reason.

If an AC solenoid plunger hangs open, the result is jokingly referred to as a "smoke test", which the solenoid is almost certain to fail. Within a few seconds, the coil heats to the point that the insulation begins to smoke. Then the insulation breaks down completely, allowing the coil to short circuit. Of course, when this occurs, the solenoid has "failed the smoke test" and must be replaced.

Plunger out

Plunger attracted in

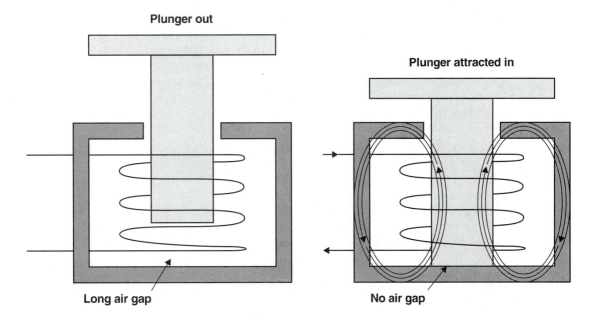

Long air gap

No air gap

a. Air gap solenoid, T-plunger type.

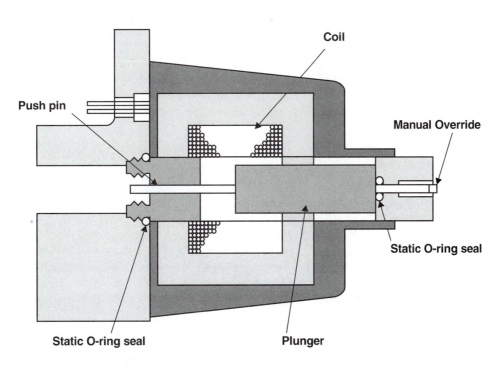

Coil

Push pin

Manual Override

Static O-ring seal

Static O-ring seal

Plunger

b. Wet armature solenoid

Figure 10.1 A solenoid valve consists of three major components: the coil, the frame (C-stack), and the plunger.

A major advantage of DC solenoids is that they do not experience this in-rush current phenomenon. Therefore, they are not subject to the high heat generation. In fact, the initial current flow is virtually identical to the holding current, so the coil really doesn't care where the plunger is. A properly applied DC solenoid should never have a problem passing the smoke test.

The Valves

Since solenoids are applied primarily to directional control valves, we don't need to spend much time discussing the mechanical design of the valves themselves. Any type of directional control valve can be designed to be operated by a solenoid.

The Marriage

The combination of the solenoid operator with the mechanical valve body results in what we call a *solenoid valve*. The purpose of the solenoid is to shift the spool in the valve to change the internal flow paths. As seen in Figure 10.2,

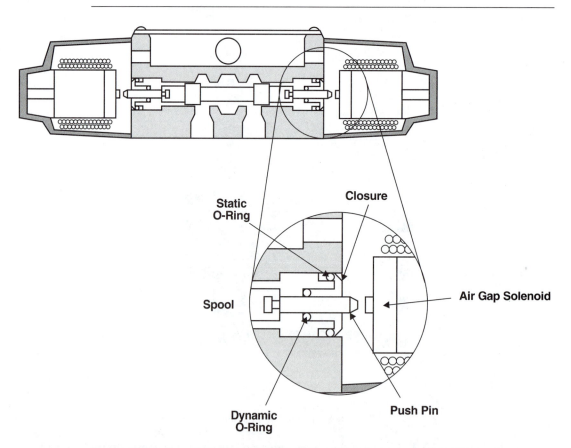

Figure 10.2 In solenoid spool actuation, the plunger pushes the pushpin when the solenoid is energized.

when the solenoid is energized, the plunger pushes the pushpin. The pushpin, in turn, pushes the valve spool to its new position. Because of the solenoid design, there can be no intermediate positions; thus, the spool is moved through its full stroke.

Since the spool position is controlled directly by the solenoid plunger, this type valve is referred to as a *direct-acting solenoid valve*. There are only three basic configurations of direct-acting valves. If the valve contains a two-position spool, it may be controlled by either one solenoid and a spring, or by two solenoids. If a three-position spool is used, then two solenoids and two springs are required. Figure 10.3 shows the graphic symbols used for these configurations.

Figure 10.3a shows a two-position spool with a single solenoid. We always consider that springs and solenoids *push* the spool (although there are some designs in which the solenoid pulls). Therefore, when the solenoid is not energized, the spring pushes the spool to its unactuated position, as shown. Energizing the solenoid causes it to push the spool to its actuated position as shown in the figure. De-energizing the solenoid allows the spring to push the spool back to its original position.

The two-position valve with the two solenoids (Figure 10.3b) really doesn't have a "normal," or unactuated, position because there is no device to "return" the spool when the solenoids are de-energized. When either solenoid is energized, it pushes the spool to the opposite position. The spool will stay in that position—even if the solenoid is de-energized—until the other solenoid is energized. Be careful in this design that both solenoids don't get energized at the same time—especially if AC solenoids are used—because you would have a smoke test in progress.

In these double-solenoid designs, it is sometimes desirable to use a detent to be sure that the spool stays in position with the solenoid de-energized. As seen in Figure 10.4, the detent mechanism is simply a notch or groove in the spool and a spring-loaded ball. The spring is very light, but it provides enough force to hold the ball in the groove and keep the spool from drifting. The force is easily overcome when the opposite solenoid is energized.

Looking back to Figure 10.3c, we see the three-position spool configuration. Two solenoids are required to push the spool to the two end positions. Two springs are required to return the spool to its center position when the solenoids are de-energized.

Solenoids have limited force capabilities, so the size of spool that they can operate directly is also limited. To overcome this problem, rather than try to build bigger, more powerful solenoids (which could cause high electrical power consumption), we use what we call a *pilot valve*. Pilot valves are normally used if the valve port size is larger than ⅜ inch.

The pilot concept here is the same as we discussed in the chapter on directional control valves, except that the pilot section is solenoid-operated instead of manually operated. Figure 10.5 shows a solenoid-operated pilot valve "piggy-backed" onto the main valve. Virtually any functional configuration is possible in pilot valves.

The symbols for pilot valves must accurately represent the valve function. To be absolutely correct, the symbol should be a composite, showing the pilot section with its actuators, the main spool with all its various parts, all internal porting, and so forth. Figure 10.6 shows such a composite ("complete") symbol.

a. SPRING OFFSET

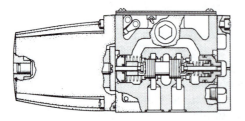
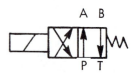

b. DETENTED

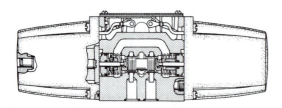
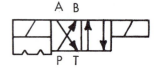

c. SPRING CENTERED

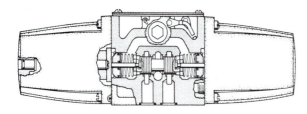
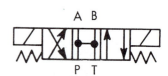

Figure 10.3 ISO symbols for the three basic configurations of direct-acting solenoid valves.
Source: Courtesy of Vickers, Inc.

While it is complex, it is also completely accurate. Such symbols are seldom used, however, because the valve function can be accurately depicted in a simplified diagram that is easier to draw and easier to read. The simplified symbol in Figure 10.6 represents the same valve as the composite, complete symbol. Notice that the main valve spool is used in the simplified drawing, but all actuators, springs, etc., are shown to represent the total valve operation.

When a pilot valve is used, there is sometimes a need to control the rate at which the main spool shifts. Since the main spool responds to the pilot section

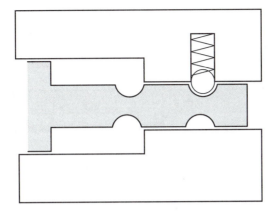

Figure 10.4 A detent mechanism keeps the spoil in position with the solenoid deenergized.

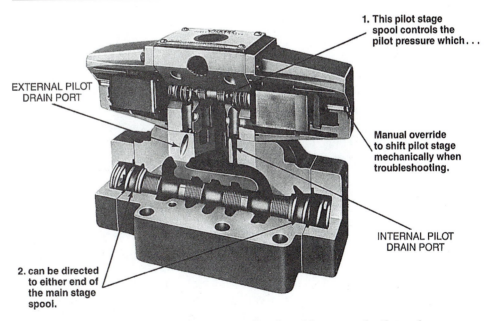

1. This pilot stage spool controls the pilot pressure which . . .

EXTERNAL PILOT DRAIN PORT

Manual override to shift pilot stage mechanically when troubleshooting.

INTERNAL PILOT DRAIN PORT

2. can be directed to either end of the main stage spool.

Figure 10.5 This cutaway shows a small solenoid-operated pilot valve mounted atop the main valve. The internal porting is visible.
Source: Courtesy of Vickers, Inc.

Pilot Drain

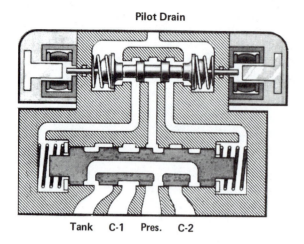

Tank   C-1   Pres.   C-2

(A). Valve Construction.

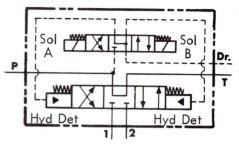

(B). Complete Graphic Symbol.

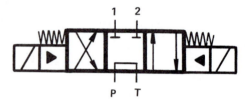

(C). Simplified Graphic Symbol.

Figure 10.6 This solenoid-controlled, pilot-operated directional valve can be represented by a composite, complete symbol or a simplified symbol.
Source: From Industrial Fluid Power, vol. 2, published by Womack Educational Publications.

in the same way that a cylinder would respond, it is logical to assume that the spool speed is a function of the flow rate, just as cylinder speed is. This is exactly the case—the lower the flow rate, the more slowly the spool shifts. To provide this flow control, a section containing two small needle valves (often referred to as *chokes*) is sandwiched between the pilot section and the main spool. The symbol for such an arrangement is shown in Figure 10.7. Note that a meter-out arrangement is used.

The Controls

The electrical control circuits for solenoid valves may be as simple as a pushbutton or a toggle switch or as complex as numerous sensors—position, pressure, temperature, and so on—providing inputs for *programmable logic controllers* (PLCs). The symbols representing the more common of these devices are shown in Appendix C. As with fluid power valve symbols, these electrical symbols represent the unactuated position of the device. They may be normally open (N.O.) or normally closed (N.C.). Note that these terms have exactly opposite meanings when applied to electrical devices and hydraulic valves. A normally closed valve has no fluid flow through the valve from the pressure port. A normally closed switch provides a current flow path through it.

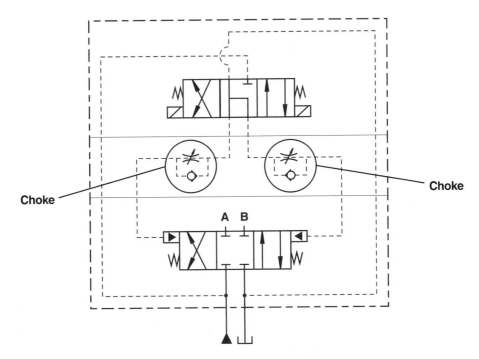

Figure 10.7 In a pilot-operated valve with choke, a meter-out arrangement is used.

Relay Logic Diagrams

While wiring diagrams and schematics are sometimes used to show the control circuits for solenoid valves, *relay logic diagrams* (sometimes called *ladder logic diagrams*) provide an easily read representation of the electrical control circuitry. The rules are relatively simple. First, the "ladder" can be either vertical or horizontal. If it is vertical, the left-hand rail is the positive, or "hot," power lead, while the right-hand rail is the negative, neutral, or common lead. If the ladder is horizontal, the top rail is the positive lead.

All switching (logic) is normally shown on the "hot" end of the rungs, with the output device (which can be a relay, solenoid, motor, light, bell, or whatever) shown as the last item on the rung before the right-hand rail. Output devices are almost never in series, but parallel outputs are common. In complex circuits, every device should be labeled to show its relationship to the overall circuit. The symbol itself shows the function of the device.

This may be somewhat confusing so far, so let's look at some examples to help clarify things a bit. Figure 10.8 is a very simple circuit to operate an electric motor. It uses a normally open toggle switch to make and break the circuit to the motor. When the switch is closed, the motor runs; when it is opened, the motor stops.

If we wanted to provide temperature protection for the motor, we would include a normally closed thermal switch in the control logic, as shown in

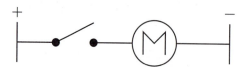

Figure 10.8 This electric motor control circuit uses a normally open toggle switch.

Figure 10.9. The logic requires that the toggle switch be actuated *and* that the thermal switch *not* be actuated in order for the motor to run.

Figure 10.10 shows a circuit for operating a solenoid valve (or any solenoid operated device) using a momentary pushbutton. When the button is pushed, the solenoid is energized; however, since a momentary pushbutton is shown, as soon as the button is released, the solenoid is de-energized. If this is not satisfactory, some other logic devices are necessary. One alternative would be to use a latching pushbutton that is mechanically held in its actuated position until the operator deactuates it. Another alternative is to use a control relay. Let's look at this type circuit.

A *control relay* is really an electrically actuated electrical switch, (Figure 10.11). It is very similar to a solenoid. It has a wire coil through which a current is passed to create a magnetic field. This field pulls in a plunger that operates the switch mechanism. The switch may open or close only one set of contacts, or it may operate several contact sets, some of which open and some of which close each time the relay is actuated. In any case, *all* contacts operated by the

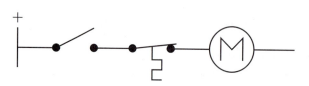

Figure 10.9 This electric motor control circuit uses a normally closed thermal switch.

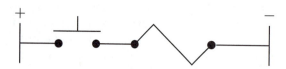

Figure 10.10 This solenoid control circuit uses a normally open momentary pushbutton.

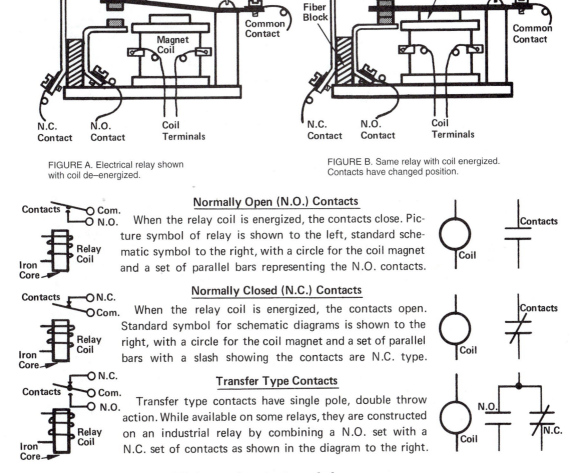

FIGURE A. Electrical relay shown
with coil de–energized.

FIGURE B. Same relay with coil energized.
Contacts have changed position.

Normally Open (N.O.) Contacts

When the relay coil is energized, the contacts close. Picture symbol of relay is shown to the left, standard schematic symbol to the right, with a circle for the coil magnet and a set of parallel bars representing the N.O. contacts.

Normally Closed (N.C.) Contacts

When the relay coil is energized, the contacts open. Standard symbol for schematic diagrams is shown to the right, with a circle for the coil magnet and a set of parallel bars with a slash showing the contacts are N.C. type.

Transfer Type Contacts

Transfer type contacts have single pole, double throw action. While available on some relays, they are constructed on an industrial relay by combining a N.O. set with a N.C. set of contacts as shown in the diagram to the right.

Figure 10.11 Relay and contact symbols.

Source: From Industrial Fluid Power, vol. 2, published by Womack Educational Publications.

relay change position every time the relay is energized or de-energized. Let's look at operating our solenoid with a relay.

Figure 10.12 is a basic *relay logic circuit* using two rungs. On the first rung we have the momentary pushbutton and the relay (labeled 1CR). The second rung contains the relay contacts (labeled 1CR-1, indicating that it is the Number 1 contact controlled by relay 1CR) and the solenoid. Pushing the pushbutton completes the logic circuit to energize the relay. Energizing the relay causes the normally open contacts to close and provides power to the solenoid. We really haven't gained much here, because as soon as the pushbutton is released, the relay is de-energized, the contacts open, and power to the solenoid is lost. To take care of this problem, a holding circuit is required.

A *holding circuit* makes use of a second set of contacts in the relay in parallel

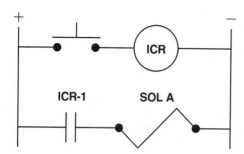

Figure 10.12 This relay circuit for operating a solenoid uses two rungs.

with the pushbutton and provides continuous power to the relay once it has been energized (Figure 10.13). Now when the pushbutton is depressed, the relay is energized and both sets of associated contacts close. The solenoid is energized through 1CR-1, while 1CR-2 provides the path to keep the relay energized, even after the pushbutton is released. (Since the relay is, essentially, keeping itself energized, this is sometimes called a *bootstrap circuit*.)

We still have a problem, though. Now we can't de-energize the solenoid, so we need yet another control element in the logic circuit for the relay. We need a normally closed pushbutton on the relay side of the holding circuit, as shown in Figure 10.14. With this pushbutton, we can break the relay circuit, open the contacts, and de-energize the solenoid.

Now let's look at the ladder logic diagram for using a limit switch to cause a hydraulic cylinder to retract automatically. The hydraulic circuit we're using is shown in Figure 10.15a. Notice that the symbol used for the limit switch is

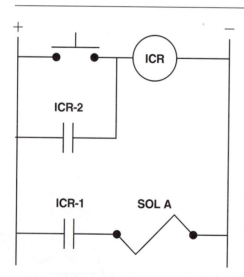

Figure 10.13 This solenoid control circuit has a holding circuit.

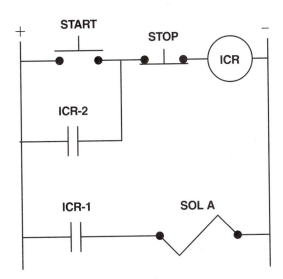

Figure 10.14 This complete ladder diagram for using a relay to operate a solenoid shows a normally closed pushbutton on the relay side of the holding circuit.

not the same as used in the ladder diagram of Figure 10.15b, but the labels show that they represent the same switch. When we push the START button, the relay is energized. This completes the holding circuit through 1CR-1 and the normally closed limit switch (1LS) and energizes the solenoid through 1CR-2. This, in turn, causes the cylinder to extend until it contacts 1LS. The limit switch opens and breaks the holding circuit. The relay is consequently de-energized, the contacts open, the solenoid is de-energized, the spring kicks the spool back to its unactuated position, and the cylinder retracts. As soon as the cylinder backs off the limit switch, the contacts close again, but the holding circuit has already been broken because the contacts (1CR-1) have opened. The only way to reestablish the circuit is to push the START button again.

Note that, without the hydraulic circuit diagram, the ladder diagram is somewhat meaningless. The corresponding symbols in the two diagrams must be labeled so that it will be obvious that they are the same device.

There is no limit to the possible combinations of hydraulic and control devices. The exercises at the end of this chapter include some circuits that will help you to develop your skills in this area. Remember that there are usually several ways to accomplish the same operation. The challenge is to find the most effective circuit.

10.3 Proportional Control Valves

A more sophisticated version of a solenoid valve is the *proportional control valve,* often referred to as an EHPV (for *electrohydraulic proportional valve*).

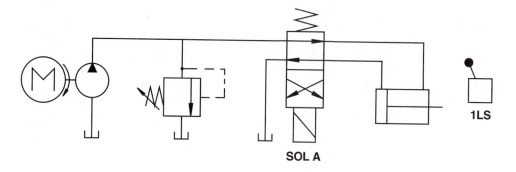

a. Hydraulic circuit.

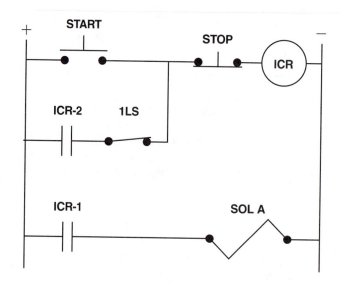

b. Ladder diagram.

Figure 10.15 This automatic cylinder retraction uses a limit switch.

The EHPV uses a proportional solenoid with an electronic controller to position the valve spool. Proportional solenoids can be used to operate pressure and flow control valves, but the most common application is in directional control valves, so we'll concentrate on this area once again.

Proportional Solenoids

With any solenoid design, the force output of the solenoid is a function of several variables of design and the current in the coil. For any specific solenoid, the force is proportional to the square of the current and inversely proportional to the square of the airgap (the distance the plunger is from its fully closed position). This is shown graphically in Figure 10.16.

If there is a spring in the valve, it represents an opposition to the plunger movement. The spring usually has a linear force-displacement curve that has a slope determined by the spring constant. If we superimpose the spring curve on the solenoid curves, we get Figure 10.17. Every point at which the spring curve intersects a solenoid force curve is a point at which the two forces are equal. At these points the plunger will stop. By increasing the current, we can increase the plunger force to drive the plunger in further by compressing the spring.

This technique is not used in standard solenoid applications because it is not needed, and because it requires additional electronics for controlling the current. Proportional solenoids, however, are based on this concept. There are some slight modifications in the solenoid design that cause the solenoid force to have a flat spot in it, so that when the solenoid force-versus-displacement curve family is plotted, we get something like Figure 10.18. When this family of curves is overlaid with the spring curve, Figure 10.19 results. The proportional solenoid is designed so that all intersections occur in the area where the solenoid force curve is flat. The area is known as the *control zone* and is usually only about 0.060 inches long. Again, every intersection of the solenoid and spring curves represents a force balance situation that causes the plunger to stop. It can be moved in or out by increasing or decreasing the control current.

The Valves

A major advantage of proportional solenoid valves over standard solenoid valves is their ability to provide directional and flow control in a single valve. When

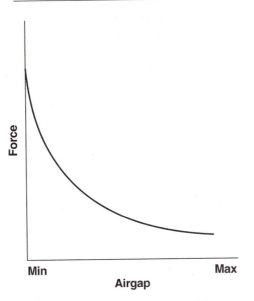

Figure 10.16 Typical solenoid force versus stroke curve at constant current.

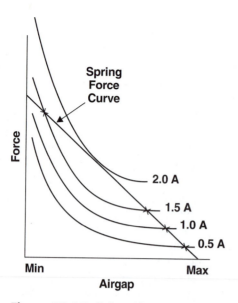

Figure 10.17 Solenoid force curves with spring force curve superimposed

a standard solenoid is used, flow control must be accomplished by some other means, as we have already discussed. An EHPV provides flow control by positioning the spool to control the port opening. This can simplify the hydraulic control circuit significantly. For example, the circuit in Figure 10.20 uses standard solenoid valves and flow control valves to provide speed control (four

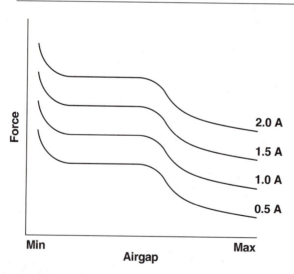

Figure 10.18 Proportional solenoid force versus displacement curves for varying current

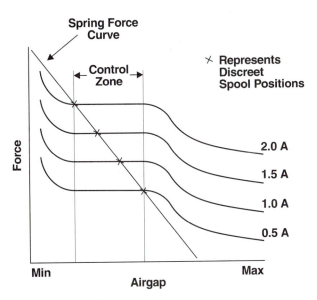

Figure 10.19 Proportional solenoid force versus displacement curves with spring force overlaid

different speeds) for a hydraulic motor. This system requires the use of three directional control valves and four flow control valves, whereas the same four-speed function can be accomplished using one proportional valve as shown in Figure 10.21.

Precise flow control requires that valve clearances be very small—on the order of 10 μm (0.0004 in)—and that very close manufacturing tolerances be observed. (One micrometer [μm] is one-millionth of a meter or about 0.00004 in.) Additional control can be achieved by using metering notches on the spool lands, as shown in Figure 10.22. The notches allow flow to begin long before the valve land actually clears the ports. Very small clearances and high-precision manufacturing enhance the flow control accuracy.

The Controls

In order to vary and accurately control the solenoid current, sophisticated electronic controls are needed. One such control unit is shown in Figure 10.23. No attempt will be made here to explain the operation of the controller; we'll just look at its use. In the drawing shown, everything inside the solid lines is contained on the controller circuitboard. The numbers along the edge represent terminal numbers on the connector block that holds the board.

All proportional solenoids operate on direct current. The power supplies of most controllers are 24 VDC. For the controller shown, the power supply is connected to terminals 30 and 32. This power is smoothed, conditioned, distributed to appropriate control points, and so on, by the electronic components on the board. A rotary *potentiometer*, connected between terminals z2, b2 and b6,

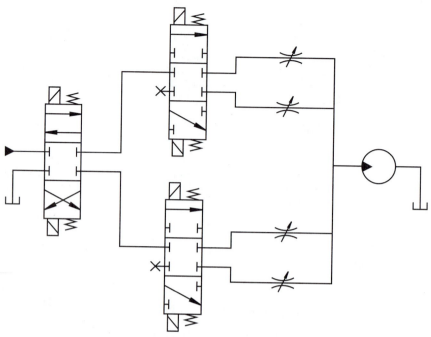

Figure 10.20 This circuit uses standard solenoid valves to provide four different hydraulic motor speeds.

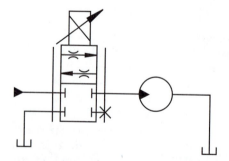

Figure 10.21 This circuit uses a proportional solenoid valve to provide four different hydraulic motor speeds.

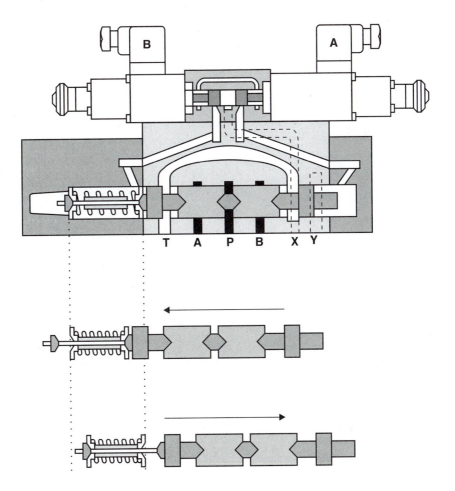

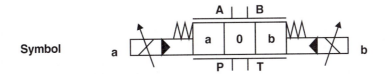

Figure 10.22 Spool configuration example, note the pilot-operated proportional directional valve with single-sided "spring centering".
Source: The Rexroth Corporation.

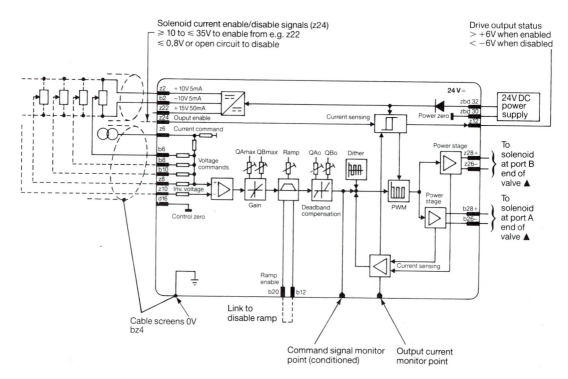

Figure 10.23 This PCB functional layout shows sophisticated electronic controls.

Source: Courtesy of Vickers, Inc.

is used to vary the command signal to the proportional solenoid. The potentiometer signal is again processed by various components before it is finally sent to one of the solenoids, which are connected to terminals z26, z28 and b26, b28. By connecting several potentiometers between terminals z2, b2 and terminals b6, b8, b10, z8, and z10, and providing appropriate switching, we can change the signal to the solenoid between four discrete levels. This allows the actuator speed and direction to be varied by simply flipping a switch. Automatic speed control can be achieved by using limit switches, timers, PLCs, or other control devices to provide switching between the potentiometers.

Proportional solenoid valves lend themselves readily to processes that require automatic speed variations (as in the motor circuit of Figure 10.20) or cyclic inputs up to about 40 Hz (in a fatigue tester, for example). Low-flow circuits can use direct-acting valves, while pilot-operated valves are used for high flow rates.

Proportional valves may use internal feedback circuits to help control spool position, but continuous feedback from the load (speed, position, pressure, etc.) is seldom used with these valves. Normally, only discrete switching devices such as limit switches, pressure switches, and the like are used to switch between the control potentiometers.

10.4 Servovalves

Another type of valve in which the output is proportional to an electrical input is the *servovalve*. Servovalves differ from proportional valves in several ways—higher performance (speed, accuracy, cycle rate, response time, etc.), method of actuation, clearances and manufacturing tolerances, electronic controls, feedback, and price, to list a few. In the early days of proportional valve technology, the differences in performance were great. Many of these differences are not as obvious now, because of advances in proportional valve technology. In fact, some high-end EHPVs have better characteristics than some low-end servovalves. Although the once sharp lines of differences have blurred, we can still consider the servovalve to be, in general, a somewhat more sophisticated device that provides considerably greater control capability than the proportional valve.

Servovalves are very high-performance units that normally use continuous feedback from the hydraulic system to provide very accurate and high-speed response to varying command signals or system disturbances. Valve position is set by a torque motor that receives control signals from an electronic unit called an *operational amplifier*. Servovalves can be used to provide pressure or flow control, but the most common application is as a combined direction and flow control valve. Once again, we'll concentrate on this last application.

Torque Motor

A *torque motor* is a small, limited-rotation device used to move the valve mechanism. Figure 10.24 shows a typical torque motor. The armature is essentially an electromagnet. Its polarity is controlled by the direction of current flowing through the coil. The pole pieces are permanent magnets, so changing the polarity of the armature induces a small amount of rotation in the appropriate direction. Through a mechanical linkage, the rotating armature moves the valve mechanism by a corresponding amount. Torque motors always operate on DC power.

The Valve

Servovalve components are very high-precision, close-tolerance devices. The spool and bore are usually hand-lapped and are often matched, noninterchangeable sets. Radial clearances between the spool and the bore are on the order of three to five micrometers. Figure 10.25 shows the possible configurations of spool lands relative to the valve ports—underlapped, overlapped, and zero-lapped or line-to-line. An underlapped spool has lands that are slightly (0.5 to 1.5 percent) narrower than the port opening so that there is a small amount of designed-in leakage when the spool is centered. Likewise, an overlapped spool is wider than the port by 0.5 to 1.5 percent. The zero-lapped spool—which is the most commonly used—has lands that are precisely the same width as the port openings. The advantage of zero-lapping is that any movement of the spool results in an immediate, proportional fluid flow. This gives a near linear response to even the smallest of command inputs around the center (or null) point.

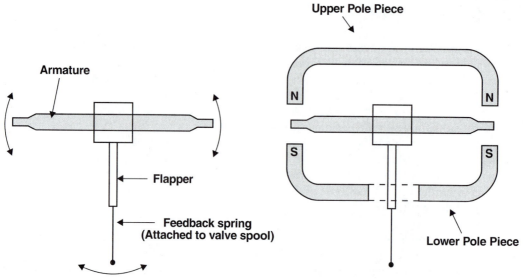

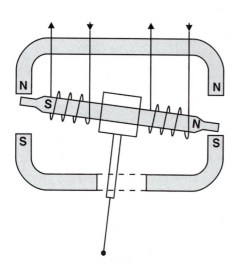

a. The armature moves only a few degrees in each direction.

b. The pole pieces are permanent magnets

c. The direction of DC current flow polarizes the armature and determines the direction of rotation. The strength of the current determines the force required to return it to its center position.

Figure 10.24 A torque motor is a small limited-rotation device used to move the valve mechanism.

left (-Y)   **right (+Y)**

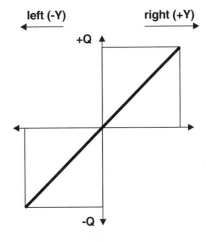

Zero-lapped or line-to-line:
There is no volumetric flow in control spool position Y=0.
There is constant volumetric flow via the control land in position |Y|>0.

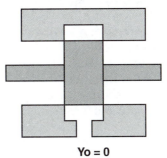

Yo = 0

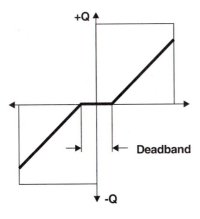

Overlapped:
Control openings remain closed in the area |Y| ≤ Yo.
There is constant volumetric flow via a control land in position |Y|>Yo.

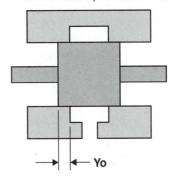

Yo

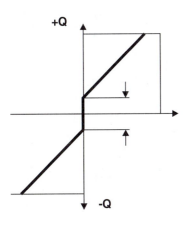

Underlapped:
There is a constant volumetric flow via both control lands in the area |Y| < Yo.
There is volumetric flow via only one control land in position |Y| ≥ Yo.

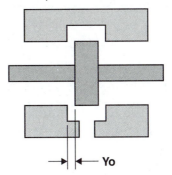

Yo

Figure 10.25 The possible configurations of spool lands are (a) line-to-line, (b) overlapped, and (c) underlapped.
Source: The Rexroth Corporation.

Only the smallest of the servovalves can be direct-acting (usually termed *single-stage*) because of the very limited output force of the torque motor. Figure 10.26 is a typical single-stage servovalve. The torque motor armature is linked directly to the valve spool so that any movement of the armature results in the opening or closing of the valve ports. The mechanical feedback spring works against the torque motor to control the position of the spool. As the spool moves, the feedback spring force is increased as a result of the movement, until finally, the feedback spring force equals the torque motor force. The resultant force balance stops the spool and holds it in position.

Most servovalves are *two-stage* units. The first stage—termed the *hydraulic amplifier*—is the equivalent of a pilot stage in other valves and controls the operation of the main spool. It is operated by the torque motor and directs pilot pressure and flow to the ends of the main spool. There are several different types of hydraulic amplifier, as shown in Figure 10.27. The spool-type first stage is very similar to the pilot operated DCVs we have discussed previously. The jet-pipe first stage uses a nozzle (jet pipe) attached to the torque motor armature to direct pressurized fluid flow into two receivers. When the jet pipe is exactly centered, the pilot flow is divided evenly between the two receivers so that the pressure is equal on each end of the main spool. The result is a force balance that stops the spool and holds it in place. Any movement of the torque motor results in rotation of the jet pipe to cause a higher pressure in one receiver than the other. The result is a force imbalance on the main spool that causes it to move. An internal feedback mechanism works against the torque motor to center the jet pipe when the commanded position is reached.

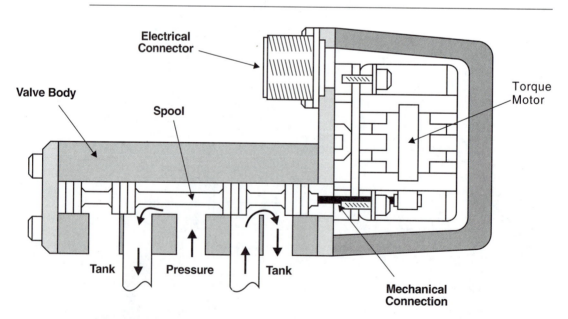

Figure 10.26 The single-stage servovalve is the only one that is direct-acting.

Source: Courtesy of Vickers, Inc.

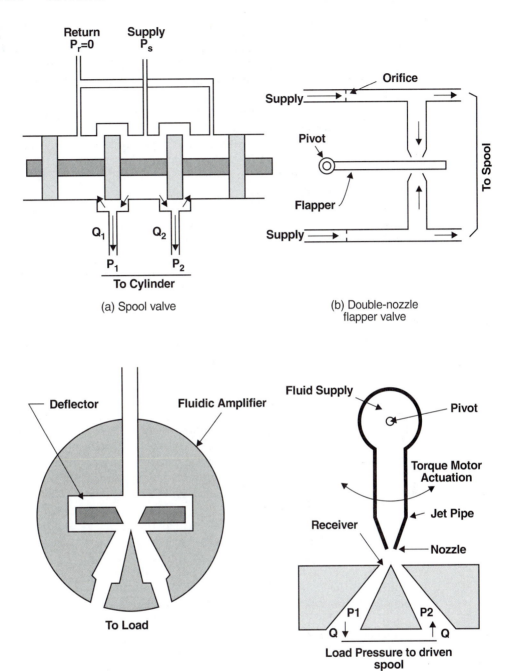

Figure 10.27 The first stage of the servovalve is the hydraulic amplifier.

A double flapper-nozzle first stage has the pilot flow split evenly between the two pilot channels. Each channel has a nozzle, as shown in the figure, which is continuously dumping fluid into an internal drain. The flapper is attached to the torque motor armature and extends into the internal drain cavity between—but very close to—the two nozzles. With the flapper centered, the flow from the two nozzles is equal and the pressure in the two pilot channels is equal. Again we have a force balance holding the main spool in place. Movement of the torque motor causes the flapper to move away from one nozzle (decreasing the pressure in the associated pilot channel) as it moves toward the other (increasing the pressure in its pilot channel). The result is force imbalance, spool movement, internal feedback, flapper centered, spool stopped—just like the other types.

Because of the very fine clearances, servovalves are extremely sensitive to contaminated fluid. Solid particles can cause jamming (often termed *silting*) and wear that can quickly render a servovalve inoperable. Therefore, the key to servovalve reliability is good fluid filtration. Always consider the manufacturer's recommended acceptable fluid contamination levels to be the *maximum* acceptable level and strive to provide much cleaner fluid for the valve.

The Controls

Controllers for electrohydraulic servovalves are relatively sophisticated when compared to those used with proportional valves. This is because the electronics must handle both an input signal and a feedback signal in order to provide the appropriate output signal to the torque motor. Controllers of this type are called *operational amplifiers* or *op amps*. Figure 10.28 shows the block diagram of a typical op amp. As with the proportional controllers, we'll make no effort here to understand *how* it works; we'll look only at what it does. Again, as with the proportional controllers, the components shown within the solid outline are contained on the printed circuitboard, and the numbers represent the terminals on the connectors.

Contained on the board are several potentiometers for adjusting various functions of the op amp. Perhaps the most important of the adjustments is the *gain* of the amplifier. If the gain is too high, the system becomes "twitchy" and could become unstable, oscillating very rapidly and even ending in structural failure. If the gain is too low, the system will be sluggish and respond very slowly to input commands or system disturbances.

Although the popular meaning of *gain* is "increase" when used in the context of system performance, gain actually defines an output in response to an input. Mathematically, we can express gain as

$$\text{Gain} = \frac{\text{Output}}{\text{Input}} \qquad\qquad (10.1)$$

For example, if we use a hydraulic motor that turns at 1200 rpm when the flow rate is 10 gpm, we can say that the gain of the motor is

$$\text{Gain} = \frac{1200 \text{ rpm}}{10 \text{ gpm}}$$

That is, the input is 10 gpm, and the output is 1200 rpm. You can see from this that the units of gain will be determined by the character of the output

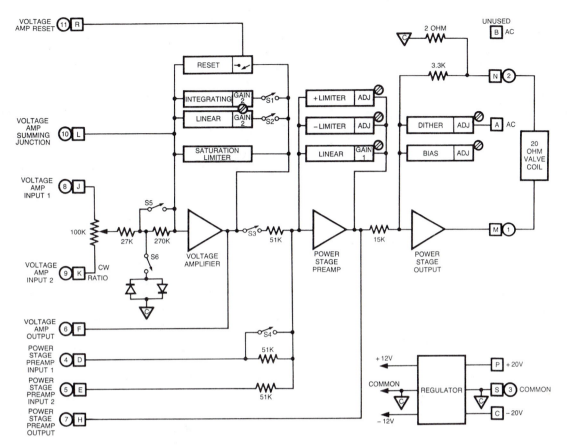

Figure 10.28 Typical operational amplifier control module.
Source: Courtesy of Vickers, Inc.

and the input. Common units in servovalve operation are milliamps per volt (mA/V), volt per volt (V/V), gallons per minute per milliamp (gpm/mA), volts per revolution per minute (v/rpm), and so on.

The System

Before we can go much further with this discussion of gain, we need an understanding of the total servo system concept. In Figure 10.29 we see a general block diagram of a typical servo system. In this case, we are controlling the rate at which the load is rotating.

The command input R (which stands for Reference signal) tells the controller how fast the load should turn. The reference signal is fed into the summing junction where it is compared to the feedback signal, F, which is an indication of how fast the load is actually turning. If these two inputs are equal—indicating that the load is turning at the commanded speed—there is no output signal from the summing junction, so the torque motor stays where it is. If

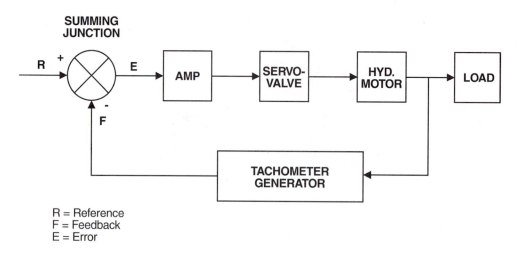

R = Reference
F = Feedback
E = Error

Figure 10.29 In this typical servo system, the common input R (for "Reference signal") tells the controller how fast the load should turn.

there is a difference between the reference and feedback signal, an error signal, E, is generated and sent to the torque motor through the amplifier stage. The torque motor moves by an amount proportional to the error signal. This changes the flow rate to the motor, changing the motor speed, hence the load speed. This change in load speed is picked up by the tach generator, which produces voltage output proportional to the load speed. This is the feedback signal, F, that is fed back to the summing junction. This type of closed-loop comparison and adjustment goes on continuously in a servo system. The gain of the system will determine how quickly and accurately the required adjustments can be made.

System gain can be calculated from a knowledge of the gains of each component in the system. Let's look at a simple example. Suppose we have the basic setup shown in Figure 10.30. In this figure, R is the reference signal, F is the feedback signal, and E is the error signal. G is the combined gain of all the components between the summing junction and the load and is termed the *forward loop gain*. H is the gain of all the components in the feedback loop from the load to the summing junction. It is termed the *feedback loop gain*. The controlled output is indicated by the letter C.

The error signal, E, is the sum of the reference and feedback signals, R and F. (Notice that F is indicated as a negative electrical polarity.) Thus, we can write

$$E = R - F \tag{10.2}$$

The feedback signal, however, is the product of the controlled output and the feedback loop gain,

$$F = C \times H \tag{10.3}$$

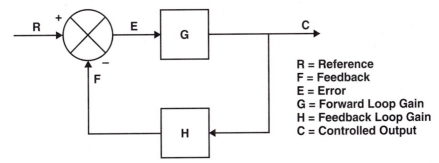

Figure 10.30 In this simple feedback loop, G is the combined gain of all the components between the summing junction and the load, and H is the gain of all the components in the feedback loop from the load to the summing junction.

so Equation 10.2 becomes

$$E = R - (C \times H) \tag{10.4}$$

The controlled output, C, is the product of the error signal, E, and the forward loop gain, G, so we can write

$$C = E \times G \tag{10.5}$$

Solving Equation 10.5 for E and substituting into Equation 10.4 gives

$$\frac{C}{G} = R - (C \times H)$$
$$C = (R - (C \times H))G$$
$$C = (R \times G) - (C \times G \times H)$$
$$R \times G = C + (C \times G \times H) = C \times (1 + (G \times H))$$

and finally,

$$\frac{C}{R} = \frac{G}{1 + (G \times H)} \tag{10.6}$$

Equation 10.6 is termed the *control ratio* of the system, which is exactly the same thing as the system's closed-loop gain, where C is the output and R is the input.

Example 10.1

The speed of a mixer in a chemical process must be accurately controlled to ensure the proper mixing of the chemicals without shearing certain polymers and reducing the strength of the end product. Find the closed-loop gain of the system shown here and determine how fast the load is rotating.

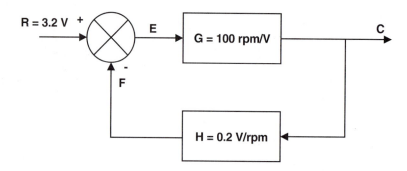

Example 10.1

Solution

From Equation 10.6,

$$\frac{C}{R} = \frac{G}{1 + (G \times H)} = \frac{100 \text{ rpm/V}}{1 + (100 \text{ rpm/V} \times 0.2 \text{ V/rpm})} = 4.76 \text{ rpm/V}$$

This means that the load speed increases 4.76 rpm for every volt of input signal. To find the load speed, solve Equation 10.6 for C:

$$C = \frac{G}{1 + (G \times H)} \times R$$

Substituting, we get

$$C = 4.76 \text{ rpm/V} \times 3.2 \text{ V} = 14.24 \text{ rpm}$$

Example 10.2

The circuit shown here is a portion of the system used to accurately position the panels of an automobile body for welding. The command signal is set to move the panel to the precise position and hold it there until the welding operation is completed. Determine the distance the cylinder will extend in response to the commanded input.

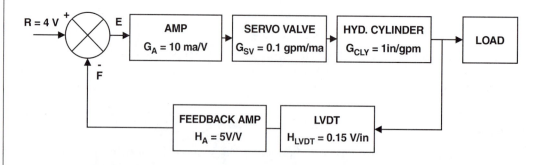

Example 10.2

Solution

We will find the system gain using Equation 10.6, but first we need to find the values of G and H to plug into that equation. The forward loop gain is the product of all the gains in that section. Thus,

$$G = G_A G_{SV} G_{CYL} = 10 \text{ mA/V} \times 0.1 \text{ gpm/mA} \times 1 \text{ in/gpm} = 1 \text{ in/V}$$

Similarly, we find H from

$$H = H_{LVDT} H_A = 0.15 \text{ V/in} \times 5 \text{V/V} = 0.75 \text{ V/in}$$

Now we can find the final position from

$$C = \frac{G}{1 + (G \times H)} \times R = \left[\frac{1 \text{ in/V}}{1 + (1 \text{ in/V} \times 0.75 \text{ V/in})} \right] \times 4 \text{ V}$$

$$= 2.28 \text{ in}$$

Some systems have *internal* (or *parallel*) *feedback loops* in addition to the main forward and feedback loops. Figure 10.31 is a block diagram of such a circuit. To find the control ratio of this system, we must first resolve the internal loop. We handle this exactly as we would if it were a major loop, using Equation 10.6:

$$\frac{C}{R}_{\text{INNER LOOP}} = \frac{G_2}{1 + G_2 H_2}$$

Now we can substitute this for the G_2 block in the figure to give Figure 10.32. This now becomes a term in the general equation for forward loop gain, so that we have

$$G = \frac{G_1 G_2 G_3}{1 + G_2 H_2}$$

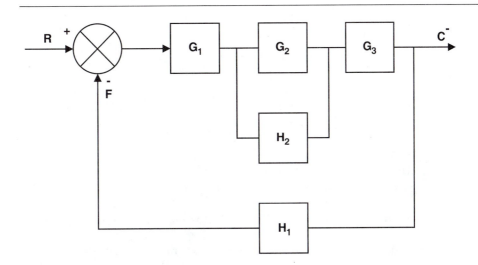

Figure 10.31 This servo system has an internal feedback loop in addition to the main forward and feedback loops.

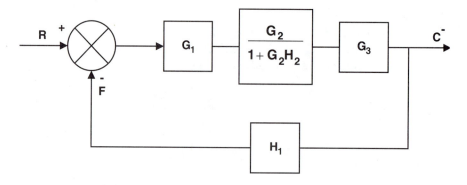

Figure 10.32 Here the internal feedback loop is resolved into forward loop gain.

which we plug into Equation 10.6 for the control ratio for the entire control loop. This gives

$$\frac{C}{R} = \frac{G}{1 + G \times H_1} = \frac{\dfrac{G_1 G_2 G_3}{1 + G_2 H_2}}{1 + \dfrac{G_1 G_2 G_3}{1 + G_2 H_2} \times H_1}$$

which finally becomes

$$\frac{C}{R} = \frac{G_1 G_2 G_3}{1 + G_2 H_2 + G_1 G_2 G_3 H_1}$$

Any circuit, regardless of its complexity, can be reduced to a single equation in this manner. You always start with the innermost loop and work outward.

10.5   Electrohydraulic Valve Symbols

The primary difference between the symbols representing electrohydraulic valves and mechanically operated valves is the *actuator symbol*. The symbology for the valve mechanism is basically the same in all cases.

Figure 10.33a shows a four-way, three-position, solenoid-operated, spring-centered, closed-center DCV. In Figure 10.33b, we see the same valve, but this time, proportional solenoids are used. Notice the differences between these two symbols—the arrow (meaning "adjustable") across the solenoid symbols, and the two lines parallel to the symbol. These lines indicate that the valve position is infinitely variable between the two ends. If metering notches are used on the spool lands, the symbol may indicate this by using arcs (similar to the orifice symbol) along the flow path arrows, as shown in Figure 10.33c.

The servovalve symbol is shown in Figure 10.33d. The actuator symbol actually represents the summing junction with the command and feedback signal inputs and the error signal output. Again, the spool is shown as having an infinite number of possible positions.

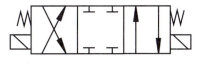

a. Solenoid valve.

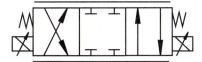

b. Proportional valve.

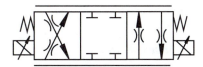

c. Proportional valve with
metering notches.

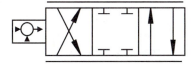

d. Servovalve.

Figure 10.33 Electrohydraulic directional control valve symbols.

10.6   Troubleshooting Tips

Solenoid Valves

Symptom	Possible Cause
Solenoid burned out	1. Spool stuck 2. Voltage too high 3. Voltage too low 4. Ambient temperature too high 5. Two solenoids energized at same time
Spool doesn't move	1. Solenoid not energized or inoperable 2. Contaminated fluid 3. Fluid viscosity too high

Proportional Solenoid Valves

Symptom	Possible Cause
Solenoid burned out	1. Ambient temperature too high
Spool doesn't move	1. Spool stuck 2. Faulty circuitboard 3. Current setting too low 4. Contaminated fluid 5. Fluid viscosity too high 6. Power supply connections reversed 7. No command signal 8. Wire loose on spool feedback loop (where applicable)
Spool moves too slowly	1. Ramp control set too high

Servovalves

Symptom	Possible Cause
Torque motor burned out	1. Current limits set too high
Spool doesn't move	1. Spool stuck 2. Contaminated fluid 3. Fluid viscosity too high 4. Faulty electronics 5. No command signal 6. Loose or broken wires
System operation jerky, vibrates or hammers—lines break	1. Gains set too high
System operation slow—won't react or hold command position	1. Gains set too low
Load position or speed drifts	1. Gains set too low 2. Valve leaking internally 3. Valve land edges worn

10.7 Summary

Electrohydraulic control valves can be classified as solenoid, proportional sole-noid, and servovalves. These three types vary significantly in their perfor-mance, complexity, cost, and contaminant sensitivity. Solenoid valves, the least expensive of the three, provide a single function only; that is, the valves are either open or closed. Thus, they are basically directional control devices. Pro-portional valves, on the other hand, can be moved to virtually any position within their spool stroke. This allows the valve to control both directional and flow control. The electronics required to provide the control of the spool position as well as the small clearances required to minimize internal leakage make the cost of a proportional valve considerably higher than a solenoid valve. However, the ability to combine flow and direction control and to provide multiple flow rates from a single valve can reduce the overall system cost by reducing the number of valves required.

Servovalves provide the highest performance capability of the three valve types, although today the high-end proportional valves are approaching the capabilities of many low-end servovalves. They provide extremely accurate control of the controlled parameters—speed, position, acceleration, pressure, or flow—and very fast response to changes in those parameters. This high degree of control and response requires sophisticated electronics, accurate transducers, and precisely manufactured valve hardware. This, in turn, means that servo systems are far more expensive than the other two valve types.

Electrohydraulics is a very large area that includes not only the valves we have discussed briefly here, but also servo pumps, servo motors, some very sophisticated transducers, digital electronics, natural frequencies of electronics, valves, and systems, programmable logic controllers, and many other topics. This chapter can serve only as a very brief introduction to the topic. For more detailed information, see the texts listed at the end of this chapter. The real world of fluid power involves extensive use of electronics, so the reader is strongly encouraged to learn as much as possible about this extremely important field.

10.8   Key Equations

1. System gain (control ratio): $\dfrac{C}{R} = \dfrac{G}{1 + (G \times H)}$ (10.6)

Review Problems

Note: The letter "S" after a problem number indicates SI units are used.

1. List the three general types of electrohydraulic valves.
2. Explain the operation of a solenoid.
3. Explain the difference between an airgap solenoid and a wet armature solenoid.
4. Explain the differences between a standard solenoid valve and a proportional solenoid valve.
5. Explain the operation of an EHPV.
6. Explain the difference between an EHPV and an electrohydraulic servovalve.
7. Explain the operation of a torque motor.
8. Explain the difference between a single-stage servovalve and a two-stage servovalve.
9. List three different types of hydraulic amplifiers.

Servovalve Gain

10. Calculate the gain for the servovalve system shown here.

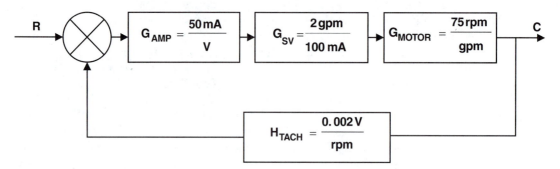

Review Problem 10–10

11. Calculate the gain for the servovalve system shown here.

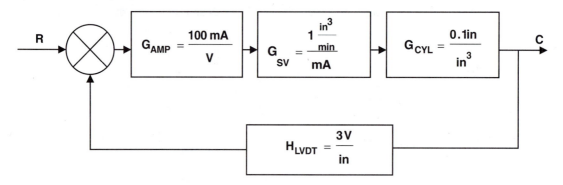

Review Problem 10–11

12. Calculate the gain for the servovalve system shown here.

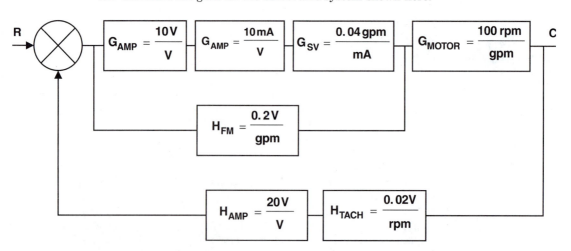

Review Problem 10.12

13. In the circuit shown here, the op amp saturates at 300 mA. How fast will the 2.2 in³ hydraulic motor rotate when the command pot is rotated 200 degrees?

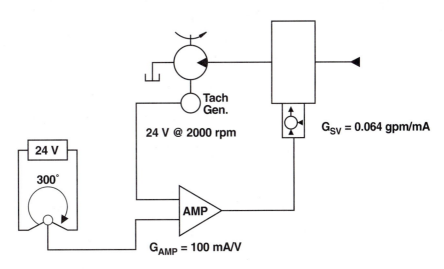

Review Problem 10.13

Circuit Practice

14. In the circuit shown here, what will happen when SW1 is closed?

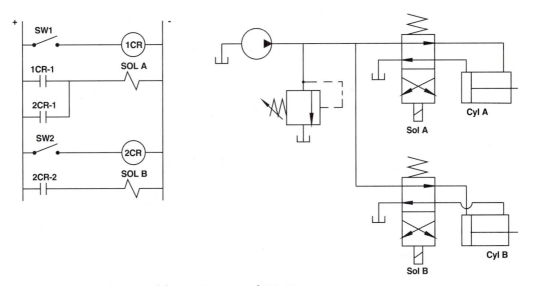

Review Problems 10.14 and 10.15

15. In the circuit shown for Review Problem 14, what will happen when SW2 is closed with SW1 open?

16. Write a complete description of the operation of the system shown here.

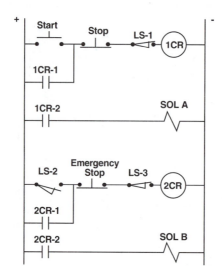

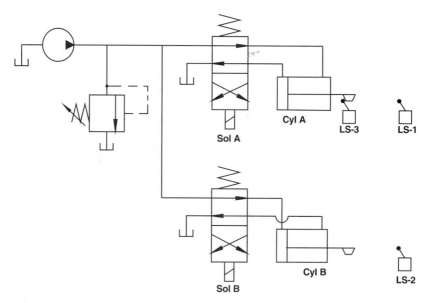

Review Problem 10.16

For the following problems, use the standard hydraulic and electrohydraulic symbols as employed in this book.

17. Draw a hydraulic circuit and the ladder diagram required to cause a single-acting hydraulic cylinder to extend to a certain point, then retract automatically. Use a single-solenoid DCV.

18. Draw a hydraulic circuit and the ladder diagram required to cause a double-acting hydraulic to extend to a certain point, then retract automatically. Use a single-solenoid DCV.

19. Draw a hydraulic circuit and the ladder diagram required to cause a double-acting hydraulic cylinder to reciprocate automatically between two points. Use a single-solenoid DCV. Be sure to include a method for stopping the action.
20. Draw a hydraulic circuit and the ladder diagram required to cause two hydraulic cylinders to operate in sequence without the use of sequence valves. The order of operation should be: Extend 1, Extend 2, Retract 2, Retract 1.

Suggested Additional Reading

1. F. Don Norvelle, *An Introduction to Electrohydraulics* (Jenks, OK: Amalgam Publishing Co., 1993).
2. Michael J. Tonyan, *Electronically Controlled Proportional Valves*, (New York: Marcel Dekker, 1985).
3. James E. Johnson, *Electrohydraulic Servo Systems*, 2d ed. (Cleveland, OH: Penton/IPC, 1982).
4. *Industrial Hydraulics Manual*, 2d ed. (Rochester Hills, MI: Vickers, Inc., 1989).
5. *The Hydraulic Trainer,* vol 2, (Main, Germany: Mannesmann Rexroth GmbH, Lohr am 1986).
6. Werner Götz, *Proportional Valves Theory and Operation*, (Stuttgart, Germany: Robert Bosch GmbH, 1984).
7. Werner, Götz, *Electro-Hydraulic Closed Loop Control Systems Theory and Application*, (Stuttgart, Germany: Robert Bosch GmbH, 1987).
8. *Design of Electrohydraulic Servo Systems for Industrial Motion Control*, (Cleveland, OH: Parker Hannifin Corporation, 1992).
9. *Handbook of Electrohydraulic Formulae*: (Cleveland, OH: Parker Hannifin Corporation, 1992).

Filtration

Outline

11.1 Introduction

It is very unusual to find an entire chapter in an introductory text devoted to the subject of filtration. It is included here to emphasize the extreme importance of *filtration* (actually, contamination control as a whole) in ensuring high reliability from virtually any modern high-pressure fluid power system.

It is widely estimated that 70 to 75 percent of all fluid power system failures result from contaminated fluid. Imagine the boon to industry as a whole if the hydraulic system failure rate were cut by 70 to 75 percent! The savings would amount to untold millions of dollars each year. These large savings would not result merely from the repair of the system and the replacement of hardware. This is a relatively small part of the cost of a failure. The real expense is in the cost of the unproductive time on the equipment—commonly called *downtime*. Downtime can cost thousands of dollars per hour. For example, one copper refinery calculated its downtime cost to be $1342.50 per *minute*. Few companies can afford much of that!

Objectives

When you have completed this chapter, you will be able to

- Discuss the importance of filtration in the life, reliability, and performance of a hydraulic system.
- Discuss the types of contamination.
- Explain the sources of contamination.
- Discuss the different types of filter media.
- Explain the filter Beta rating.

- Calculate filter efficiency based on the Beta rating.
- Discuss the possible locations for filters and their advantages and disadvantages.
- Explain how filters capture particles.
- Explain the selection criteria for filters.
- Discuss the standard terminology associated with filters.

11.2   Types of Contamination

In the strictest sense of the word, anything in a system that is not supposed to be there can be termed *contamination*. A partial list follows:

- Dirt
- Metal particles
- Water
- Air
- Chemicals (other than the fluid formulation)
- Acid
- Heat
- Biological growths
- Radiation
- Cigarette butts

While all of these are (or at least have the potential to be) detrimental to a hydraulic system, the two that are the most harmful are heat and "dirt." Here, "dirt" includes all types of solid particles. We'll discuss the heat problem when we get to the section on heat exchangers (Chapter 13). In this chapter, we'll deal with the dirty subject of particulate contamination.

As a quick review, let's consider once again the relative size of the particles involved. Particle size is measured in micrometers (μm)—usually referred to as *microns*. A micron is one millionth of a meter or, in the U.S. Customary system, about 0.000039 inch. Thus, one inch is about 25,400 microns. The unaided eye can detect items as small as 40 microns—approximately the diameter of a blond human hair. The clearances in the components of modern, high-pressure hydraulic systems are frequently less than 10 microns, and seldom exceed 25 microns.

11.3   Sources of Contamination

There are many sources of dirt. First is the manufacturer of the individual components that make up the system. Airborne dirt particles, machine cuttings, poor deburring techniques, and even the assembly of the piece parts can leave a tremendous number of particles in each component.

As these components are assembled into a system, the problem is compounded. Mating of threads generates new particles, most of which are pushed into the components or trapped in the pipes, hoses, or tubing. The further the assembly progresses, the more difficult the cleaning procedure becomes, because there are more places for the particles to hide or become trapped. If cleaning has not been carried out in small steps along the way, the cleaning of the fully assembled system can be difficult, time consuming, and expensive. It also may not be successful.

Once the user acquires the component, or the new system, it is possible to cause a great deal of damage and generate a tremendous number of particles unless great care is used. The first problem can occur if the user assumes that new oil is clean oil. New oil, especially that from drums and bulk storage, can be extremely dirty. It can contain dirt, metal particles, rust, water, and just about anything else. New oil should always be filtered before use through filters that are *at least* as good as those in the system, preferably even better. This will usually require the use of a filter unit that can be used to clean the fluid in the drum or bulk storage before use.

Improper startup procedures can actually destroy a system in a matter of minutes. *Never* fill the reservoir of a new sytem with oil and start up the system at full load. A proper startup begins with ensuring that every component is filled with filtered oil before the GO button is pushed. This includes filter housings and cylinders, but especially pumps and motors. Those rotating elements must never be operated dry. The manufacturer will usually provide an instruction sheet that includes the best method for filling the pump or motor.

Once everything is filled with clean oil, the system should be run with no load for a very short time, then shut down and checked over. Before starting, relief valves should be set at their minimum settings, and all valves adjusted to keep restrictions at a minimum and thus ensure minimum pressure drops.

If all seems in order after this initial "bump," the system can be restarted— still under no load—and run for an extended period. Gradually, in small increments, pressures can be increased, flow rates adjusted, and so on, giving adequate time for opposing surfaces to wear in and mate properly.

This break-in run should be done with fine filtration installed. Obviously, a careful watch on the filter pressure drop is required so that as the element begins to clog, the system can be shut down and the filters changed. These break-in filters should be replaced before the system is put into service.

If there is any doubt about the cleanliness of the system as it is received (and there probably will be), it is usually a good idea to remove as many of the system components as possible and install hoses or tubing in their places. The pump can then be started up, and the reservoir, pump, and plumbing flushed thoroughly. One by one, the components can be reinstalled and flushed, beginning as close to the pump as possible and progressing through the system, so that the trash from one component won't be pushed into one that has already been cleaned. It may even be desirable to use an external filter system connected downstream of the most recently connected component during this type of operation.

Once the system is in operation, the maintainers often become their own worst enemies, because they often contribute to the contamination problem due to a lack of understanding. One of the worst problems is the so-called "preventive maintenance" program that calls for time changes of components. Components that are working correctly should not be removed, because the removal and replacement of the component generates large quantities of debris that can damage the system. It is far better to develop a "predictive mainte-nance" program based on periodic analysis of the fluid and various performance parameters. In this way, unnecessary component removals can be avoided, and the components can be used to the end of their useful lives while minimizing the danger of a catastrophic failure.

Other problems include (1) the "shotgun" approach to maintenance ("I don't know which component is bad, so I'll change them all"), (2) leaving connections open during the maintenance actions, and (3) failure to properly maintain filters. Oil servicing is a major problem. Pouring "new" oil into the reservoir from a dirty bucket should be a criminal offense, yet it happens every day.

While the system is operating, dirt can enter through reservoir breathers and rod wipers seals (or *scrapers*). In most industrial system, reservoirs are open to the atmosphere and are allowed to "breathe" as the volume of oil in the tank changes due to the movement of actuators. As air is "inhaled," dirt, moisture, chemicals, etc, are inhaled with it. Techniques for controlling this unwanted inhalation of contaminants will be discussed in Chapter 13.

The wiper seals on cylinder rods can also be a major point of entry for atmospheric contaminants. As the rod extends, a fine film of oil is carried out into the atmosphere on its surface. Dirt, moisture, and other airborne contaminants settle on this oil film. Much of it is carried back into the cylinder because the wiper seal fails to do its job. This can eventually lead to a large buildup of dirt in the cylinder, leading to rod pressure seal failures, and wear and internal leakage around the piston. One possible way to reduce this problem is to use pleated "boots" made of a rubber-like material to cover the rod as it extends. However, the best cure for this problem is to minimize the entry (termed *ingression*) past the wiper seal by using the seals that show low contaminant ingression rates in the SAE J1185 test.

In the SAE test, a wiper seal is installed on a rod that reciprocates inside a dust box. Standard AC Fine Test Dust—a standard test contaminant consisting primarily of sand particles ranging from 0 to 80 μm in a consistent size distribution—is circulated by fans inside the box. As the rod passes through the wiper seal (as if it were retracting), a spray ring washes it down with oil. The oil is periodically collected, and the weight of dirt per unit volume of oil is determined by a gravimetric analysis. The results of the test are reported as milligrams of dirt per liter of oil per *cycle-meter* of rod travel, or the length of stroke multiplied by the number of strokes.

Figure 11.1 shows some typical results of the SAE J1185 test. It is readily seen that there is a vast difference in wiper seal performance. Seals with high ingression rates (numbers 4 and 5, for example) can cause some serious problems that will result in low reliability and perhaps even a catastrophic failure.

As any mechanical system operates, some amount of wear is going to occur. When the fluid contains solid particles, the wear rate is accelerated. Contamination is self-perpetuating and self-accelerating—the more there is, the more there is. The only way to break the cycle is to deal with the sources and provide sufficient filtration to remove the accumulated contaminants.

In 1976 the Fluid Power Research Center (FPRC) at Oklahoma State University conducted a study to determine the rate at which solid contaminants (from both internal and external sources) entered operating fluid power systems. The study included aircraft, industrial installations, and several types of mobile equipment. The results were staggering! It was found that 10^8 particles, 10 micrometers in size and larger, entered the average system every minute.[1]

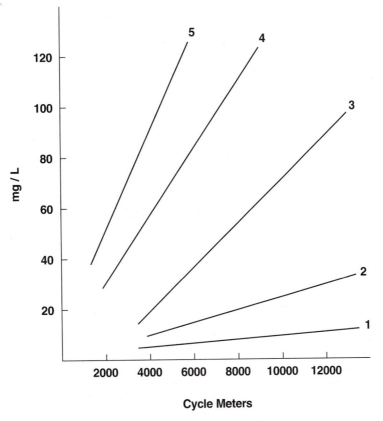

Figure 11.1 Typical results of the wiper seal ingression test (SAE J1185).

That's 100,000,000 particles every 60 seconds! It's easy to see why good contamination control is important. Without it, the fluid in the average system would turn to sludge in a very short time.

11.4   Filters

In discussing *filters*, we need to touch on several characteristics that describe various physical attributes of these devices. These include configuration, media (the material that traps the particles), and the micron rating of the filter. A discussion of filter capture mechanisms is included, also.

Filter Types

The vast majority of filters used in fluid power systems are either spin-on (canister) or cartridge types. The *spin-on filter* (Figure 11.2) looks like the canister filters used on most automobiles as crankcase oil filters. In this type of unit, the filter element (the part that does the filtering) is contained in a

Figure 11.2 Many filters used in fluid power systems are spin-on or canister filters.

throwaway housing. The filter head is plumbed into the fluid line. It may contain a bypass valve, some sort of filter condition indicator, and perhaps internal check valves to prevent loss of fluid from the system when the canister is removed. It also serves as the base on which the canister is mounted. As with the automotive filters, the entire canister is removed and discarded, and a new canister is installed.

Figure 11.3 shows a cartridge-type filter. This assembly consists of the filter head similar to the spin-on type, but instead of a disposable canister, there is a filter bowl (a *keeper*), and a cartridge (filter element) that is inserted into the assembly. The cartridge may be disposable, or it may be reusable after cleaning. To change this filter, you remove the bowl, take out the old element, insert the new one, and reinstall the bowl. You should clean the bowl and fill it with clean oil (if possible) before you reinstall it.

Filter Media

Many materials can be used as *filter media*. The ancient Romans used clay filters for their drinking water. Today, the most efficient filters available (at least from a contaminant removal standpoint) are made of a type of clay. These are called Fuller's earth or diatomaceous earth filters. They consist basically of *diatoms* (fossils of microscopic animal life) that remove not only particles, but also acids from the fluid. These filters are expensive and somewhat delicate, so their use is rather limited. More common media include woven and sintered metals, paper, and non-woven inorganic fibers.

Metals are often used in applications where high differential pressure across the filter element can be expected. They are made either by weaving very fine wire strands (as may be seen in screen wire) or by laying the strands across

Figure 11.3 In the cartridge type filter, the cartridge may either be disposable or reusable after cleaning.
Source: Courtesy of Schroeder Industries.

one another and applying heat and force to cause them to "weld" together in a process called *sintering*. Sintered metal spheres may also be used.

Wire screen (sieve) media are usually referred to by "mesh" or "sieve number" rather than by pore size. This defines the number of openings per linear inch of material. The number and size of openings in this material is governed by American Society for Testing and Materials standard ASTM E 11. Table 11.1 shows the standard mesh (sieve) numbers and the "size" of the openings for each. The openings in the standard media are square.

Most metal-media filters are cleanable. That is, rather than throwing the element away, you use various techniques to remove the particles from it so it can be reused. This sounds like a very good idea, and it is, if it is done properly. If you choose cleanable filters, there are three important things to remember. First, the finer the element, the more difficult it will be to clean. In fact, most elements cannot be cleaned completely; some dirt will be left in the element at the end of the cleaning process. This means that it will take less dirt to clog it the next time it is used, which is the second point. It is not unusual for more dirt to be left in the media after each cleaning, so that the useful life is shorter each time.

The third thing to remember is that these reusable filters won't last forever. Obviously, from the discussion above, if the useful life is shorter after each

Table 11.1 Standard Wire Mesh Sizing (ASTM E 11)

Number	Mesh	Average Opening Size	
		μm	in
6	6	3350	0.1330
7	7	2800	0.1112
8	8	2360	0.0937
10	9	2000	0.0794
12	10	1700	0.0675
14	12	1400	0.0556
16	14	1180	0.0468
18	16	1000	0.0397
20	20	850	0.0337
25	24	710	0.0282
30	28	600	0.0238
35	32	500	0.0199
40	35	425	0.0169
45	42	355	0.0141
50	48	300	0.0119
60	60	250	0.0099
70	65	212	0.0084
80	80	180	0.0071
100	100	150	0.0060
120	115	125	0.0050
140	150	106	0.0042
170	170	90	0.0036
200	200	75	0.0030
230	250	63	0.0025
270	270	53	0.0021
325	325	45	0.0018
400	400	38	0.0015
450	—	32	0.0013
500	—	25	0.0010
635	—	20	0.0008

cleaning, there comes a time when it isn't worth reusing. More importantly, however, the element may be damaged by the process used to clean it. If acid is used to destroy the particles, it may also damage the media. If ultrasonic vibration is used, parts of the metal may eventually fail due to fatigue. Fluid pulsing and backflushing may also cause fatigue. Even if the element survives the cleaning process, it may not survive the handling and shipping. Before any reusable filter is reinstalled, it should be subjected to a fabrication integrity test (to be discussed later in this chapter).

The mainstay of the filter media world since the early 1970s has been paper. It is usually impregnated with resin or fiberglass for additional strength. It is normally pleated (folded like an accordion), then rolled into a cylinder, bonded to end caps of some sort, and usually (but not always) has a center support tube and, occasionally, an outside support tube. Figure 11.4 shows a typical paper element. These elements are not cleanable. At the end of their installed

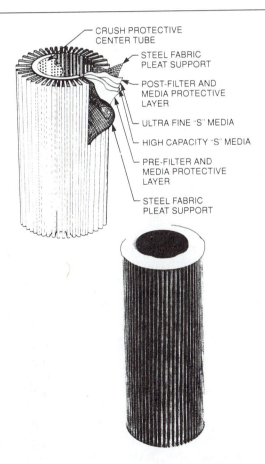

CRUSH PROTECTIVE
CENTER TUBE

STEEL FABRIC
PLEAT SUPPORT

POST-FILTER AND
MEDIA PROTECTIVE
LAYER

ULTRA FINE "S" MEDIA

HIGH CAPACITY "S" MEDIA

PRE-FILTER AND
MEDIA PROTECTIVE
LAYER

STEEL FABRIC
PLEAT SUPPORT

Figure 11.4 Paper filter elements consist of a pleated paper medium, often with other supporting materials, around a protective center tube. An outer metal protective cover is often used.
Source: Courtesy of Schroeder Industries.

life, they are discarded and replaced with a new element. The discarded elements should not simply be thrown in the trash barrel, since disposal companies can recover the metals and oils from these elements in an environmentally safe manner.

A relative newcomer to filter media are the inorganic materials. These are synthetic, usually non-woven materials that use extremely-small-diameter fibers to make up the media. The term "non-woven" means that, like paper, they are a mass of randomly laid fibers rather than a screen or woven cloth. Their main advantage over paper is that the fiber size can be much smaller, and the fiber diameter can be accurately controlled. The smaller the diameter of the fibers, the more densely they can be packed. This results in smaller openings between fibers, and, as a result, finer filtration.

There are obviously other filter media. You may be familiar with some we haven't discussed. The three discussed here—metal, paper, and man-made fibers—are used in the vast majority of fluid power filters. You may see them combined with magnets in some instances to enhance the removal of ferromagnetic particles. They may also be combined with hygroscopic materials to remove trace moisture from mineral based fluids.

Capture Mechanisms

Almost invariably, when we think about how filters remove particles from a fluid (if we think about it at all), we visualize a fine-mesh screen with the dirty fluid flowing into it and clean fluid flowing out of it, with all the dirt left on the screen. In this visualization, the dirt particles are removed simply because they are too large to go through the holes in the screen. This type of filtration, usually termed *surface filtration,* is an important mechanism in fluid power filters, but it is only one of several mechanisms involved in modern filters.

Before discussing mechanisms, let's look at the structure of the media itself. Figure 11.5 is a magnified photograph of a modern, non-woven filter media. You can see that there is no order to its structure—the fibers lay randomly, and there is no regular pattern of pores or pore sizes. Furthermore, the media obviously has a finite depth, and there are no apparent flow paths straight through it. Consequently, it is termed a *depth-type* media. We say that it has *tortuous* flow paths, meaning paths full of curves and turns, (not meaning "causing pain," as in listening to a professor's bad jokes, for instance.)

With this media, we can readily imagine two capture mechanisms. The first, as we have already discussed, is *surface capture* (see Figure 11.6). Some particles are simply too large to enter any of the surface openings. They collide with the media and simply sit there. As more and more particles are captured in this manner, they form a cake that, to some extent, improves the characteristics of the filter.

The second is a similar mechanism, except that it happens inside the media instead of on its surface. The particle enters the media, but encounters a barrier inside it. Thus, it is captured in the depth of the media. This is aptly termed a *depth capture* mechanism (again refer to Figure 11.6).

Some other mechanisms are not so readily apparent, because they are the results of physical phenomena rather than physical barriers. One of these is

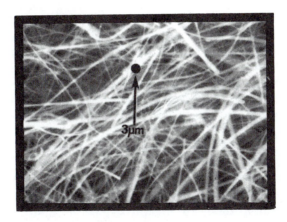

Figure 11.5 In a non-woven filter media, the fibers lay randomly.
Source: Pall Corporation, East Hills, NY.

hydrostatic capture (Figure 11.6). In this case, a particle somehow finds its way into an irregularity, or pocket, in the media and is held there by hydrostatic forces. Still another mechanism involves *electrostatic* or *Van der Waals attractions* between the particle and the media. These latter mechanisms are relatively weak. They are very susceptible to disruption by flow surges, vibration, impulses, or even a change in the relative humidity.

A final mechanism is one that has been termed *brickworking* or *log jamming*.

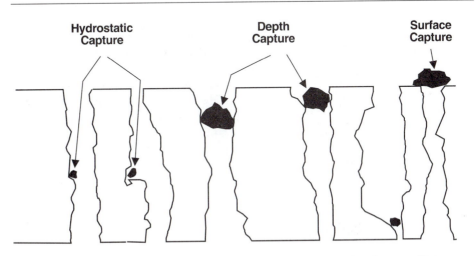

Figure 11.6 Three capture mechanisms are possible in depth-type filters: surface, depth, and hydrostatic.

This is a surface mechanism in which particles that are small enough to enter the opening in the media interfere with one another so that none of them get in. It is similar to pouring sand through a fine screen. If you pour the sand onto the screen slowly, the majority of it will go through. If you dump the whole bucketful on at one time, very little will go through. This is also a weak mechanism that is easily disturbed—just shake the screen.

Filter Ratings

If you were to purchase a new pump for your fluid power system and happened to read the instructions before you installed it, you might come across a statement similar to this: "To ensure best performance and longer pump life, use 25-micron filtration." How would you react to this statement? If you were conscientious, you would probably set out to find a 25-micron filter, but what would you actually be looking for, and how would you know when you found it? The truth is that such terminology has no universally accepted definition.

In general, the *micron rating* of a filter—that is, the size of particles it is supposed to remove—is presented in one of three ways:

- Absolute rating
- Nominal rating
- Beta rating

Absolute Rating: The absolute rating of a filter should define the particle size where 100 percent of the particles that size and larger are trapped by the filter. Unfortunately, it doesn't work that way in today's depth-type filters. The original definition of *absolute* was the largest hard spherical particles that would pass through the filter under specified conditions. Today, many filter manufacturers have their own definition of absolute and do their own testing. As a result, you as a consumer have no way to compare filters based on their "absolute" rating.

Nominal Rating: Originally, the nominal rating of a filter defined the size of hard, spherical particles at which 98 percent of the particles would be captured. That term has been so badly abused that most standards organizations discourage its use. Again, it provides no basis for comparing filters.

Beta Rating: The Beta rating, developed at the FPRC at Oklahoma State University in the late 1960s and early 1970s does provide a basis for comparing filters. Basically, the Beta rating is defined as the ratio of the number of particles of a given size entering a filter to the number of particles that same size leaving the filter. Since this is the only method currently available for comparing filters in a reliable way, let's look at the rating method in detail.

11.5   Beta Rating

The method used for determining the *Beta rating* of a full-flow fluid power filter is defined in the International Organization for Standardization (ISO) Standard ISO 4572, *Multipass Filtration Test*. Briefly, the test requires that

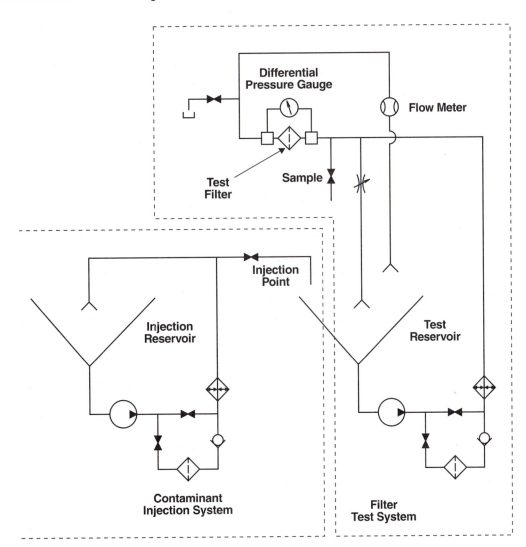

Figure 11.7 The Multipass Filter Test gives the Beta rating. This is a schematic of the test stand used.

NOTE: Reservoir shapes emphasize the requirement for conical design.

the test filter be installed in a test system as shown in Figure 11.7. A steady fluid flow rate is established that is compatible with the filter type, size, and material. Fluid contaminated with AC Fine Test Dust (ACFTD) is transferred from the injection reservoir to the test reservoir at a controlled rate so that the total amount of contaminant injected is known at all times. The fluid from the test reservoir passes through the filter, circulates through the system, and returns to the test reservoir. As it returns, it carries with it any contaminant that was not captured by the filter. This contaminant is pumped back to the filter from the test reservoir—hence the name *multipass*.

At several predetermined points in this procedure—based on the pressure drop across the filter element—fluid samples are extracted from the flow stream upstream and downstream of the filter. An automatic particle counter is used to assess the numbers of particles of selected sizes in these samples. The Beta rating for any particle size is calculated from Equation 11.1:

$$\beta_x = \frac{n_u}{n_d} \tag{11.1}$$

where

β_x (Beta sub x) = Beta rating at $x\mu$m

n_u = number of particles $\geq x\mu$m per milliliter of fluid upstream of the filter.

n_d = number of particles $\geq x\mu$m per milliliter of fluid downstream of the filter.

Particle counts are per milliliter of fluid. This Beta rating is calculated for each test point for several different particle sizes, x. The minimum ratio for each particle size is specified as the Beta rating for that size. They are expressed, for example, as $\beta_5 = 3$, $\beta_{10} = 12$, $\beta_{15} = 25$, and so on.

As you may have already deduced, the test stand used in conducting the multipass test is a very critical part of the process. The stand must provide for precise control of the contamination level being injected into the test system. It must be constructed so that it will not generate particulate that can find its way into the test system. Likewise, it must not have any particle traps that would remove particles from the test system. It must also provide for accurate control of the flow rate through the system as well as the fluid temperature.

Likewise, the fluid sampling and particle counting techniques must be carefully controlled so that extraneous contamination will not be introduced into the fluid samples. Sample container cleanliness is especially critical.

An example of the results of a multipass filtration test is shown in Table 11.2. The first column in the table indicates the point during the test that the samples were taken. The first sample was taken two minutes after the test began. The next was taken when 10 percent of the maximum allowable pressure drop across the filter was reached, and so on. The table then shows the number of particles in the upstream and downstream samples for each particle size at each sample point. These particle counts are in the rows labeled "UP" and "DOWN," respectively. Notice that at each test point, the Beta rating has been determined for 10, 20, 30, 35, 40, and 50 μm.

As the differential pressure (that is, the pressure drop across the filter) increases, the Beta ratings vary. The manner in which they vary depends on several factors related to the filter media and construction. For some filters, Beta may increase as the pressure drop increases. For others, it may decrease. In still others, there may be an initial increase followed by a decrease or a decrease followed by an increase. Whatever the case, the lowest Beta (meaning the poorest performance) for each reported particle size is the one reported. This is shown as the "minimum Beta" in the table.

Looking at the calculation of the Beta rating, a couple of things become immediately obvious. First, if Beta equals 1, there is the same number of particles of that size upstream and downstream of the filter—none have been removed. This means that the filter is totally useless for that particle size. The second is that the better job the filter is doing, the higher the Beta rating will be.

Table 11.2 Sample Results of Multipass Filtration Test (ISO 4572)

		Particle Distribution Analysis (Particles per Milliliter)					
Sample Pt		$> 10~\mu m$	$> 20~\mu m$	$> 30~\mu m$	$> 35~\mu m$	$> 40~\mu m$	$> 50~\mu m$
Initial		10.90	0.80	0.00	0.00	0.00	0.00
	Up	6705.00	413.70	90.87	56.13	35.87	16.27
	Down	5642.00	182.30	12.07	4.00	1.00	0.00
2 Min	Beta	1.19	2.27	7.53	14.03	35.87	—
	Up	75771.00	2872.00	234.00	100.00	54.00	22.00
	Down	74525.00	2663.00	178.00	58.67	18.67	2.67
10%	Beta	1.02	1.08	1.31	1.70	2.89	8.24
	Up	82579.00	3747.00	320.00	126.00	55.33	19.33
	Down	80700.00	3333.00	216.00	68.67	22.67	4.67
20%	Beta	1.02	1.12	1.48	1.83	2.44	4.14
	Up	75312.00	3428.00	293.30	120.70	60.00	16.00
	Down	67622.00	2783.00	193.30	62.00	19.33	1.33
40%	Beta	1.11	1.23	1.52	1.95	3.10	12.03
	Up	57392.00	2829.00	276.00	120.70	57.33	22.67
	Down	54223.00	2418.00	180.70	61.33	23.33	3.33
80%	Beta	1.06	1.17	1.53	1.97	2.46	6.81
Minimum Beta		1.02	1.08	1.31	1.70	2.44	4.14

Note: ACFTD Capacity (grams): Apparent 65.82, retained 52.01.

Example 11.1

Let's consider a practical application problem. Say we have a servovalve that has been found to give the required performance and reliability only if the number of particles 5 μm and larger in the fluid entering the valve is less than 25 per milliliter. A particle count reveals that the system fluid actually has 875 particles in that critical size range. In order to protect the servovalve from the highly contaminated fluid, what β_5 rating is required?

Solution

Since the valve can tolerate only 25 particles, the job of the filter in this system must be to trap and retain at least 850 of the 875 particles that enter it with each milliliter. The required β_5 rating is found by dividing the actual particle count in the influent fluid by the allowable particle count in the effluent. Thus,

$$\beta_5 = \frac{875}{25} = 35$$

Therefore, if you were specifying a filter to protect the servovalve, you would call for a filter with a β_5 of at least 35. It would be wise to specify a higher Beta rating, for example $\beta_5 = 50$.

Before the development of the Beta concept, many filter manufacturers would use the dirt removal efficiency of their filters in their advertising. Unfortunately, as with the different methods of describing the filter micron ratings, the efficiency ratings were obtained in different ways. This resulted in considerable difficulty in comparing filters made by different manufacturers. While particle-removal efficiency is the ultimate test of a filter, the confusion surrounding efficiency determinations caused researchers at the Fluid Power Research Center to purposely avoid the term "efficiency" as they developed the Beta concept. It is possible, however, to use the Beta rating to determine a filter's efficiency in removing any specified particle size. The equation used for this is simply

$$\eta_x = \frac{\beta_x - 1}{\beta_x} \tag{11.2}$$

where

$$\eta = \text{efficiency}$$
$$x = \text{particle size}$$

Example 11.2

Fluid samples taken upstream and downstream of a filter contain 10,890 and 565 particles larger than 10 μm, respectively. Find the β_{10} rating and the removal efficiency at 10 μm.

Solution

The β_{10} rating is found as before.

$$\beta_{10} = \frac{10,890}{565} = 19.3$$

The efficiency at which particles 10 μm and larger are removed can now be calculated as

$$\eta_{10} = \frac{\beta_{10} - 1}{\beta_{10}} = \frac{19.3 - 1}{19.3} = 0.95$$

The Beta rating provides a method for a meaningful comparison between filters produced by different manufacturers. This method of filter rating allows the user to specifically define what the filter is to do. It is a tremendous improvement over the terms used previously (absolute, nominal, etc).

You must remember, however, that the multipass test is conducted under somewhat idealized conditions. This, of course, is true of most standards. While it is desirable to reproduce the exact conditions under which the test is to operate, defining those conditions is virtually impossible, because everyone's application and conditions are different. Therefore, most standard tests depart from reality and move toward test conditions that are controllable, repeatable, and reproducible. While the end result, hopefully, will allow a reasonable comparison of articles tested under such controlled conditions, it may not demonstrate exactly how the article will perform in the "real world."

This is the case with the multipass test for rating filters. The test parameters call for a constant temperature, steady flow rate, and a constant rate of contaminant ingression using AC Fine Test Dust. These conditions are seldom found in actual usage. Usually, the fluid temperature varies throughout the work day, the flow is more likely to be surging and pulsing than steady, and ingression rates may vary with environmental conditions, wear rates, and duty cycles.

AC Fine Test Dust is not found in many systems. All of these things may adversely affect the ability of a filter to remove and retain particles. This means that the Beta rating obtained on a multipass test stand is unlikely to be obtained on an operating system.

In the light of this, you are probably ready to question the usefulness of the Beta rating. The Beta technique is, in fact, valid for ranking the performance of filters "on the job" in that a filter that performs well on the multipass stand will do better on the job than one that performed poorly on the stand.

Recall that in Chapter 4, when we were discussing the contaminant sensitivity of hydraulic pumps, we said that the Omega rating of the pump was related to the Beta rating of the filter required to ensure a 1000-hour lifetime for the pump. You can see now that selecting a pump with a low Omega rating (consequently, a high contaminant tolerance) can significantly ease the system filtration requirement. This should not be considered a license to skimp on filtration. There is no excuse for poor filtration.

11.6   Additional Considerations

It is unfortunate that with filters, as with political candidates, getting the right answer to one question doesn't automatically imply that all other answers will be correct. In addition to finding a filter with the correct Beta rating, there are several other factors that must be considered. Included in these are the flow capacity, pressure drop, and dirt-holding capacity.

Flow Capacity

Filter manufacturers publish data for their filters which includes, among other things, the maximum flow for which a given filter is suitable. This flow rate is based on the filter type, media, size, and so forth. It should never be exceeded. It becomes critical, therefore, to determine accurately the maximum flow rate that will be experienced by each filter. It may be well above the pump output, as the following example shows.

Example 11.3

Consider the system shown here. The pump output is 10 gpm. The cylinder has a 4 in bore and a 2 in rod. What is the minimum flow capacity of the return line filter?

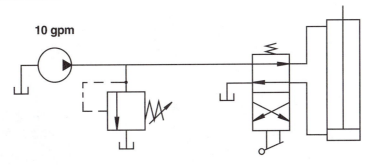

10 gpm

Example 11.3

Solution

We approach this problem by considering the relationships between flow rates, areas, and piston velocities. Let's start by looking at the extension stroke. With our pump producing 10 gpm, we can find the extension speed of the cylinder from Equation 2.13:

$$Q = Av$$

or

$$v_{ext} = \frac{Q}{A_{piston}}$$

$$= \frac{10 \text{ gal/min} \times 231 \text{ in}^3/\text{gal} \times \text{ min/60 sec} \times \text{ft/12 in}}{12.56 \text{ in}^2}$$

Therefore,

$$v_{ext} = 0.26 \text{ ft/sec}$$

How does this affect the return flow rate from the rod end of the cylinder? We find out by using the same equation, but now the area of concern is the annular area of the piston minus the rod area. Obviously, the velocity of the rod side of the piston is the same as that of the blind side of the piston (at least, we hope so). Plugging this into our equation for flow rate and converting the gpm gives

$$Q_{return} = (A_{piston} - A_{rod}) \times v$$

$$= (12.56 \text{ in}^2 - 3.14 \text{ in}^2) \times \frac{0.26 \text{ ft}}{\text{sec}} \times \frac{12 \text{ in}}{\text{ft}} \times \frac{\text{gal}}{231 \text{ in}^3} \times \frac{60 \text{ sec}}{\text{min}}$$

Therefore,

$$Q_{return} = 7.63 \text{ gpm}$$

This is somewhat lower than our pump is pushing into the blind end. It shows simply that we are displacing a smaller volume of fluid from the rod end than we are putting into the blind end for the same amount of movement.

Now let's reverse our valve and retract the cylinder. We're still using our full pump output of 10 gpm. Therefore, our retraction speed is

$$v_{return} = \frac{Q}{A_{piston} - A_{rod}}$$

$$= \frac{10 \text{ gal/min} \times 231 \text{ in}^3/\text{gal} \times (\text{ft/12 in} \times \text{min/60 sec})}{12.56 \text{ in}^2 - 3.14 \text{ in}^2}$$

Therefore,

$$v_{return} = 0.34 \text{ ft/sec}$$

The return flow is now coming from the blind end of the cylinder. We find it as before, except we now use the full face area of the piston. We find that

$$Q_{return} = A_{piston} \times v$$

$$= 12.56 \text{ in}^2 \times 0.34 \text{ ft/sec} \times \text{gal/231 in}^3 \times 12 \text{ in/ft} \times 60 \text{ sec/min}$$

Therefore,

$$Q_{\text{return}} = 13.31 \text{ gpm}$$

which is somewhat greater than our pump output. Therefore, if we sized our filter for the pump flow rate of 10 gpm, we could be in trouble.

We could reach the same conclusions by considering the ratio of the blind-side and rod-side areas of the piston. Here, our piston area is 12.56 in², while the rod-side annulus area was 9.42 in², a ratio of 4 to 3, or 1.333. On the extension stroke, then, the pump flow and the return flow would be in that same ratio. Thus, on extension, the return flow (from the rod end) would be

$$1.333 = \frac{10}{Q}$$

or

$$Q = \frac{10}{1.333} = 7.5 \text{ gpm}$$

(The difference in this flow rate and the one calculated previously is in rounding off certain factors.)

To determine the return flow on retraction (from the blind end), we take the ratio of the rod-end area to the piston area. That is a ratio of 9.42 to 12.56 in², or 3 to 4, which is 0.75. We then find Q as before.

$$.75 = \frac{10}{Q}$$

or

$$Q = \frac{10}{.75} = 13.33 \text{ gpm}$$

Again, rounding off accounts for the slight difference.

Pressure Drop

The term *pressure drop* (often called the Δp) has been widely misunderstood and misused. The term actually refers to the loss in pressure from the inlet to the filter to the outlet from the filter. That is easily understood. The problem comes in defining the locations of the inlet and the outlet. If we measure the pressure drop from the inlet port of the filter assembly to the outlet port, then we get the *assembly pressure drop*. This includes the pressure drop through the ports, housing, filter media, and so on.

If we are using a cartridge type filter, we may be interested in the pressure drop across the element itself as well as the entire assembly. This could be of interest if we are considering the purchase of a replacement element or in selecting a filter for the pump suction line.

It is generally considered that finer filtration automatically implies a higher pressure drop, but this is not a valid generality. Pressure drop across a filter media is affected by both the "fineness" of the media (often referred to as the "tightness" of the media) and the flow density through the media. *Flow density* is defined as the flow rate per unit area and is usually presented in gallons per minute per square foot of media.

Now, if we generalize that, for a specified area of media, a tighter filter will have a higher pressure drop, we will usually be correct. Note, however, that two filter cartridges of the same "size" may not necessarily have the same area. For example, while two paper cartridges may have the same diameter and length (giving them the same cylindrical area), one may have more and deeper pleats than the other. This will give it a greater surface area, thus reducing its flow density for a given flow rate. This will result in a lower pressure drop for a given media. In filters using synthetic media, finer filtration may be achieved at lower pressure drops. The pores are smaller, but there are more of them.

Filter manufacturers provide pressure drop versus flow rate data for each of their filters. Figure 11.8 is an example of such data. This information will normally include fluid specifications (viscosity, temperature, etc) and the conditions under which the data were obtained. You will need to consider the filter's performance in your system in light of the test parameters.

Another pressure drop of interest is the *terminal pressure drop*, which is specified by the manufacturer and is the maximum pressure differential at which the filter should remain in service. At this point, it is time to change the filter because it has clogged (or loaded up) with particles to the point that its useful life is finished. In many filters, once this terminal pressure drop is reached, the media begins to distort slightly, allowing more and larger particles to pass through it. This is often seen in the results of a multipass test as a decrease of the filter Beta ratings at one or more particle sizes.

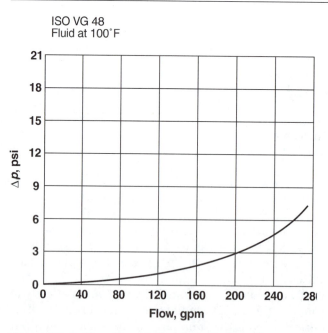

Figure 11.8 The filter P-Q curve is provided by the manufacturer.

The *collapse/burst pressure* is the differential pressure at which the media distorts significantly or actually ruptures. This results in an abrupt drop in the differential pressure, the immediate release of many of the previously retained dirt particles, and a catastrophic loss of filter efficiency. This point is well above the terminal pressure drop. Therefore, in a well maintained system, catastrophic filter failure should never occur.

One other area in which some confusion is likely to occur is in discussing the maximum operating pressure of the filter. The filter *assembly* (head, bowl, connectors, etc) must be structurally strong enough to withstand the maximum operating pressure experienced in the portion of the system in which it is installed. The filter *element*, however, needs to withstand only the pressure *drop* that occurs as fluid flows through it. With the exception of this differential pressure, the element experiences the same pressure on both sides of the media. Thus, it is the differential pressure, and not the system pressure that is of concern when speaking of the element. In some exceptional cases, the element may be required to withstand a differential pressure equal to the full system pressure without collapsing or bursting.

Many filter assemblies provide some sort of device to warn the user when the pressure drop across the element approaches the limit (terminal) pressure. These devices normally measure or sense the differential pressure, although some will sense the pressure increase in the housing or bowl due to the clogging of the filter.

Among those devices that sense the differential pressure are differential pressure gauges GO/NO GO indicators (which are actually pressure gauges with green (GO) and red (change the filter) color bands rather than numbers), pressure switches or other transducers that initialize some sort of warning (light, buzzer, etc), and pop-up indicators provide a visible warning. Of these, the pop-up has traditionally been the least reliable. Older types may experience one of several failure modes, including:

1. False activation due to cold startups (temporary high pressure drop due to high viscosity).
2. False activation due to pressure or flow surges during system operation.
3. Activation due to impact.
4. Failure to operate due to contaminant jamming, corrosion, rust, etc.

Manufacturers have used a number of design variations to prevent these problems. Design changes such as bimetallic springs, time delay devices, and protective caps to exclude dirt and moisture have improved their reliability significantly. Figure 11.9 shows several different types of indicators.

Many filter assemblies are also equipped with bypass valves that open up at a set differential pressure. The purpose of these valves is to port fluid around the clogged filter element to prevent its subsequent collapse or burst. When the valve opens, the system is subjected to unfiltered fluid. This is potentially disastrous to any system, especially if the situation is not quickly corrected by changing the filter cartridge.

Changing filters based on the time of use is seldom cost effective. It usually results in the removal of the filter when there is a significant amount of useful life remaining, or after it has passed its useful life, perhaps even after it has collapsed or burst.

Figure 11.9 Filter clogging indicators may be visual or electrical.
Source: Courtesy of Schroeder Industries.

Figure 11.10 shows the ISO symbols for filters. Figure 11.10a is the basic filter symbol. As with all other symbols, only the function is represented; therefore, the same symbol represents all filters regardless of type or placement in the system. Figure 11.10b represents a filter with a differential pressure indicator, while Figure 11.10c represents a filter with both an indicator and a bypass valve.

Dirt-Holding Capacity

The ability of a filter to retain solid particles is often termed its "dirt-holding capacity." This is an important characteristic of a filter, and may very well be the deciding factor in the choice between filters with similar Beta ratings and pressure drops.

The capacity of a filter element is a natural piece of information that results from the multipass filtration test (Beta rating test). In that test, contamination in the form of AC Fine Test Dust is added to the test system at a constant rate. Therefore, the amount (weight in milligrams) of ACFTD that has been introduced into the system can be determined at any time during the test. It is a simple matter, then, to report the amount of contamination that was added to the test system in order to bring the filter to the terminal pressure drop. This is, in fact, done as a part of the multipass test. It is reported as the *capacity* or the *apparent capacity* of the filter. This is shown at the bottom of Table 11.1

Unfortunately, the determination of the filter capacity in this way can be misleading, because the "apparent" capacity is the amount of ACFTD *added to the system* and not necessarily the amount actually *captured by the filter*. Most filters will allow at least a small percentage of the test contaminants to pass through them. These particles simply continue to be pumped around the circuit by the system pump and are suspended in the fluid. By the time the terminal pressure drop is reached, a substantial amount of the injected dirt may be running around in the fluid rather than being retained in the filter. You can see from this that a poor filter may look very good from an apparent capacity point of view. It may require a large injection of ACFTD to bring it

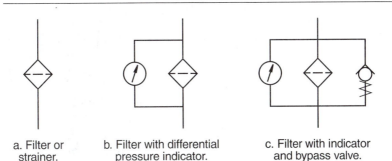

a. Filter or   b. Filter with differential   c. Filter with indicator
   strainer.   pressure indicator.   and bypass valve.

Figure 11.10 ISO filter symbols, like all other symbols, represent only the function.

to the terminal pressure drop, but much of that dirt is in the test fluid rather than in the filter.

This problem has prompted some test laboratories to report a *retained capacity,* determined by evaluating the test fluid gravimetrically to weigh the dirt in the test system fluid. The difference between this weight and the total weight added (the apparent capacity) is the actual weight of the dirt retained in the filter. There may be a significant difference. Beware!

Other Filter Tests

Several other filter tests may be of interest to the potential user.

1. *Fabrication Integrity Test—ISO 2942.* This test, also termed the *bubble-point test*, is designed primarily for cartridge type elements. Its purpose is to determine if there are any flaws in the element that will allow unfiltered fluid to pass through it. Such flows—if they exist—are commonly found at the ends of the element where the end caps are bonded to the media, at the seam where the two ends of the media have been bonded together, or in the creases of the pleats where cracks may occur. Of course, pinhole leaks, punctures, tears, and other flaws may occur anywhere in the media due to manufacturing defects, rough handling, or deterioration due to age. Regardless of the location or reason for the flaw, allowing unfiltered fluid to pass through a filter is highly undesirable.

This test is simple to conduct and is highly recommended for every cartridge element before it is installed in the system. All that is required for an abbreviated version of this test is a vat of isopropyl alcohol large enough to allow immersion of the entire element to a depth of about 3 inches, a low-pressure air source with a pressure regulator and gage, and rubber stoppers for each end of the cartridge—one with a hole to allow the insertion of an air hose as shown in Figure 11.11. With the stoppers and air hose in place, the element is immersed in the alcohol. The air pressure is slowly increased until bubbles begin to come from the element. If there is a flaw, the bubbles will appear as a stream coming from that flaw. If there is no flaw, bubbles will appear uniformly and nearly simultaneously over the entire surface of the filter. If a flaw is detected, no matter how small, do not use the element. If the element is suitable for use, allow the alcohol to evaporate completely before installing the element. The air pressure required in this test is very low—a few inches of water, at most.

2. *Collapse/Burst Resistance Test—ISO 2941.* This test is normally conducted by the manufacturer or a testing laboratory rather than by the user. It is usually performed in conjunction with the multipass test. When the terminal pressure drop is reached, the injection of the contaminant is continued until the element is either ruptured or collapsed by the differential pressure.

3. *Material Compatibility Test—ISO 2943.* Obviously, it is necessary that the materials in a filter be compatible with the fluid in which it will be used. This test involves soaking the filter element in the designated fluid at an elevated temperature for 72 hours. At the end of that time, several tests, including the fabrication integrity, multipass, and collapse/burst are conducted to determine the effects of the fluid.

4. *Media Migration—SAE J806.* This test is conducted by the manufacturer or a test laboratory to determine if the media releases any of its material in

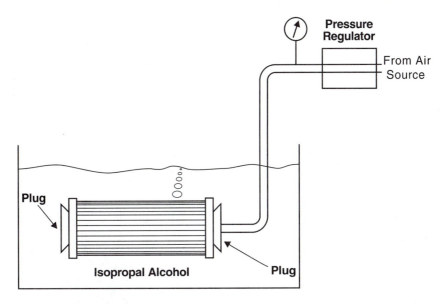

Figure 11.11 The fabrication integrity test is recommended for every cartridge element.

use. The test is for the filter material itself rather than for the release of previously captured contaminants.

11.7   Filter Placement

System Filtration

Filters may be placed at numerous locations throughout a system (Figure 11.12). The location (one or more) depends on the function the filter must accomplish. In some locations, a single component will be protected. In others, protection will be afforded to the entire system.

In many systems, filtration begins in the reservoir. In some cases, a magnet is placed inside the reservoir to attract ferromagnetic particles. A *suction strainer* is frequently found inside the reservoir. Notice that this is referred to as a *strainer* and not as a filter. The reason for this is that this device is often a metal screen media—usually a 100-mesh screen. That amounts to a 149-micron pore size. While this seems extremely large, remember that as this strainer becomes sufficiently clogged to choke off the fluid flow, pump cavitation and subsequent pump failure could result. Filters of this type are often referred to as *rock stoppers*, because their purpose is simply to keep the "big stuff" from getting into the pump. Because these strainers are inside the tank, they are often forgotten and not changed as often as they should be.

The suction strainer is frequently supplemented, or even replaced, by a *suction-line filter*. Because of its location, it usually cannot be a fine filter,

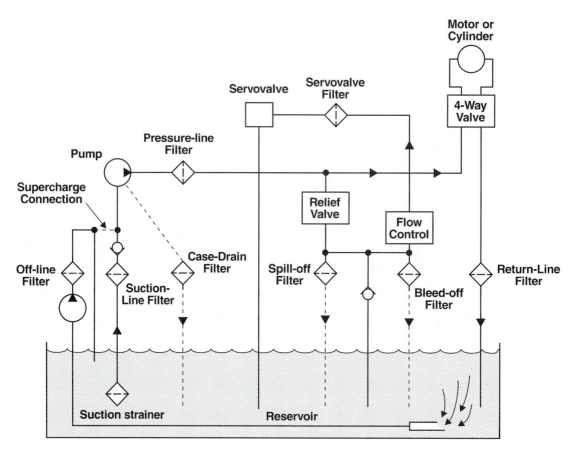

Figure 11.12 Filter locations depend on the function desired.

although it may be somewhat tighter than the strainer. These filters must be monitored carefully, because a clogged element could lead to cavitation failure of the pump.

Probably the most common filter location is in the pressure line just after the pump outlet. In this location, all fluid from the pump is filtered before it gets into the system. This is a good strategy, because the pump is usually the worst generator of wear particles in the entire circuit. Removal of these particles before they get into the system can prevent serious wear problems on other components.

Pressure-line filters have one major shortcoming that is either unrecognized or frequently overlooked. Pressure spikes or flow surges will significantly reduce the efficiency of any filter by driving particles through the media. The changes in flow forces caused by the surges act almost like a hammer, causing the particles to spread the media and go through. This sometimes results in "clouds" of particles escaping from the media. These high concentrations of particles are potentially more harmful to components than a steady input of lower concentrations. Pressure-line filters are frequently subjected to these harmful spikes and surges.

A common practice is to provide *dedicated filters* for especially contaminant-sensitive components such as servovalves. These filters may be much finer than the other filters in the system.

Return-line filters are often the most effective filters in the operating portion of the system. Virtually all system flow passes through them. Since pressure and flow surges tend to be somewhat less dramatic than in the pressure line, and because the flow velocity is lower, there is less diminishment of filter efficiency in this location. Return-line filters are frequently finer than those in most other locations, but care must be taken to prevent the buildup of a back-pressure in the return line.

The drain lines, bypass lines, and other component lines that return directly to the reservoir often are filtered to ensure that those components don't add contaminants to the system.

Off-Line Filtration

In addition to the filters that are placed in the operating system, *off-line filters*—such as the one shown on the left-hand side of Figure 11.1—can be used very effectively. These filters—also called *bypass filters* or *kidney loops*—can be extremely effective because they are usually very fine and are not subject to the spikes and surges of the system filters. An off-line filtration system is independent of the system operation because it has its own pump and filters. Fluid is taken from the reservoir, passed through the filter at a constant flow rate (the slower the better), and returned to the reservoir or dumped into the system pump suction line.

The performance of these filters can be spectacular. With a high-Beta filter and a flow rate designed to filter the total reservoir capacity approximately ten times per hour, the fluid can be "polished" to an extremely clean state. Remember, though, that *off-line filters do not take the place of the normal system filters*. They do not have the ability to protect the system from a catastrophic failure or a sudden ingression of dirt from outside the system. Don't install an off-line filter and throw away your system filters. That is flirting with disaster!

11.8   Water Removal Filters

Small amounts of water can be removed from the system fluid by specially designed filters that may be chemically treated with a hygroscopic chemical or contain some material such as wood chips that tends to absorb and hold water. Other filter housings contain heating elements to evaporate water. Caution must be used with these heater-type filters to avoid overheating the fluid.

11.9   Contamination Control Program

In this chapter, we have discussed the extremely important subject of filtration. Good filtration can extend the life of systems and components and increase up-time and productivity while decreasing operating costs. However, filtration is only one phase of an effective contamination control program. Prevention of contamination, detection of contamination, and the proper selection of components to minimize the effects of contamination are also required.

Prevention addresses methods of keeping contamination out of the system. It includes recognizing the sources of contamination and taking the proper steps to eliminate the problem from those sources. It also includes the use of the proper fluid (correct viscosity, additives, temperature ranges, etc) as well as fluid maintenance. This means starting with the right fluid, cleaning it before putting it into the system, and checking it regularly for water, acids, oxidation, and viscosity.

Detection implies some sort of fluid monitoring program in which fluid samples are extracted from the system on a regular basis and analyzed for indications of degradation. These analyses should include, at a minimum, particle counting, acid level, water content, and viscosity.

Extracting fluid samples is critical to the success of any fluid monitoring program. The sample must be representative of the fluid in the system; therefore, every effort must be made to prevent adding contaminants to the sample. The most common problem here is the container in which the fluid is collected. Use only containers that have been specially cleaned for this purpose. Regardless of how carefully you wash the canning jar, soft drink bottle, or whatever, it is very unlikely that it will be so clean that it will not significantly change the results of a particle count.

Another problem in sampling is the location from which the sample is extracted. The system reservoir is a poor choice, since contaminants tend to collect there, and it will likely be far dirtier than the rest of the system. A well-designed fluid power system should include sampling valves located in active flow lines throughout the system.

We have already discussed the *contaminant sensitivity of components*. When designing a new system or specifying replacement components, low Omega ratings (that is, good contaminant tolerance) should be required where possible. In this way, if the system should become contaminated for any reason, the component should be damaged less than those with higher Omega ratings.

The details of establishing and conducting a total contamination control program is beyond the scope of this text. You should, however, be aware of its importance and make it a priority in any fluid power work you might undertake.

11.10   Troubleshooting Tips

Symptom	Possible Cause
Filter pop-up indicator pops	1. Filter clogged 2. Cold startup 3. Flow surges
Filter pop-up indicator doesn't pop, even though filter is clogged	1. Wrong indicator—that is, Δp too high 2. Button jammed

Symptom	Possible Cause
Filter element clogs too frequently	1. It's doing its job 2. High rate of external ingression 3. High rate of internal wear
High fluid contamination levels—high component wear rate	1. Beta rating too low 2. Wrong element installed—doesn't fit housing 3. Element installed incorrectly 4. No element installed 5. Media damaged 6. Media collapsed

11.11 Summary

Contamination control is the key to system reliability and life. There are many different types of contamination—solid particles, heat, water, chemicals, gases—which have detrimental effects on the system and the fluid. Contamination can enter the system from external sources and it can be generated within the system.

A good contamination control program will include prevention, detection, and removal of contamination, and the selection of components with good contamination tolerance. The removal portion of the program requires that good filters be placed where they will provide the greatest protection for the system components. Filters should be specified by their Beta ratings to ensure that they are capable of providing the needed protection.

In fluid power system design, operation, and maintenance, always be guided by this concept: Keep it clean; keep it cool.

11.12 Key Equations

Beta rating: $\beta = \dfrac{n_u}{n_d}$ (11.1)

Beta efficiency: $\eta = \dfrac{\beta - 1}{\beta}$ (11.2)

Reference

1. R. K. King, R. K. Tessmann, and L. E. Bensch, "Ingression Rates—Actual Field Measurement," *Basic Fluid Power Research Program*, vol 10, Fluid Power Research Center, Oklahoma State University, 1976.

Review Problems

Note: The letter "S" after a problem number indicates SI units are used.

1. List some common sources of particle contamination in hydraulic systems.

2. Which component is usually the worst generator of wear particles in a hydraulic system?
3. What percent of fluid power system failures can be attributed to contaminated fluid?
4. List several types of filter media. Which is the most commonly used?
5. What is meant by the "mesh" of a wire screen type media?
6. What is a micron? How does it compare to an inch?
7. Explain the difference between surface-capture and depth-capture mechanisms.
8. Define the following filter rating terms:

    a. Absolute
    b. Nominal
    c. Beta rating

9. List several possible locations for filters in fluid power systems.
10. List several factors that must be considered in the selection of filters for fluid power systems.

Beta Ratings

11. In a multipass test, the particle counts for particles ≥ 5 μm are 3000 and 200 for the upstream and downstream samples, respectively. What is the β_5 of the filter?
12. Complete the following table:

Particle size (μm)		**5**	**10**	**15**	**20**	**25**
Particle counts	UP	65000	14000	3000	1200	100
	DOWN	12000	200	20	3	1
Beta Rating						

13. Which of the filters represented by the following Beta ratings is the most efficient?

    a. $\beta_{10} = 0.98$
    b. $\beta_{10} = 1.0$
    c. $\beta_{10} = 1.4$
    d. $\beta_{10} = 15$

14. Determine the efficiencies represented by the following Beta ratings:

    a. $\beta_5 = 7$
    b. $\beta_{15} = 30$
    c. $\beta_{10} = 5$
    d. $\beta_{10} = 75$
    e. $\beta_{10} = 100$
    f. $\beta_{10} = 200$

Flow Rates

15. A cylinder has a 4 in bore and a 1 in rod. Determine the flow rate out of the rod end during extension if the input flow rate is 10 gpm.
16. For the cylinder of Review Problem 15, what is the flow rate out of the blind end on retraction if the input flow rate is 10 gpm?
17. A cylinder has a 4 in² piston and a 2 in² rod. Determine the output flow rates on extension and retraction if an input flow rate of 6 gpm is used for both directions.

Fluid Conduits

Outline

12.1 Introduction

Fluid is conveyed between components by *fluid conduits*—a generic term for pipe, tubing, and hoses. In this chapter, we'll take a brief look at each of these types of conduits as well as the fittings and connectors required in their use. We will also examine methods for determining the pressure losses that result from fluid flow through fluid conduits and fittings.

Objectives

When you have completed this chapter, you will be able to

- Explain the standard sizing schemes for pipes, tubing, and hose.
- Discuss hose construction and specifications.
- Discuss the different threads used to make conduit connections.
- Select conduits based on flow rate and pressure requirements.
- Discuss common pipe, hose, and tubing fittings and connectors.
- Calculate the pressure losses associated with flow through fluid conduits.

12.2 Pipes

Let's begin with pipes. *Piping* is distinguished from hydraulic *tubing* by its extreme rigidity and thick, heavy walls. Pipe is normally constructed of high-tensile-strength mild steel. As a result, it can be used where very high pressures are expected. Pipe also lends itself well to stationary, industrial applications as well as to applications involving rugged or hostile environments and strong vibration.

Table 12.1 Recommended Fluid Velocities

Line Type	Velocity Range	
	ft/sec	(m/s)
Pump Suction	2–4	(0.6–1.2)
Return	10–15	(3–4.5)
Medium Pressure (to 2000 psi, or 13.87 MPa)	15–20	(4.5–6)
High Pressure (above 2000 psi, or 13.8 MPa)	20–30	(6–9)

Pipe Sizing

In specifying pipe sizes, we must be concerned with both the internal diameter and the wall thickness. The *internal diameter* (ID) is important because the fluid velocity at any given flow rate depends on that parameter. Recall Equation 2.13.

$$v = \frac{Q}{A}$$

where A is the flow area.

The fluid velocity throughout a circuit is important because of losses and inefficiencies that may result from friction within the conduit. The higher the velocity is, the greater the losses for a given fluid. As a rule of thumb, velocities in various parts of the system should be on the order of those listed in Table 12.1. The relationship between fluid velocity and pressure losses will be examined in detail in Section 12.5.

Example 12.1

A hydraulic pump is producing 20 gpm (76 Lpm). It has been determined that the fluid velocity in the pressure line of the system should be between 20 and 25 ft/sec (6 to 7.6 m/s). Determine the minimum and maximum pipe diameters that could be used.

Solution

From equation 2.13, we can find the flow areas required for the limiting velocities.

$$A_{MIN} = \frac{Q}{v_{MAX}}$$

$$= \frac{20 \text{ gal/min} \times \text{min/60 sec} \times \text{ft}^3/7.48 \text{ gal} \times 144 \text{ in}^2/\text{ft}^2}{25 \text{ ft/sec}}$$

Therefore,

$$A_{MIN} = 0.26 \text{ in}^2$$

or 1.65 cm².

Likewise,

$$A_{\text{MAX}} = \frac{Q}{v_{\text{MIN}}}$$

$$= \frac{20 \text{ gal/min} \times \text{min/60 sec} \times \text{ft}^3/7.48 \text{ gal} \times 144 \text{ in}^2/\text{ft}^2}{20 \text{ ft/sec}}$$

so that

$$A_{\text{MAX}} = 0.32 \text{ in}^2$$

or 2.1 cm^2.

Note, however, that these answers are in areas, while the problem concerned the inside diameters. We can easily find the diameters, through, using the equation for the area of a circle:

$$A = \frac{\pi D^2}{4}$$

so that

$$D = 2\sqrt{A/\pi}$$

For this problem, then,

$$D_{\text{MIN}} = 2\sqrt{A_{\text{MIN}}/\pi} = 2\sqrt{0.26/\pi} = 0.57 \text{ in}$$

or 1.45 cm, and

$$D_{\text{MAX}} = 2\sqrt{0.32/\pi} = 0.64 \text{ in}$$

or 1.63 cm.

The wall thickness of the conduit is of concern because the material of which the conduit is constructed must withstand the maximum system pressure without rupturing, splitting, or bursting. There are two methods for calculating the minimum wall thickness required. The method used depends on the ratio of the thickness of the material (t) to the inside diameter (D) of the conduit. Calculations for thin-walled conduits ($t/D \leq 0.1$) are made by the *Barlow formula for hoop stress*, while the *Lame hoop stress equation* is used for thick-walled conduits. ($t/D > 0.1$).

Barlow's formula is

$$S_u = \frac{p \times D}{2t} \qquad\qquad (12.1)$$

where

S_u = ultimate tensile strength (psi, kPa, or MPa).
p = maximum system pressure (psi, kPa, MPa).
D = inside diameter (in, cm, or mm).
t = wall thickness (in, cm, or mm).

The maximum allowable hoop stress (S_{ALL}) is found by applying a safety factor (N) between 4 and 8, depending on the application, so that

$$S_{\text{ALL}} = \frac{S_u}{N}$$

Thus, Barlow's formula becomes

$$S_{ALL} = \frac{S_u}{N} = \frac{p \times D}{2tN} \tag{12.2}$$

Solving this for wall thickness gives

$$t = \frac{p \times D \times N}{2S_u} \tag{12.3}$$

Example 12.2

Determine the minimum wall thickness for a thin-walled pipe ($D = 2$ in) for a system pressure of 2000 psi (13.8 MPa). The material is SAE 1020 carbon steel with an ultimate tensile strength of 60,000 psi (41.4 MPa). Use a safety factor of 6.

Solution

From Equation 12.3,

$$t = \frac{p \times D \times N}{2S_u} = \frac{2000 \text{ lb/in}^2 \times 2 \text{ in} \times 6}{2 \times 60,000 \text{ lb/in}^2} = 0.20 \text{ in}$$

or 0.51 cm. Since t/D = 0.2/2 = 0.1, the use of the thin-wall equation is justified.

Example 12.3

A pipe is used in a system that has a maximum pressure of 15 MPa (2175 psi). The pipe has a 7.5 cm (2.95 in) ID and an ultimate tensile strength of 400 MPa (58,000 psi). Find the minimum wall thickness that can be used. Use a safety factor of 6.

Solution

Again, we use Barlow's equation and solve for t.

$$t = \frac{p \times D \times N}{2S_u}$$

$$= \frac{15,000 \text{ kPa} \times 7.5 \text{ cm} \times 6}{2 \times 400,000 \text{ kPa}} = 0.84 \text{ cm}$$

or 0.33 in. Since t/D = 0.112 > 0.1, the thin-wall equation is not suitable.

For thick-walled pipe (t/D > 0.1), the Lame equation is used. Here, we will actually use the ASTM modified form of the Lame equation, which is written as follows.

$$t = \frac{p \times r_i}{(f \times E) - 0.6p} \tag{12.4}$$

where

t = wall thickness (in, cm, or mm).
r_i = inside radius (in, cm, or mm).
p = pressure (psi or MPa).
f = allowable stress (psi, kPa, or MPa).
E = weld joint efficiency (usually = 1).

To convert this information to terms more consistent with those used in the Barlow equation, let's make the following substitutions:

$$\frac{D}{2} = r_i$$

where D = inside diameter; and

$$f = S_{ALL}$$

where allowable stress equals

$$\frac{S_u}{N}$$

and E = 1

$$t = \frac{p \times D}{2(S_{ALL} - 0.6p)} \tag{12.5}$$

Example 12.4

Recalculate the wall thickness required for Example 12.3 using the equation for a thick-walled pipe.

Solution

In this case, we use Equation 12.5. Thus,

$$t = \frac{p \times D}{2(S_{ALL} - 0.6p)}$$

Here

$$S_{ALL} = \frac{S_u}{N} = \frac{400 \text{ MPa}}{6} = 66.7 \text{ MPa}$$

or 69 MPa.

Substituting, we get

$$t = \frac{15\text{MPa} \times 7.5 \text{ cm}}{2(66.7\text{MPa} - (0.6 \times 15\text{MPa}))}$$

$$= 0.97 \text{ cm}$$

or 0.38 in.

A very logical question to ask at this point is, "How do I know whether to use Equation 12.3 or Equation 12.5?" Before that question can be answered, we need to have a look at the listing for standard pipe sizes shown in Appendix D. An examination of these tables discloses an interesting phenomenon. Notice that the first column is headed "Nominal Pipe Size." The word "nominal" here simply means designation or name. Next, notice that there are headings such as "Schedule 40," etc. "Schedule" refers indirectly to the thickness of the pipe wall, with Schedule 40 being the thinnest and Schedule 160 the thickest. Other schedules can be found in complete tables.

Now look at the columns headed "Inside Diameter" under each schedule and note that the IDs are all different for any given nominal pipe size. Also note that this ID almost never matches the nominal pipe size; nor does the outside diameter match. Thus, a one-inch pipe is a one-inch pipe in name only, not in actual measurement, in much the same way that a 2 × 4 board is only a nominal size.

To help us determine whether we assume thin-walled (Equation 12.3) or thick-walled (Equation 12.5), apply the t/D test to a few sizes. Doing this, we find that for Schedule 40 pipes, $t/D > 0.1$ for all nominal pipe sizes 1¼ in or smaller, so those sizes must be considered thick-walled. Anything larger can be considered thin-walled. The same exercise for Schedule 80 pipe shows that 3 in nominal and smaller-pipes are considered thick-walled, while 3½ in and larger are thin-walled.

It would be very fortunate indeed if the diameter and wall thickness you calculated for a specific requirement actually existed. Usually, however, after making your calculations, you must refer to listings such as Appendix D or to manufacturers' standard listings to determine what is available. Then you must select the nominal pipe size and schedule that most nearly meets your requirements. Let's look at an example of how this can be done.

Example 12.5

Select the proper standard pipe size that would limit the fluid velocity to 25 ft/sec (7.6 m/s) in a system operating at 30 gpm (114 Lpm) and 3000 psi (20.7 MPa). The tensile strength of the pipe is 60,000 psi (414 MPa). Use a safety factor of 6.

Solution

We must determine both the pipe ID and the wall thickness to satisfy the operating parameters. We again use Equation 2.13 to find the flow area:

$$A = \frac{Q}{v} = \frac{30 \text{ gal/min} \times \text{ft}^3/7.48 \text{ gal} \times \text{min}/60 \text{ sec} \times 144 \text{ in}^2/\text{ft}^2}{25 \text{ ft/sec}}$$

$$= 0.385 \text{ in}^2$$

or 2.5 cm²,

so that the ID is

$$D = 2\sqrt{\frac{A}{\pi}} = 2\sqrt{\frac{0.385}{\pi}} = 0.70 \text{ in}$$

or 1.8 cm.

Since this is smaller than the 2 in Schedule 40 pipe, we must consider it to be thick-walled and use Equation 12.5 to calculate the wall thickness.

$$t = \frac{p \times D}{2\,(S_{\text{ALL}} - 0.6p)}$$

$$S_{\text{ALL}} = \frac{S_u}{6} = \frac{60,000 \text{ psi}}{6} = 10,000 \text{ psi}$$

or 69 MPa. Therefore,

$$\therefore t = \frac{3000 \text{ lb/in}^2 \times 0.7 \text{ in}}{2\,(10,000 - (0.6 \times 30000 \text{ lb/in}^2))} = 0.13 \text{ in}$$

or 0.33 cm.

These calculated values represent the minimum diameter and wall thickness that are acceptable. Referring to Appendix D, we find that a ¾ in Schedule 40 pipe has an acceptable ID, but the wall is too thin; however, the ¾ in Schedule 80 pipe has an acceptable ID (0.742 in) as well as an acceptable wall thickness (0.154 in). Thus, we would choose the ¾ in Schedule 80 pipe for this application.

Pipe Fittings

Piping systems are assembled using various types of *pipe fittings*—couplings, unions, elbows, tees, reducers, etc. Figure 12.1 illustrates several different types of fittings.

Two different types of threads are commonly used on pipes and pipe fittings in the United States. Water pipes and other "nonindustrial" piping will normally utilize the American or National Standard Taper Thread, better known as the NPT (for National Pipe Taper) thread. The fluid power industry standards call for the use of the Dryseal American or National Pipe Taper Thread where piping is used. A more familiar name for this thread is the NPTF (for National Pipe Taper Fuel) thread.

Figure 12.2 is an illustration of these two thread types. Notice that the mechanisms of these two devices are completely different. The NPT thread is sealed and secured mechanically by an interference fit between the flanks of

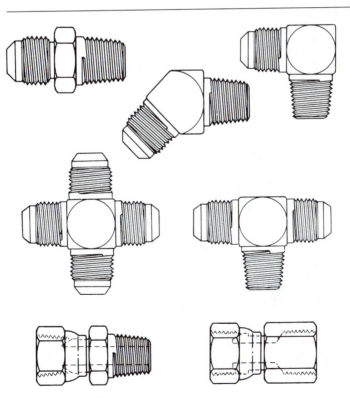

Figure 12.1 Piping systems are assembled using various types of fittings.
Source: Courtesy of Aeroquip Corporation.

USA Standard, NPT (National Pipe Taper)

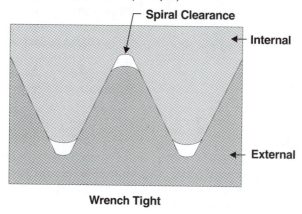

Wrench Tight

USA Standard Dryseal, NPTF (National Pipe Taper Fuel)

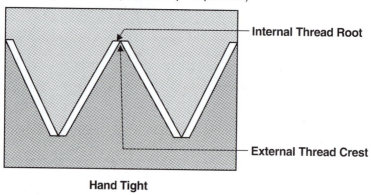

Hand Tight

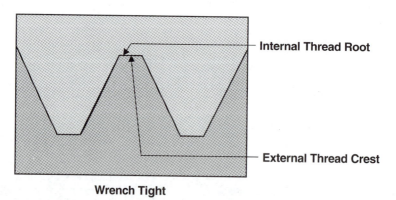

Wrench Tight

Figure 12.2 Two different pipe threads are commonly used: NPT for nonindustrial uses and NPTF for the fluid power industry.

the mating threads. Notice, though, that the crests of the threads do not contact the roots of the opposing thread. This leaves a spiral clearance that results in a leakage path through the mated threads. To achieve a complete seal, some type of thread sealant ("pipe dope") must be used. Teflon tape is often used for this purpose, although it is actually a lubricant, not a sealant. Teflon tape can represent a major problem if it is not used properly, because small strings of it can be cut off during tightening and find their way into valves, orifices, and so on. Note: *NPT threads should never be used in fluid power system piping*.

This problem is not present in the NPTF thread. When this thread is only hand-tightened, the crests of one thread are in contact with the roots of the other. Wrench-tightening results in the deformation of the crests until there is total, intimate contact between all parts of the mating threads. This eliminates the leakage path.

There is a major problem with the NPTF threads, however. The deformation that allows the seal to be made occurs on the first tightening, and it is permanent. In order to ensure a good seal after a thread has been removed and reassembled, additional deformation must occur. While this may be possible once or twice, repeated disassembly and reassembly will almost invariably result in a leaking fitting. Even when one new thread section is used, the previous deformations of the other are likely to cause leaks. For this reason, the use of pipe and pipe threads in fluid power applications should be avoided wherever possible. The leaks that have resulted from the use of pipe threads has done untold damage to the industry because users have become fed up with having to deal with leaks.

12.3   Tubing

In many applications, *tubing* provides a totally acceptable (and often preferred) alternative to the use of piping. In addition to being much less prone to leakage, tubing offers the advantages of being much lighter in weight and easily bent to preclude the need for fittings to change directions. The lighter weight is especially attractive for certain applications (aerospace, for example). The elimination of most fittings further reduces the system weight and also eliminates possible leak sources. Tubing has the disadvantage, though, of being less "abuse-resistant" than pipe.

In fluid power systems, tubing materials include aluminum, copper, steel, and stainless steel. Aluminum may be used for some low-pressure systems or in the return lines of high-pressure systems where very low pressures are experienced. Note: *Copper tubing should never be used in fluid power applications*. Both alloy and stainless steels possess the mechanical properties needed for most fluid power applications, although stainless steel is required where corrosion from the fluid or the environment may be a hazard. Plastic tubing is sometimes used for pump suction lines and drain lines.

Tubing material, manufacturing, and strength standards are covered in SAE specifications J356 and J524 through J529. High-strength alloy tubing is covered by American Iron and Steel Institute standards AISI 4130 and 8630.

Because tubing is relatively thin-walled, it lends itself readily to being bent into the desired shape. The bending operation must be done carefully, however, to prevent kinking or flattening of the tubing. Normally, a tube-bending tool

Figure 12.3 This tube-bender can be used to bend tubing to any angle up to 90 degrees.

Source: Courtesy of Imperial Eastman, 6300 W. Howard St., Chicago, Ill.

such as the one shown in Figure 12.3 is used. This device allows the tubing to be bent to any angle up to 90 degrees. Other tools can provide greater bend angles. In addition to this special tool, tube inserts such as springs or rubber rods are often used to help preserve the circular cross section of the tubing. Sand is sometimes used, but it is not recommended for fluid power applications. The bend radius should be no less than six times the internal diameter of the tubing.

Tubing Sizing

In specifying the size of tubing to be used for a particular application, we must again be concerned with the internal diameter (velocity considerations) and wall thickness (strength). Unfortunately, the standards and methods for describing tubing sizes are different from those for pipes.

Appendix E lists some standard tubing sizes. The first column, which is labeled "Tube O.D.," is the actual outside diameter of the tubing. The second column lists the wall thickness, while the third column shows the internal diameter of the tube. It is necessary to specify the tubing OD and wall thickness after you have calculated the required ID and wall thickness.

Example 12.6

Determine the steel tubing size required for a pressure line operating at 2000 psi (13.8 MPa) and 10 gpm (37.85 Lpm). Use a safety factor of 6, and assume a tensile strength of 55,000 psi (37.9 MPa).

Solution

We will need to calculate both the internal diameter and the wall thickness. Looking first at the ID, we know from Table 12.1 that the recommended flow velocities range is between 15 and 20 fps (4.5 and 6 m/s). Using this information in Equation 2.13, we find that the minimum acceptable diameter (based on the maximum acceptable flow velocity) is

$$A_{MIN} = \frac{Q}{v_{MAX}}$$

$$= \frac{10 \text{ gal/min} \times \text{min/60 sec} \times \text{ft}^3/7.48 \text{ gal} \times 144 \text{ in}^2/\text{ft}^2}{20 \text{ ft/sec}}$$

$$= 0.16 \text{ in}^2$$

or 1.03 cm².

Remember that this is the area, and we want diameter. Thus,

$$D_{MIN} = 2\sqrt{\frac{A_{MIN}}{\pi}} = 2\sqrt{\frac{0.16 \text{ in}^2}{3.14}} = 0.45 \text{ in}$$

or 1.15 cm.

When calculating the wall thickness required for tubing applications, we will take the conservative approach and assume thick-walled tubing. Thus, we need to use Equation 12.5.

$$t = \frac{p \times D}{2(S_{ALL} - 0.6p)} = \frac{(2000 \text{ lb/in}^2 \times 0.45 \text{ in})}{2(9167 \text{ lb/in}^2 - (0.6 \times 2000 \text{ lb/in}^2))}$$

$$= 0.056 \text{ in}$$

or 0.143 cm = 1.43 mm.

Next, we go to Appendix E to find a standard tubing size that would work in our system. Notice that the minimum wall thickness requirement can be satisfied in all sizes larger than a ¼ in tube. Therefore, we need only to find the smallest size that will give us the ID required. We find that we can use ⅝ inch tubing with a 0.065 in (1.65 mm) wall thickness to satisfy our requirements.

To check to see that the minimum velocity of 15 fps (4.5 m/s) is satisfied by this ID, we return to Equation 3.12, where we find that

$$v_{MIN} = \frac{Q}{A_{MAX}}$$

$$A = \frac{\pi D^2}{4} = \frac{\pi \times 0.495 \text{ in}^2}{4} = 0.192 \text{ in}^2$$

Therefore,

$$v_{MIN} = \frac{10 \text{ gal/min} \times \text{min/60 sec} \times \text{ft}^3/7.48 \text{ gal} \times 144 \text{ in}^2/\text{ft}^2}{0.192 \text{ in}^2}$$

$$= 16.7 \text{ fps}$$

or 5.1 m/s, which is acceptable.

As you might expect, larger diameters of tubing have lower allowable working pressures unless the wall thickness is increased. As the diameter increases,

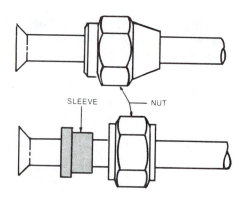

Figure 12.4 The top flared fitting is a two-piece type, and the bottom one is a three-piece type.
Source: Courtesy of Hydraulics and Pneumatics Magazine.

the internal cylindrical area increases. This means that the force exerted on the walls is increased.

Tube Fittings

Tubing systems are assembled with *fittings* that are attached to the tubing ends, but not by threading, the way pipe fittings are attached to pipe. These fittings are classified as either flared or flareless, depending on the method used to attach them to the tube. A *flared fitting*, shown in Figure 12.4, consists of a sleeve and a packing nut that are slipped over the end of the tubing. The end of the tubing is then "flared" with a *flaring tool* as shown in Figure 12.5.

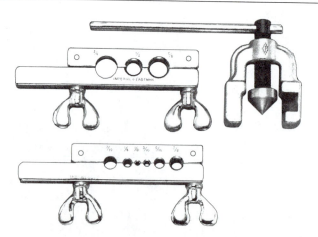

Figure 12.5 Flaring bars and a flaring tool are used to create the "flare."
Source: Courtesy of Imperial Eastman, 6300 W. Howard St., Chicago, Ill.

This flare seats on the cone-shaped surfaces on the mating connector (Figure 12.6). The two conical surfaces form a fluid seal when the packing nut is securely tightened.

Two different flare angles are used in the United States. The SAE 37 degree flare, formerly known as the Joint Industrial Conference (JIC) flare, is used in fluid power applications. The SAE 45 degree flare is commonly used in automotive and air conditioning applications, but not in fluid power systems. The 37 degree flare fittings are not compatible with 45 degree flare fittings, and the two configurations are not interchangeable.

An advantage of flared fittings is that they are relatively easy to connect. Because there is no biting or compression of the tube, most of the tightening can be done by hand. Only the final tightening requires a wrench, and then only to turn the nut a fraction of a turn. No more than about 200 lb · in of torque is required for assembly.

Flared fittings are very vibration resistant. They tend to remain tight and relatively leak-free even when subjected to severe vibration. They also have the reputation for being "blowout proof." Although the fittings may leak under some conditions, very rarely will the tubing blow out of the fitting. Thus, catastrophic failure of flared tubing is virtually unknown.

Two major considerations in the use of flaring are the hardness of the tube material and the wall thickness. Hard tubing may be very difficult to flare. It may either split during the flaring operation, or be cold-worked enough to fail after a short time in use. The wall thickness is of concern because a thicker material causes both the sealing surface and the thread engagement to be reduced (Figure 12.7). Thus, both ductility and wall thickness are important considerations when determining whether flaring is acceptable for a given tube type. For steel tubing, for example, the acceptable wall thickness range for successful flaring is from 0.035 to 0.109 inch.

Flared fittings are commonly used with system pressures up to 2000 psi (13.8 MPa). Above that point, *flareless fittings* are usually employed. These may be of several types, ranging from a simple ferrule through fittings applied

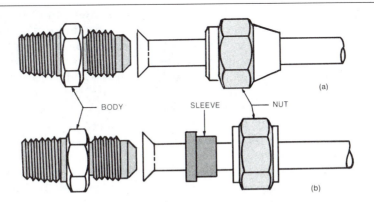

Figure 12.6 The flare seats on the mating adaptor, or connector. The top fitting (a) is a two-piece type, and the bottom one (b) is a three-piece type.
Source: Courtesy of Hydraulics and Pneumatics Magazine.

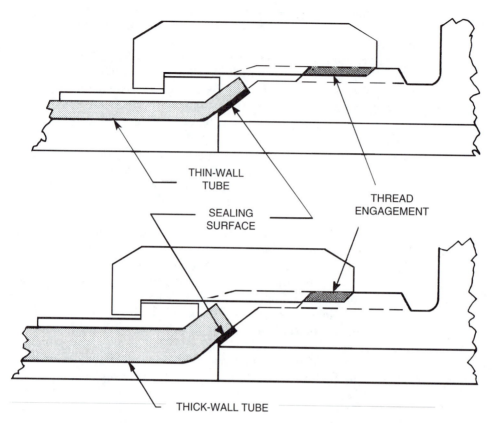

Figure 12.7 With a flared fitting, heavier wall tubing tends to reduce sealing surface and thread engagement.
Source: Courtesy of Hydraulics and Pneumatics Magazine.

using cryogenic techniques. The *ferrule* fitting, probably the simplest, consists of a ferrule with carefully chamfered edges and a packing nut. The straight end of the tubing is inserted into the connector, which is internally chamfered to receive the ferrule. When the packing nut is tightened, the mating surfaces form the seal while the ferrule compresses and grips the tubing.

A group of somewhat more sophisticated fittings are termed *bite-type* because of the mechanism by which they grip the tubing. Some of these are shown in Figure 12.8 This is a compression-type mechanism which causes the sleeve to bite into the tube surface as the nut is tightened. These fittings vary in complexity, depending on design and application. Tube wall strength is of major concern when considering using bite-type flareless fittings. If the wall is too thin or the material is too soft, the tube may distort or collapse completely as the sleeve begins to bite into the outer surface. For steel tubing, the thickness range for using bite-type fittings is 0.065 to 0.120 inch.

Another group of flareless fittings are the *swaged* fittings, which require a swaging machine to deform the materials of both the fitting and the tubing to

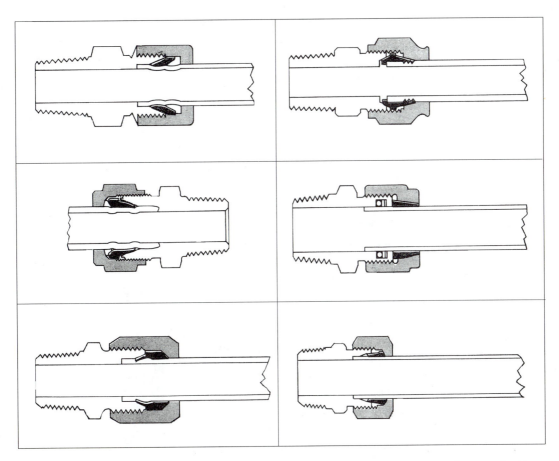

Figure 12.8 The compression types of flareless fittings vary in complexity. The sleeve bites into the tube surface as the nut is tightened.
Source: Courtesy of Hydraulics and Pneumatics Magazine.

form a very strong mechanical grip. The swaging may be either internal or external—that is, the fitting may be deformed onto the tube from the outside, or the tube may be deformed into the fitting from the inside. (Figure 12.9) Other types of fittings may be brazed or welded onto the tubing.

Probably the most technologically sophisticated fitting is the cryogenic fitting (Figure 12.10). These *cyrogenic sleeve* fittings are made of a nickel-titanium alloy material with some peculiar temperature-related characteristics. The fitting is originally machined somewhat undersize. When the fitting is to be used, it is immersed in liquid nitrogen. When its temperature stabilizes, a mandrel is drawn through it to give it a larger internal diameter. It is then slipped over the tubing (still chilled), positioned properly, and allowed to warm up. Its peculiar characteristic is that when it warms up, it returns to exactly the same dimensions as before it was chilled. The deformation caused by the mandrel is only temporary. The result is that it shrinks back to its original,

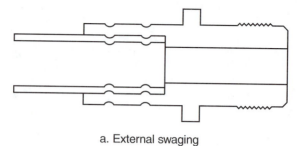

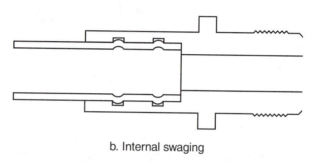

a. External swaging

b. Internal swaging

Figure 12.9 Fittings can be permanently attached to tubing by either internal or external swaging.

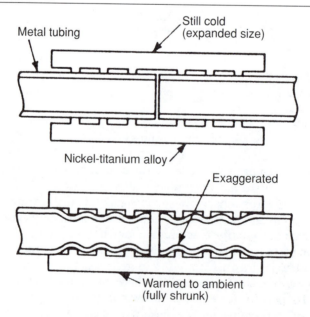

Metal tubing

Still cold
(expanded size)

Nickel-titanium alloy

Exaggerated

Warmed to ambient
(fully shrunk)

Figure 12.10 A cryogenic sleeve fitting stays expanded until warmed.

undersized internal diameter and creates an extremely strong joint. This joint is so reliable and leak-free that it is commonly used in assembling hydraulic systems for submarines. It is also used on the 8000 psi (55 MPa) experimental aircraft hydraulic systems.

The disadvantages of this fitting design are that it is relatively expensive and requires special handling equipment because of the cryogenic temperatures.

Regardless of the type of fitting used, the fitting must accomplish two functions—it must mechanically contain the forces due to the internal pressures, and it must seal to prevent leakage.

Mechanical forces are exerted on the internal surfaces of the tubing by the fluid pressure acting on the tubing walls. The greater the tubing diameter, the more surface area there is exposed to the pressure; therefore, the forces are greater as the inside diameter increases. The fitting must provide sufficient gripping force to prevent the tube from blowing out of the fitting.

The writers of *Hydraulics and Pneumatics* magazine list four categories of holding and sealing problems:

- difficult-to-hold, easy-to-seal
- difficult-to-seal, easy-to-hold
- difficult-to-hold and seal
- easy-to-hold and seal

An example of a difficult-to-hold, easy-to-seal application is a high pressure system using a fluid that has a large molecular structure such as a petroleum based hydraulic oil. This application would be suitable for a flared fitting for pressures up to 2000 psi. For higher pressures or for thick-walled tubing, a bite-type flareless fitting could be used. For very high pressures, the cryogenic-type fitting might be required.

The water-containing fluids operating at low pressures fall into the difficult-to-seal, easy-to-hold category. The low viscosity of the water makes it prone to leakage, even at low pressure. Bite-type fittings are suitable for these applications.

High pressure systems (3000 psi and higher) with small molecular structures constitute the difficult-to-seal, difficult-to-hold application. Cryogenic fittings are best suited for this application, although some of the better bite-type fittings may also be satisfactory.

Unfortunately, the use of tube fittings cannot completely guarantee against leaks, because even if the tube fitting itself does not leak, the tubing must, at some point, be connected to the system components. The connectors (fittings) used for this task typically are adaptors that have one tube-compatible end and one pipe-threaded end, usually NPTF for fluid power applications. This leads us to the same old problem.

In recent years, a new attachment technique, termed the SAE *O-ring boss (ORB)* fitting has become available (Figure 12.11), which replaces the NPTF thread. It consists of a straight threaded male stem with an O-ring and threaded female part with a machined bevel at the thread entrance. When the fitting is tightened, the O-ring compresses and forms the fluid seal between the boss on the male portion and the bevel on the female portion. The threads form the mechanical connection, but play no part in the sealing. This type of connector is not interchangeable with NPT or NPTF threads and must be specified when ordering components.

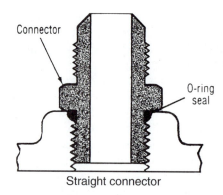

Figure 12.11 SAE straight thread O-ring port fittings are favored as replacements for pipe thread fittings by many manufacturers and users of hydraulic components.
Source: Courtesy of Hydraulics and Pneumatics Magazine.

Another method to attach tubing to components is the use of a *flanged conduit* and a four-bolt *split flange* (Figure 12.12). The conduit fitting consists of a circular, flat-faced flange with a groove that contains an O-ring. The component port, rather than being threaded, is simply a smooth-bore opening surrounded by a flat machined surface against which the O-ring seals. Threaded holes are spaced to accept the mounting bolts. The O-ring and conduit flange are held in contact with the port by attaching the two halves of the split flange and tightening the mounting bolts.

The SAE split flange system is normally used on larger-diameter connections because it is easy to install and provides easy orientation of the fitting prior to final torquing of the mounting bolts. It has very high pressure capabilities, and seldom leaks when properly assembled.

12.4   Hoses

Where there is a great deal of vibration or relative movement between parts of the system, *hoses* are commonly used. Hoses allow a great deal of flexibility in the routing of the fluid conduit, but provide no rigidity as do pipes or tubes.

Hose Specifications

Hydraulic hoses used in fluid power systems must conform to SAE Standard J517. This standard specifies both the hose construction and radial dimensions. Fluid power hoses are termed *100R-series hoses*. Figure 12.13 shows the family

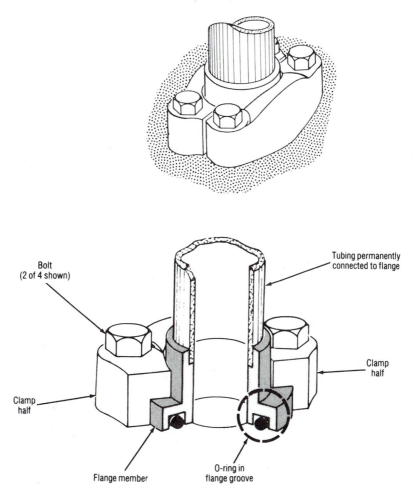

Figure 12.12 SAE split-flange O-ring port fittings are widely used on heavy equipment such as tractors and bulldozers.
Source: Courtesy of Hydraulics and Pneumatics Magazine.

of 100R hoses. Note that all 100R hoses have three basic elements—and inner tube, a reinforcement, and protective cover. The design and materials used vary, depending on the intended use. Hoses are designated by different numbers (100R1, 100R2, etc.) depending on materials and construction.

The materials normally used for the inner tube are synthetic rubber or thermoplastic. The reinforcement is provided by one or more layers of wire or textile braid. The outer cover may be either synthetic rubber or thermoplastic. Various materials may be used between the layers to help anchor the materials together and prevent chaffing. In addition to the outer protective cover, extra protection may be provided by an external, braided wire cover.

Hose sizes are normally designated by dash numbers that indicate the nominal inside diameter in sixteenths of an inch. For instance, a half-inch hose would be termed a —8 (dash eight), while a —5 hose would have an internal

SAE 100R1

Type A — This hose shall consist of an inner tube of oil resistant synthetic rubber, a single wire braid reinforcement, and an oil and weather resistant synthetic rubber cover. A ply or braid of suitable material may be used over the inner tube and/or over the wire reinforcement to anchor the synthetic rubber to the wire.

Type AT — This hose shall be of the same construction as Type A, except having a cover designed to assemble with fittings which do not require removal of the cover or a portion thereof.

SAE 100R2

The hose shall consist of an inner tube of oil resistant synthetic rubber, steel wire reinforcement according to hose type as detailed below, and an oil and weather resistant synthetic rubber cover. A ply or braid of suitable material may be used over the inner tube and/or over the wire reinforcement to anchor the synthetic rubber to the wire.

Type A — This hose shall have two braids of wire reinforcement.

Type B — This hose shall have two spiral plies and one braid of wire reinforcement.

Type AT — This hose shall be of the same construction as Type A, except having a cover designed to assemble with fittings which do not require removal of the cover or a portion thereof.

Type BT — This hose shall be of the same construction as Type B except having a cover designed to assemble with fittings which do not require removal of the cover or a portion thereof.

SAE 100R3

The hose shall consist of an inner tube of oil resistant synthetic rubber, two braids of suitable textile yarn, and an oil and weather resistant synthetic rubber cover.

SAE 100R4

The hose shall consist of an inner tube of oil resistant synthetic rubber, a reinforcement consisting of a ply or plies of woven or braided textile fibers with a suitable spiral of body wire, and an oil and weather resistant synthetic rubber cover.

SAE 100R5

The hose shall consist of an inner tube of oil resistant synthetic rubber and two textile braids separated by a high tensile steel wire braid. All braids are to be impregnated with an oil and mildew resistant synthetic rubber compound.

SAE 100R6

The hose shall consist of an inner tube of oil resistant synthetic rubber, one braided ply of suitable textile yarn, and an oil and weather resistant synthetic rubber cover.

SAE 100R7

The hose shall consist of a thermoplastic inner tube resistant to hydraulic fluids with suitable synthetic fiber reinforcement and a hydraulic fluid and weather resistant thermoplastic cover.

Figure 12.13 Fluid power hoses are termed SAE 100R-Series hoses.

Source: Reprinted with permission from SAEJ746, © 1992, Society of Automotive Engineers, Inc.

SAE 100R8

The hose shall consist of a thermoplastic inner tube resistant to hydraulic fluids with suitable synthetic fiber reinforcement and a hydraulic fluid and weather resistant thermoplastic cover.

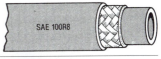

SAE 100R9

Type A — This hose shall consist of an inner tube of oil resistant synthetic rubber, 4-spiral plies of wire wrapped in alternating directions, and an oil and weather resistant synthetic rubber cover. A ply or braid of suitable material may be used over the inner tube and/or over the wire reinforcement to anchor the synthetic rubber to the wire.

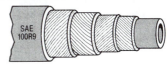

Type AT — This hose shall be of the same construction as Type A, except having a cover designed to assemble with fittings which do not require removal of the cover or a portion thereof.

SAE 100R10

Type A — This hose shall consist of an inner tube of oil resistant synthetic rubber, 4-spiral plies of heavy wire wrapped in alternating directions, and an oil and weather resistant synthetic rubber cover. A ply or braid of suitable material may be used over the inner tube and/or over the wire reinforcement to anchor the synthetic rubber to the wire.

Type AT — This hose shall be of the same construction as Type A, except having a cover designed to assemble with fittings which do not require removal of the cover or a portion thereof.

SAE 100R11

This hose shall consist of an inner tube of oil resistant synthetic rubber, 6-spiral plies of heavy wire wrapped in alternating directions and an oil and weather resistant synthetic rubber cover. A ply or braid of suitable material may be used over the inner tube and/or over the wire reinforcement to anchor the synthetic rubber to the wire.

SAE 100R12

This hose shall consist of an inner tube of oil resistant synthetic rubber, 4-spiral plies of heavy wire wrapped in alternating directions, and an oil and weather resistant synthetic rubber cover. A ply or braid of suitable material may be used over or within the inner tube and/or over the wire reinforcement to anchor the synthetic rubber to the wire. [HP]

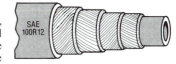

Figure 12.13 Continued.

diameter of ⁵⁄₁₆ in. An exception to this is the 100R5 series, for which the dash number is the nominal outside diameter.

Because hoses are constructed of elastomeric materials, they tend to respond to internal pressure by changing their dimensions. Depending on the hose design and material, a hose may shorten as much as 4 percent or even lengthen as much as 2 percent. These dimensional changes could result in stresses that can lead to failure of the hose-fitting assembly if the hose is not properly installed. (We will discuss installation later in this chapter.) In some designs, dimensional changes can be avoided by a properly constructed wire braid reinforcement laid at an angle of 54 degrees, 40 minutes to the axis of the hose.

Table 12.2 is an excerpt from a hose manufacturer's brochure giving application information about its SAE 100R-series hose. Notice that hoses are pressure rated according to size, and that the larger the hose, the lower its rated working

Table 12.2 Specifications for SAE 100 R-Series Hoses.

SUMMARY OF SAE J517 100R-SERIES HOSE MAXIMUM OPERATING PRESSURE $\frac{PSI}{MPa}$

Nominal Hose I.D. Size, In	100R1	100R2	100R3	100R4	100R5	100R6	100R7	100R8	100R9	100R10	100R11	100R12	100R13	100R14
1/8														1500/10.3
3/16	3000/20.7	5000/34.5	1500/10.3		3000/20.7	500/3.4	3000/20.7	5000/34.5		10 000/68.9	12 500/86.2			1500/10.3
1/4	2750/19.0	5000/34.5	1250/8.6		3000/20.7	400/2.8	2750/19.0	5000/34.5		8750/60.3	11 250/77.6			1500/10.3
5/16	2500/17.2	4250/29.3	1200/8.3		2250/15.5	400/2.8	2500/17.2							1500/10.3
3/8	2250/15.5	4000/27.6	1125/7.8			400/2.8	2250/15.5	4000/27.6	4500/31.0	7500/51.7	10 000/68.9	4000/27.6		1500/10.3
13/32	2250/15.5				2000/13.8									1000/6.9
1/2	2000/13.8	3500/24.1	1000/6.9		1750/12.1	400/2.8	2000/13.8	3500/24.1	4000/27.6	6250/43.1	7500/51.7	4000/27.6		800/5.5
5/8	1500/10.3	2750/19.0	875/6.0		1500/10.3	350/2.4	1500/10.3	2750/19.0						800/5.5
3/4	1250/8.6	2250/15.5	750/5.2	300/2.1			1250/8.6	2250/15.5	3000/20.7	5000/34.5	6250/43.1	4000/27.6	5000/34.5	800/5.5
7/8	1125/7.8	2000/13.8			800/5.5									800/5.5
1	1000/6.9	2000/13.8	565/3.9	250/1.7			1000/6.9	2000/13.8	3000/20.7	4000/27.6	5000/34.5	4000/27.6	5000/34.5	800/5.5
1-1/8					625/4.3									600/4.1
1-1/4	625/4.3	1625/11.2	375/2.6	200/1.4					2500/17.2	3000/20.7	3500/24.1	3000/20.7	5000/34.5	
1-3/8					500/3.4									
1-1/2	500/3.4	1250/8.6		150/1.0					2000/13.8	2500/17.2	3000/20.7	2500/17.2	5000/34.5	
1-13/16					350/2.4									
2	375/2.6	1125/7.8		100/0.7					2000/13.8	2500/17.2	3000/20.7	2500/17.2	5000/34.5	
2-3/8					350/2.4									
2-1/2		1000/6.9		62/0.4							2500/17.2			
3				56/0.4	200/1.4									
4				35/0.2										

Minimum burst of 100R hoses is at least 4 times operating pressure

pressure. This reflects the fact that as the hose diameter increases, the internal effective area increases (as with tubing). Thus, any given pressure causes a greater force as the hose area increases.

Hose Couplings

Hose *couplings* can be divided into two major groupings—permanently attached or reusable. *Permanently attached* couplings—as the name implies—become an integral part of the hose assembly when properly installed. They normally consist of an end connection that has a stem inserted into the hose and an outer sleeve or shell that slides over the outside of the hose. The shell is then permanently deformed by a controlled crushing or crimping action (termed *swaging*) so that it grips the hose and stem (Figure 12.14). The connecting fitting can be of virtually any design—pipe thread, 37 or 45 degree flare; male or female; straight, 45 or 90 degree elbow, and so on.

Hose assemblies with permanently attached couplings can be purchased from manufacturers in standard configurations, or they may be assembled in the "field" by the user. Field assembly offers the advantage of allowing a user to keep few, if any, hose assemblies on hand. All the user needs is an assortment of required hoses and couplings. When a new assembly is needed, he or she cuts the required length of hose and installs the required couplings, using a crimping machine designed for that purpose. This can save considerable time, storage space, and money in stocking and inventory control.

Note that hoses should be assembled only by trained personnel using the correct equipment and techniques. Field-assembled hoses should be pressure-tested before they are installed on the equipment.

When an assembly containing a permanent coupling is discarded, the coupling goes with it. In order to reduce the expense of purchasing replacement couplings, and also to make assembly of hoses and couplings a more easily accomplished field exercise, *reusable* couplings are often used (Figure 12.15).

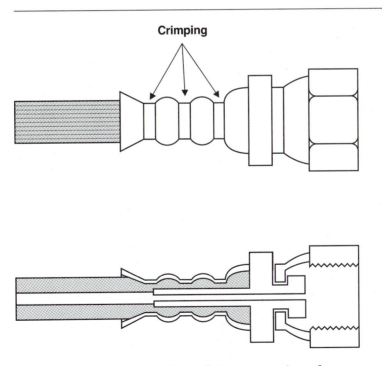

Figure 12.14 Permanent hose fittings are crimped, or swaged, to grip the hose and stem.

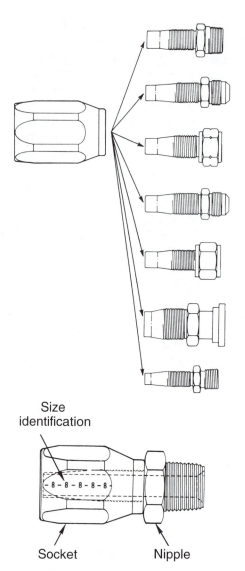

Figure 12.15 Reusable hose fittings are either screwed together as these are or a clamping mechanism is used to secure the coupling to the hose.
Source: Courtesy of Aeroquip Corporation.

These are either screwed together or a clamping mechanism is used to secure the coupling to the hose. The screw-type coupling, like the permanent couplings, has both an insert (nipple or stem) and an outer shell (socket or sleeve). The shell usually has internal threads that allow it to be screwed onto the outer surface of the hose, as well as threads to receive the mating threads from the stem. The stem, which has a long taper, is inserted into the hose and screwed into the threads of the shell. The hose is compressed between the stem and the shell to provide both gripping and sealing. There are many configurations of screw-type assemblies, depending on the application and manufacturer. Some of these couplings are designed to fit over the outer protective covering of thin cover hoses. Other types require that the cover be removed (a procedure called *skiving*), especially in thick-walled hoses.

Clamp-style reusable couplings use an inserted stem and a two-piece outer shell that is clamped around the hose and secured by two or more nuts and bolts. This type is more expensive than the screw-type, but it is easier to install and may be the only type that can be used on some large hoses.

Assembly of any reusable coupling should be done with caution and strictly in accordance with the manufacturer's instructions. Pressure-testing of the assembly is certainly a reasonable precaution when possible.

Hose coupling and fitting requirements are specified in SAE standard J343.

A popular type of hose coupling is the quick connect (or quick disconnect) coupling. These couplings are used to allow rapid assembly or disassembly of hose circuits. *Quick connects* consist of two components—a socket and a plug (Figure 12.16). When the plug is inserted into the socket, a locking mechanism activates to lock the two pieces together. This is commonly a spring-loaded collar that holds a set of recessed balls in the socket into a groove in the plug. An internal O-ring seal is usually used to prevent leakage.

Quick connects may contain a check valve in both the socket and the plug so that both units are sealed when disconnected. When connected, the check valves are pushed open to allow free flow through the coupling. This configuration is termed a *two-way shutoff* coupling (Figure 12.16).

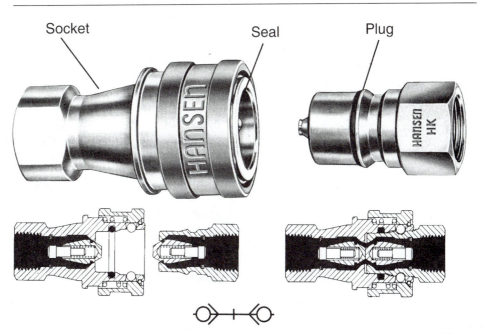

Figure 12.16 Two-way shutoff quick disconnect (or quick connect) coupling involves a socket and a plug.

Source: Courtesy of Tuthill Corp., Hansen Coupling Div.

Another configuration is a *one-way shutoff* in which one end (usually the socket) is sealed when disconnected, but the other end is open. The third configuration has no check valves, so both ends are open when disconnected. This is termed a *straight-through* coupling. The symbols for all three configurations are shown in Figure 12.17. The two-way shutoff is the most commonly used coupling for hydraulic applications.

In liquid flow applications, the flow capacity of quick couplings is expressed as a C_v rating. As with flow control valves, this rating refers to the flow rate (gpm) of water at 60°F that will flow through the coupling with a pressure drop of 1 psi.

Hose Testing

The Society of Automotive Engineers (SAE) has issued three documents concerning the testing of hoses and hose assemblies. We have already mentioned the first of these—SAE J517—when we discussed the 100R-series specifications. The second is SAE J343—Test Procedures for SAE 100R-Series Hydraulic Hose and Hose Assemblies. It contains more than 20 tests for hoses and assemblies. Included among these are checks for internal and external dimensions, dimensional changes due to internal pressure, proof and burst pressure, physical properties of the hose materials, and impulse testing.

The third document is not actually a standard, but rather an SAE Information Report that details the procedures for conducting a recommended, but not

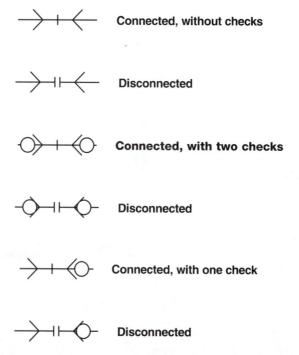

Connected, without checks

Disconnected

Connected, with two checks

Disconnected

Connected, with one check

Disconnected

Figure 12.17 Symbols for quick disconnect (quick connect) couplings

required, flex-impulse hose test. It is SAE J1405—Flex-Impulse Test Procedures for Hydraulic Hose Assemblies.

There are several things that you should remember when specifying or purchasing hose assemblies.

1. There is no such thing as an "SAE approved hose or hose assembly." SAE issues standards by which hoses and other devices may be manufactured and/ or tested. SAE does not test hardware or issue approvals.

2. The requirements of SAE J343 cover only the initial qualification of a design or manufacturing technique. It is not a continuing test requirement.

3. The only continuing tests the hose manufacturer is required to conduct are contained in SAE J517. They include the following:

- Dimensions
- Proof pressure
- Change in length due to pressure
- Burst pressure
- Visual examination

Hose Specification

We have already seen that the SAE standards cover the physical construction, materials, and test requirements of hoses, fittings, and hose assemblies. Unfortunately, conformance to SAE standards is voluntary. Customers therefore are well advised to specify only hoses and hose assemblies that meet the requirements of the applicable SAE standards.

Hose assemblies must be compatible with the hydraulic fluid being used and the maximum operating temperatures of the fluid, as well as being sized to satisfy pressure and flow requirements.

Although hoses and fittings from different manufacturers may meet all SAE specification requirements, they may not be mutually interchangeable. A very common failure that occurs when using hoses and fittings from different manufacturers for field assemblies (whether permanent or reusable hardware is used) is hose blow-off, where the hose separates from the hardware. This is usually the result of pressure impulses. To avoid this problem, hoses and fittings should be matched for performance. The safest approach is to use hoses and fittings from the same manufacturer and assemble them in strict adherence to that manufacturer's procedures.

Hose Installation

While hoses provide a great deal of flexibility in routing, they must be installed carefully in order to avoid subjecting them to stresses or abrasion that could cause them to ,fail in operation. Figure 12.18 illustrates some typical hose installations, with examples of proper and improper techniques.

Abrasion is a very subtle danger in hose installations, because damage to the outer cover may appear to be superficial or merely cosmetic when, in fact, major damage to the hose structure may have resulted. Of course, the protective covering does not normally have any significant effect on the strength of the hose. If, however, abrasion allows moisture or chemicals to reach the reinforcing wire or fiber braid, the braid could be degraded, and this could lead to premature

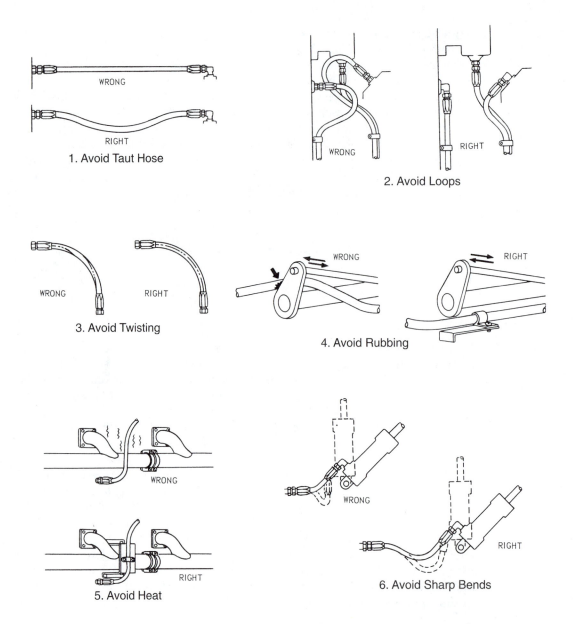

Figure 12.18 Hoses must be installed carefully in order to avoid subjecting them to stresses or abrasion.

Source: Reprinted by permission of Deere and Company, © 1987, Deere and Company. All rights reserved.

hose failure. Therefore, hoses should be routed and clamped to eliminate the abrasion hazard. Particular points of caution are sharp edges on which the hose could rub and moving parts of the machine that could crimp, cut, or abrade the hose.

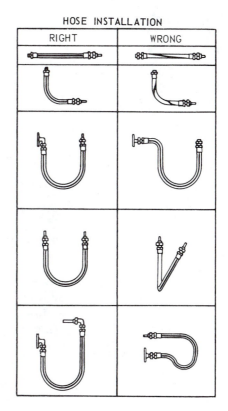

Figure 12.18 Continued.

When routing and attaching hoses, be sure that the hoses are not twisted. Twisting causes internal stresses that could lead to failure of the hose at the fitting or separation of the plies (or layers) of the hose, leading to subsequent leaks or even catastrophic blowouts. To help you avoid this problem, all hoses are marked with what is termed a *lay line*. It may be nothing more than a stripe painted along the hose, or it may be a "label" listing the manufacturer, hose type, pressure rating, and so forth. Whatever the case, always be sure that the lay line is straight after installation. If it is spiraling around the hose, disaster may be waiting somewhere along that spiral path.

12.5 Pressure Losses in Conduits

Earlier in this chapter, the problem of pressure losses in conduits was discussed briefly. In this section, we'll look at this problem in detail and present methods for determining the magnitude of these losses.

Bernoulli's Equation

Bernoulli's equation is a highly simplified approach to analyzing the fluid flow through a series pipeline. It is based on the assumption that there are no energy losses as the fluid flows between point 1 and point 2; therefore, the total energy possessed by the fluid is the same at each point, even though the form of the energy may have changed. It assumes that the fluid possesses energy in three forms—pressure energy, potential energy (due to its elevation), and kinetic energy (due to its velocity). Since the equation assumes no losses, the sum of these three energies is constant. Thus, we can write

Pressure Energy + Potential Energy + Kinetic Energy = Constant

Mathematically, we express these pressure energy as

$$\text{Pressure Energy} = \frac{w \times p}{\gamma}$$

where

$$w = \text{fluid weight.}$$
$$p = \text{pressure.}$$
$$\gamma = \text{specific weight.}$$

Potential energy is expressed as

$$\text{Potential Energy} = w \times z$$

where

$$w = \text{fluid weight.}$$
$$z = \text{elevation above a datum.}$$

Kinetic energy is expressed as

$$\text{Kinetic Energy} = \frac{w \times v^2}{2g}$$

where

$$w = \text{fluid weight.}$$
$$v = \text{fluid velocity.}$$
$$g = \text{acceleration of gravity.}$$

To express the energy possessed by w pounds of the fluid, we can write

$$\frac{w \times p}{\gamma} + (w \times z) + \frac{w \times v^2}{2g} = \text{Constant}$$

or, in considering two points (1 and 2),

$$\frac{w \times P_1}{\gamma} + (w \times z_1) + \frac{w \times v_1^2}{2g} = \frac{w \times p_2}{\gamma} + (w \times z_2) + \frac{w \times v_2^2}{2g}$$

Since we are considering a constant weight of fluid, the w in the equation cancels to give us

$$\frac{p_1}{\gamma} + z_1 + \frac{v_1^2}{2g} = \frac{p_2}{\gamma} + z_2 + \frac{v_2^2}{2g} \tag{12.6}$$

This is the traditional form of the Bernoulli equation. Note that each term has the units of length (feet or meters). Let's look at a couple of examples to illustrate its use.

Example 12.7

Determine the pressure at point 2 in the series pipeline shown here. Assume no losses between the two points. The fluid is water (specific weight = 62.4 lb/ft³, or 9.8 kN/m³).

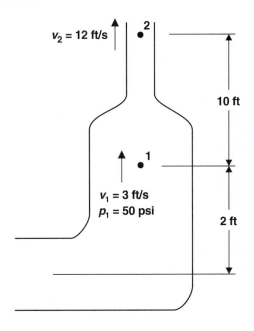

Example 12.7

Solution

First, we write the full form of Bernoulli's equation *in the direction of flow:*

$$\frac{p_1}{\gamma} + z_1 + \frac{v_1^2}{2g} = \frac{p_2}{\gamma} + z_2 + \frac{v_2^2}{2g}$$

Next, we solve the equation for the pressure at point 2:

$$p_2 = p_1 + \gamma \left[z_1 - z_2 + \frac{v_1^2 - v_2^2}{2g} \right]$$

We know everything we need to know for this equation, so we can substitute directly into it:

$$p_2 = 50\ \text{lb/in}^2 + 62.4\ \text{lb/ft}^3 \left(2\ \text{ft} - 12\ \text{ft} \right.$$

$$\left. + \frac{(3\ \text{ft/sec})^2 - (12\ \text{ft/sec})^2}{(2)\,(32.2\ \text{ft/sec}^2)} \right) \frac{\text{ft}^2}{144\ \text{in}^2}$$

$$= 50\ \text{lb/in}^2 - 5.24\ \text{lb/in}^2 = 44.76\ \text{psi}$$

or 309 kPa.

This 5.24 psi (36 kPa) loss in pressure between the two points is due to the increases in elevation and fluid velocity.

Example 12.8

The pipeline shown here is carrying hydraulic fluid with a specific gravity of 0.87. The pipe is horizontal. Assuming no losses, determine the pressure at Point 2.

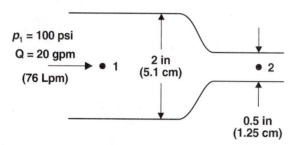

Example 12.8

Solution

Again, we write Bernoulli's equation in the direction of flow. Thus,

$$\frac{p_1}{\gamma} + z_1 + \frac{v_1^2}{2g} = \frac{p_2}{\gamma} + z_2 + \frac{v_2^2}{2g}$$

Because the pipe is horizontal, $z_1 = z_2$, so these terms cancel one another, leaving

$$\frac{p_1}{\gamma} + \frac{v_1^2}{2g} = \frac{p_2}{\gamma} + \frac{v_2^2}{2g}$$

Now we can solve the equation for p_2.

$$p_2 = p_1 + \gamma \left(\frac{v_1^2}{2g} - \frac{v_2^2}{2g} \right)$$

We don't know v_1 or v_2, but we can find them using $Q = Av$. Thus,

$$v_1 = \frac{Q}{A_1} = \frac{(20 \text{ gal/min})(231 \text{ in}^3\text{/gal})(60 \text{ sec/min})}{3.14 \text{ in}^2}$$
$$= 2.04 \text{ ft/sec } (0.62 \text{ m/s})$$

and $v_2 = \dfrac{Q}{A_2} = \dfrac{(20 \text{ gal/min})(231 \text{ in}^3\text{/gal})(60 \text{ sec/min})}{0.196 \text{ in}^2}$
$$= 32.71 \text{ ft/sec } (9.96 \text{ m/s})$$

also,

$$\gamma = \text{Sg} \times \gamma_{\text{water}} = 0.87 \times 62.4 \text{ lb/ft}^3 = 54.29 \text{ lb/ft}^3$$

or 8.53 kN/m³.

Putting these into our equation for p_2 gives

$$p_2 = 100 \text{ lb/in}^2 + \frac{54.29 \text{ lb/ft}^3}{144} \left(\frac{(2.04 \text{ ft/sec})^2 - (32.65 \text{ ft/sec})^2}{(2)(32.2 \text{ ft/sec}^2)} \right)$$

$$= 100 \text{ lb/in}^2 - 6.22 \text{ lb/in}^2 = 93.78 \text{ psi } (647 \text{ kPa})$$

or 647 kPa.

In this case, the loss in pressure is due to the increase in velocity. This illustrates the so-called Bernoulli principle—as the velocity of a fluid in a pipeline increases, the pressure in the pipeline decreases.

General Energy Equation

Although the Bernoulli equation is very convenient for numerous applications, its requirement that there be no energy losses is not very realistic for most flow situations. There are energy losses due to friction between the fluid and the pipe, as a result of changing the direction of the flow through elbows, valves, and other fittings, and due to the fluid viscosity. Thus, to be more realistic, we must add a term to the Bernoulli equation to account for these losses. To complete our quest for realism, we must also include terms to account for the energy added by the pump and energy removed by actuators. We will use the following symbology:

$$h_A = \text{energy added.}$$
$$h_R = \text{energy removed by actuators.}$$
$$h_L = \text{energy lost.}$$

These values have the unit of length (feet or meters) and are usually referred to as *head*. (The units actually are ft · lb/lb or N · m/N, referring to the energy per unit weight of fluid, but the weight terms are normally canceled out, leaving only feet or meters.) The resulting *general energy equation* becomes

$$\frac{p_1}{\gamma} + z_1 + \frac{v_1^2}{2g} + h_A - h_R - h_L = \frac{p_2}{\gamma} + z_2 + \frac{v_2^2}{2g} \qquad (12.7)$$

Example 12.9

For the circuit shown here, determine the pressure at the pump outlet while the cylinder is extending. The pressure at the cylinder inlet is 750 psi (5.2 MPa). The head loss between the pump outlet and the cylinder inlet is 40 ft (12.2 m). The cylinder is 20 ft (6.1 m) above the pump. The fluid velocity is 15 ft/sec (4.5 m/s). The specific weight of the fluid is 55.7 lb/ft³ (8.76 kN/m³).

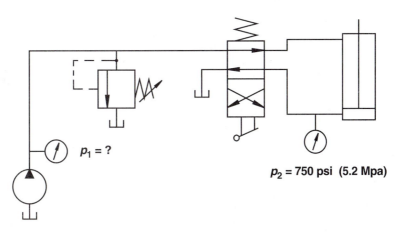

$p_1 = ?$

$p_2 = 750$ psi (5.2 Mpa)

Example 12.9

It is essential that the general energy equation be written in the direction of the flow; therefore, we put point 1 at the pump outlet and point 2 at the cylinder inlet. The equation, then, is written

$$\frac{p_1}{\gamma} + z_1 + \frac{v_1^2}{2g} + h_A - h_R - h_L = \frac{p_2}{\gamma} + z_2 + \frac{v_2^2}{2g}$$

Since the pipe size and fluid velocity is constant, $v_1 = v_2$. Also, since the pump and the cylinder are outside the portion of the system defined by points 1 and 2, h_A and h_R are zero for this example. Our equation now reduces to

$$\frac{p_1}{\gamma} + z_1 - h_L = \frac{p_2}{\gamma} + z_2$$

Solving this for p_1 gives

$$p_1 = p_2 + \gamma \left(z_2 - z_1 + h_L \right)$$

Substituting the values into this equation, we get

$$p_1 = 750 \text{ lb/in}^2 + 55.7 \text{ lb/ft}^3 \times (20 \text{ ft} - 0 \text{ ft} + 40 \text{ ft}) \, \frac{\text{ft}^2}{144 \text{ in}^2}$$

$$= 750 \text{ lb/in}^2 + 23.2 \text{ lb/in}^2 = 773.2 \text{ psi}$$

or 5.3 MPa.

Laminar and Turbulent Flow

Flow through conduits may be smooth and well organized, or they may be very disorganized and unpredictable. The organized flow is termed *laminar*, while the disorganized is termed *turbulent*. The flow regime that is likely to be found in a pipeline can be predicted by calculating the *Reynolds number* of the flow. This is done by using Equation 12.8:

$$N_R = \frac{v \times D}{v} \tag{12.8}$$

where

N_R = Reynolds number (dimensionless).
v = fluid velocity (ft/sec or m/sec).
D = pipe ID (ft or m).
v(nu) = kinematic viscosity (from Appendix F) (ft²/sec or m²/sec).

Experiments have shown that the flow regimes are generally defined by the Reynolds number as follows:

$$N_R < 2000 : \text{laminar}$$
$$N_R > 4000 : \text{turbulent}$$

The range between 2000 and 4000 is termed the *transition* or *critical* regime and is totally unpredictable.

As we will see in some examples to follow, it is desirable to keep the fluid velocity low enough to ensure that the flow is laminar in order to minimize head losses.

Example 12.10

An ISO VG 46 fluid at 104°F is flowing through a 2 in (5.1 cm) pipe at 30 ft/sec (9 m/s). Is the flow laminar or turbulent?

Solution

From Appendix F, we see that the kinematic viscosity of the fluid is 4.95 $\times$ 10^{-4} ft²/sec (4.6×10^{-5} m²/s). Using Equation 12.8, we find that the Reynolds number is

$$N_R = \frac{v \times D}{\nu} = \frac{30 \text{ ft/sec} \times 2 \text{ in} \times \text{ft/12 in}}{4.95 \times 10^{-4} \text{ ft}^2/\text{sec}} = 10{,}101$$

This is greater than 4000; thus, the flow is turbulent.

Darcy Equation

The value of h_L in the general energy equation is found from the *Darcy equation* (Equation 12.9):

$$h_L = f \times \frac{L}{D} \times \frac{v^2}{2g} \quad \frac{m^2/s^2}{m/s^2} = m \tag{12.9}$$

where

f = friction factor.
L = pipe length.
D = pipe ID.
v = flow velocity.
g = acceleration of gravity.

The friction factor must be determined, but the method used to find it depends upon whether the flow is laminar or turbulent. If the flow is laminar, the friction factor is found from Equation 12.10:

$$f = \frac{64}{N_R} \tag{12.10}$$

For turbulent flow, the task of determining the friction factor is somewhat more involved. While several empirical equations have been developed for this purpose, the quickest and most convenient method is to use the graph shown in Figure 12.19, commonly referred to as the *Moody diagram*. Two parameters are needed—the Reynolds number and the relative roughness.

The Reynolds number scale across the bottom of Figure 12.21 is a logarithmic scale. It begins on the left-hand edge at 6×10^2, or 600. The line marked 10^3 is 1000; the line marked 2 is 2×10^3, or 2000. Then comes 3×10^3, or 3000, 4×10^3, or 4000, and so on to 10^4, which is 10×10^3, or 10,000. The scale then repeats, starting at 2, but this time it means 2×10^4, or 20,000.

Notice the diagonal line in the upper left-corner labeled "Laminar Flow." This is the plot of Equation 12.10. It ends at $N_R = 2 \times 10^3$, which is the upper limit for laminar flow. The dashed projection of the line is a guess at what might be happening in the transition region between 2000 and 4000. Since this region is so unpredictable, efforts to determine friction factors are futile, so we'll not bother with it.

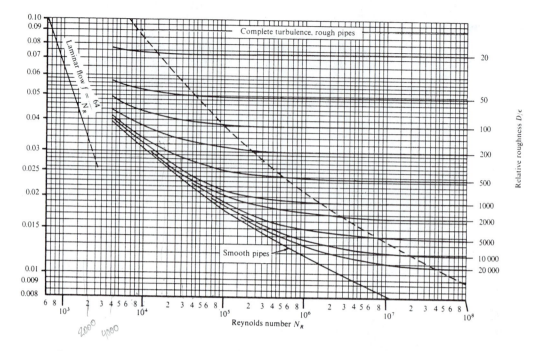

Figure 12.19 The Moody diagram for determining the friction factor in turbulent flow employs two parameters, the Reynolds number and the relative roughness.

Source: Reproduced courtesy of Richard Pao.

The relative roughness is found from Equation 12.11:

$$\text{Relative Roughness} = \frac{\text{Pipe Diameter}}{\text{Surface Roughness}}$$

or

$$RR = \frac{D}{\varepsilon} \qquad\qquad\qquad \textbf{(12.11)}$$

where

$$RR = \text{relative roughness.}$$
$$D = \text{pipe diameter.}$$
$$\varepsilon \text{ (epsilon)} = \text{surface roughness.}$$

Relative roughness is dimensionless. The design value of surface roughness for the steels commonly used in fluid power systems is 1.5×10^{-4} ft (4.6×10^{-5} m).

To find the friction factor, first locate the curve representing the calculated value of relative roughness on the right-hand scale. Follow that curve until it intersects the calculated value of the Reynolds number. From that point, go horizontally to the left-hand scale and read the value for the friction factor.

Example 12.11

Use the Moody diagram to determine the friction factors for a Reynolds number of 20,000 and a relative roughness of 1000.

Solution

Locate the curve for the relative roughness of 1000 on the right-hand scale, and follow it until it intersects the Reynolds number (20,000 or 2×10^4). Now go straight across to the left-hand axis and read the value of f, which is approximately 0.0275.

The values of f determined from the Moody diagram are always approximations. Different users can get different values, especially when the relative roughness values fall between those shown on the scale. All you can do in this case is to imagine that a curve for that value exists and that it follows the same profile as the existing curves. Then use your imaginary curve in the same way that we used the 1000 value in Example 12.10.

Equivalent Length

Valves and fittings in the piping system also cause pressure losses. The easiest way to calculate these losses is to use the *equivalent length* method to estimate the effect of a valve or fitting by treating it as if it were an additional length of pipe. Table 12.3 lists some common devices and their equivalent length values, which are given as the length-to-diameter (L_e/D) ratios so that they can be used directly in the modification of the Darcy equation shown in Equation 12.12:

$$h_L = f \times \frac{L_e}{D} \times \frac{v^2}{2g}$$ (12.12)

Table 12.3 Equivalent Length Values

Device	Equivalent Length L_e/D
Check valve	150
90° standard elbow	30
45° standard elbow	16
Close return bend	50
Standard tee-run	20
Standard tee-branch	60

Source: The Crane Company, Chicago, Il

Example 12.12

Find the head loss due to a standard 90 degree elbow for a friction factor of 0.030 and a flow velocity of 8 ft/sec (2.4 m/s).

Solution

From Table 12.3, the equivalent length value for a 90 degree standard elbow is 30. Using Equation 12.12, we find the head loss is

$$h_L = f \times \frac{L_e}{D} \times \frac{v^2}{2g}$$

$$= \frac{(0.030)\,(30)\,(8\ \text{ft/sec})^2}{(2)\,(32.2\ \text{ft/sec}^2)} = 0.89\ \text{ft}$$

or 0.27 m.

Putting It Together

Now let's look at an example problem that will bring all of this together.

Example 12.13

For the system shown here, determine the pressure at the pump outlet. The fluid is ISO VG 68 at 104°F (40°C). It has a specific gravity of 0.9. The commercial steel pipe is ½ in Schedule 40 and is 30 ft (9.14 m) long. All fittings are standard. The hydraulic motor inlet is 12 ft (3.66 m) above the pump. The flow rate is 60 gpm (414 Lpm), and the pressure at the motor inlet is 2500 psi (17.2 MPa).

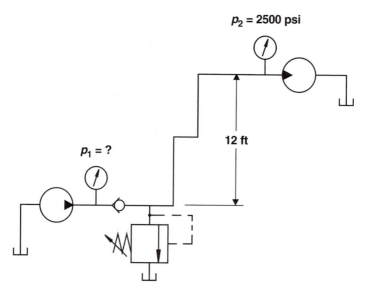

Example 12.13

Solution

First, we write the general energy equation in the direction of the flow.

$$\frac{p_1}{\gamma} + z_1 + \frac{v_1^2}{2g} + h_A - h_R - h_L = \frac{p_2}{\gamma} + z_2 + \frac{v_2^2}{2g}$$

Since there are no pumps or motors between points 1 and 2, h_A and h_R are both zero. Since the pipe size is the same throughout, $v_1 = v_2$. Thus, our equation reduces to

$$\frac{p_1}{\gamma} + z_1 - h_L = \frac{p_2}{\gamma} + z_2$$

Solving for the pump outlet pressure, we get

$$p_1 = p_2 + \gamma \left(z_2 - z_1 + h_L \right)$$

We find the specific weight to be

$$\gamma = 0.9 \times 62.4 \ \text{lf/ft}^3 = 56.2 \ \text{lb/ft}^3 \ (8.83 \ \text{kN/m}^3)$$

We already know p_2, z_1, and z_2, but we must use the Darcy equation to find h_L. We need the total head loss, which will include the pipe, the ball-type check valve, and the four elbows, so we can write

$$h_L = \left(f \times \frac{L}{D} \times \frac{v^2}{2g} \right)_{\text{pipe}} + \left(f \times \frac{L_e}{D} \times \frac{v^2}{2g} \right)_{\text{c.v.}} + \left(4f \times \frac{L_e}{D} \times \frac{v^2}{2g} \right)_{\text{elbows}}$$

or

$$h_L = f \left[\left(\frac{L}{D} \right)_{\text{pipe}} + \left(\frac{L_e}{D} \right)_{\text{c.v.}} + 4 \left(\frac{L_e}{D} \right)_{\text{elbows}} \right] \frac{v^2}{2g}$$

Now we need to find the Reynolds number, but to do so, we first need to calculate the velocity. From Appendix D, we see that the area of a ½ in Schedule 40 pipe is 0.304 in^2 (1.956 cm^3), so our velocity is

$$v = \frac{Q}{3.12A} = \frac{60}{3.12 \times 0.304} = 63.3 \ \text{ft/sec}$$

or 19.3 m/s.

Again, we use Appendix D to find the pipe diameter, which is 0.622 in (0.052 ft). Thus, using Equation 12.8, we find that the Reynolds number is

$$N_R = \frac{v \times D}{\nu} = \frac{63.3 \ \text{ft/sec} \times 0.052 \ \text{ft}}{7.3 \times 10^{-4} \ \text{ft}^2\text{/sec}}$$

$$= 4509$$

Since this is in the turbulent flow regime, we must use the Moody diagram to find f. So now we need to find the relative roughness. The surface roughness is 1.5×10^{-4} ft (4.6×10^{-5} m); thus,

$$RR = \frac{D}{\varepsilon} = \frac{0.052 \ \text{ft}}{1.5 \times 10^{-4} \ \text{ft}} = 346.6$$

From the Moody diagram, we estimate the value of f as 0.041. Using the equivalent length values from Table 12.3, we find the total head loss to be

$$h_L = 0.041 \left(\frac{30 \ \text{ft}}{0.052 \ \text{ft}} + 150 + 4(30) \right) \frac{(63.3 \ \text{ft/sec})^2}{(2) \ (32.2 \ \text{ft/sec}^2)}$$

$$= 2160 \ \text{ft}$$

or 658 m.

From this, we find the pressure at the pump outlet to be

$$p_1 = 2500 \text{ lb/in}^2 + 56.2 \text{ lb/ft}^2 \ [12 \text{ ft} - 0 \text{ ft} + 2160 \text{ ft}] \ \frac{\text{ft}^2}{144 \text{ in}^2}$$

$$= 3348 \text{ psi}$$

or 23.1 MPa.

You should be aware that the velocity and head loss in this example are very high and would constitute a poor design, a high energy waste, and a high heat generation. The pipe size used here was much too small, but was chosen intentionally for the purpose of demonstrating the use of the Moody diagram for such calculations. In real design situations, you should attempt to achieve laminar flow where possible.

A Quicker Way

The analytical method we have just discussed for finding the pressure losses in pipelines is accurate but can be very time-consuming. A method that is much less accurate but provides a reasonable first estimate involves the use of tables similar to Tables 12.4 and 12.5. Such tables are available from pipe manufacturers and in various handbooks concerning fluid flow.

Table 12.4 is based on an ISO VG 46 hydraulic oil having a specific gravity of 0.9. It is for Schedule 40 commercial steel pipe only.

Table 12.4 Pressure Loss in Schedule 40 Pipe (psi per foot of pipe)

Nominal Pipe Size (in)	Flow Rate (gpm)						Equivalent Length (feet)	
	5	10	15	20	25	30	Elbow (90 Deg)	Tee (Branch)
$\frac{1}{8}$	10.7460	21.4830	88.9200	154.7010	230.9400	317.1600	0.7	1.3
$\frac{1}{4}$	3.2040	6.4224	9.5940	33.3450	49.7250	69.0300	0.9	1.8
$\frac{3}{8}$	0.9360	1.9080	2.8620	3.8070	10.9080	15.7140	1.2	2.5
$\frac{1}{2}$	0.3780	0.7524	1.1250	1.5030	1.8720	2.2590	1.6	3.1
$\frac{3}{4}$	0.1260	0.2448	0.3663	0.4896	0.6093	0.7317	2.1	4.1
1	0.0468	0.0927	0.1395	0.1863	0.2322	0.2799	2.6	5.2
$1\frac{1}{2}$	0.0081	0.0168	0.0252	0.0336	0.0419	0.0504	4	8.1
2	0.0031	0.0061	0.0092	0.0123	0.0155	0.0189	5.2	10.3
$2\frac{1}{2}$	0.0015	0.0030	0.0045	0.0061	0.0076	0.0091	6.2	12.3
3	0.0006	0.0013	0.0019	0.0025	0.0032	0.0038	7.7	15.3
4	0.0002	0.0004	0.0006	0.0009	0.0011	0.0013	10.1	20.1

Note: This table is based on VG 46 fluid, Sg = 0.9.

Table 12.5 Pressure Loss in Schedule 40 Pipe (kPa per meter of pipe)

Nominal Pipe Size (in)	Flow Rate (Lpm) 5	10	25	50	75	100	Equivalent Length (meters) Elbow (90 Deg)	Tee (Branch)
$\frac{1}{8}$	65.183	130.365	325.914	655.258	3473.553	5763.526	0.23	0.43
$\frac{1}{4}$	19.419	38.837	97.093	194.186	745.327	1264.797	0.30	0.59
$\frac{3}{8}$	5.717	11.433	28.583	57.165	85.748	284.743	0.40	0.83
$\frac{1}{2}$	2.256	4.513	11.282	22.563	33.845	45.126	0.53	1.02
$\frac{3}{4}$	0.735	1.471	3.677	7.354	11.030	14.707	0.69	1.35
1	0.280	0.560	1.400	2.799	4.199	5.598	0.86	1.72
$1\frac{1}{2}$	0.050	0.100	0.251	0.502	0.753	1.005	1.32	2.67
2	0.018	0.037	0.092	0.185	0.277	0.370	1.72	3.40
$2\frac{1}{2}$	0.009	0.018	0.045	0.091	0.136	0.182	2.11	4.06
3	0.004	0.008	0.019	0.038	0.057	0.076	2.54	5.05
4	0.001	0.003	0.006	0.013	0.019	0.026	3.33	6.64

Note: This table is based on VG 46 fluid, Sg = 0.9.

The right-hand side of the table lists the equivalent lengths for various tees and elbows in length of pipe. These values are simply added to the total pipe length, then multiplied by the pressure drop per foot or meter of pipe.

Example 12.14

Use Table 12.4 to estimate the pressure loss in 15 ft of 2 in Schedule 40 pipe for a flow rate of 25 gpm.

Solution

Locate the pipe size on the left and the flow rate across the top. The row and column intersect at 0.0155 psi. This is the pressure loss per foot of pipe. To find the total loss, multiply this by the pipe length. This gives a total loss of

$$0.0155 \text{ psi/ft} \times 15 \text{ ft} = 0.23 \text{ psi}$$

Example 12.15

Use Table 12.5 to estimate the pressure loss in 10 m of 1 in Schedule 40 pipe for a flow rate of 75 Lpm.

Solution

From Table 12.5, we see that the pressure loss in the pipe is 4.199 kPa per meter of pipe. Therefore, the loss in 10 m is

$$4.199 \text{ kPa/m} \times 10 \text{ m} = 41.99 \text{ kPa}$$

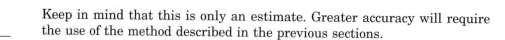

Keep in mind that this is only an estimate. Greater accuracy will require the use of the method described in the previous sections.

Example 12.16

A system has 25 ft of 1 in Schedule 40 pipe and four 90 degree elbows. Estimate the total pressure loss for a flow rate of 15 gpm.

Solution

Table 12.4 shows that the pressure loss per foot of pipe is 0.1395 psi. The equivalent length for a 90 degree elbow is 2.6. The four elbows give a total equivalent length of 10.4 ft. Adding this to the actual pipe length gives a total length of 35.4 ft. The pressure loss, therefore is

$$0.1395 \text{ psi/ft} \times 35.4 \text{ ft} = 4.94 \text{ psi}$$

12.6 Troubleshooting Tips

Symptom	Possible Cause
Fluid leak at joints	1. NPT threads 2. NPTF threads reused 3. Loose fitting 4. Split flare or cracked tubing due to overtightening 5. Mating of different threads 6. O-ring damaged (SAE fittings)
Hose ruptured or pulled from fitting	1. Excessive pressure 2. Pressure spikes 3. Improperly assembled hose fitting 4. Hose incompatible with system fluid 5. Hose damaged externally 6. Hose improperly installed (twisted, kinked, bend radius too small) 7. Wrong series hose
Pump damaged by cavitation or starvation	1. Suction line too long 2. Suction line too small 3. Loose or leaking connections in suction line 4. Too many fittings, elbows, etc., in suction line 5. Viscosity too high
Insufficient pressure at load	1. Pressure line too long 2. Pressure line too small 3. Too many fittings, elbows, etc., in pressure line 4. Viscosity too high

12.7 Summary

While pipe, tubing, or hoses can be used to convey fluid throughout a fluid power system, careful attention must be paid to selecting the best conduit for a particular application. Piping, for example, despite its disadvantages, is the practical (in fact, only) choice for some cases.

The sizing of pipe, tubing, or hoses depends on pressures and flow rates that dictate the wall thickness and internal diameter to be used. Compatibility and temperature sensitivity must also be considered, especially for hoses. In all cases, installation and routing must receive special attention to ensure that unnecessary failures are not experienced due to stresses or abrasion.

12.8   Key Equations

Wall thickness (thin-walled):
$$t = \frac{p \times D \times N}{2S_u} \tag{12.3}$$

Wall thickness (thick-walled):
$$t = \frac{p \times D}{2(S_{ALL} - 0.6p)} \tag{12.5}$$

Bernoulli's equation
$$\frac{p_1}{\gamma} + z_1 + \frac{v_1^2}{2g} = \frac{p_2}{\gamma} + z_2 + \frac{v_2^2}{2g} \tag{12.6}$$

General energy equation:
$$\frac{p_1}{\gamma} + z_1 + \frac{v_1^2}{2g} + h_A - h_R - h_L \tag{12.7}$$

$$= \frac{p_2}{\gamma} + z_2 + \frac{v_2^2}{2g}$$

Reynolds numbers:
$$N_R = \frac{v \times D}{\nu} \tag{12.8}$$

Darcy equation:
$$h_L = f \times \frac{L}{D} \times \frac{v^2}{2g} \tag{12.9}$$

Friction factor (laminar flow):
$$f = \frac{64}{N_R} \tag{12.10}$$

Relative roughness:
$$RR = \frac{D}{\epsilon} \tag{12.11}$$

Review Problems

Note: The letter "S" after a problem number indicate SI units are used.

General

1. Explain the meaning of "one inch" as related to pipes, hoses, and tubing.
2. Explain the difference between NPT and NPTF threads and the SAE O-ring boss (ORB) fittings.
3. List and describe several methods for attaching fittings to tubing.
4. Describe the differences between reusable and permanent hose fittings.
5. What does the "dash number" (-8, for example) mean when related to hose sizing?
6. What is the SAE series designation for hydraulic hoses?
7. What is the flare angle used for hydraulic fittings?
8. List the fluid velocity ranges for suction, pressure, and return lines.
9. What tubing material must never be used for fluid power applications?
10. List the three parts of a hose.

Pipe Sizing

11. A pressure line carries fluid flow at 40 gpm. The system pressure is 1500 psi. What is the smallest Schedule 40 pipe that can be used in this application based on the guidelines in Table 2.1?
12. A pump is producing a flow of 60 gpm. Find the smallest Schedule 40 pipe that can be used for the suction line in this system based on the guidelines in Table 2.1.
13S. A system operating at 20 MPa has a flow rate of 140 Lpm. Using Table 2.1 as a guide, find the smallest Schedule 40 pipe that can be used for the pressure line.
14S. A system requires 80 Lpm pump output. Determine the smallest Schedule 40 pipe that can be used for the suction line using the guidelines of Table 2.1.

Wall Thickness

15. Determine the minimum wall thickness for a thin-walled pipe ($D = 1.5$ in) for a system pressure of 1750 psi. The pipe material has an ultimate tensile strength of 60,000 psi. Use a safety factor of 6.
16S. A 5 cm ID pipe is used in a system in which the pressure is 20 MPa. The pipe has an ultimate tensile strength of 350 MPa. Using a safety factor of 6, determine the minimum wall thickness that can be used.

Reynolds Number

17. Fluid is flowing through a 1 in Schedule 40 pipe at 30 gpm. The fluid is an ISO VG 32. Calculate the Reynolds number in the pipe.
18. An ISO 68 fluid is being pumped at 60 gpm. Determine the smallest Schedule 40 pipe that can be used to maintain laminar flow.
19S. Find the Reynolds number for an ISO VG 68 oil flowing through a 2 in Schedule 40 pipe at 150 Lpm.

Pressure Loss

20. An ISO VG 46 fluid (Sg = 0.85) at 104°F is flowing through a 2 in Schedule 80 pipe at 45 gpm. Calculate the pressure loss through 100 ft of the pipe.
21. A 2 in Schedule 80 pipe carries ISO VG 46 oil (Sg = 0.9) at 125 gpm. The oil temperature is 104°F. Calculate the pressure loss through 100 ft of pipe.
22S. A 1 in Schedule 40 pipe is carrying ISO VG 100 oil (Sg = 0.9) at 40°C. The flow rate is 250 Lpm. Calculate the pressure loss per meter of pipe.
23S. The flow rate in Review Problem 22S is increased to 1000 Lpm. Calculate the pressure loss per meter of pipe.

System Problems

24. In the system shown here, the 3 in (7.6 cm) cylinder must raise the load at 14 in/sec (53 cm/sec). The pressure drop across the directional control valve is 22 psi (150 kPa) in each direction. There is a total of 14 ft (4.25 m) of pressure line and 12 ft (3.7 m) of return line. All pipe is 1½ in Schedule 40 steel. Calculate the pump outlet pressure required to raise the load. Account for all valves and fittings. The fluid is an ISO VG 32 oil at 104°F (40°C). The piping, cylinder, and pump are on the same level.

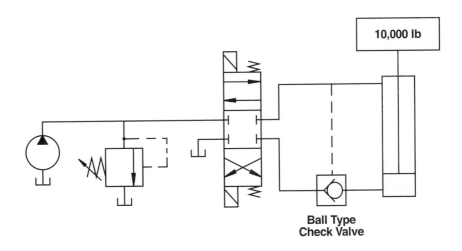

10,000 lb

Ball Type
Check Valve

Review Problem 12–24

Ancillary Devices

Outline

13.1 Introduction

The term *ancillary devices* seems to imply items that are perhaps nice to have but are not really important to the operation of the system. The devices listed in this chapter, however, serve a very important role in circuits where they are utilized. Reservoirs, accumulators, and intensifiers, along with seals and instrumentation, are included in the grouping; and be aware that other authors might very well devote a separate chapter to any one of these devices. Their inclusion in this chapter under *ancillary devices* is not meant to demean their significance.

Objectives
When you have completed this chapter, you will be able to

- Discuss basic reservoir construction and functions.
- Discuss the various types of accumulators.
- Discuss typical accumulator applications and the advantages offered by each.
- Explain intensifiers and their applications.
- Discuss the more commonly used types of seals for fluid power applications.
- Discuss the more common types of fluid power instrumentation and how each device works.
- Explain how heat exchangers are used in fluid power systems.
- Calculate the heat dissipation rate of heat exchangers.

13.2 Reservoirs

To most of us, a *reservoir,* such as the one shown in Figure 13.1, is simply a tank in which to store hydraulic fluid: fluid flows from the reservoir to the

447

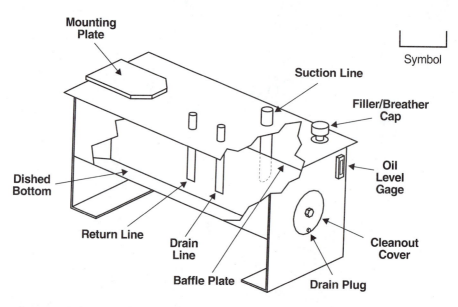

Figure 13.1 This is a typical industrial hydraulic reservoir configuration.

pump, circulates through the system, and returns to the tank whence it came. In truth, however, the reservoir actually plays several major roles, including

- Heat exchanger (to cool the fluid).
- Filter (to remove solid particles).
- Deaerator (to remove air and gases).
- Dehydrator (to remove water).

So holding fluid is only one of a reservoir's roles, and it may not be the most important one, depending on the system design. Let's consider each of these functions and see how they are accomplished in this tank with no moving parts.

Heat Exchange

Remember, in our discussion of valves we found that any energy we put into hydraulic fluid that we do not subsequently remove in the form of useful work is manifested in the form of heat. Thus, every time we encounter a pressure drop across a valve (or any other device), we encounter an increase in fluid temperature. If we don't dissipate that heat somewhere in the system, we end up with an ever increasing temperature spiral that will eventually lead to an unacceptable loss of viscosity, and eventually to a complete chemical breakdown of the fluid.

As we will see later in this chapter, we could use a *heat exchanger* specifically designed for this purpose. Another way to get rid of unwanted heat is through the reservoir walls. Ideally, of course, the system should be designed to prevent the losses that lead to temperature rise, but some losses are inevitable, and we must design our systems to deal with the resultant temperature rises.

The dissipation of heat from the fluid in the reservoir results from all three heat-transfer mechanisms—convection, conduction, and radiation. Heat transfer by *convection* requires a significant circulation or movement of the liquid through the regions of different temperatures. Convection may be natural or forced. *Natural convection* occurs because warmer fluid is less dense and rises and is replaced by cooler fluid next to the wall. *Forced circulation* results when a pump (or fan or blower in the case of air) is used to move the fluid. *Conduction* is a heat transfer mechanism that occurs on a molecular level within the liquid or gas. It is the result of collisions or intimate contact between the molecules. Thermal *radiation* is the transfer of heat by means of electromagnetic radiation. The amount of heat radiated depends on the configuration of the surface, the surface temperature, and the material and finish of the surface. By a combination of these mechanisms, heat passes from the fluid into the tank walls and then into the atmosphere (assuming that the atmospheric temperature is cooler than the fluid temperature, since heat moves from the high-temperature area to the low-temperature area).

The efficiency of this operation depends on two parameters in the reservoir design—the total surface area exposed to the atmosphere and the length of time that the fluid remains in the reservoir.

The rate at which heat is transferred from a surface is measured in British Thermal Units per square foot per hour (Btu/(ft$^2 \cdot$ hr)) or watts per square meter (W/m^2). Thus, the larger the reservoir, the more square feet of surface area there are for heat dissipation. The effectiveness of the heat transfer is also enhanced by direct contact of the fluid with the walls of the tank.

The rate at which heat is dissipated (transferred) from the surface of a reservoir is calculated from

Heat Transfer Rate = Heat Transfer Coefficient $\times$ Area
$\times$ Temperature Difference Between Fluid and Air

$$q = C \times A \times \Delta T \qquad \qquad \textbf{(13.1)}$$

where

q = heat transfer rate (Btu/hr or W).

C = heat transfer coefficient (Btu/hr·ft^2·°F or W/m^2·°C)

A = total fluid contact area (ft^2 or m^2).

ΔT = temperature difference (°F or °C).

The total fluid contact area is the total surface area of the reservoir that is in contact with the fluid. The bottom of the reservoir can be included in this total area only if it is high enough off the floor to allow free circulation of air beneath it. Table 13.1 lists some typical values of C. Equation 13.1 can also be used to estimate the heat transfer rate for heat exchangers. In this case, the total outside surface area of the pipes carrying hydraulic fluid is used for A.

Table 13.1 Typical Values of Heat Transfer Coefficients

Situation	Heat Transfer Coefficient, C	
	Btu/hr·ft²·°F	W/m²°C
Tank inside machine or with little air circulation around it	2 to 5	10 to 25
Steel tank sitting in unrestricted air flow	5 to 10	25 to 50
Steel tank with natural airflow directed toward it (not forced)	10 to 13	50 to 65
Steel tank with forced air circulation, or oil-to-air heat exchanger	25 to 60	125 to 300
Oil-to-water heat exchanger	80 to 100	400 to 500

Example 13.1

A fluid reservoir is mounted so that air circulates freely around it as shown here. The fluid is 1.5 ft (0.46 m) deep in the tank. The fluid temperature is 130°F (54°C). The air temperature is 70°F (21°C). Find the maximum heat transfer rate from the reservoir.

Solution

Since we're looking for the maximum heat transfer rate, we'll use $C = 10$ from Table 13.1. We need to determine the surface area of the tank in contact with the oil. First calculate the side walls. The areas are

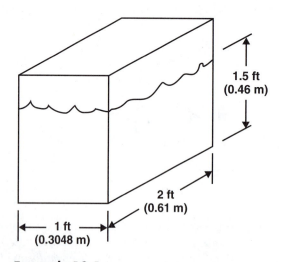

1.5 ft (0.46 m)

2 ft (0.61 m)

1 ft (0.3048 m)

Example 13.1

$$A = 2 \text{ ft} \times 1.5 \text{ ft} \times 2 \text{ Walls} = 6 \text{ ft}^2$$

or 0.56 m². Now calculate the end walls:

$$A = 1 \text{ ft} \times 1.5 \text{ ft} \times 2 \text{ Walls} = 3 \text{ ft}^2$$

or 0.28 m². Finally, calculate the bottom:

$$A = 1 \text{ ft} \times 2 \text{ ft} = 2 \text{ ft}^2$$

or 0.19 m².

the total contact area, therefore, is 11 ft² (1.03 m²). Using Equation 13.1, we get

$$q = C \times A \times \Delta T = \left(10 \frac{\text{Btu}}{\text{hr} \cdot \text{ft}^2 \cdot °\text{F}}\right) \times 11 \text{ ft}^2 \times (130 - 70)°\text{F}$$

$$= 6600 \text{ Btu/hr}$$

or 1.9 kw.

The transfer of heat from the fluid to the dissipating surfaces is not instanta-neous. Therefore, the fluid must be allowed to remain in contact with the walls for as long as possible. This is a function of the length of time the fluid remains in the reservoir—termed *dwell* or *residence* time. In order to increase dwell time, many reservoirs are built with a series of *baffles* around which the fluid must travel. The tortuous path provided by these baffles prevents the fluid from traveling directly from the return line to the pump inlet line. It also promotes mixing of the fluid, which improves the heat transfer situation. Figure 13.2 illustrates some baffle configurations.

An oversized reservoir offers another advantage in that there is simply much more fluid, so that the average bulk fluid temperature increases at a much lower rate than it would with a smaller quantity of fluid. This, coupled with

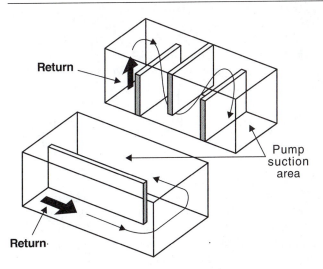

Figure 13.2 Baffles increase dwell time.

the increased surface area and the lower turnover rate of the fluid leads to much lower overall temperatures.

You may well be asking what constitutes this "oversized" reservoir. That actually depends on your source of reference. Most system specifications call for the reservoir to contain three times the pump flow rate. If the pump operates at 5 gpm, then the reservoir capacity should be 15 gallons. The mobile equipment industry tends to use a 1 to 1 ratio; a 5 gpm pump would require a 5 gallon reservoir. Other requirements vary according to application. It might be safe to say, then, that an oversized reservoir might be on the order of 10 to 1—10 gallons of fluid for each gpm of output.

To determine whether the reservoir has sufficient surface area to dissipate the heat generated by the system operations, we can compare the heat generation rate of the system to the heat dissipation rate of the reservoir.

Example 13.2

A hydraulic system has an overall efficiency of 65 percent. It uses a 50 gpm (189 Lpm) fixed displacement pump that operates at a constant 1000 psi (6.9 MPa). The reservoir holds 100 gal (378.5 L) of oil and has a total surface area of 40 ft^2 (3.72 m^2). Its heat transfer coefficient is 8 Btu/(ft^2 · hr · °F) (45.4W/m^2°C). Assume no heat dissipation through components and piping. Can the reservoir dissipate enough heat to maintain the fluid temperature below 130°F (54°C) if the air temperature is 80°F (27°C)?

Solution

First, we need to find the heat generation rate of the system. The total hydraulic horsepower produced by the pump is found from Equation 2.11

$$HHP = \frac{p \times Q}{1714} = \frac{1000 \times 50}{1714} = 29.2 \text{ HP}$$

or 21.8 kw. Since the system efficiency is 65 percent, the remaining 35 percent is turned into heat, giving us a heat generation rate of

$$HGR = 0.35 \times 29.2 \text{ HP} \times 42.4 \frac{\text{Btu}}{\text{min} \times \text{HP}} \times \frac{60 \text{ min}}{\text{hr}}$$
$$= 26,000 \text{ Btu/hr}$$

or 7.62 kW. Next, we used Equation 13.1 to determine the amount of heat the reservoir can dissipate at the maximum temperature difference (130 − 80 = 50°F, or 28°C).

$$q = C \times A \times \Delta T = \left(8 \frac{\text{Btu}}{(\text{hr} \cdot \text{ft}^2 \cdot \text{°F})}\right) \times 40 \text{ ft}^2 \times 50\text{°F}$$
$$= 16,000 \text{ Btu/hr}$$

or 4.6 kW.

Since the heat generation rate is greater than the heat dissipation rate, the temperature cannot be maintained at 130°F (54°C). To maintain this temperature, some other means of heat dissipation—such as a heat exchanger—must be used. We'll discuss heat exchangers later in this chapter. If you were using a system such as this one, you would be well advised to look for ways to reduce the system losses rather than resorting to the use of a heat exchanger.

Otherwise, you lose twice—you pay for the wasted energy to put the heat in, and you pay for the heat exchanger and its operation to take the heat out.

Another heat exchange function that might take place in the reservoir is the reverse of what we have just been discussing. That is, we might want to heat the fluid rather than cool it. This situation could arise during the startup of a system that has been exposed to low temperatures for an extended time. The cold fluid may not be pumpable, so it may be necessary to heat it to lower its viscosity. This is usually accomplished by electrical heating elements immersed in the fluid, although in some very large plant installations, steam or hot water pipes are sometimes run through the tanks.

If electric heaters are used, care must be taken not to "cook" the fluid that contacts the heating elements. This results in degradation of the fluid and the creation of a coating of burned fluid on the elements that severely reduces their heating efficiency. As a rule of thumb, the heat output of intank heating elements should not exceed 3 watts per square inch of heating element surface.

Sedimentation

A reservoir acts as a filter when solid particles are allowed to settle to the bottom as the fluid passes through. Unfortunately, not all of the particles will settle out of the fluid, even if it remains at rest over very long periods of time. For settling to occur, the force of gravity must overcome buoyant forces and viscous drag. Relatively large, dense particles (and remember that we are talking on a microscopic scale when discussing particle size) will tend to settle out fairly rapidly, while small particles (smaller than 10 microns, say) may never settle out, especially in a high-viscosity fluid.

The longer the dwell time of the fluid in the reservoir, the more opportunity there is for *particle drop-out*. This constitutes another factor favoring oversized reservoirs, although one might argue—and validly so—that good filtration (internal and external) should remove the large particles before they get to the reservoir. It can also be pointed out that an off-line filter loop will do far more to clean the fluid in the reservoir than sedimentation could ever do.

Deaeration

Reservoirs that are at atmospheric pressure provide an opportunity for deaeration, the release of air that may have been trapped in the fluid for any reason. Air in fluids may take three different forms:

- Dissolved
- Entrained
- Free

As we discussed in Chapter 4, most hydraulic oils will contain about 10 percent air by volume under atmospheric pressure. This *dissolved air* is contained in the "empty spaces" between the fluid molecules, and does not increase the volume of the fluid.

As we exceed the amount of air that can be *dissolved* in a fluid, we begin to encounter *entrained air*. We see this in the form of bubbles which are 0.001 to 0.030 in (25 to 750 μm) in diameter. Fluids with entrained air tend to look cloudy or foamy. Generally, entrained air is attempting to get out of the fluid.

If the fluid is allowed to sit undisturbed for a time, the bubbles will usually agglomerate into larger bubbles of free air, rise to the surface and escape. Most modern hydraulic fluids contain antifoaming additives to help in releasing the air from the fluid. The rate at which this process will occur depends on the size of the bubble and the viscosity of the fluid (Stoke's Law), as well as the air pressure above the fluid (Henry-Dalton Law).

The implication of *Stoke's Law* is that the smaller the bubble, the longer it takes to rise to the surface in any given fluid. The rise rate is also inversely proportional to the viscosity of the fluid; therefore, the more viscous the fluid, the longer it will take any given bubble to rise to the surface. Regardless of the situation, it is obvious that some finite amount of time is required to allow air bubbles to escape from the oil. This, in turn, means that a more efficient deaeration process will be realized if the fluid flows through the reservoir slowly. In an undersized reservoir, or in one that is not adequately baffled, bubbles in the oil can be swept back into the pump inlet line before they can rise to the surface. As discussed in Chapter 4, this can lead to pump damage and eventual failure due to *pseudo-cavitation*—a process in which collapsing air bubbles allow high-velocity fluid to impinge on pump surfaces. The result is that metal particles are dislodged, contaminating the fluid and destroying critical pump clearances.

The *Henry-Dalton Law* says that the amount of air that can be dissolved in a fluid is directly proportional to the air pressure above the fluid:

$$h = h_0 \times \left(\frac{p}{p_0}\right) \tag{13.2}$$

where

h = actual air content (in^3 or cm^3).
h_0 = air content at standard day barometric pressure (in^3 or cm^3).
p = actual pressure (absolute) (psia, in Hg, or kPa abs).
p_0 = standard pressure (14.7 psia, 29.92 in Hg, or 101 kPa abs).

Example 13.3

One gallon (3.785 L) of hydraulic oil will contain 23 in^3 (375 cm^3) (or about 10 percent of its volume) of dissolved air at 14.7 psia (101 kPa abs). How much air will it hold at 200 psia (1380 kPa abs)?

Solution

Using Equation 13.2,

$$h = h_0 \left(\frac{p}{p_0}\right)$$

$$= 23 \text{ in}^3 (200 \text{ psia}/14.7 \text{ psia})$$

$$= 312.9 \text{ in}^3 (5132 \text{ cm}^3)$$

or 5.13 L. This answer is amazing! At 200 psia, we can actually have more than a gallon of air in a gallon of oil. What's even more amazing is that we won't even know it's there until we reduce the pressure below 200 psia. As the pressure is reduced, the air will be released and may cause a great deal of foaming as it comes out of the solution.

Example 13.4

What would happen if we exposed the oil from the previous example to a vacuum of 14.96 in Hg (38 cm Hg)?

Solution

Again, we use Equation 13.2.

$$h = h_0 \left(\frac{p}{p_0} \right)$$

$$= 23 \text{ in}^3 \left(\frac{14.96 \text{ in Hg}}{29.92 \text{ in Hg}} \right)$$

$$= 11.5 \text{ in}^3$$

or 189 cm^3.

This means that the fluid could hold only half as much air in solution. Therefore, half of the dissolved air would evolve from the solution and form air bubbles that would rise to the surface and dissipate.

Free air is seen as large "globs" of air in the system. This condition is most commonly found in cylinders, accumulators, and elevated points in the system where the fluid velocity is low enough for entrained air bubbles to agglomerate and form large air pockets.

While dissolved air is expected and generally acceptable, air in any form can have detrimental effects in the system. In Chapter 3, we discussed fluid oxidation and saw that this phenomenon requires oxygen—the presence of air in any form will suffice. We have also discussed the advantages of using liquids instead of gases—primarily the near-incompressibility of liquids. This advantage is lost with highly aerated fluids. Fluid column stiffness is lost because the air in the fluid is compressible, and, in fact, must be compressed before any other work can be done. This causes system response to be slow and spongy. Finally, aerated fluid causes pseudo-cavitation in pumps, valves, and so on. This can cause severe damage and rapidly degrade the components to absolute uselessness.

In a reservoir that is open to atmospheric pressure, as are most industrial reservoirs, we can remove only the entrained and free air, even during the very long dwell times or even when the system is shut down. If the reservoir is pressurized, the release of air will be restricted by the Henry-Dalton Law.

Creating a vacuum in the reservoir will significantly accelerate the release of free and entrained air and promote the dissolution of the dissolved air. An advantage of this practice is that the fluid can act as an air "sponge," readily absorbing any air that might be bleeding into the system or introduced by bad maintenance practices. This (reservoir vacuum) is seldom seen in other than laboratory situations, however, because it is more expensive and troublesome than standard breathing reservoirs. Reservoir vacuum could also be detrimental as far as pump suction is concerned and could lead to pump cavitation as discussed in Chapter 4.

A design that is sometimes seen, however, is a reservoir containing a sloped screen through which the fluid flows as it moves from the return line to the pump inlet line. Air bubbles tend to adhere to the screen and thus be removed from the flow stream. Eventually, they agglomerate, forming large bubbles that rise to the surface and dissipate. The most effective arrangement seems to be number 100 to number 400 wire mesh inclined about 20 degrees to the horizontal. Air removal efficiencies in excess of 95 percent have been experienced, but again, dwell time is very important, because the removal efficiency is dependent on the velocity of the fluid as it passes through the screen.

Dehydration

During periods of system downtime, when the fluid is standing in the reservoir, water in fluids that are less dense than water will tend to separate from the oil and settle to the bottom where it can be drained off. This separation will occur only if the water and oil have not been emulsified.

As with the separation of air from the oil, the separation of water is not instantaneous. Little separation is likely to occur as long as the fluid is moving through the reservoir unless very large amounts of water are present.

Alternative Devices

Some of the functions we have discussed are performed fairly efficiently by the reservoir. For instance, there is little question that the large surface area of the reservoir walls (especially one that is intentionally oversized) dissipate heat efficiently, although the efficiency is severely reduced on hot, still days. Deaeration is also accomplished fairly efficiently, especially in large reservoirs with long fluid dwell times. This function is normally restricted by the fact that the reservoir is usually at atmospheric pressure, or higher, so that none of the dissolved air is removed. The percentage of free and entrained air removed is a strong function of the fluid dwell time.

It is often found, however, that other, dedicated devices can perform these functions more efficiently than the reservoir. While there is no question that a great deal of dirt ends up upon the bottom of the tank (all you need to do is look), good filters perform this function far more efficiently.

Water removal is another function that can be performed more efficiently by devices specifically designed for that purpose (refer to Chapter 11). A properly designed water separator removes water from the oil continuously, while the system is operating. It does not require quiescent fluid to be effective, and the water is removed in an active rather than passive operation.

Heat exchangers, while viewed by some as unnecessary when large reservoirs are used, can be utilized very effectively. In some cases, the bulk or weight of an oversized reservoir is undesirable. The thermal demands of the system may require a heat exchanger in some cases.

Air-oil separators, or *deaerators*, offer some advantages over reservoirs in the removal of air from the fluid. One such device—the Separate-Aire manufactured by Seaton-Wilson (Figure 13.3)—utilizes a vacuum arrangement to remove a significant amount of dissolved air from operating fluid systems. While this may not be necessary in normal industrial systems, it can be very important in applications such as aircraft flight control systems, where sluggish or spongy operation could mean disaster.

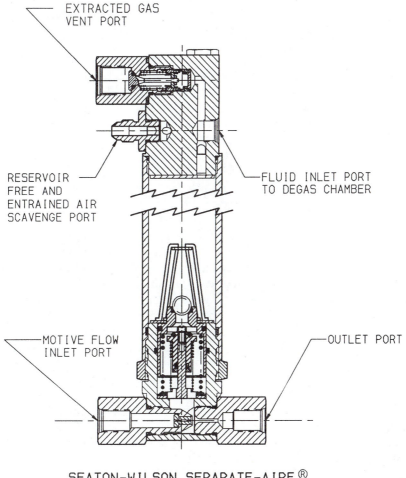

SEATON-WILSON SEPARATE-AIRE®
FLUID DEGASSER
U.S. PATENT NO. 3969093

Figure 13.3 The Seaton-Wilson deaerator utilizes a vacuum arrangement to remove dissolved air.

Source: Seaton-Wilson, a BEI Electronics Company (Seaton-Wilson Air-Oil Separator, Patent No. 3,273,313).

Reservoir Construction

Figure 13.1 depicts a standard industrial reservoir along the lines recommended by the Joint Industrial Conference (JIC). Some of the features we have already mentioned can be seen in this drawing; for instance, the baffle plate. In this design, a single plate runs the length of the tank. There is a drain plug in the bottom that can be used to drain off settled water as well as to empty the entire reservoir. Both ends contain plates that can be removed to facilitate cleaning the entire tank. Notice that the tank is standing well clear of the floor to allow air circulation to enhance heat transfer.

Breathers

A *breather* reservoir is intended for industrial applications and operates at atmospheric pressure. It gets its name from the fact that it inhales and exhales as the fluid level changes due to system operation. As we discussed in Chapter 11, atmospheric dirt is one of the major sources of fluid contamination. Therefore, it is imperative that good filtration be provided to clean the incoming air.

In addition to the removal of airborne dirt, some breathers are designed to remove the moisture from incoming air. These *desiccant* breathers (Figure 13.4) contain both a particulate filter and a water-absorbent agent that extracts water vapor from air as it is drawn into the unit.

An alternative method of preventing the ingress of dirt and moisture is to prevent the inhaling of atmospheric air. This can be done with a so-called *pressurized* breather, which contains a vacuum breaker to allow an initial influx of air when the system is first started up and fluid is pulled from the reservoir. In subsequent operation, as fluid returns to the tank, a relief valve prevents the air from being expelled from the tank. Rather, it is compressed and will be exhausted only if the relief valve setting (which can be from 3 to as much as 25 psig) is exceeded. Caution must be used when adding a pressurized breather to an existing reservoir to ensure that the tank structure can withstand the internal pressure. Otherwise, a good idea can go very sour.

In the section on accumulators, we will see yet another way to prevent the ingestion of dirt by the reservoir.

The ideal situation, of course, would be to use a completely sealed reservoir that has no communication with the environment. Unfortunately, this is seldom done because of the expense and weight involved in providing sufficient strength to withstand the internal pressures that might result.

Strainers

A *strainer* is often attached to the pump inlet line. As discussed previously, these are necessarily very coarse elements, seldom providing better than 150 micron filtration. Their main purpose is to prevent trash in the reservoir from finding its way into the pump.

If such a strainer is used, two very important points must be remembered. First, the strainer should always be covered by at least 3 inches of fluid. Otherwise, a vortex could form that would allow air to be pulled into the pump. Second, these strainers must be changed periodically. If they are allowed to clog up, they can severely restrict the fluid flow to the pump, resulting in pump cavitation. Unfortunately, strainers in the tank are unseen, so they are forgotten more often than not. The reservoir design must accommodate the maintenance task of changing the strainer. It simply isn't practical to build a unit that requires extensive disassembly or the removal of large quantities of oil to get to the strainer. The strainers in such units don't get changed very often.

Pump Inlet and Return Lines

The locations of the *pump inlet line* and the system *return line* (and any other lines bringing fluid into the tank) can be critical in the ability of the tank to

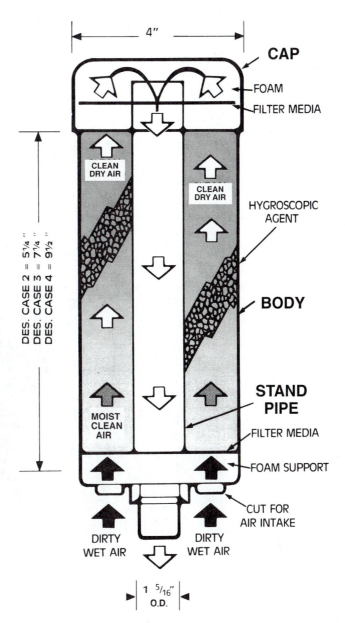

Figure 13.4 A dessicant breather removes moisture as well as dirt from incoming air.
Source: DES-CASE Corporation, P.O. Box 729, Goodlettsville, TN 37070.

perform all the jobs that are thrust upon it. We've already seen that the fluid dwell time is very important. Therefore, we should arrange the inlet and return lines to provide the longest and most tortuous path possible. They should be at opposite ends of the tank or on the same end but on the opposite sides of a longitudinal baffle. The pump inlet line must be near, but never on, the bottom

Figure 13.4 Continued.

of the tank. If it were on the bottom, dirt that has settled out of the fluid might be picked up when the pump starts. Return lines must always be below the surface of the fluid, preferably extending to near the bottom of the tank. (An exception to this is some drain lines that must terminate above the surface to prevent unacceptable back-pressures.) The addition of a *diffuser* on the end of the return line will promote the release of air and generate mixing of the returning fluid with that in the tank to aid in heat transfer.

Covers

Very often (but not always), the top of the reservoir will be a heavy cover that is bolted to the side walls. It is very common for the pump and drive motor, some valves, the filters, and so on, to be mounted on this cover. In other cases, it is simply the lid. Whatever the case, it is important that a good gasket be provided for the cover to prevent the entrance of atmospheric air. It is also important that the cover be kept on the reservoir and bolted down securely to maintain the integrity of that gasket. Otherwise, the door is opened for major contamination problems.

A well designed and maintained reservoir can eliminate many potential problems. A poorly designed or maintained reservoir can spell disaster in an otherwise near perfectly designed system. It is so simple that its importance is often overlooked.

Be aware that the example reservoir shown in Fig. 13.1 is only one of a multitude of possible designs. Figure 13.5 shows other designs commonly found in industrial applications. There probably is no such thing as a common reservoir design for mobile applications. In aircraft, tractors, and other mobile equipment, reservoirs are usually put wherever they can be fit in conveniently and the space available often dictates the shape.

13.3 Accumulators

An *accumulator* is a fluid energy storage device, similar to a capacitor in an electrical circuit. In discussing accumulators, the most common application is

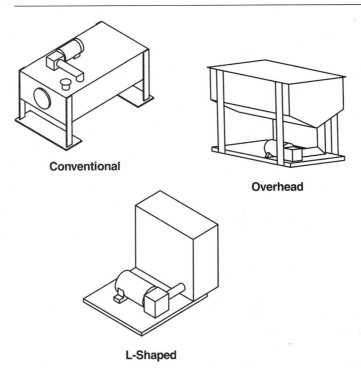

Conventional

Overhead

L-Shaped

Figure 13.5 Many industrial reservoir configurations are possible.

deemed to be pulsation damping. While this is an important application, we will see as we work through this section that it is only one of many uses for these versatile devices. Surprisingly, it is also the thing that they do least well. They are better suited for several other jobs in fluid power systems. *Caution! A charged accumulator means that the system is pressurized.* All accumulator circuits should be designed so that the fluid side of the accumulator can be discharged to prevent injury to personnel or damage to equipment prior to beginning any maintenance actions.

Accumulator Types

There are several different styles of accumulators. *Weight-loaded* accumulators (Figure 13.6) are basically hydraulic cylinders with a constant load. They have the advantage of supplying their entire volume of fluid at a constant pressure—something no other type of accumulator can do. In spite of this unique advantage, weight-loaded accumulators are not common in fluid power applications because of their physical size and the weights required to provide significant pressure. They are usually designed and fabricated for a specific application and custom built to meet the requirements of that application. They are used for stationary applications where size and weight are not at a premium.

On the other end of the scale (size-wise) are the *spring-loaded* accumulators (Figure 13.7). These devices depend on a spring to exert force on a piston in order to provide pressure. As the accumulator is charged with fluid from the pressure line, the spring is compressed. The amount of compression depends on the system pressure. As with any compression-type spring, the more it is compressed, the more force (hence, pressure) is required to compress it more.

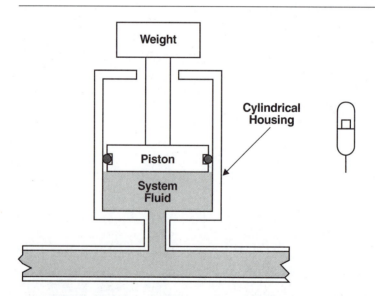

Figure 13.6 Weight-loaded accumulators are basically hydraulic cylinders with a constant load.

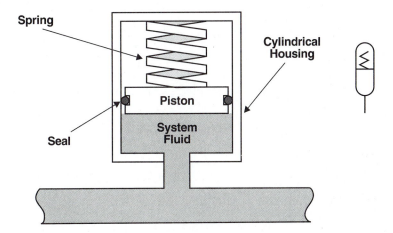

Figure 13.7 Spring-loaded accumulators depend on a spring to exert force on a piston in order to provide pressure.

When there is a demand for the fluid stored in the accumulator (which means that the system pressure is reduced to something less than the pressure in the accumulator), the spring force on the piston will push fluid from the accumulator and into the system. As the spring extends, its force decreases. As a result, the pressure of the fluid remaining in the accumulator also decreases. If we remove so much fluid from the accumulator that the pressure (resulting from the spring force) falls below the required system pressure, the remaining fluid is useless for system operation.

Example 13.5

Consider the spring-loaded accumulator shown in Figure 13.7 that has a piston diameter of 2 in (5 cm). The spring in the accumulator has a spring constant of 500 lb/in (875 N/cm). The maximum compression of the spring is 6 in (15.2 cm). Find:

    a. The volume of fluid in the accumulator if the spring is compressed 3 in (7.6 cm).

    b. The accumulator pressure when the spring is fully compressed.

    c. The usable volume of fluid if the minimum system pressure requirement is 700 psi (4.8 MPa).

Solution

    a. The amount of fluid in the accumulator is found by multiplying the piston area by the length occupied by the fluid.

Thus

$$\text{Vol} = A \times L = 3.14 \text{ in}^2 \times 3 \text{ in} = 9.42 \text{ in}^3$$

or 155 cm^3.

b. The maximum pressure is found from

$$p = \frac{F}{A}$$

In this case, the force is provided by the spring and can be calculated from Equation 8.1.

$$F = k \times d$$

Thus,

$$F = k \times d = 500 \text{ lb/in} \times 6 \text{ in} = 3000 \text{ lb}$$

or 13.4 kN. From this, we find that the maximum pressure in the accumulator is

$$p = \frac{F}{A} = \frac{3000 \text{ lb}}{3.14 \text{ in}^2} = 955 \text{ psi}$$

or 6.6 MPa.

c. The maximum volume of oil in the accumulator occurs when the spring is fully compressed. As calculated in part "b," this occurs when the pressure is 955 psi (6.6 MPa). When the pressure drops to 700 psi (4.8 MPa) (the lower limit of useful pressure), the spring pushes the piston to a lower position. We find this new position by finding first the force due to that pressure

$$F = p \times A = (700 \text{ lb/in}^2)(3.14 \text{ in}^2) = 2198 \text{ lb}$$

or 9.8 kN, and then the distance the spring is compressed as a result of that force,

$$d = \frac{F}{k} = \frac{2198 \text{ lb}}{500 \text{ lb/in}} = 4.4 \text{ in}$$

or 11.2 cm.

The usable volume of oil is the volume between the fully compressed position (6 in) and this lower position, which is determined by the minimum allowable pressure. Thus,

$$V_{\text{USABLE}} = V_{\text{MAX}} - V_{\text{MIN}} = A(d_{\text{MAX}} - d_{\text{MIN}})$$
$$= 3.14 \text{ in}^2 (6 \text{ in} - 4.4 \text{ in}) = 5 \text{ in}^3$$

or 82 cm³.

A third type of accumulator is the hydropneumatic accumulator. These devices use a compressed gas to provide the force that causes the fluid pressure. Nitrogen is normally used to charge accumulators for two reasons. First, it is a "dry" gas that contains no moisture. This is important because any rust or corrosion inside the accumulator shell could cause severe weakening and eventually structural failure.

The second reason for using nitrogen is that, being an inert gas, it will not support combustion. If oil (or oil vapors) find their way into the gas side of the accumulator, a situation resembling a diesel cylinder could result. If the accumulator were charged with air, there could occur the proper combination of air, fuel (the oil), and pressure to cause a diesel-type ignition. The results could be disastrous. In either of these cases, the failing accumulator could explode like a bomb.

Like the weight and spring-loaded accumulators, the hydropneumatic *piston accumulator* (Figure 13.8) resembles a hydraulic cylinder except that there is no rod. There is, however, a cylinder barrel and a piston that separates the hydraulic fluid from the gas. As the pump forces fluid into the accumulator, the piston is pushed upward, compressing the gas that is trapped because of a closed valve. When a demand on the system causes the fluid pressure to drop, the gas expands and pushes fluid out of the accumulator. As with the spring-loaded accumulator, there is a limited volume of fluid that can be used because of the minimum pressure requirements of the system. This volume is calculated in a similar manner as before except that gas compression rather than spring compression must be considered.

The compression of a gas results in heating the gas. During the compression, one of three things will happen to the heat:

1. All of the heat will be removed from the gas so that its temperature remains constant. This is termed an *isothermal* process.
2. None of the heat will be removed from the gas so that its temperature rises to some maximum possible level. This is termed an *adiabatic* process.
3. Part of the heat is removed from the gas so that the gas temperature rises but does not reach the maximum as in an adiabatic process. This is termed a *mixed* thermal process. In the normal operation of an accumulator, this mixed process usually occurs.

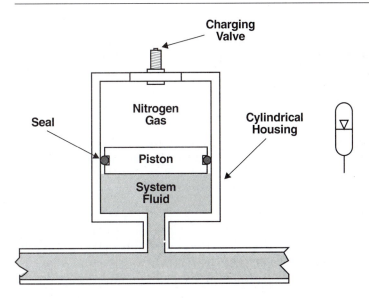

Figure 13.8 Piston accumulators resemble hydraulic cylinders except that there is no rod.

The exact thermal process in this mixed area is difficult to predict because it is very dependent on the speed of the compression or decompression, environmental conditions, and so forth. As a result, little effort is made to determine the exact response of the gas charge as the accumulator operates. Instead, it is normal to look at both the isothermal and adiabatic reactions with the understanding that the real situation will lie somewhere between the two extremes.

The amount of usable oil in the accumulator, then, is a function of the pressure of the gas charge. This pressure depends on both the amount of compression and the thermal process involved. The "gas laws" that determine the reaction of the gas to the different processes will be discussed in detail in Chapter 14.

When a gas is compressed at constant temperature (isothermally), the initial and final pressures and volumes are related by Equation 13.3. This equation is referred to as *Boyles' Law*.

$$p_1 \times V_1 = p_2 \times V_2 \qquad \qquad \textbf{(13.3)}$$

For the adiabatic case, where the final pressure is the result of both the compression and heating, this relationship is given by

$$p_1 \times V_1^{1.4} = p_2 \times V_2^{1.4} \qquad \qquad \textbf{(13.4)}$$

In both cases, the pressures must be in absolute values.

The nitrogen is introduced into the accumulator through a charging valve located at the top of the gas chamber. This is done while the oil chamber is completely empty (or at least unpressurized so that the piston can force the oil out). The gas pressure is increased to a predetermined level termed the *gas precharge*. This precharge pressure is usually one-third to one-half the normal operating pressure of the system. Oil is then introduced into the oil chamber at a pressure dictated by the system load. The oil continues to flow into the accumulator until the fluid system pressure reaches the relief valve setting. During this process, the gas pressure is being increased by the compression process.

The amount of usable oil in the accumulator is determined by finding the volume required to raise the pressure in the gas chamber from the minimum usable fluid pressure to the fluid system relief valve pressure.

Example 13.6

Consider a piston-type accumulator as shown here. The gas precharge (initial pressure with no fluid in the accumulator) is 200 psi (1380 kPa). Determine the usable volume of oil the accumulator can provide if the oil pressure must be between 500 and 700 psi (3450 and 4826 kPa). Consider both isothermal and adiabatic processes.

Solution

The 700 psi (4826 kPa) pressure represents the upper limit of both the gas and oil pressures. Therefore, the first step is to determine the amount of oil required in the accumulator to cause that pressure to exist.

In the isothermal situation, we use Equation 13.3:

$$p_1 \times V_1 = p_2 \times V_2$$

Right now, we're looking at gas volumes, but we will relate these to oil volumes in just a minute.)

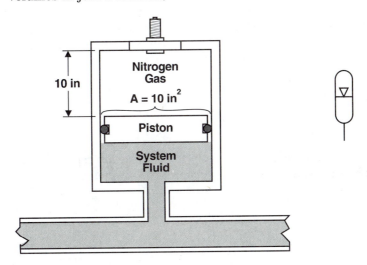

Example 13.6

The pressures in the problem statement are gage pressures, so we convert these to absolute pressures:

$$p_1 = 200 + 14.7 = 214.7 \text{ psia}$$

or 1480 kPa.

$$p_2 = 700 + 14.7 = 714.7 \text{ psia}$$

or 4928 kPa.

The initial gas volume is

$$V_1 = 10 \text{ in}^2 \times 10 \text{ in} = 100 \text{ in}^3$$

or 1640 cm.

Therefore, Equation 13.3 gives us the final gas volume as

$$V_2 = \frac{p_1 \times V_1}{p_1} = \frac{214.7 \text{ psia} \times 100 \text{ in}^3}{714.7 \text{ psia}}$$
$$= 30 \text{ in}^3$$

or 492 cm^3.

This means that the volume of the gas charge has been reduced by 70 in^3 (1148 cm^3). This also represents the amount by which the volume of oil in the accumulator has been *increased*.

To find the usable volume, we now determine how much oil would give us a gas pressure (and, consequently, an oil pressure) of 500 psi (gage) (3450 kPa). Again, we resort to Equation 13.3. This time, p_2 is 514.7 psia (3550 kPa abs); therefore,

$$V_2 = \frac{p_1 \times V_1}{p_2} = \frac{214.7 \text{ lb/in}^2 \times 100 \text{ in}^3}{514.7 \text{ lb/in}^2} = 41.7 \text{ in}^3$$

or 684 cm^3.

Remember that this is our "final" gas volume at 500 psig (3450 kPa). The oil volume then is $100 - 41.7 = 58.3$ in³ (956 cm³).

So, at 700 psi (4826 kPa) we have 70 in³ (1148 cm³) of oil in the accumulator. At 500 psi (3450 kPa), this had reduced to 58.3 in³ (956 cm³). This volume is useless to us because the next drop we take out is below 500 psi (3450 kPa). Therefore, the useful volume of oil in the accumulator is the difference between these two volumes, or

$$V_{\text{USABLE}} = V_{700} - V_{500} = (70 - 58.3) \text{ in}^3 = 11.7 \text{ in}^3$$

or 192 cm³.

In the adiabatic case, we use the same procedure, but utilize Equation 13.4. Thus, at maximum pressure, the gas volume is

$$p_1 \times V_1^{1.4} = p_2 \times V_2^{1.4}$$

or

$$V_2^{1.4} = \frac{p_1 V_1^{1.4}}{p_2} = \frac{214.7 \text{ lb/in}^2 \times 100^{1.4} \text{ in}^3}{714.7 \text{ lb/in}^3}$$

$$V_2 = 42.4 \text{ in}^3$$

or 695 cm³. This means that our oil volume is $100 - 42.4 = 57.6$ in¹ or 945 cm¹. At 500 psi, we get

$$V_2^{1.4} = \frac{214.7 \text{ lb/in}^2 \times 100^{1.4} \text{ in}^3}{514.7 \text{ lb/in}^2}$$

$$= 53.6 \text{ in}^3$$

or 879 cm³, so that the oil volume is $100 - 53.6 = 46.4$ in³ (761 cm³). In this case the usable oil volume is $57.6 - 46.4 = 11.2$ in³ (184 cm³).

In a real world application, the actual usable volume would normally be somewhere between these two values.

By far the most common type of accumulator is the *bladder* type, which uses a rubber or elastomeric bladder to contain the gas charge (Figure 13.9). In preparing the accumulator for use, the system is shut down. The accumulator bladder is then precharged with nitrogen. The precharge pressure will be according to the manufacturer's recommendation for the particular application. As the system pressure builds, the accumulator fills with oil and the gas is compressed. The oil pressure and the gas pressure are always equal. When there is a system demand, the system pressure drops, and the gas pressure pushes oil from the accumulator. A variation on the bladder is the *diaphragm* type shown in Figure 13.10.

The accumulator shown in Figure 13.11 is termed a *gas-over-oil* accumulator. No provision is made to separate the gas from the oil in this type. They are normally used only in special applications, usually where large volumes of oil are needed. A major problem with this type of unit is that there is a direct interface between the high-pressure gas and the oil. As a result, large quantities of gas can be dissolved in the oil (Henry-Dalton Law). This gas can come out

Figure 13.9 Bladder-type accumulators use a rubber or elastomeric bladder to contain the gas charge.
Source: Courtesy of Greer Hydraulics Service.

of solution in low-pressure areas (the reservoir, for instance), leading to the many gas-related problems that are damaging to the system.

Accumulators are commonly specified by type and volume. The volume may be specified in cubic inches, gallons, or liters. Because they are pressure vessels, their design, manufacture, and maintenance are governed by the pressure vessel codes of the American Society of Mechanical Engineers (ASME). The codes require regular inspection of specific accumulator types by certified inspectors.

Table 13.2 is an example of the data tables describing accumulator performance. This table compares the pressurized oil volume with the gas precharge pressure for a one-gallon, bladder-type accumulator. It shows the oil volume for both isothermal and adiabatic processes, and allows us to estimate both the total volume of oil in the accumulator and the usable volume it contains.

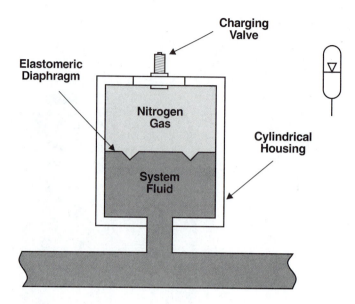

Figure 13.10 Diaphragm-type accumulators are a variation on the bladder type accumulator.

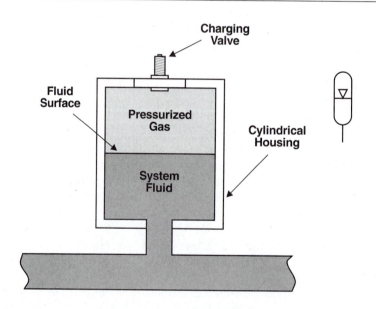

Figure 13.11 Nonseparated gas charged (gas-over-oil) accumulators are used where large volumes of oil are needed.

Table 13.2 Adiabatic/*Isothermal* Accumulator Performance Chart (231 in³ Accumulator)

Legend — Upper Row: Adiabatic; Lower Row: *Isothermal*. Cell values are Liquid Volume (in.³) Stored in Accumulator.

Gas Precharge—PSI (gage)	Type	Operating Pressure—PSI (gage) 100	200	300	400	500	600	700	800	900	1000	1100	1200	1300	1400	1500	1600	1700	1800	1900	2000	2100
100	Adiab.		86.6	113	144	158	168	175	182	186	190	192	196	198	200	202	204	206	207	209	210	211
	Isoth.		*112*	*154*	*174*	*187*	*196*	*202*	*207*	*211*	*214*	*216*	*218*	*220*	*222*	*223*	*224*	*225*	*226*	*227*	*227*	*228*
200	Adiab.			57.4	90.0	112	126	138	147	155	161	166	170	174	178	181	184	186	188	190	192	194
	Isoth.			*76.6*	*116*	*141*	*157*	*168*	*178*	*184*	*190*	*195*	*198*	*202*	*204*	*207*	*209*	*211*	*213*	*214*	*215*	*216*
300	Adiab.				43.4	71.4	91.1	105	118	127	136	143	148	153	157	162	165	169	172	174	177	179
	Isoth.				*58.5*	*94.0*	*118*	*134*	*148*	*158*	*166*	*173*	*176*	*184*	*188*	*191*	*194*	*197*	*199*	*202*	*203*	*205*
400	Adiab.					34.2	58.8	77.3	92.0	103	114	121	128	135	141	145	149	153	157	160	163	165
	Isoth.					*46.7*	*78.5*	*101*	*118*	*132*	*143*	*151*	*159*	*165*	*171*	*175*	*179*	*183*	*186*	*189*	*191*	*194*
500	Adiab.						28.5	50.2	67.0	80.5	91.8	102	110	117	123	128	134	138	142	146	149	152
	Isoth.						*39.3*	*67.5*	*88.6*	*105*	*119*	*130*	*139*	*146*	*153*	*159*	*164*	*169*	*173*	*176*	*179*	*182*
600	Adiab.							24.6	43.6	58.8	72.1	83.2	92.4	101	108	114	120	126	130	132	136	140
	Isoth.							*33.8*	*59.0*	*78.8*	*95.0*	*108*	*119*	*128*	*136*	*143*	*149*	*154*	*159*	*164*	*168*	*171*
700	Adiab.								21.7	38.6	53.0	65.1	75.5	84.6	92.6	99.5	106	112	117	121	125	129
	Isoth.								*29.9*	*52.5*	*71.1*	*86.3*	*99.4*	*110*	*119*	*127*	*134*	*141*	*146*	*151*	*155*	*160*
800	Adiab.									19.1	35.0	48.0	59.3	69.4	78.1	85.8	92.5	99.8	105	110	114	119
	Isoth.									*26.2*	*47.7*	*64.5*	*79.4*	*91.9*	*102*	*111*	*119*	*127*	*133*	*139*	*144*	*148*
900	Adiab.										17.4	31.6	43.6	54.7	63.9	72.5	80.0	86.8	92.8	98.5	104	108
	Isoth.										*24.1*	*43.2*	*59.4*	*73.3*	*84.9*	*95.5*	*104*	*112*	*120*	*126*	*132*	*137*
1000	Adiab.											15.7	28.7	40.5	50.9	59.5	67.8	75.0	81.5	87.5	93.0	98.0
	Isoth.											*21.5*	*39.5*	*55.0*	*68.2*	*79.6*	*89.7*	*98.4*	*106*	*113*	*120*	*125*
1100	Adiab.												14.2	26.8	37.4	47.2	55.9	63.4	70.4	76.9	82.6	88.0
	Isoth.												*19.8*	*36.6*	*50.8*	*63.9*	*74.7*	*84.4*	*93.1*	*101*	*108*	*114*
1200	Adiab.													13.3	24.8	35.0	44.4	52.1	59.8	66.5	72.8	78.5
	Isoth.													*18.6*	*34.2*	*47.7*	*60.0*	*70.2*	*79.8*	*88.2*	*95.7*	*103*
1300	Adiab.														12.3	23.1	32.5	41.0	49.6	56.4	63.1	69.1
	Isoth.														*17.1*	*31.8*	*44.6*	*55.9*	*66.3*	*75.5*	*83.9*	*91.1*
1400	Adiab.															11.6	21.7	30.8	39.0	46.3	53.5	59.8
	Isoth.															*15.9*	*29.9*	*42.2*	*53.0*	*62.7*	*71.9*	*80.0*
1500	Adiab.																10.6	20.2	28.9	36.9	44.4	51.9
	Isoth.																*15.0*	*28.0*	*39.8*	*50.1*	*59.8*	*68.5*

Upper Row: Adiabatic
Lower Row: *Isothermal*

Liquid Volume (in.³) Stored in Accumulator

Source: Courtesy of Parker Hannifin Corporation.

Example 13.7

An accumulator circuit is designed to provide a maximum accumulator pressure of 2100 psi. The normal operating pressure of the system is 1600 psi. Use Table 13.2 to determine the usable volume of oil in the accumulator between those two pressures if the gas precharge is 500 psi. Repeat the exercise for 1000 and 1500 psi precharges. Consider both isothermal and adiabatic processes.

Solution

From Table 13.2, we see that at a precharge of 500 psi (3450 kPa) and an operating (system) pressure of 2100 psi (14.5 MPa), the accumulator contains 152 in^3 (2493 cm^3) of oil for the adiabatic case. For an isothermal process, it contains 182 in^3 (2985 cm^3) of oil. At 1600 psi (11 MPa), it contains 134 and 164 in^3 (2198 and 2690 cm^3) for the two cases. The usable volumes, then, are

$$152 - 134 = 18 \text{ in}^3$$

or 295 cm^3 (adiabatic), and

$$182 - 164 = 18 \text{ in}^3$$

or 295 cm^3 (isothermal).

If we increase the precharge to 1000 psi (6.9 MPa), the maximum volumes are 98 and 125 in^3 (1607 and 2050 cm^3) at 2100 psi (14.5 MPa), and 67.8 and 89.7 in^3 (1112 and 1471 cm^3) at 1600 psi (11 MPa). Usable volumes are now

$$98 - 67.8 = 30.2 \text{ in}^3$$

or 495 cm^3 (adiabatic), and

$$125 - 89.7 = 35.3 \text{ in}^3$$

or 579 cm^3 (isothermal). or

Note that, although the total volumes of oil were lower at each pressure, the usable volumes almost doubled.

Looking at the 1500 psi (10.3 MPa) precharge, our volumes are only 51.9 and 68.5 in^3 (851 and 1123 cm^3) at 2100 psi (14.5 MPa), and 10.6 and 15 in^3 (174 and 246 cm^3) at 1600 psi (1.1 MPa). The usable volumes are now up to

$$51.9 - 10.6 = 41.3 \text{ in}^3$$

or 677 cm^3 (adiabatic), and

$$68.5 - 15 = 53.5 \text{ in}^3$$

or 877 cm^3 (isothermal).

The point of this exercise is to demonstrate that, as the gas precharge pressure goes up, the *total* volume of oil that the accumulator can hold goes down; however, the *usable* volume (which is usually the important aspect) will increase.

Typical Accumulator Applications

Two particularly useful accumulator applications are in supplementing pump flow rate and in maintaining system pressure. These uses are easily illustrated by the following examples.

Example 13.8

In a particular manufacturing process, three hydraulic cylinders are required to extend in sequence, then retract simultaneously in a 12 second cycle as shown here. Determine the horsepower savings that result from using an accumulator in the circuit.

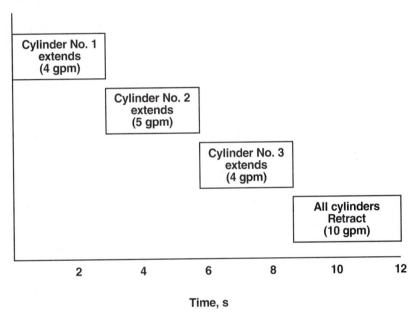

Time, s

Example 13.8

Solution

If a fixed displacement pump is used for this cycle, its capacity will be determined by the retraction action, which requires 10 gpm (37.85 Lpm). This action requires only 3 seconds, or 25 percent of the total cycle. For the remaining 9 seconds, the excess pump capacity will be dumped across the relief valve. As we've already seen, this is costly in wasted energy and generates considerable heat.

If the system pressure is 1500 psi (10.3 MPa), the horsepower requirement for this operation (ignoring losses) would be

$$HHP = \frac{p \times Q}{1714} = \frac{1500 \times 10}{1714} = 8.75\,HP$$

or 6.53 kW.

If we install an accumualtor to provide a portion of the flow for the retraction of the cylinder, a smaller pump could be used. To calculate the size pump required, let us designate the pump flow as X gpm. During the extension of the first cylinder, 4 gpm (15.1 Lpm) is being used to extend that cylinder. The remainder is being used to charge the accumulator. Therefore, the volume flow rate going into the accumulator is $(X - 4)$ gpm, or $(X - 15.1)$ Lpm. Similarly, for the second and third cylinders, the accumulator is charged at the rates of $(X - 5)$ and $(X - 4)$ gpm, or $(X - 18.9)$ and $(X - 15.1)$ Lpm, respectively. During the retraction cycle, the accumulator is discharged. The discharge rate plus the pump flow rate must provide the 10 gpm (37.85 Lpm) required for retraction, that is, $(10 - X)$ gpm, or $(37.85 - X)$ Lpm.

If we consider that we will use the total volume put into the accumulator during the extension cycle for the retraction cycle, then we can write a sum of the flow rates as

$$(X - 4) + (X - 5) + (X - 4) = 10 - X$$

Solving for X gives

$$4X = 23$$

$$X = 5.75 \text{ gpm}$$

or 21.76 Lpm. Thus, a 5.75 gpm (21.76 Lpm) pump and an accumulator will do the same job that required a 10 gpm (37.85 Lpm) pump working alone. The pump horsepower requirement would consequently be reduced to

$$HHP = \frac{p \times Q}{1714} = \frac{1500 \times 5.75}{1714} = 5.03 \text{ HP}$$

or 3.75 kW. This represents a 42.5 percent saving in horsepower input (ignoring losses). Such savings can be significant over the long run.

The volume of oil required from the accumulator can be determined from the parameters of the retraction cycle. Without the accumulator, the 10 gpm (37.85 Lpm) pump could retract all three cylinders in 3 seconds. Thus, the total volume of oil required is

$$V = \left(10 \frac{\text{gal}}{\text{min}}\right)\left(\frac{\text{min}}{60 \text{ sec}}\right)\left(\frac{231 \text{ in}^3}{\text{gal}}\right)(3 \text{ sec}) = 115.5 \text{ in}^3$$

or 1894 cm³. Our 5.75 gpm (21.76 Lpm) pump, in 3 seconds, will produce

$$V_{PUMP} = \left(5.75 \frac{\text{gal}}{\text{min}}\right)\left(\frac{\text{min}}{60 \text{ sec}}\right)\left(\frac{231 \text{ in}^3}{\text{gal}}\right)(3 \text{ sec})$$

$$= 66.4 \text{ in}^3$$

or 1089 cm³. The accumulator would be required to supply only the remaining 115.5 − 66.4 = 49.1 in³ (805.2 cm³) for the retraction cycle. The 1 gallon accumulator of the previous example would satisfy this requirement.

Example 13.9

The hydraulic cylinder shown here is required to extend in 3 seconds to push a workpiece into the workplace. This requires a 20 gpm (76 Lpm) pump. The cylinder then clamps the piece while a 55 second machining process is completed. The cylinder then retracts in 2 seconds. The pump pressure throughout the stationary portion of the cycle is 1000 psi (6895 kPa). Compare the horsepower needed to operate this system when using only the system pump and when using an accumulator and a small pump.

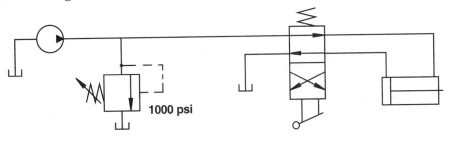

1000 psi

a. Cylinder circuit without accumulator.

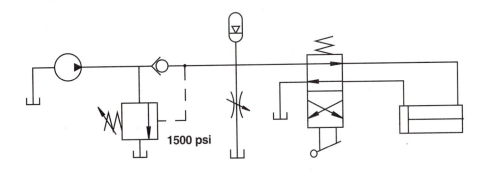

1500 psi

b. Cylinder circuit with accumulator and unloading valve.

Example 13.9

Solution

The hydraulic horsepower required for this operation is

$$HHP = \frac{p \times Q}{1714} = \frac{1000 \times 20}{1714} = 11.7 \text{ HP}$$

or 8.7 kW. If there is no leakage in the system, then the entire pump horsepower is being dumped across the relief valve for 55 seconds of this 1 minute cycle, causing a heat generation of 455 Btu (8 kW) during each cycle.

If we add an accumulator to the system along with an unloading valve that opens at 1500 psi (10.3 MPa) (as shown in the illustration), then considerable savings can be realized. If the system is designed properly, the accumulator alone can be used to operate the cylinder, while the pump's only purpose is to charge the accumulator. In this case, using the same 20 gpm (76 Lpm) pump, we would have the pump operating at 1500 psi (10.3 MPa) for a maximum of 5 seconds to recharge the accumulator. (Actually, it would operate at lower pressures for most of that time, reaching 1500 psi (10.3 MPa) only at the instant that the accumulator became fully charged.) For the remaining 55 seconds of the cycle, the unloading valve would be open and the pump flow would be dumping back to tank at low pressure, perhaps 30 psi (207 kPa). Calculating the horsepower requirements for this system, we get, during recharging

$$HHP = \frac{p \times Q}{1714} = \frac{1500 \times 20}{1714} = 17.5 \text{ HP}$$

or 13.1 kW for 5 seconds, and

$$HHP = \frac{30 \times 20}{1714} = 0.35 \text{ HP}$$

0.26 kW for 55 seconds.

   Again, this will bring about considerable energy savings over time. Note, also, that because we have 55 seconds during which to recharge the accumulator, a much smaller pump and prime mover could be used to accomplish the task. Since the cylinder requires 20 gpm (76 Lpm) for a total of 5 seconds during the cycle, the accumulator discharges only 1.67 gal (6.3 L) of fluid. Since the recharge time is 55 seconds, a pump flow rate of only 0.03 gallons per second, or 1.82 gpm (6.89 Lpm) would be required to have the accumulator completely recharged and ready for the next cycle.

   If we use a 2 gpm (7.6 Lpm) pump to be on the safe side, we would have a pump horsepower requirement of 1.75 HP (1.3 kW) for essentially the entire 1 minute cycle.

Accumulators can also be used to provide emergency power to ensure safe shutdown in case of power failure, clamp or lock devices (brakes, for instance) when equipment is shut down, or to provide a "one-shot" operation when power has been lost.

   Remember, when the accumulator is charged, the system may be pressurized. Use extreme caution when performing maintenance and making adjustments on systems on which accumulators are used. Be sure that system pressure is discharged.

   As mentioned earlier, accumulators are also used frequently to damp surges or pressure spikes in the system. When used in this application, the gas precharge must be relatively low—usually about one-fourth of the system operating pressure. This low precharge can cause a soft or spongy operation of the system that may be unacceptable in certain applications.

   A variation on the normal accumulator is a device called a *desurger* (Figure 13.12). It is designed to especially for surge damping. The desurger damps pressure pulsations by combining throttling orifices with the energy-absorbing action of an expandable sleeve surrounded by compressed gas. This arrangement dissipates the kinetic energy change that is created when the velocity of

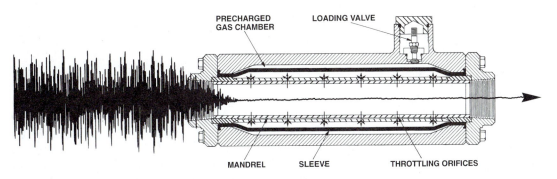

Figure 13.12 A desurger designed for surge damping.

Source: Nixco Manufacturing Company.

the flowing fluid suddenly changes, as normally occurs when a valve is suddenly opened or closed.

Structurally, a desurger consists of a *mandrel* (or inner pipe section) perforated with throttling orifices. Around the mandrel is a synthetic rubber sleeve. A case and two caseheads clamp the mandrel and seal the sleeve in place to form a gas chamber around the outside of the sleeve. An integral loading valve is used to fill the chamber with gas, usually nitrogen. It is installed as if it were a section of pipe so that all the liquid flowing through the system will

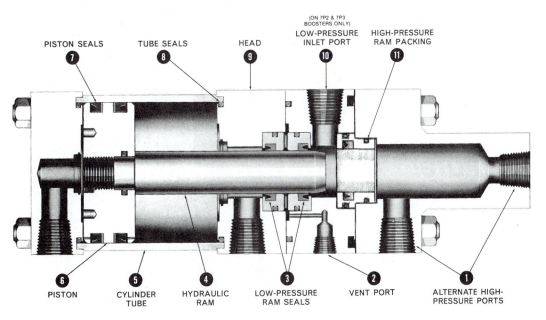

Figure 13.13 An intensifier makes use of the phenomenon of pressure intensification.

Source: Courtesy of Coltec Industries, Hammond, Ind.

flow through the desurger. The unit should be sized so that it offers negligible resistance to flow.

13.4 Intensifiers

In Chapter 6, we discussed the *pressure intensification* that can occur on the rod end of a hydraulic cylinder. This happens whenever the piston is being extended and there is a flow restriction from the rod end. The intensification is the result of the smaller active area on the rod side of the piston and is in proportion to the area ratio of the piston and rod sides of the piston.

An *intensifier* is a device that makes use of this pressure intensification phenomenon by using a large piston on the input end and a small piston within a small-bore cylinder on the output end (Figure 13.13). These devices can be used where small quantities of fluid are required at a pressure higher than the normal system operation.

Example 13.10

Consider the intensifier shown here. The maximum pressure available at the inlet is 500 psi (3.4 MPa). Find the maximum output pressure available.

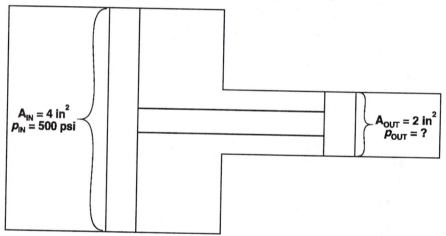

Example 13.10

Solution

The output pressure is found from the input pressure and the ratio of the two areas. Thus,

$$P_{OUT} = P_{IN} \times \frac{A_{IN}}{A_{OUT}}$$

$$= 500 \text{ lb/in}^2 \times \frac{4 \text{ in}^2}{2 \text{ in}^2} = 1000 \text{ psi}$$

or 6895 kPa. Remember that this is the *maximum* output capacity. The actual output would depend on the output load demand as it does with any other fluid power output device. The actual capability can be reduced by leakage or internal friction.

Example 13.11

In the system shown here, an intensifier is used to convert low-pressure shop air into the hydraulic pressure required to operate a clamping cylinder. The clamp must be moved 2 in (5 cm) to make it close properly. Determine the input pressure and intensifier stroke length required for proper clamp operation. A 3000 lb (13.4 kN) force is needed to close the clamping device.

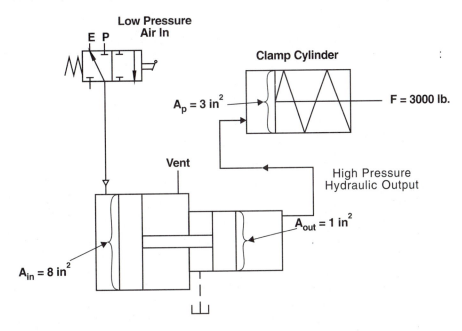

Example 13.11

Solution

The first step is to determine the pressure required to operate the clamping cylinder.

$$P_{CLAMP} = \frac{F}{A_p} = \frac{3000 \text{ lb}}{3 \text{ in}^3} = 1000 \text{ psi (6895 kPa)}$$

This is the same as the pressure output required from the intensifier.

$$P_{OUT} = P_{CLAMP} = 1000 \text{ psi (6895 kPa)}$$

From this we can find the required input pressure based on the ratio of the two ends of the intensifier.

$$P_{IN} = P_{OUT} \times \frac{A_{OUT}}{A_{IN}}$$

$$= 1000 \text{ lb/in}^2 \times \frac{1 \text{ in}^2}{8 \text{ in}^2}$$

$$= 125 \text{ psi}$$

or 862 kPa.

To find the stroke requirement, we can consider the output end of the intensifier to be just another hydraulic cylinder. This gives us a two-cylinder system similar to those used to demonstrate multiplication of force in Chapter 2. Thus, we can find the stroke by considering the volume of oil transferred from the intensifier to the clamping cylinder. We can use the equation,

$$\text{Vol}_{\text{INT}} = \text{Vol}_{\text{CYL}}$$
$$A_{\text{OUT}} \times \text{Stroke}_{\text{INT}} = A_{\text{CYL}} \times \text{Stroke}_{\text{CYL}}$$
$$\text{Stroke}_{\text{INT}} = \frac{A_{\text{CYL}} \times \text{Stroke}_{\text{CYL}}}{A_{\text{OUT}}}$$
$$= \frac{3 \text{ in}^2 \times 2 \text{ in}}{1 \text{ in}^2} = 6 \text{ in}$$

or 15.2 cm. Thus, the intensifier must stroke 6 in (15.2 cm) to move the clamping cylinder the required 2 in (5 cm).

13.5 Seals

There are two broad categories of *seals*—static and dynamic. *Static* seals are those that are used between non-moving parts. They are often referred to as *gaskets*, although gaskets (which are usually flat and made of either fibrous or metallic materials) are only one type of static seal. Other types of materials (elastomeric, rubber, plastic, Teflon, etc.) and other shapes (O-rings, squares, tees, etc.) may be used, depending on design requirements or simply preference.

Dynamic seals are used to prevent leakage between parts that move relative to one another. Among the more common applications are piston and rod seals in cylinders, shaft seals in pumps and motors, and valve seals. While there are often many seal cross sections (O-ring, tee, U-cup, chevron, etc.) that may satisfy a particular requirement, compatibility between seal material and the fluid, and the pressure capabilities of the seal must always be primary considerations in seal selection.

Figure 13.14 shows several different seal cross sections. Note that many terms that are used in describing seals come from these cross-sectional shapes.

Regardless of the shape of a seal, proper functioning requires that it deform in response to pressure in order to close off leakage paths. Consider, for example, a typical O-ring seal installation on the piston of a hydraulic cylinder (Figure 13.15). The cross section of the uninstalled seal is, for practical purposes, perfectly round (a). When the O-ring is installed, it is deformed, or preloaded, to establish the sealing line or band (b). This deformation is commonly termed *squeeze*. The amount of squeeze varies from 10 percent to a maximum of 35 percent of the O-ring's cross-sectional diameter. Installing the piston into the cylinder barrel may lead to further deformation of the seal.

As pressure is applied to the cylinder, the seal will be forced to the unpressurized end of the seal groove, called the *gland*, where it begins to deform due to the fluid pressure (c). This deformation results in a high-pressure contact between the seal and the two moving surfaces to prevent the leakage of fluid between them.

A common problem in this type of sealing is the *extrusion* of the seal (d), in which the pressure causes some of the seal to be pushed into the space between

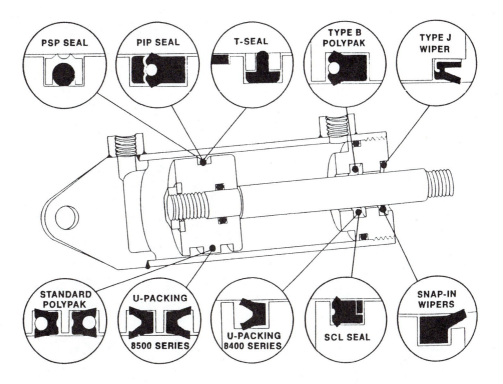

Figure 13.14 Dynamic seal cross-sections vary in configuration, but all seals deform in response to pressure in order to close off leakage paths.
Source: Courtesy of Parker Hannifan Corporation.

the piston and the barrel. Extrusion, in itself, is really not the problem, but it sets the stage for a process known as *nibbling* wherein the extruded bit of the seal is pinched off by small radial movements of the piston. This problem can be prevented by the use of some sort of anti-extrusion device such as a backup ring (e).

Compatability of the seals with the system fluid is a very important design consideration. If the seals are not compatible, seal *degradation* and subsequent failure are likely to occur. Degradation may take the form of excessive swell of the seal material. Some swell (about 5 percent) is desirable to ensure a tight seal, but excess swell can cause unacceptable seal friction or tearing of the seal as the parts move. Other forms of degradation including shrinkage, softening to a gummy texture, embrittlement, permanent set, crumbling, or dissolving. In addition to losing the sealing function, many of these seal failures cause large particles of the seal material to become contaminants in the system fluid. These particles are frequently large enough to plug orifices, plug the lubricating ports in the pistons of piston pumps, and jam valve spools and poppets.

The most commonly used seal materials for petroleum based fluids are *elastomers* such as fluorocarbon, nitrile, or polysulphide. Buna-N, a nitrile based product, is the most commonly used seal material for petroleum oils.

a. Uninstalled.

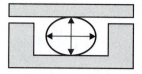
b. Installed (squeezed).

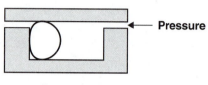
c. Pressurized.

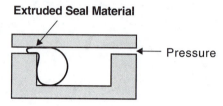
d. Extrusion due to pressurization.

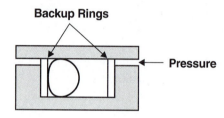
e. Backup rings prevent extrusion.

Figure 13.15 O-ring seal installation requires deformation called squeezing of 10 to 35 percent of its cross-sectional diameter. Backup rings are often used to prevent seal damage that results from extrusion.

Phosphate ester fluids require the use of butyl seals. Other fluids require still different elastomers. Possibly the greatest across-the-board compatibility is Viton, which performs acceptably with a wide variety of fluids. There are literally hundreds of elastomer compounds, however, so it is always best to check the compatability of a specific seal material with your specific fluid before installing the seal.

The seals we have been discussing are used primarily to keep fluid in. Another family of seals are used for keeping liquids, air, and dirt out of the system. These exclusion devices are usually termed *wiper* or *scraper* seals. As a cylinder rod extends, it collects on its extended surface a large amount of environmental contaminants. The wiper seal has the task of scraping these contaminants from the rod as it retracts so that they can't enter the system. As mentioned during the discussion of contamination control, some of these devices work very well. Others do a poor job, contributing significantly to system contamination problems.

An effective way to reduce the contamination problem associated with cylinder rods is to use *protective boots* (Figure 13.16). Boots are available for rods with 0.5 to 10 in (1.27 to 25.4 cm) diameters and virtually any length. Their convolutions allow them to "accordion" as the cylinder extends and retracts. Many different materials are available, so boots can be purchased that are compatible with virtually any chemicals and fluids, as well as a wide range of temperatures.

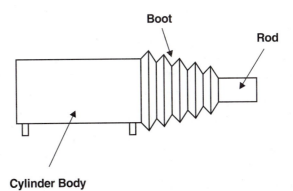

Figure 13.16 A protective boot helps reduce contamination problems associated with cylinder rods.

Boots should be inspected regularly for cracks and abrasion. If contaminants—especially chemicals and water—enter through openings in the boot and become trapped, serious rust or corrosion problems may result.

13.6   Instrumentation

The primary *instrumentation* found in fluid power systems is pressure gages, flow meters, and temperature gages.

Pressure Gages

Pressure gages can be either mechanical or electromechanical. Most gages seen today are mechanical, but the electromechanical devices are gaining in popularity because they provide digital readouts. Their electrical outputs can also be fed directly into computer control systems, servo feedback systems, or maintenance data recorders.

Mechanical gages commonly use one of the following devices to translate pressure into a gage reading:

- Bourdon tube
- Bellows
- Metallic diaphragm

Bourdon tube gages may use either a C-shaped tube or a spiral tube. Figure 13.17 shows a Bourdon gage with a C-shaped tube. The tube itself has a flattened cross section, with one end open to fluid pressure and the other end sealed. As the pressure in the tube increases, the tube attempts to straighten out. This mechanical movement produces movement of the pointer over the graduated scale on the gage face. Other gage designs may use either helical or spiral Bourdon tubes, but in both cases, the movement of the pointer results from the mechanical changes in the tube as the pressure changes.

For low-pressure measurements (below 12 psi, or 83 kPa), metallic bellows or metallic diaphragms are usually used. A *bellows-type* gage is shown in Figure 13.18. Fluid pressure causes the bellows to expand. Then the linkage attached to both the bellows and the pointer causes the pointer to respond to the bellows' movements.

Diaphragm gages operate in a similar manner. Instead of relying on the movement of a bellows, a thin sheet of metal is used as a diaphragm plate. This plate deforms elastically in response to fluid pressure, and the pointer mechanism responds to these small deformations.

Electromechanical pressure gages sometimes make use of a *linear variable differential transformer (LVDT)*. An LVDT (Figure 13.19) consists of a primary coil and two secondary coils and produces a voltage output proportional to the displacement of a magnet free to move within the coils. The inductance of the two secondary coils to the primary coil changes as the core is moved away from the center position. This movement results from the deflection of a standard mechanical-type gage mechanism (Bourdon tube, bellows, etc.) Instead of moving a pointer, the device pushes or pulls on the LVDT core. The electrical

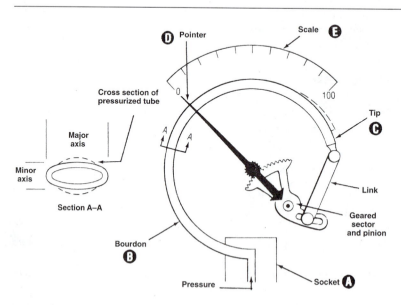

Figure 13.17 Bourdon tube gages may use either a C-shaped tube as shown here or a spiral tube.

Source: Courtesy of Hydraulics and Pneumatics Magazine.

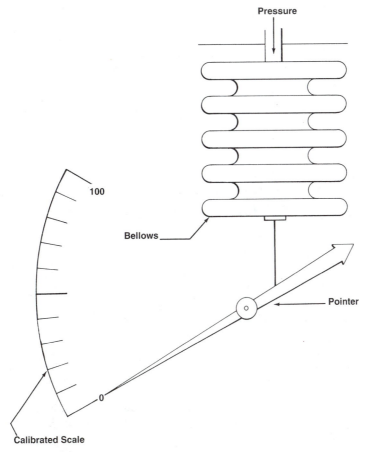

Figure 13.18 A bellows-type pressure gage is often used for low-pressure measurements.

Source: Courtesy of Hydraulics and Pneumatics Magazine.

output from the LVDT is easily converted into whatever form desired for the particular application.

A second type of electronic pressure gage utilizes strain gage transducer technology coupled with high-tech electronics. As with the LVDT units, the output can be either a digital readout or a data signal transmitted to a remote location. These gages are often battery-powered and can replace mechanical gages directly. *Strain gage* pressure gages have a wide variety of applications ranging from 0 to 15 psi (0 to 100 kPa) to 0 - 10,000 psi (0 - 70 MPa). The accuracy of the units is usually from 0.25 to 0.5 percent of the full-scale output.

A more recent development also makes use of a strain gage type of sensing unit. The strain gage for this device, however, consists of piezoelectric resistors in a Wheatstone bridge strain gage array mounted on the surface of a silicone semiconductor chip. The back side of the chip is then etched to shape the chip into a thin silicone diaphragm. The chip is mounted on a backing plate with

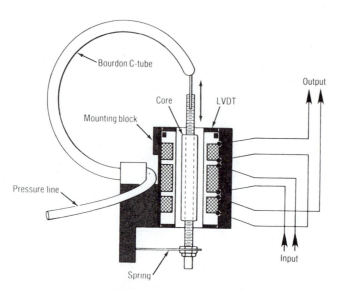

Figure 13.19 Electromechanical pressure gages sometimes use a linear variable differential transformer (LVDT).
Source: Courtesy of Hydraulics and Pneumatics Magazine.

the etched back vented to the atmosphere. Electrical leads are attached to the legs of the Wheatstone bridge and passed through holes in the backing plate. After these holes are sealed, the entire unit is mounted in a cavity in the end of a pressure probe. The cavity is then filled with a silicone oil and sealed with a stainless steel diaphragm. When the probe is installed, fluid pressure exerted on the stainless steel diaphragm is transmitted by the silicone oil to the chip. The resulting deflection of the chip causes a change of resistance in the piezo-electric resistors and, through the associated electronics, gives a pressure readout.

Pressure gages are normally calibrated by use of a *deadweight tester*. This device uses calibrated weights supported by a high-precision hydraulic cylinder to provide a known force. Pressure is built up in the cylinder by the use of a hand pump. When the pressure is sufficient to raise and hold the calibrated weights, the test gage pressure is read. This reading should correspond to the pressure required to support the known load. If it is not, then the gage reading should be adjusted to that pressure, usually by turning a calibration screw on the gage face or on the back of the case.

Flowmeters

There are basically three different principles involved in measuring fluid flow. *Flow meters* are accordingly grouped as

- Differential pressure meters
- Positive displacement meters
- Velocity meters

Differential pressure flow meters consist of two parts, regardless of the specific design of the device. The primary element causes a change in the kinetic energy of the flow system, and this causes a differential pressure across that portion of the device. The secondary element is the part that senses this pressure differential and provides a relative readout. While the readout is actually the pressure drop across the primary element, rather than the flow rate, the scale on the meter is graduated to indicate the flow rate.

A very popular type of differential pressure flow meter is the *orifice* type (Figure 13.20). Although they find little use in fluid power systems, most other facets of industry use orifices extensively. An orifice flow meter is a very simple device consisting of a flat plate with a precision hole drilled in it. The orifice plate is installed in the pipe between two flanges. Pressure taps are installed upstream and downstream of the plate. Lines from these taps are run to a differential manometer or other differential pressure reading device. The actual flow rate is proportional to the differential pressure reading.

Other differential pressure flow meter types include

- Flow tubes
- Flow nozzles
- Pitot tubes
- Elbow meters
- Target meters
- Variable-area meters

Variable-area flow meters (Figure 13.21), also termed *rotameters*, are commonly used in fluid power applications, especially on test benches and flow benches. The rotameter consists of a tapered tube containing a float. Fluid flowing through the tube causes the float to rise until the differential pressure between the upper and lower surfaces exactly balances the weight of the float.

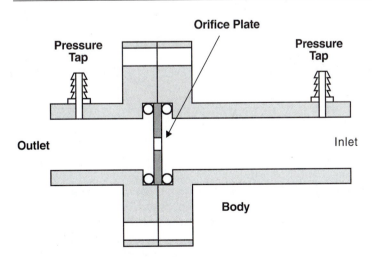

Figure 13.20 Orifice flowmeters are a popular choice for many applications outside the fluid power industry.

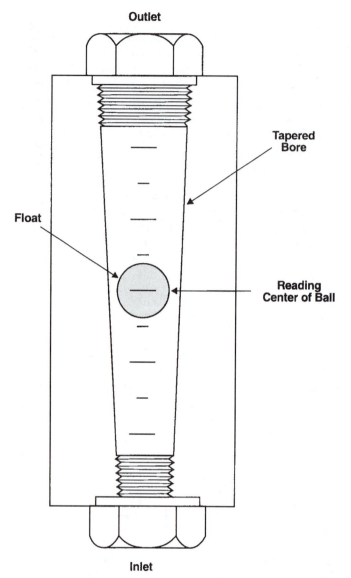

Figure 13.21 Variable-area flowmeters, also called rotameters, are commonly used in fluid power applications.

At that point, the flow rate can be read on the scale. Since most rotameters use a tube made of glass or other transparent, but low-strength material, they are usually limited to relatively low-pressure applications.

A variation on the typical rotameter is the *Hedland Flowmeter* shown in Figure 13.22. This device uses an annular orifice formed by a piston and a tapered cone instead of a float as is normally used. With no flow, the piston is held in place at the base of the cone by a calibrated spring. Flow through the meter results in a differential pressure across the piston orifice that moves the

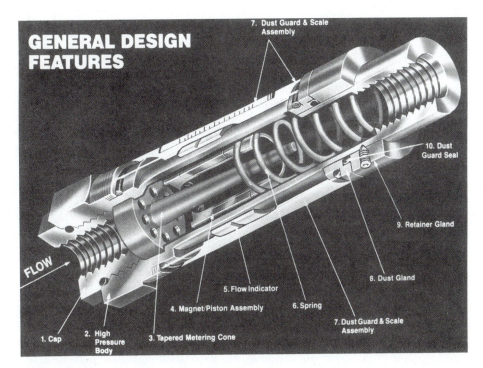

Figure 13.22 The Hedland flowmeter uses an annular orifice formed by a piston and a tapered cone instead of a float.
Source: Hedland Flow Meters, Racine, Wis.

piston against the spring. Piston movement and the orifice area are proportional to the flow rate. The flow tube, instead of being glass or plastic, is a nonmagnetic metal. The indicator ring on the outside of the body is magnetic and follows a magnet, which is part of the piston assembly. The black line on the indicator ring is read against the scale located on the transparent dust cover. Unlike the float-type meters that must be installed vertically, this meter operates properly in any position.

Positive-displacement flowmeters act in a manner similar to a hydraulic motor. They mesure flow by separating the fluid into accurately measured volumes as it flows through the metering section. A counting device keeps track of the number of volumes passed per unit time. This information is then converted to flow rate. There are four flowmeters in this category:

- Reciprocating piston meters
- Oval-gear meters
- Nutating-disk meters
- Rotary-vane meters

Velocity meters use the primary element to respond to the velocity of the fluid flow. The secondary elements convert this velocity to flow rate. They include

- Turbine meters
- Vortex meters
- Electromagnetic meters
- Ultrasonic meters

The most widely used of these types in fluid power applications are the *turbine meters*. The primary element in a turbine meter is a multi-bladed turbine that is installed in the flow pipe. The turbine turns at a speed proportional to the fluid velocity. The rotational speed is sensed by a counting device that produces an electrical pulse for each blade passage sensed. These pulses are then converted to flow rate based on the particular turbine design and flow area. Turbine meters can have very high accuracy, but they must be properly calibrated and installed. They may be very susceptible to contamination that can cause bearing damage as well as wear on the turbine blades.

Temperature Gages

As we have seen in earlier chapters, fluid temperature has a significant effect on the operation, reliability, and life of a hydraulic system. Therefore, it is desirable to have instrumentation for monitoring the fluid temperature. This means keeping track of the temperature of the fluid in the reservoir. Consequently, a *temperature gage*, or *probe* (usually a thermocouple) is inserted into the reservoir and provides either a readout (dial or digital) or some type of warning device such as a light or beeper to indicate an out-of-limits situation.

In some cases, the temperature associated with the operation of an individual component or branch of a circuit is to be monitored. This requires the location of temperature sensors at, or relatively near, the point of interest. Ports for inserting temperature probes may prove more practical in these locations than permanent installations of probes.

The external temperature of a device or conduit can sometimes be used to provide significant information about heat generation. In these cases, the use of *temperature-sensitive patches* may be useful. These patches—which are mounted on an adhesive backing so that they can be stuck onto the surfaces in question—change color when a limit temperature is exceeded. A more sophisticated technique—*infrared thermography*—can sometimes be used successfully to locate "hot spots" in a large system.

On-line Contamination Monitors

In an effort to prevent damage and wear degradation to contaminant-sensitive components, *on-line contamination monitors* are sometimes used. One such device is a magnetic plug that is mounted in the system, usually in the reservoir. This plug includes an electrical switching device activated by the metal particles that collect on the surface. One such device uses a set of closely spaced but open contacts. When the opening is bridged by metal particles, the electrical circuit is completed, and a warning is activated. Another device uses a capacitive bridge that is sensitive to the metal particles collected on the surface grid. When the quantity of particles collected exceeds a predetermined amount, an alarm is activated.

In other on-line applications, laser scanners and ultrasonic transmitter-receivers (similar to sonar) are used to detect solid particles as they are carried through a sensing zone by the system fluid. These devices can be used to provide continuous reporting on particle sizes and numbers, or simply to activate an alarm when specified limits are exceeded.

13.8   Heat Exchangers

The term *heat exchanger* is usually considered to mean a "cooler". In fact, a heat exchanger may be either a heater or a cooler. A cooler is required when the heat generation rate of a system (from either internal or external sources) exceeds the heat dissipation rate from the reservoir, piping, and hardware. On the other hand, a heater may be required to raise the fluid to some minimum temperature prior to startup, prevent freezing or separation of water-containing fluids, or help to maintain an acceptable operating temperature when the system is exposed to extreme cold.

Coolers

Coolers for hydraulic fluids normally use either water or air as the heat transfer medium. These two types are often referred to as *oil-to-water* and *oil-to-air* coolers. Oil-to-water (or simply water) coolers are commonly found in stationary applications were there is a ready source of cooling water. They are generally more efficient and effective than air coolers. They can be used to obtain lower fluid temperatures than air coolers, and are normally more effective in providing a constant hydraulic fluid temperature over the long term. This constant hydraulic fluid temperature aspect results from the near constant year-round, day or night temperature of the water (whether from mains or wells) that is used in such coolers.

Figure 13.23 is a drawing of a typical water cooler. This type of cooler is termed a *shell-and-tube* cooler. Water flows through the shell and around the tube bundle that carries the oil. Baffles are used to cause the cooling water to flow essentially perpendicular to the tube bundle. This provides a more efficient temperature gradient than could be achieved from having the water flow parallel to the tube bundle. As with reservoirs, the rate of heat flow is directly proportional to the temperature difference. Therefore, the cooler the water at any point in the heat exchanger, the more effectively it will remove the heat from the hydraulic fluid.

Shell-and-tube exchangers are usually either single-pass or double-pass units. In the *single-pass* units, the two fluids usually enter at opposite ends. In the *double-pass* units, the cooling water enters and leaves the shell through ports at the same end. The hydraulic fluid usually enters at that end, also, but exits through the opposite end. Numerous other configurations may be found, especially in very large industrial systems.

In addition to the temperature differential between the water and the hydraulic fluid, the amount of heat transfer also depends on the heat transfer coefficients between the surfaces of the tube and the two liquids, the thermal conductivity and thickness of the tube material, the total area of the tube in contact with each fluid, and any corrosion, deposits, or fouling of the tubes. Detailed and exact analysis of heat exchanger performance is a complex subject

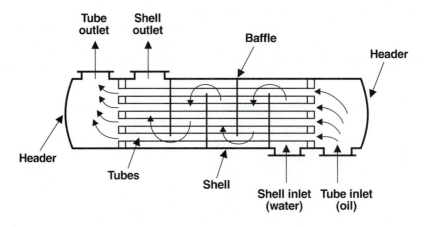

a. Shell-and-tube heat exchanger
(single-pass)

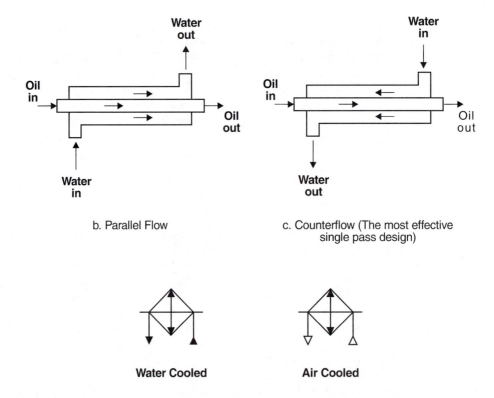

b. Parallel Flow

c. Counterflow (The most effective
single pass design)

Water Cooled   **Air Cooled**

Figure 13.23 One type of heat exchanger is a shell-and-tube cooler.

best left to texts on heat transfer; however, Equation 13.1 can be used to
estimate the performance of heat exchangers. Suppliers and manufacturers of

heat exchangers can normally perform detailed analysis for you or supply you with graphs or tables that will allow you to determine the heat dissipation capacity of any specific unit. This information will usually be given in Btu per minute or hour. By evaluating the heat generation rate of a system, you can select an appropriate heat exchanger. Remember, though, that it is always better to prevent the heat generation than to simply deal with removing the heat.

Oil-to-air (or air) coolers operate in the same way as the radiator on an automobile. The hydraulic fluid is directed through a series of small tubes in order to provide a large area for heat dissipation. Air is forced across the tubes, usually by a fan or blower. All of the conditions that are factors in the effectiveness of water coolers apply to air coolers also. The heat differential is often a problem, especially in units that are outdoors. Since the air temperature usually depends on the time of day as well as upon the season, the heat dissipation capacity of these units can be very inconsistent. Air coolers are generally limited to those applications in which the desired fluid temperature is at least 10°F (5.6°C) above the air temperature.

Fluid Heaters

A second heat-exchanger-type device is used for heating fluid. This may be accomplished by using hot water in a shell-and-tube heat exchanger. In most cases, though, *fluid heating* is accomplished in the reservoir. The units used in this case are usually termed *tank heaters*.

One common type of tank heater is a *tubing* or *pipe bundle* that carries hot water, steam, or hot air through the reservoir. A second type is an electrical heating element called an *immersion heater*. Both types can be effective, but both types also have potential shortcomings. For instance, if a leak develops in a tubing bundle, the fluid could be seriously contaminated with air or water.

The electrical immersion heater, if not properly sized, can overheat the oil that comes in contact with it, causing scorching, rapid oxidation and severe deterioration of the oil. Solutions to this problem include the use of low-watt-density heating elements and stirrers to ensure that the fluid circulates instead of remaining in contact with the heater.

Heat Exchanger Controls

In a well-designed hydraulic system, it is unlikely that the use of a heat exchanger—either a cooler or a heater—will be required at all times. In fact, it may not even be desirable. Therefore, some type of control should be incorporated to ensure that the exchanger is used only when it is needed. There are several methods used to operate such controls, most of which are triggered by fluid temperature. (A few such controls are operated by timers without regard to actual fluid temperature.)

One type of control uses a temperature-sensing electrical switch to open or close a solenoid valve to direct the fluid through or around the exchangers as required. A variation of this is a solenoid-operated shutoff valve that controls the flow of cooling water. The air cooler equivalent of this is a temperature switch to turn the cooling fan on or off.

Another alternative is a valve constructed of materials that open or close the flow path through the heat exchanger by expansion and contraction in response to the fluid temperature. Some of these valves are *proportional flow* devices, which means that the amount of flow passing through the valve is proportional to the fluid temperature. In these valves, only enough flow is directed through the cooler to maintain the bulk fluid temperature at the required level.

Similar techniques can be used to turn heaters on and off or to control the flow of hydraulic fluid, water, steam, or air to keep the fluid heated to the required temperature. In most applications, heating is required only to facilitate startup. Once the fluid has reached the required temperature, system operation will usually provide enough heating to keep it warm.

13.9 Summary

With the exception of reservoirs and seals, it is likely that many systems will not include any of the devices mentioned in this chapter. In fact, some hydrostatic transmissions don't even include a reservoir. There are many applications, however, in which accumulators and intensifiers can either simplify system design or save energy. In some cases, heat exchangers—either heaters or coolers—are necessary if the system is to function properly. Many systems don't have pressure gages, and very few have flowmeters and temperature gages; however, a well-designed system, even if it does not actually include instrumentation, will include readily accessible and easily used ports in which these devices can be used for troubleshooting and setup.

13.10 Key Equations

Heat transfer rate from
a reservoir:
$$q = C \times A \times \Delta T \tag{13.1}$$

Henry-Dalton Law:
$$h = h_0 \left(\frac{p}{p_0} \right) \tag{13.2}$$

Boyles' Law:
$$p_1 \times V_1 = p_2 \times V_2 \tag{13.3}$$

Adiabatic compression:
$$p_1 \times V_1^{1.4} = p_2 \times V_2^{1.4} \tag{13.4}$$

Review Problems

Note: The letter "S" after a problem number indicates SI units are used.

General

1. List and define several functions that a reservoir performs in addition to holding oil.
2. What three factors determine the rate at which air bubbles will be released from a liquid?
3. List and describe the three forms in which air can exist in a liquid.

4. How much air (percent by volume) is normally dissolved in hydraulic fluid at atmospheric pressure?
5. Why are baffles used in reservoirs?
6. List and describe the three methods of "loading" accumulators.
7. List and discuss the three most common applications for accumulators.
8. Define "adiabatic" and "isothermal."
9. Describe the operation of an intensifier.
10. What is meant by seal "squeeze"?
11. List and describe the two basic types of oil coolers.

Heat Transfer

12. A steel reservoir with a heat transfer coefficient of 12.5 contains hydraulic fluid to a depth of 2.5 ft. The reservoir is a 2 × 3 ft rectangular shape. The air temperature around the reservoir is 85°F, while the fluid temperature is 125°F. Air circulates freely under the reservoir. Find the rate at which heat is dissipated from the reservoir.
13. A reservoir with a heat transfer coefficient of 5 is placed in a closed room where the temperature reaches 110°F. The fluid temperature must not exceed 130°F. To maintain the fluid at this temperature, the heat dissipation rate must be 800 Btu/min. How much surface area is required?
14. An oil-to-water heat exchanger with an effective surface area of 3000 in² is required to dissipate 50 HP of lost energy in order to maintain the oil at 125°F. What water temperature is required if the heat transfer coefficient is 90?
15S. The depth of oil in a rectangular fluid reservoir (0.75 × 1.5 m) is 1.0 m and has a temperature of 60°C. The reservoir has a heat transfer coefficient of 40. The ambient temperature is 35°C. Find the heat transfer rate. The reservoir sits flat on the floor.
16S. An oil-to-air heat exchanger is required to dissipate 15 kW of waste heat to maintain the system fluid at 55°C. The heat exchanger has a surface area of 4 m² and a heat transfer coefficient of 140. Determine the maximum air temperature at which the required temperature can be maintained.

Air Content

17. A certain fluid will hold 9 percent air by volume at standard atmospheric pressure. Determine

    a. The volume of air (in³) in 1 gallon at 100 psig.
    b. The volume of air (in³) in 1 gallon at 5 psia.

18S. A hydraulic fluid contains 10 percent air by volume at standard atmospheric presure. Find

    a. The volume of air (cm³) in 1 liter at 350 kPa gage.
    b. The volume of air (cm³) in 1 liter at 35 kPa absolute.

Accumulators

19. An accumulator has a total volume of 600 in³. It is precharged with nitrogen to 1200 psi. Assume an adiabatic process. Determine the amount of oil the accumulator will contain when the system pressure is

    a. 1500 psi
    b. 2250 psi
    c. 3000 psi

20. Repeat Review Problem 20 using an isothermal process.
21S. A 6.0 L accumulator is precharged with nitrogen to 7.5 MPa. Assume an

isothermal process. Determine the amount of oil the accumulator will contain when the system pressure is

    a. 10 MPa
    b. 20 MPa
    c. 30 MPa

22S. Repeat Review Problem 21S using an adiabatic process.

System Problems

23. In the system shown here, the 6 in³ hydraulic motor produces a 250 ft·lb output at 500 rpm. The motor has volumetric and mechanical efficiencies of 0.9 and 0.92, respectively. The flow coefficient of the flow control valve is 0.5. The oil has a 0.9 specific gravity and a specific heat of 0.45 Btu/lb°F. There is a 20 psi pressure drop through the directional control valve in each direction. The pressure line and the return line have pressure drops of 50 psi and 10 psi, respectively. The pump output is 20 gpm. The total system contains 100 gal of oil.

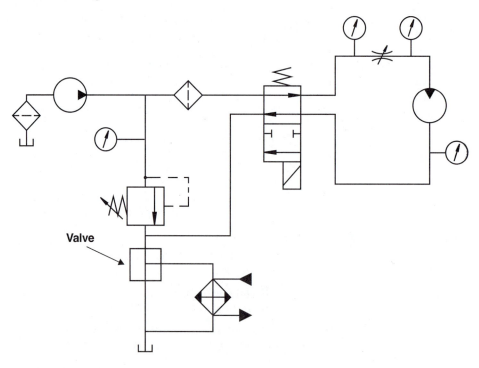

    When the fluid temperature reaches 130°F, the diverter valve shifts to direct the return oil flow through the heat exchanger. The heat transfer coefficient of the exchanger is 100. The incoming water temperature is 60°F.

    The oil temperature at system startup is 80°F. Assume that the reservoir has an average heat dissipation rate of 2500 Btu/hr. The piping has an average dissipation rate of 1000 Btu/hr. Find

    a. The time after startup when the diverter valve opens to direct flow through the heat exchanger.
    b. The heat exchanger area required to maintain the oil temperature at 130°F.

Chapter 14

Pneumatics: Basic Concepts and Air Preparation

Outline

14.1   Introduction

To this point, our discussions have been limited to the use of liquids as the power transmission medium in the discipline we've called hydraulics. In many applications, the use of gas instead of liquid makes more sense. Systems using gas (whether it be air, nitrogen, or any other gas) are called *pneumatic* systems.

Much of the previous material concerning hydraulics is directly applicable to pneumatics; however, there are also considerable differences between the two.

Objectives

When you have completed this chapter, you will be able to

- Explain the behavior of gases in pneumatic systems.
- Explain the basic gas laws.
- Discuss the differences between absolute and gage temperatures and pressures.
- Explain the operation of different types of flowmeters and calculate the effects of pressure on air flow rates.
- List and discuss the different types of air compressors.
- Discuss compressor controls.
- Complete the calculations necessary to size a compressor for a given application.
- Explain the purpose of a receiver and calculate receiver volumes for various applications.

- Discuss the effects of moisture on pneumatic systems and explain the meaning of moisture-related terms.
- List and explain the operation of different types of dryers for compressed air systems.
- Explain the need for, and the operation of, filters, regulators, lubricators, and silencers.

14.2   General Information on Pneumatics

Pneumatics is defined as the science and technology that deal with the laws governing compressed air flow. There are systems where this definition is expanded to include other gases, but we'll keep it simple here by considering only air as our working medium.

There are several advantages of pneumatics over other power transmission media. For example, the atmosphere is the source of the media, so no purchase orders, storage devices, or delivery times are involved. Compressed air is safe to use in almost any environment; it doesn't present a fire hazard, is not adversely affected by temperature or radiation, doesn't mess up equipment or the floor if it leaks, doesn't get your hands dirty, presents little in the way of toxic hazards, and can store large amounts of energy in small, portable containers. There is a large variety of simple, reliable, and relatively inexpensive pneumatic tools available, and small compressors can be used to provide large volumes of stored energy for intermittent service.

Unfortunately, there are also some significant disadvantages. These include the spongy or springy effect due to compressibility, the high cost of compressing and conditioning the air, lack of positive speed control, the noise and danger of escaping air, the shrapnel effect than can occur if a pressure vessel ruptures, and the difficulty of accurately calculating or predicting the performance of systems (especially large ones).[1] Nevertheless, pneumatics is the obvious choice for many applications involving relatively low output power requirements (relative to typical hydraulic power capabilities) or systems operating at less than 200 psi. Later in the chapter, we'll look at some typical applications.

14.3   Gas Laws

Before we begin to look at the performance and sizing of pneumatic components, we need to gain some understanding of the behavior of air as it is compressed. We covered this topic briefly when we discussed accumulators in Chapter 13, but we need to cover them in more detail for our pneumatics applications. As mentioned earlier, accurately predicting the behavior of air is a bit difficult. This is because we base all our calculations for air on the behavior of a perfect gas. Air—especially under the conditions found in most pneumatic systems—comes close to being a perfect gas, so we are reasonably justified in our use of the perfect gas analogy. Argon and neon, which are monatomic gases, are very nearly perfect gases.

Absolute Temperature and Pressure

In the calculations we'll be making in the following sections, we must use *absolute* temperatures and pressures. These differ from the temperature and pressure values used in our common applications, which are based on things that are familiar to us but are essentially quite arbitrary. In the Fahrenheit temperature scale, for instance, water freezes at 32°F and boils at 212°F. In the Celsius scale, these values are 0°C and 100°C, respectively. The two temperature scales are related by the equation

$$°C = \frac{5}{9}(°F - 32) \tag{14.1}$$

When discussing pressures, we commonly use *gage* pressure, which is based on the current ambient (atmospheric) pressure. In this scale, we arbitrarily say that the atmospheric pressure is 0 psig (the "g" meaning gage) and measure all pressure relative to that point. In the SI system, atmospheric pressure is 0 kPag (1 psi = 6.895 kPa).

When we make calculations concerning gas behavior, we can no longer rely on these arbitrary systems. A simple example will show why this is true. Suppose you were compressing a gas that was initially at atmospheric pressure (0 psig), and let's say that you had a pressure ratio of 10 to 1. In other words, you were increasing the pressure by a factor of 10. What would the final pressure be? Simply multiply the initial pressure (p_1) by the pressure ratio (r) to get the answer (p_2).

$$p_2 = p_1 \times r = 0 \times 10 = 0 \text{ psi}$$

That obviously isn't the correct answer!

Absolute pressure is based on a perfect vacuum. This perfect vacuum has a pressure of 0 psia or 0 kPa abs. In the absolute scale, pressure is *always* measured positive relative to a perfect vacuum. There can be no such thing as -5 psia. Gage pressure is related to absolute pressure by Equation 14.2.

$$p_{abs} = p_{gage} + p_{atm} \tag{14.2}$$

where p_{atm} is the current, local atmospheric pressure. Standard atmospheric pressure at sea level (based on 59°F (15°C) and 36 percent relative humidity) is 14.7 psia or 101 kPa abs. In the gage system, pressures can be negative relative to atmospheric pressure. These negative pressures are commonly termed *vacuums*. It is obvious, however, that the range of negative pressures is extremely limited, because you can't have anything less than a perfect vacuum. If the atmospheric pressure were 14.5 psia, and you calculated a pressure of -25 psig, you should recognize immediately that you had an error somewhere in your calculations, because, physically, such a pressure would be impossible. Figure 14.1 compares the absolute and gage pressure scales.

In these situations, we must base our calculations on an absolute, rather than an arbitrary, reference. For temperature calculations, we use *absolute zero*. This is the temperature at which all molecular activity ceases. We call temperatures measured from this point *absolute temperatures*. In the U.S. Customary system of measurements, we express absolute temperature in degrees Rankine (°R). Degrees Fahrenheit is converted to degrees Rankine by Equation 14.3.

$$°R = °F + 460 \tag{14.3}$$

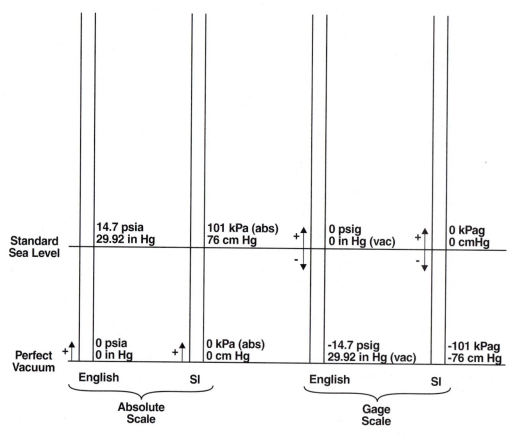

Figure 14.1 The gage pressure scale is based on arbitrary measurements, whereas the absolute pressure scale is based on a perfect vacuum.

In the SI system, absolute temperature is expressed in degrees Kelvin (°K).

$$°K = °C + 273 \qquad (14.4)$$

Figure 14.2 compares the four temperature scales.

Boyle's Law

When compressing air, three factors must be considered—volume, pressure, and temperature. All of the gas laws consider the relationships of the factors based on specified conditions. *Boyle's Law* relates volume and pressure changes that occur in a compression process in which the temperature is held constant (termed an *isothermal* process). This is expressed as

$$p_1 \times V_1 = p_2 \times V_2 \qquad (14.5)$$

The subscripts refer to the intial and final states. The pressures must be absolute rather than gage.

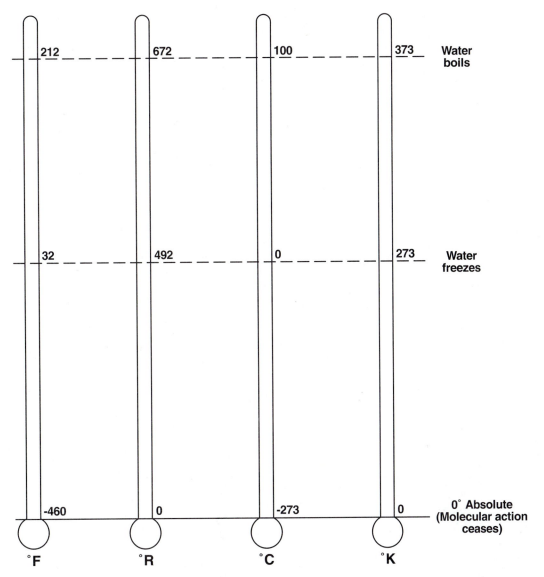

Figure 14.2 The four temperature scales are Fahrenheit (°F), Rankine (°R), Celsius (°C), and Kelvin (°K).

Example 14.1

Air at standard atmospheric pressure is contained in a cylinder with a volume of 24 in³ (394 cm³) as shown here. If the piston is pushed into the cylinder to reduce the volume to 7 in³ (115 cm³), what will the pressure gage read after the compression? Consider only an isothermal compression.

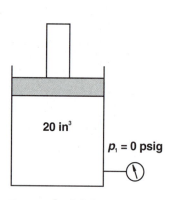

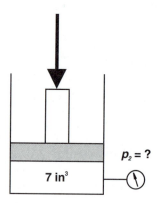

Example 14.1

Solution

First, we must convert gage pressure to absolute pressure using Equation 14.2.

$$p_1 = p_{abs} = p_{gage} + p_{atm}$$
$$= 0 + 14.7 \text{ psia}$$

Therefore,

$$p_1 = 14.7 \text{ psia}$$

or 101 kPa abs. Now, we solve Equation 14.5 for p_2 and plug in the values for V_1, V_2, and p_1.

$$p_2 = \frac{p_1 V_1}{V_2}$$

$$= \frac{14.7 \text{ psia} \times 20 \text{ in}^3}{7 \text{ in}^3} = 42 \text{ psia}$$

or 289 kPa abs. Notice that this is the absolute pressure. Gages, however, usually read gage pressure, so we must use Equation 14.2 to convert back into psig.

$$p_2 = p_{gage} = p_{abs} - p_{atm}$$
$$= 42 \text{ psia} - 14.7 \text{ psia}$$

Therefore,

$$p_2 = 27.3 \text{ psig}$$

or 188 kPag.

Charles's Law

Charles' Law applies to the situation in which temperature and volume are allowed to change while pressure is held constant. Equation 14.6 is used in this situation. Again, the subscripts refer to the initial and final states. Temperature must be expressed as an absolute value.

$$\frac{T_1}{T_2} = \frac{V_1}{V_2} \quad (p = \text{constant}) \qquad \qquad \textbf{(14.6)}$$

Example 14.2

In the container shown here 25 in³ (410 cm³) of air is initially at 70°F (21°C). If the air is heated to 250°F (121°C), find its final volume, assuming that the pressure remains constant.

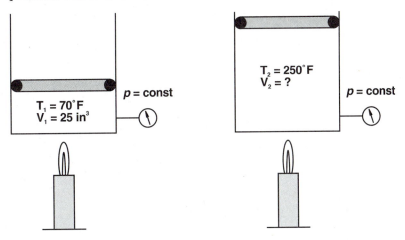

Example 14.2

Solution

In this case, we use Equation 14.6 to calculate the final volume, V_2. The equation requires the use of absolute temperatures, so we convert from °F to °R using Equation 14.3.

$$T_1 = 70°F + 460 = 530°R \, (294°K)$$

$$T_2 = 250°F + 460 = 710°R \, (394°K)$$

Now, solve Equation 14.6 for V_2 and plug in the values:

$$V_2 = \frac{V_1 T_2}{T_1}$$

$$= \frac{25 \text{ in}^3 \times 710° \text{ R}}{530° \text{ R}}$$

Therefore,

$$V_2 = 33.49 \text{ in}^3$$

or 549 cm³.

Gay-Lussac's Law

The third situation is one in which the volume is held constant while temperature and pressure are varied. *Gay-Lussac's Law,* Equation 14.7, addresses this situation.

$$\frac{p_1}{T_1} = \frac{p_2}{T_2} \qquad\qquad \textbf{(14.7)}$$

This equation tells us that pressure and temperature are directly related. Both must be in absolute measurements.

Example 14.3

The air in the container shown here is initially at 100°F (37.8°C) and 30 psig (207 kPa). If the air is heated to 250°F (121°C), find the final pressure shown on the gage. Atmospheric pressure is 14.7 psia (101 kPa abs).

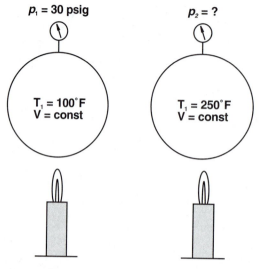

$p_1 = 30$ psig $p_2 = ?$

$T_1 = 100°F$
$V = const$

$T_1 = 250°F$
$V = const$

Example 14.3

Solution

Using Equations 14.2 and 14.3, we convert to absolute temperatures and pressures.

$$T_1 = 100°F + 460 = 560°R$$

or 311°K.

$$T_2 = 250°F + 460 = 710°R$$

or 394°K.

$$p_{1\,abs} = p_1 + p_{atm} = 30 \text{ psig} + 14.7 \text{ psia} = 44.7 \text{ psia}$$

or 308 kPa abs.

Now solve Equation 14.7 for p_2 and plug in the values.

$$p_2 = \frac{p_1 \times T_2}{T_1} = \frac{44.7 \text{ psia} \times 710°R}{560°R}$$

$$= 56.67 \text{ psia}$$

or 391 kPa abs. The pressure gage will read

$$p_2 = 56.67 \text{ psia} - 14.7 \text{ psia} = 41.97 \approx 42 \text{ psig}$$

or 290 kPag.

General Gas Law

If we combine Boyle's, Charles', and Gay-Lussac's laws into one general expression, we get what is often called the *General Gas Law*:

$$\frac{p_1 \times V_1}{T_1} = \frac{p_2 \times V_2}{T_2} \tag{14.8}$$

Unlike the other equations, none of the factors is held constant. Pressures and temperatures must be absolute.

Example 14.4

The air in a cylinder is initially at 700°F (371°C) and 450 psig (3103 kPa) and occupies a volume of 13 in³ (213 cm³). If its temperature is increased to 1000°F (538°C) while its volume is reduced to 5 in³ (82 cm³), what is the final gage pressure? Atmospheric pressure is 14.7 psia (101 kPa abs).

Solution

As before, we must convert to absolute temperatures and pressures.

$$p_{1\,abs} = p_{1\,gage} + p_{atm} = 450 \text{ psig} + 14.7 = 464.7 \text{ psia}$$

or 3204 kPa abs.

$$T_{1R} = T_{1F} + 460 = 700°F + 460 = 1160°R$$

or 644°K.

$$T_{2R} = T_{2F} + 460 = 1000°F + 460 = 1460°R$$

or 811°K. Now solve Equation 14.8 for p_2 and substitute these values.

$$p_{2\,abs} = \frac{p_1 \times V_1 \times T_2}{T_1 \times V_2} = \frac{464.7 \text{ psia} \times 13 \text{ in}^3 \times 1460°R}{1160°R \times 5 \text{ in}^3}$$

$$= 1520.7 \text{ psia}$$

or 10.5 MPa. Therefore, the gage would show

$$p_2 = p_{2\,abs} - p_{atm}$$

$$= 1520.7 \text{ psia} - 14.7 \text{ psia}$$

$$= 1506 \text{ psig}$$

or 10.4 MPa.

With this understanding of the behavior of air under various circumstances, we will have a better feel for what is happening as the components in the system operate. Now we're ready to look at the components.

Gas Flow Measurement

The flow of gases can be measured in many different ways, using most of the same instruments that were discussed in Chapter 13 for use with hydraulic fluid flow measurements. The most commonly used air flow device is the *rotameter*—a vertical, tapered tube containing a ball or float. The drag on the ball caused by the air flowing through the tube creates a force that lifts the ball. As the tube widens, the air velocity (and consequently the drag force on the ball) decreases. At a point proportional to the air flow, the drag force is balanced by the gravitational force on the ball, and it remains stationary. The flow rate can then be read from the graduations on the tube.

Rotameters normally show flow in standard cubic feet either per minute (scfm) or per hour (scfh). In SI units, the volume is either liters or cubic meters. If the air is exhausted from the rotameter into the atmosphere at standard atmospheric pressure, the flow rate read on the scale is correct. However, this is seldom the case; instead the air leaves the rotameter and enters the system where the pressure is higher than standard atmospheric pressure, so the density of the air is greater than "standard." The reading on the rotameter is therefore not standard cubic feet per minute, but rather "high density cubic feet per minute"—a lower value than at standard pressure. To find the actual flow rate, the "observed" flow rate (as read from the rotameter) must be corrected to standard conditions. This is done by Equation 14.9

$$Q_A = Q_0 \sqrt{\frac{p_A}{p_S}}$$ **(14.9)**

where

Q_A = actual flow rate.

Q_0 = observed flow rate.

p_A = absolute pressure at the outlet of the rotameter.

p_S = standard atmospheric pressure (14.7 psia or 101 kPa abs).

Example 14.5

A flowmeter shows a flow rate of 150 scfh (4.2 cmh), but the pressure at the outlet of the meter is found to be 80 psig (552 kPa). What is the actual flow rate through the meter?

Solution

For us to use Equation 14.9, p must be in absolute terms; therefore,

$$p_A = 80 + 14.7 = 94.7 \text{ psia}$$

or 653 kPa abs. From Equation 14.9, then, we find that the actual flow rate is

$$Q_A = Q_0 \sqrt{\frac{p_A}{p_S}} = 150 \sqrt{\frac{94.7}{14.7}} = 381 \text{ scfh}$$

or 10.7 cmh.

In extreme cases, temperature compensation may also be required. In most applications, however, the range of temperatures involved has relatively little effect on air density, so they are usually ignored.

14.4 Compressors

Many of the components found in pneumatic systems have direct counterparts in hydraulic systems. Most components peculiar to pneumatic systems are involved in preparing the air for use by linear and rotary actuators. Figure 14.3 depicts a typical pneumatic system. We'll use this to guide our walk through the components, beginning with the compressor.

Compressor Types

Air *compressors* come in many varieties and can be classified in many ways. One classification groups them by the method used to impart energy to the

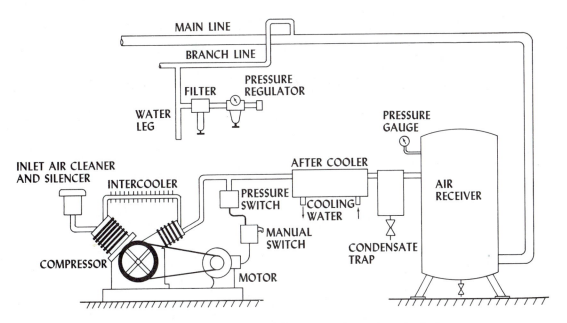

Figure 14.3 A typical industrial pneumatic system includes a compressor and air preparation equipment.
Source: Courtesy of Hydraulics and Pneumatics Magazine.

air: dynamic or positive displacement. These groups are similar in concept to the nonpositive and positive displacement hydraulic pumps.

In the *dynamic compressors*, pressure rise results from an increase in kinetic energy that is subsequently converted to static pressure as the high-velocity air enters the exit diffuser. This is accomplished by accelerating the air through a series of compression stages. The most common dynamic compressor types are centrifugal and axial.

Figure 14.4 shows a typical *centrifugal compressor*. Air enters the impellor at the eye and is accelerated as it moves outward toward the tips of the vanes. After leaving the impellor, the air enters a diffuser. If a single-stage compressor is used, the diffuser directs the air toward the storage tank (receiver). If a multistage compressor is used, the air is directed into the eye of the next compressor stage. Multiple stages are used to provide higher pressures. In general, a two-stage compressor will go to about 65 psi (450 kPa), a four-stage unit will reach around 150 psi (1000 kPa), while 350 to 400 psi (2400 to 2750 kPa) can be obtained from a five-stage unit. Centrifugal compressors rotate at 20,000 to 100,000 rpm.

Axial compressors (Figure 14.5) are designed so that the air flow is basically parallel to the axis of the unit. These units are particularly well suited for very high volumes of air at relatively low pressures. Pressure capabilities seldom exceed 90 psi (625 kPa) except in aircraft engine applications.

The airflow delivery of both centrifugal and axial compressors are inversely proportional to their outlet pressures (which are the result of resistance to

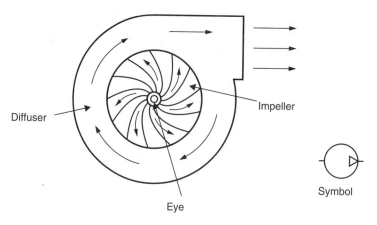

Diffuser

Impeller

Eye

Symbol

a. Centrifugal Compressor.

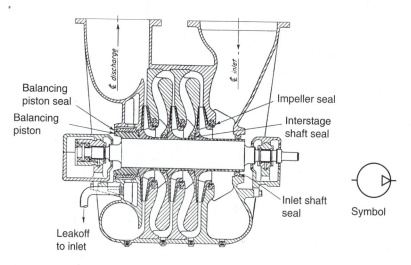

Balancing
piston seal

Balancing
piston

ℓ discharge

ℓ inlet

Impeller seal

Interstage
shaft seal

Inlet shaft
seal

Leakoff
to inlet

Symbol

b. Vertical section drawing showing typical multistage centrifugal compressor.

Figure 14.4 Centrifugal compressors may be either single stage or multistage.

Source: Reprinted with permission. *Compressed Air and Gas Handbook*, Compressed Air and Gas Institute.

flow). This means that if the outlet pressure increases, the delivery decreases. *Positive displacement compressors*, on the other hand, produce a constant delivery at a given operating speed regardless of the outlet pressure.

The most common types of positive displacement compressors are *rotary* (vane and screw) and *reciprocating* (piston). A *vane compressor* (Figure 14.6) operates in the same manner as the vane pump used in hydraulic systems. The axis of the rotor, containing the sliding vanes, is offset from the axis of the pressure ring. Thus, as the rotor turns, the volumes of the individual

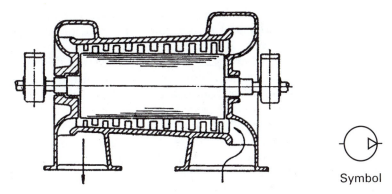

Symbol

Figure 14.5 A multistage, single-flow axial compressor is well suited for very high volumes of air at low pressures.
Source: Reprinted with permission. *Compressed Air and Gas Handbook*, Compressed Air and Gas Institute.

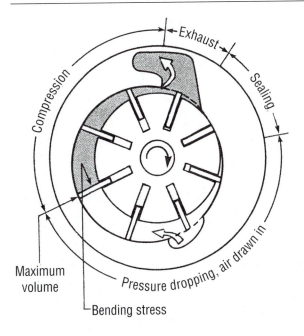

Figure 14.6 A typical rotary vane compressor has oil injected during compression to cool heat of compression. Air from vane and screw compressors is delivered to an oil separator, where liquid oil is removed.
Source: Courtesy of Hydraulics and Pneumatics Magazine.

chambers alternately increase and decrease. The increasing volume creates a low-pressure area, and air is pushed into the chamber from the inlet port. This air is carried around the ring to the point at which the volume begins to decrease. This decreasing volume compresses the air. It is then expelled through the outlet port. Vane compressors are capable of pressures to about 150 psi (1000 kPa) with power outputs ranging from less than 1 HP to as much as 500 HP (0.75 to 375 kW).[2]

A *screw compressor* is shown in Figure 14.7. Again, the similarity to the hydraulic pump counterpart is obvious. There are numerous variations on the screw compressor design. Typically, its maximum pressure capability is around 125 psi (850 kPa). The power range is from 7 HP to 300 HP (5 to 225 kW).[3]

A *reciprocating air compressor* (Figure 14.8) is similar in appearance to the piston in an automobile engine. The reciprocating motion of the piston is achieved by the rotation of the driven crankshaft. This rotation is translated into linear motion of the piston by the piston rod.

As the piston is pulled down in the bore, a vacuum results. The intake valve opens and air fills the chamber. As the crank passes bottom dead center, the piston begins to travel upward, and the intake valve closes. This results in compression of the air in the cylinder. As the piston nears the top of its stroke, the outlet valve is opened, and the compressed air is discharged. A single-stage compressor of this type is typically capable of supplying systems between 40 and 100 psi (275 to 700 kPa). Higher pressures usually require two-stage compressors.

A two-stage reciprocating compressor is shown in Figure 14.8. The first stage operates exactly like a single-stage compressor. As the air leaves the first stage,

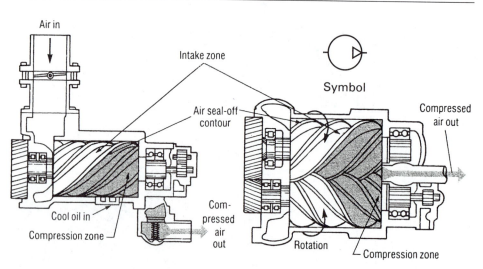

Figure 14.7 A typical rotary screw compressor. Cooled oil is injected into air at beginning of compression zone to remove heat of compression as pressure increases.

Source: Courtesy of Hydraulics and Pneumatics Magazine.

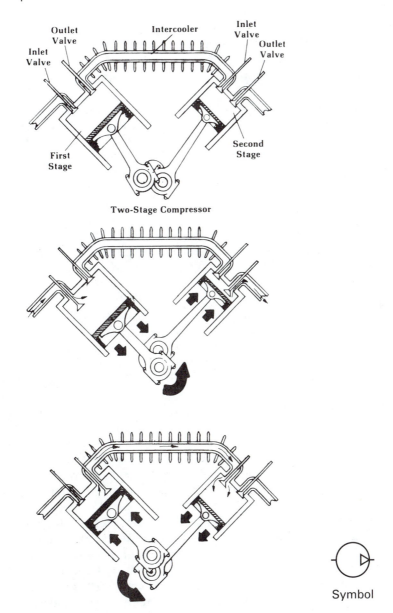

Figure 14.8 A reciprocating compressor is similar in appearance to the piston in an automobile engine. A two-stage compressor like the one shown here is required for higher pressures.
Source: Courtesy of Parker Hannifin Corporation.

however, it is directed through the intercooler to the second stage for further compression. The *intercooler* is a heat exchanger used to cool the air between the stages. The first stage of compression causes a considerable temperature rise, and the air must be cooled before it enters the second stage. This requirement stems from practical temperature limits on the compressor hardware and lubricants as well as the practicalities of the requirement to handle the tools and equipment that will be driven by the compressed air. Since the pressure of the air and its temperature are related, this cooling process reduces the pressure of the air delivered to the second stage, but increases the air density.

Once this cooled air enters the second stage, the process of the first stage is repeated. The air that is discharged from the second stage is at higher pressure and temperature (although not as high as it would have been without the intercooler). The higher pressure is achieved in a two-stage compressor at lower energy cost than it could have been with a single-stage unit.

Compressor Controls

In most applications, it is neither necessary nor desirable for the compressor to continuously supply compressed air to the system. Therefore, some technique must be used to either unload the compressor or control its output. There are several methods for doing this.

For piston-type compressors up to 25 HP (18 kW) that are driven by electric motors, a common technique is to *cycle* the motor on and off. This is done by properly positioned pressure switches that sense the maximum and minimum pressure limits for the system. For instance, the high-pressure switch might stop the motor when the pressure reaches 125 psi (850 kPa). When the pressure drops to, say, 90 psi (625 kPa), the low-pressure switch starts the motor again. This method is satisfactory for systems where the compressor duty cycle is typically 10 minutes on, 10 minutes off. Frequent cycling must be avoided to prevent harmful overheating of the electric motor. Unfortunately, frequent cycling drives up electricity costs because of the high current draw when starting an electric motor (referred to as *starting surge* or *locked-rotor amp draw*). Compressors operating on *start-stop cycles* incorporate some type of unloading mechanism so the motor is not required to start under full load.

For compressors where more frequent cycling is expected or where it is not practical to start and stop the prime mover, methods to unload the compressor are frequently used so that the prime mover will continue to run. These techniques might use the signal from the high-pressure switch to hold the inlet valve open, vary the amount that the inlet port opens (called *inlet throttling*), or hold the inlet valve closed so no air can enter the cylinder. These systems might also use a bypass valve that remains open to bleed the compressed air to the atmosphere. An electric clutch or variable-speed drive might also be employed between the prime mover and the compressor.

Compressor Sizing

A compressor must be sized to provide both sufficient pressure and sufficient airflow to operate the system efficiently. Failure to properly size the unit can prove costly in both energy consumption and personnel costs due to lost production efficiency.

Several factors must be considered in order to analyze flow properly. These include average airflow, maximum and minimum airflows, maximum and minimum pressures, duty cycle, and the condition of the air. Table 14.1 shows the air consumption of typical construction tools. Notice that these values are based

Table 14.1 Air Requirements of Various Tools

Tool	Free Air, cfm at 90 psig, 100% Load Factor
Grinders, 6″ and 8″ wheels	50
Grinders, 2″ and $2\frac{1}{2}$″ wheels	14–20
File and burr machines	18
Rotary sanders, 9″ pads	53
Rotary sanders, 7″ pads	30
Sand rammers and tampers,	
1″ × 4″ cylinder	25
$1\frac{1}{4}$″ × 5″ cylinder	28
$1\frac{1}{2}$″ × 6″ cylinder	39
Chipping hammers, weighing 10–13 lb	28–30
Heavy	39
Weighing 2–4 lb	12
Nut setters to $\frac{5}{16}$″ weighing 8 lb	20
Nut setters $\frac{1}{2}$ to $\frac{3}{4}$″ weighing 18 lb	30
Sump pumps, 145 gal (a 50-ft head)	70
Paint spray, average	7
Varies from	2–20
Bushing tools (monument)	15–25
Carving tools (monument)	10–15
Plug drills	40–50
Riveters, $\frac{3}{32}$″–1″ rivets	12
Larger weighing 18–22 lb	35
Rivet busters	35–39
Wood borers to 1″ diameter weighing 4 lb	40
2″ diameter weighing 26 lb	80
Steel drills, rotary motors	
Capacity up to $\frac{1}{4}$″-4 weighing $1\frac{1}{4}$–4 lb	18–20
Capacity $\frac{1}{4}$″ to $\frac{3}{8}$″ weighing 6–8 lb	20–40
Capacity $\frac{1}{2}$″ to $\frac{3}{4}$″ weighing 9–14 lb	70
Capacity $\frac{7}{8}$″ to 1″ weighing 25 lb	80
Capacity $1\frac{1}{4}$″ weighing 30 lb	95
Steel drills, piston type	
Capacity $\frac{1}{2}$″ to $\frac{3}{4}$″ weighing 13–15 lb	45
Capacity $\frac{7}{8}$″ to $1\frac{1}{4}$″ weighing 25–30 lb	75–80
Capacity $1\frac{1}{4}$″ to 2″ weighing 40–50 lb	80–90
Capacity 2″ to 3″ weighing 55–75	100–110

Cubic Feet of Air Per Minute Required By Sandblast

Nozzle Diameter	Compressed Air Gage Pressure (psig)			
	60	70	80	100
$\frac{1}{16}$″	4	5	5.5	6.5
$\frac{3}{32}$″	9	11	12	15
$\frac{1}{8}$″	17	19	21	26
$\frac{3}{16}$″	38	43	47	58
$\frac{1}{4}$″	67	76	85	103
$\frac{5}{16}$″	105	119	133	161
$\frac{3}{8}$″	151	171	191	232
$\frac{1}{2}$″	268	304	340	412

Source: Compressed Air and Gas Institute, *Compressed Air and Gas Handbook*, 5e, © 1989, p. 215. Adapted by permission of Prentice Hall, Englewood Cliffs, New Jersey.

on a pressure of 90 psi (625 kPa) *at the tool*. Most pneumatic tools are designed to operate at their best efficiency at this pressure; notice, however, that this is not "system" pressure, but pressure at the tool connector.

Now, let's look a minute at the airflow requirement. For instance, the heavy-duty chipping hammer requires 25 to 30 scfm (standard cubic feet per minute). What does this "scfm" actually mean, and how does it relate to ambient air? In pneumatic parlance, we come across several terms that are used to define air flow rates. These include cubic feet per minute (cfm), actual cfm (acfm), free cfm (fcfm), and standard cfm (scfm). While they are all intended to convey the same information, they really don't. For example, cfm says nothing at all about the conditions (temperature and pressure) of the air, and acfm and fcfm refer to the atmospheric, or "free," air at the inlet to the compressor. These terms are therefore not very useful for calculations unless you know the actual temperature and pressure of the air at the specific time in question. In almost every case, these three terms are actually incorrect references to the fourth term—standard cfm, or scfm. When scfm is used, the meaning is very specific—the flow rate of air at standard temperature and pressure (68°F, 14.7 psia, and 36 percent relative humidity). This is the only term that should be used when referring to compressed air flow rates.

The scfm required to operate a tool can be determined by using the General Gas Law (Equation 14.8). Since we are interested in the initial volume of air at standard conditions, we can solve this equation for V_1:

$$V_1 = \frac{p_2 \times V_2}{p_1} \times \frac{T_1}{T_2} \qquad \textbf{(14.10)}$$

The temperature term is often ignored in this equation. This is usually justified by the idea that the air can't be hot enough to make the tool too hot to handle. Therefore, it shouldn't be much higher than the standard temperature.

Example 14.5

A single-acting pneumatic cylinder has a 1.5 in (3.8 cm) bore and an 8 in (20 cm) stroke. Find the volume of air that the compressor must supply to operate at 20 cycles per minute at 100 psig (690 kPa). The air temperature in the cylinder is 90°F (32°C). The ambient temperature is 68°F.

Solution

The volume of the cylinder when it is completely extended (and therefore filled with 100 psig air) is

$$V = A \times S = 1.77 \text{ in}^2 \times 8 \text{ in} = 14.16 \text{ in}^3$$

or 232 cm³. To cycle at 20 cycles/min requires a total volume of

$$V_T = 14.16 \text{ in}^3 \times 20 \text{ cycles}$$

$$= 283.2 \text{ in}^3 \times \frac{1 \text{ ft}^3}{1728 \text{ in}^3}$$

$$= 0.164 \text{ ft}^3$$

or 4636 cm³, or 4.6 L for each minute of operation.

Remember that this air is at 100 psig, which means that we still need to calculate the standard volume. To do this, we must convert temperatures and pressures to absolute values.

$$p_{2abs} = p_2 + 14.7 = 114.7 \text{ psia}$$

or 791 kPa abs.

$$T_{1abs} = T_1 + 460 = 68 + 460 = 528°R$$
$$T_{2abs} = T_2 + 460 = 90 + 460 = 550°R$$

or 305°K.

Using Equation 4.10 to calculate the volume of air needed, we get

$$V_1 = \frac{p_2 \times V_2}{p_1} \times \frac{T_1}{T_2}$$

$$= \frac{114.7 \text{ psia} \times 0.164 \text{ ft}^3/\text{min}}{14.7 \text{ psia}} \times \frac{528°R}{550°R}$$

Therefore,

$$V_1 = 1.23 \text{ ft}^3$$

or 34.5 Lpm. If we ignore the temperature term in Equation 14.10, we get

$$V_1 = \frac{p_2 \times V_2}{p_1} = \frac{114.7 \text{ psia} \times 0.164 \text{ ft}^3/\text{min}}{14.7 \text{ psia}}$$

$$= 1.25 \text{ ft}^3$$

or 35.4 L for each minute of operation.

So we see that the temperature had very little influence on this case. The higher T_2, the more influence it will have; however, T_2 will seldom exceed 200°F (93°C) in any application. In this worst-case situation, there would be a difference of 20 percent in the calculations.

Before we can determine the power required to drive the compressor to satisfy system requirements, we need some additional definitions and equations. First, we can, with some degree of accuracy, assume that the compression process is *adiabatic*, meaning that there is no heat transferred from the air as it is being compressed, so the final pressure is the result of the decreased volume *and* the increased temperature. (The opposite of adiabatic is *isothermal*, or constant temperature, which means that all of the heat of compression is removed.) We use the adiabatic relationships in the following equations:

$$\frac{p_2}{p_1} = \left(\frac{V_1}{V_2}\right)^k \tag{14.11}$$

$$\frac{p_2}{p_1} = \left(\frac{T_2}{T_1}\right)^{\frac{k}{k-1}} \tag{14.12}$$

and

$$\frac{T_2}{T_1} = \left(\frac{p_2}{p_1}\right)^{\frac{k-1}{k}} \tag{14.13}$$

where k is the ratio of the specific heat at constant pressure (c_p) to the specific heat at constant volume (c_v). This value is 1.4 for air in our calculations. Note that temperature and pressure must be in absolute units in these equations.

Equation 14.14, again based on the assumption of the adiabatic process, is used to determine the theoretical power needed to drive the compressor:

$$HP = \frac{p_1 \times Q}{230} \times \frac{k}{k-1} \times \left[\left(\frac{p_2}{p_1} \right)^{\frac{k-1}{k}} - 1 \right] \qquad (14.14)$$

or

$$P = \frac{p_1 \times Q}{60,000} \times \frac{k}{k-1} \times \left[\left(\frac{p_2}{p_1} \right)^{\frac{k-1}{k}} - 1 \right] \qquad (14.14a)$$

where p is in absolute units and Q is flow rate in scfm or m³/min. The actual power would depend on the overall efficiency of the compressor:

Example 14.6

Determine the theoretical and actual horsepower required to drive a compressor that delivers 300 scfm (8.4 m³/min) at 100 psig (689.5 kPa). Its overall efficiency is 0.70.

Solution

$$p_2 = 114.7 \text{ psia}$$

$$\frac{k}{k-1} = \frac{1.4}{1.4-1} = 3.5$$

$$\frac{k-1}{k} = 0.286$$

Therefore, from Equation 14.14, the theoretical horsepower is

$$HP_T = \frac{14.7 \text{ lb/in}^2 \times 300 \text{ ft}^3/\text{min} \times 3.5}{230} \times \left[\left(\frac{114.7}{14.7} \right)^{0.286} - 1 \right]$$

$$= 53.7 \text{ HP}$$

or 40 kW. The actual horsepower is

$$HP_A = \frac{HP_T}{\eta_O} = \frac{53.7}{0.7} = 76.7 \text{ HP}$$

or 57.2 kW.

14.5 Receivers

A *receiver* is essentially a tank for storing air under pressure. In addition to this storage function, however, it performs several other functions that improve the overall performance of the system and reduce the operating and maintenance costs of the compressor. Included among these other functions is *pulse damping*, in which the receiver helps to eliminate the pulses generated by the compressor (especially reciprocating compressors) so as to ensure a smooth, steady airflow and a constant pressure. It also allows pressurized air to be stored in sufficient volumes to allow the compressor to be unloaded or shut down for significant portions of the system operating cycle. A properly sized receiver will allow the use of a much smaller compressor than would be necessary if the compressor had to supply the peak flow demands of the system on its own. By allowing the compressor to idle or be shut down for much of the time, maintenance and operating costs are reduced considerably. Also, moisture condenses in the receiver, so it should be drained regularly.

The primary advantage of the receiver is its ability to have a large quantity of compressed air ready to meet the high flow demands of the system. This means that a relatively small compressor can be used to charge the receiver during low demand times so that the receiver can provide the flow to satisfy the peak demands. Just as we saw in hydraulic accumulators, though, the entire volume of the receiver cannot be used. This is because, as the air is discharged into the system, the pressure in the receiver goes down. Therefore, the receiver is charged to some maximum pressure that is higher than the system requires. It is discharged until, at some point, it drops below the minimum pressure needed to operate the system.

To size a receiver, decisions must be made concerning the flow the receiver must provide, the length of time it must provide the flow, and the upper and lower pressure limits. The lower limit will be determined by the minimum pressure required to operate the system. The upper limit will usually be set by safety considerations. Most states require that receivers meet the requirements of the American Society of Mechanical Engineers (ASME) codes governing construction and maintenance of pressurized containers. Since any container that is pressurized by a gas can represent a serious safety hazard if it explodes, only ASME-approved receivers should be used.

An important aspect of receiver safety is the required *relief valve*. This valve provides protection for the vessel in case the pressure switches controlling the compressor operation fail to operate properly. Failure of the high-pressure switch would allow the pressure in the receiver to continue to build until some component in the system fails–the prime mover, a coupling, the pressure vessel structure, or the compressor itself. The relief valve is designed to prevent this *pressure vessel failure*, termed, "rapid dissemination of the containment vessel."

Because the relief valve is exposed to moisture and dirt in the air, it should be manually operated periodically to be certain that it is not rusted or jammed shut. Some insurance companies require two relief valves on pressure vessels. Figure 14.9 shows several types of relief valves. Notice that each includes some type of manual operator on the top.

Equation 14.15 can be used to size a receiver based on the predetermined factors just discussed.

$$V = \frac{14.7t\,(Q_R - Q_c)}{p_1 - p_2} \qquad\qquad \textbf{(14.15)}$$

or

$$V = \frac{101t\,(Q_R - Q_c)}{p_1 - p_2} \qquad\qquad \textbf{(14.15a)}$$

where V = receiver volume (ft^3, m^3).

   t = time during which Q_R is needed (minutes).

   Q_c = compressor delivery in standard volume (scfm, m^3/min).

   Q_R = flow demand in standard volume (scfm, m^3/min).

   p_1 = maximum pressure (psi, kPa).

   p_2 = minimum pressure (psi, kPa).

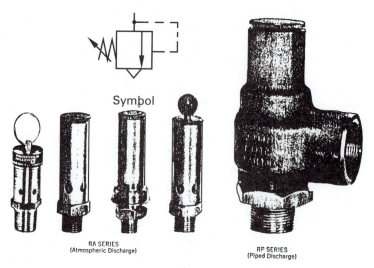

Symbol

RA SERIES
(Atmospheric Discharge)

RP SERIES
(Piped Discharge)

Figure 14.9 All types of relief valve configurations include some sort of manual operator on top.

Source: Courtesy of Fairchild, Industrial Products Division.

Example 14.7

Determine the size receiver required to provide a flow of 30 scfm (0.84 m³/min) at 90 psig (620 kPa) for 5 minutes without any compressor delivery. The receiver can be changed to 125 psig (862 kPa).

Solution

Using these values in Equation 14.15 gives

$$V = \frac{14.7t\,(Q_R - Q_c)}{p_1 - p_2}$$

$$= \frac{14.7 \text{ psia} \times 5 \text{ min} \times (30 \text{ scfm} - 0 \text{ scfm})}{(125 - 90) \text{ psi}}$$

$$= 63 \text{ ft}^3$$

or 1.78 m³. Notice that it is not necessary to convert to psia since we are finding the *difference* between the two pressures.

Although this is the actual physical volume required for this application, it is common to add about 50 percent to the calculated volume to allow for overload demands and future increases in system requirements. Based on this, a receiver volume of about 95 ft³ (2.7 m³) would probably be specified.

Receivers are usually not custom built, but are selected from a limited number of standard sizes, listed in table 14.2. We would chose the 96 ft³ receiver for this application.

Table 14.2 Standard Receiver Sizes

Diameter		Length		Volume	
in	cm	ft	m	ft³	m³
14	35.5	4	1.2	4.5	0.13
18	45.7	6	1.8	11.0	0.31
24	61.0	6	1.8	19.0	0.54
30	76.2	7	2.1	34.0	0.96
36	91.4	8	2.4	57.0	1.61
42	106.7	10	3.0	96.0	2.72
48	121.9	12	3.7	151.0	4.27
54	137.2	14	4.3	223.0	6.31
60	152.4	16	4.9	314.0	8.89
66	167.6	18	5.5	428.0	12.11

Source: Compressed Air and Gas Institute, *Compressed Air and Gas Handbook*, 5e, ©1989, p. 238. Adapted by permission of Prentice Hall, Englewood Cliffs, New Jersey.

Example 14.8

If the 96 ft³ (2.72 m³) receiver from the previous example is being supplied continuously by a compressor delivering 15 scfm (0.4 m³), how long could it operate the system before the pressure dropped to 90 psig (620 kPa)?

Solution

Solving Equation 14.15 for t gives

$$t = \frac{V(p_1 - p_2)}{14.7\,(Q_R - Q_C)}$$

Plugging in the system values, we get

$$t = \frac{96 \text{ ft}^3 \times (125 - 90)\,\text{psi}}{14.7 \text{ psi} \times (30 - 15)\,\text{ft}^3/\text{min}}$$

$$= 15.2 \text{ min}$$

If the system is used for a longer time, the pressure will drop below 90 psig (620 kPa) and the efficiency of the system will suffer.

14.6   Aftercoolers

Looking back to Figure 14.6, we see that, between the compressor and the receiver, there is a device called an *aftercooler*. This device has two purposes. First, as its name implies, it cools the hot air coming from the compressor before it enters the receiver. Second, it is the first step to *conditioning* the air, or preparing it for use in the system, before it reaches the tools and other system components. Conditioning the air means removing moisture and solid contaminants, removing or injecting lubricants as necessary, regulating the pressure, reducing noise, and, of course, cooling the air. The total conditioning job requires several different components, which we'll discuss as we follow through the system shown in Figure 14.6. Actually, we've already discussed one of them—the *receiver* provides a place where moisture can condense and dirt can drop out of the air.

Like the intercooler, the aftercooler is basically a heat exchanger used to extract some of the heat of compression from the air. It does this by passing cooling air or water over the aftercooler as the compressed air is flowing through it. Figure 14.10 shows a typical shell and tube-type aftercooler. The compressed air flows through the shell or casing, while water flows through the internal tubes to extract the heat from the air.

Before we look at the way in which the moisture is removed from the air, let's discuss *relative humidity*, defined as the ratio of the amount of moisture actually contained in the air to the amount of moisture (water vapor) it would contain if it were saturated. This is a function of the air temperature; the warmer the air, the more moisture it can hold. Thus, for any given quantity of water vapor, if the air temperature goes up, the relative humidity goes down. As the temperature goes down, the relative humidity goes up.

Another relevant factor is *dew point*, defined as the temperature at which the air is saturated with moisture—in other words, where relative humidity

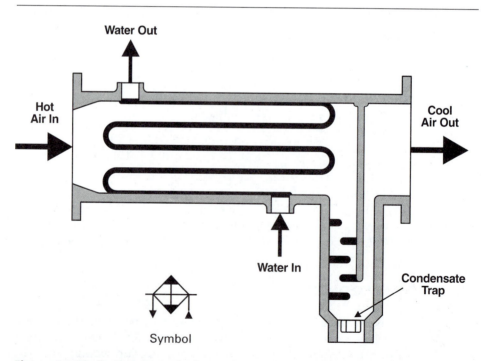

Figure 14.10 In this typical aftercooler configuration, air flows through the shell while water flows through the internal tubes to extract the heat from the air.

is 100 percent. The dew point can be reached by either increasing the actual moisture content or by lowering the temperature. In either case, moisture in the air begins to condense when the dew point is reached. Fog is a familiar example of saturated air; it occurs when the air temperature reaches the dew point. Interestingly, rain does not necessarily mean that the air is saturated (100 percent relative humidity) at ground level. It only means that the dew point has been reached at the altitude of the clouds.

In pneumatic systems, there are two different dew points to consider. The *atmospheric* dew point is the temperature at which the air becomes saturated at atmospheric pressure. The *pressure* dew point is the temperature at which the air becomes saturated at the system operating pressure.

We can use the graph in Figure 14.11 to determine the pressure dew point for an operating system. Suppose we have an atmospheric dew point of 50°F and a system operating pressure of 100 psig. To find the pressure dew point, find 50°F on the bottom axis of the graph. Follow that line upward until it intersects the 100 psi curve, then go horizontally to the right-hand axis and read the pressure dew point (about 117°F). This shows us that if the air temperature in the system remains above 117°F, there will be no condensation.

If the temperature is given in °C, then the process begins along the top of the graph. For example, if the atmospheric dew point is 10°C, and the system pressure is 7 kPa, find 10°C along the top axis. Follow the 10°C line downward until it intersects the 7 kPa curve, then go horizontally to the left to find the pressure dew point (about 47°C).

Note that in both of these examples, the pressure dew point is higher than the atmospheric dew point. That is always the case. We can also see from the graph that as the system pressure decreases, the pressure dew point decreases. For instance, suppose we start with a system pressure of 100 psig (7 kPa) and a dew point of 80°F (27°C) and reduce the pressure to 40 psig (2.8 kPa). To find the pressure dew point at the lower pressure, find 80°F on the right-hand axis and go horizontally until the 80°F line intersects the 100 psi curve. Next, go straight down until you intersect the 40 psig curve. Then go horizontally to the right-hand axis again and read the new pressure dew point (about 58°F). For the 27°C example, find 27°C on the left-hand axis and go horizontally to the 7 kPa curve. Go straight down from there to the 2.8 kPa curve, then horizontally back to the left-hand axis to read about 14.5°C.

Moisture is undesirable in pneumatic systems. It causes plumbing to rust, degrades the performance of filters, interferes with lubricants by either washing them away or displacing them on moving surfaces, causes rust and corrosion, and freezes in cold weather. Therefore, it is necessary to remove or significantly reduce the amount of moisture in the air before it accumulates to the point where it can cause problems.

The air coming from the compressor has a very high moisture content per cubic foot of air. The actual weight of water per cubic foot of air depends on the amount of water in the ambient air and the pressure ratio used. From the graph in Figure 14.12, we see that saturated air at 100°F and 14.7 psia contains about 2.85 lb of water per 1000 ft^3 of air. If we compress the air to 125 psig, then cool it back to its original 100°F, the maximum amount of moisture it can hold is only about 0.28 lb per 1000 ft^3. The remaining 2.57 lb of water (over 90 percent!) condenses and falls out of the air to be drained away at the separator.

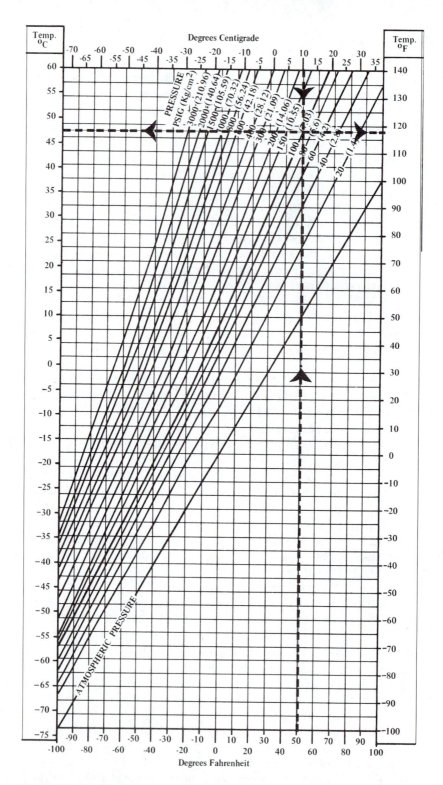

Figure 14.11 The pressure dew point can be determined from this graph according to the atmospheric pressure and the system operating pressure.
Source: Courtesy of Van Air Systems, Inc., Lake City, Pa.

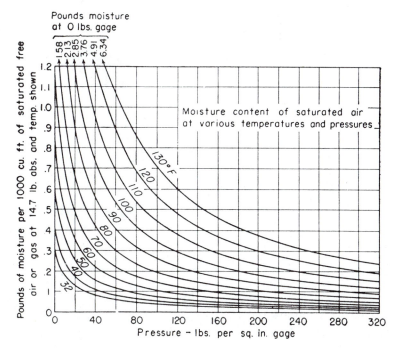

Figure 14.12 Moisture content of saturated air at various temperatures and pressures.

Source: Compressed Air and Gas Institute, *Compressed Air and Gas Handbook*, 5e, © 1989, p. 468. Adapted by permission of Prentice Hall, Englewood Cliffs, New Jersey.

This is what occurs in the aftercooler. In the real world, however, a 90 percent removal rate is not likely, because we may not be able to get the temperature back to ambient, and because the pressure drops as the air is cooled. Removal efficiencies of up to 85 percent can be achieved, though. Add to this the fact that significant percentages of the dirt and oil vapors also are left in the separator drain, and we see that the aftercooler is an important, and relatively inexpensive, first step in conditioning the air.

Example 14.9

A 100 scfm system operates 24 hours per day at an average pressure of 120 psi (825 kPa). The ambient air is saturated and is at 80°F (27°C). Determine the volume of water in gallons that passes through the compressor in one day.

Solution

From Figure 14.12, we see that the moisture content of the ambient air is 1.58 lb (0.72 kg) per 1000 ft³ (28.3 m³). At 100 scfm (2.8 m³/min), the rate at which moisture enters the system is

$$\text{Entry Rate} = \frac{1.58 \text{ lb}}{1000 \text{ ft}^3} \times \frac{100 \text{ ft}^3}{\text{min}}$$

$$= \frac{0.158 \text{ lb}}{\text{min}}$$

or $\dfrac{0.072 \text{ kg}}{\text{min}}$

In one 24-hour day, the total amount of moisture entering the system is

$$\text{Daily Rate} = \frac{0.158 \text{ lb}}{\text{min}} \times \frac{60 \text{ min}}{\text{hr}} \times \frac{24 \text{ hr}}{\text{day}}$$

$$= 227.5 \text{ lb/day}$$

or 103 kg/day.

Water weighs about 8.34 lb/gal (or 1 kg/L). That gives us

$$\frac{227.5 \text{ lb}}{\text{day}} \times \frac{\text{gal}}{8.34 \text{ lb}} = 27.28 \text{ gal/day}$$

or 103.25 L/day.

That's a lot of water to be put into a pneumatic system!

At an 85 percent removal rate, the aftercooler would remove 23.18 gal (87.7 L) of that, leaving about 4.1 gal/day (15.5 L/day) to get into the system. If dry air is required by the system, a dryer can be installed downstream of the receiver.

14.7 Dryers

The term *dry air* is relative. When applied to pneumatic systems, it really means that enough moisture has been removed from the air to ensure that the air will not become saturated as it moves through the system. That is, the relative humidity is so low that the dew point will not be reached, and condensation will not occur.

Several different types of *dryers* exist. Some overcompress and expand the air, some refrigerate it, and others use absorptive or adsorptive processes. The latter three are the more common for industrial applications. Figure 14.13 shows a *refrigeration dryer*. The wet air entering the dryer is passed through a heat exchanger where it is chilled by the liquid refrigerant. This is similar to the process used in the aftercooler, except that the minimum temperature in the aftercooler is limited by that of the cooling water or air, which in turn limits its effectiveness. In a refrigeration dryer, on the other hand, the temperature of the refrigerant can be as low as 32.4°F and can lower the temperature of the compressed air to about 35°F. Referring to the graph in Figure 14.12, we see that this would lower the moisture content of our 120 psig air from Example 14.9 to about 0.04 lb per 1000 ft^3. This means that we would have removed about 95 percent of the moisture from the air. After the air leaves the dryer and enters the system, as long as its temperature remains higher than that reached in the dryer, no condensation will occur. Thus, any moisture in the air will remain in the form of vapor. Water vapor in the air does no damage to components, because it passes through them with the air and never becomes liquid.

In many industrial applications, when the air leaves the refrigeration dryer,

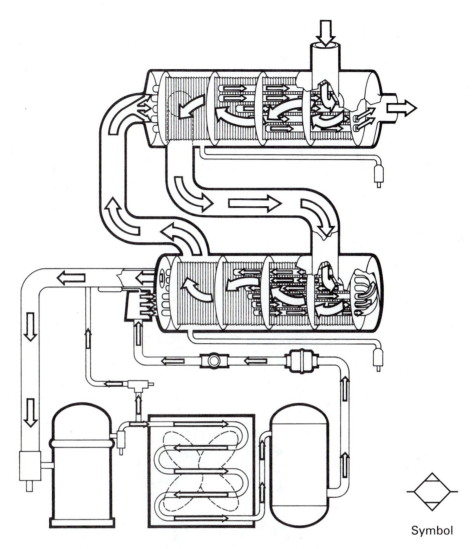

Figure 14.13 Refrigeration dryers can lower the dew point of compressed air to about 35°F.
Source: Courtesy of Van Air Systems, Inc., Lake City, Pa.

Symbol

it passes immediately into a *reheater* to bring its temperature up to the temperature of the building in which it is installed. This is done to prevent condensation on the *outside* of the piping throughout the building.

In *adsorptive dryers*, the water vapor is physically absorbed into the chemical structure of the chemical agents in the dryer. The dryer may contain chemicals such as dehydrated chalk and magnesium perchlorate which remain solid throughout the process. Lithium chloride and calcium chloride are also used. These chemicals—termed *deliquescent drying agents*—are initially solids, but liquify as they absorb water. The resultant liquid drips through a screen into a drain where it is subsequently removed (Figure 14.14). These units can provide dew points as low as $-40°F$ ($-40°C$).

Adsorption dryers remove water by using a desiccant material such as silica gel or activated alumina. The moisture is captured and held in the pores of these materials. This process is purely mechanical; there is no chemical change in the desiccant. In many industrial applications, regenerative arrangements such as the one shown in Figure 14.15 are used. By switching the four-way valve, untreated air is directed through one dryer, while dry system air is flowed through the other one to dry the desiccant for reuse. Other regenerative dryers use heaters to dry the desiccant. In small applications, the desiccant is simply replaced periodically.

14.8   Filters

We discussed solid contamination to some extent in Chapter 11, so there's no need to go over that again. Just be aware that solid contaminates cause the same types of problems in pneumatic systems that they cause in hydraulic systems. Therefore, they need to be removed, and the sooner the better.

In most systems, contaminant removal begins at the inlet to the compressor.

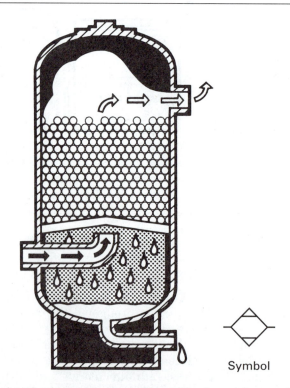

Symbol

Figure 14.14 An absorptive dryer can provide dew points as low as $-40°F$.
Source: Courtesy of Van Air Systems, Inc., Lake City, Pa.

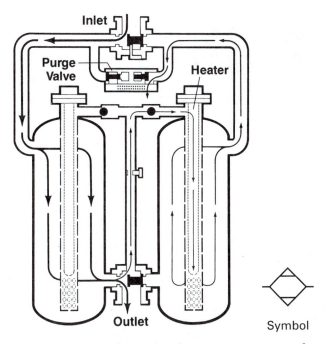

Figure 14.15 Regenerative adsorptive dryers remove water by using a desiccant material.
Source: Courtesy of Van Air Systems, Inc., Lake City, Pa.

The use of a good *intake filter* can significantly reduce the amount of dirt that is allowed to enter the compressor. This helps to extend the life of the compressor as well as the rest of the system. Intake filters are usually pleated paper types, similar to automotive air filters.

Air line filters are installed in the air line downstream of the dryer/separator near the connection for a tool or actuator. As shown in Figure 14.16, an air line filter consists of a head with inlet and outlet ports, a deflector, a shroud, a baffle plate, a filter element, and a bowl with a drain. As the air enters the bowl, the deflector plate causes it to swirl. Centrifugal force throws entrained water droplets and solid particles to the wall of the bowl. They fall to the bottom of the bowl to be drained off. The shroud ensures that the air flow is distributed evenly over the filter element. The baffle creates a "quiet zone" in the bottom of the bowl to prevent re-ingestion of the moisture and particles. The smaller particles that weren't separated by the swirling action are removed as the air flows through the filter element.

Notice that the filter we've discussed here actually removes both dirt and water, so it is more accurately termed a *filter separator*. The symbols shown in Figure 14.16 reflect that dual function.

These units must be serviced periodically. They must be drained (unless they include an automatic drain), and the filter element must be replaced.

Caution: Most clear bowls on filter separators are made of a polycarbonate plastic. They must not be used in applications where the pressure exceeds

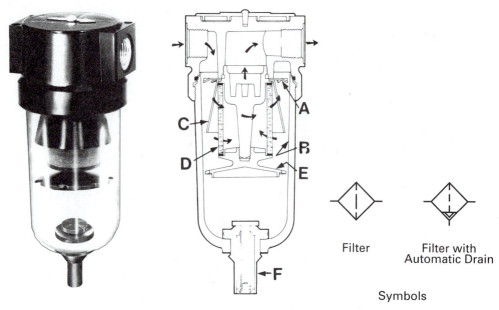

Figure 14.16 Air line filters are installed downstream of the dryer/separator and near the connection for a tool or actuator.
Source: Courtesy of Parken Hannifin Corporation.

250 psi (1725 kPa) or the temperature exceeds 150°F (65°C). They must not be exposed to chemicals such as ketones, solvents, paint thinners, chlorinated hydrocarbons, phosphate esters, or certain alcohols. These chemicals will cause fine cracks to develop (termed *crazing*) and will untimately result in the bowl's exploding under pressure. A cracked, or crazed, bowl must be removed from service immediately. If any of these adverse conditions exist, metal bowls should be specified.

Caution: Never loosen or remove the bowl from a filter when the system is pressurized. The compressed air can turn the bowl into a dangerous projectile.

14.9   Lubricators

Lubricators are used to provide lubricating oil to components such as air cylinders and air motors. There are several different approaches to providing this lubrication, but the most common is the *air line mist lubricator*. Figure 14.17 shows two typical ones. The *differential-pressure* type in Figure 14.17a is the more common of the two. In this device, the orifice in the main air passage causes a small pressure drop—usually 2 to 3 psi (15 to 20 kPa). This means that the air pressure on the oil surface is slightly higher than the pressure at the outlet of the needle valve. We've already discussed the operation of needle valves in hydraulic applications, so you can readily understand how this unit works. The differential pressure and the valve setting determine the rate at

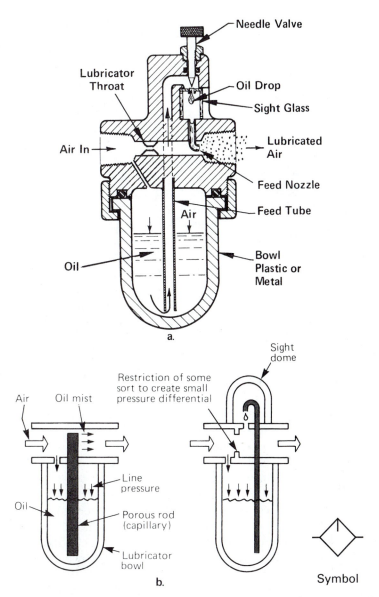

Needle Valve

Lubricator
Throat

Oil Drop

Sight Glass

Air In

Lubricated
Air

Feed Nozzle

Feed Tube

Air

Oil

Bowl
Plastic or
Metal

a.

Air   Oil mist

Restriction of some
sort to create small
pressure differential

Sight
dome

Line
pressure

Oil

Porous rod
(capillary)

Lubricator
bowl

b.

Symbol

Figure 14.17 In air line lubricators, mist lubricant can be generated with wicking or differential pressure.

Source: a. From *Industrial Fluid Power*, Vol.1. Published by Womack Educational Publications. b. A *Design News* illustration.

which oil is injected into the air stream. As the oil drips from the feed nozzle, the high-velocity airstream atomizes the drops so that they can be carried downstream as a mist. The oil mist itself does little in the way of lubricating as long as it remains airborne. The lubrication results from the oil being deposited on the surfaces of the components. This re-deposition as a liquid is termed *reclassifying*.

The rate at which the oil is injected is adjusted by the needle valve and observed through the sight glass. There are no hard, fast rules concerning the injection rate. In fact, rules of thumb vary from one drop every 3 minutes to one drop for every 20 scfm (about 0.5 m³/min) of airflow. The fact is that every system has different requirements, so you must determine for yourself the proper rate for your systems. While it is difficult to determine exactly what constitutes "enough" lubrication, "too much" is fairly easy to identify. Oil dripping from exhaust ports is one indication. Oil pooling in the piping is another.

The Occupational Safety and Health Administration (OSHA) regulations limit the acceptable concentration of airborne oil mist to 5 mg per cubic meter of air. This level is easily exceeded in a plant operation where several lubricators are used. Where the lubrication requirements of the components dictate high oil content in the system air, it will probably be necessary to reclassify the oil in the system exhaust air. This can be done reasonably effectively by standard air line filters; however, if total reclassification is necessary, a special *coalescing filter* must be used. Such units are capable of removing oil droplets as small as 0.3 μm and are up to 99.9 percent effective in removing the oil from the air. Note that these filters can also be used in-line to provide oil-free air for chemical, medical, dental, and food processes or any process where oil is prohibited.

A variation on the differential-pressure lubrication is the *recirculating* or *micromist lubricator*. It operates in the same basic manner as the unit we've just discussed, but uses a baffle plate or other technique to separate the oil droplets according to their size. Droplets smaller than about 0.2 μm are allowed to enter the air stream, while droplets larger than that size are returned to the bowl. The primary advantage of the recirculating lubricator is the distance that the oil mist can be carried through the system.

The droplets from a standard lubricator begin to precipitate from the airstream and reclassify in a very short distance. Therefore, the closer the lubricator is placed to the components they are supposed to lubricate, the more effective they are. Ten feet is usually the maximum effective distance. The micromist from a recirculating lubricator, on the other hand, can be carried several hundred feet to the components to be lubricated. Because only a small portion of the oil flowing through the needle valve actually gets into the airstream, the flow rate must be increased significantly (by a factor of 20 or more) to provide the same amount of lubrication.

The *capillary-type lubricator* shown in Figure 14.17b uses a sintered bronze rod to wick the oil into the airstream. The pressure on the oil surface forces the oil into the pores of the rod. Capillary action carries the oil up into the airstream where it is blown from the rod as a mist. The amount of oil delivered is a function of the pressure and the velocity of the airstream. The oil flow cannot be adjusted as it can in the differential-pressure type.

An empty lubricator cannot serve its purpose in the system; therefore, lubricators must be serviced on a regular basis. Some can be refilled through a special servicing valve while the system is operating, while other designs require that the bowl be removed for refilling. The type oil used must be of low viscosity so it can be atomized readily. It must also be compatible with system components, seals, and the process for which the system is used.

Caution: Never loosen or remove the lubricator bowl while the system is pressurized. The compressed air could turn the bowl into a dangerous projectile.

It should be noted that lubricators are not required for all pneumatic systems.

In fact, for many applications—instrument air, medical and dental instruments, for example—absolutely no oil can be tolerated. In many other applications, however, the trend is to eliminate the requirement for lubricators by using either permanently lubricated materials or materials that, due to their composition or finish, have very low friction. This trend has been prompted by at least three important considerations. First, the elimination of lubricators can reduce the initial and operating costs of the system. Second, if the components don't require lubrication, there's no chance of an unfilled lubricator causing component damage. The third reason is the OSHA acceptable limit on oil mist in the air, as previously discussed.

14.10 Regulators

The pressure *regulator* in a pneumatic system serves the same basic purpose as the pressure-reducing valve in a hydraulic system. The air pressure at the work station may be higher than the pressure required to do the job, so it is necessary to reduce and maintain the downstream pressure at the proper level. Air pressure regulators (Figure 14.18) consist of a valve body, with inlet and outlet ports, a poppet and seat, a pressure-sensing device (either a piston or a diaphragm, as in this case), and an adjustment screw that controls the compression of the main spring. A pressure gage is usually included. When the downstream pressure is lower than the regulator setting, the spring pushes the diaphragm downward. This, in turn, pushes the poppet wide open to allow virtually unrestricted air flow through the regulator. As the downstream pressure increases, the increase is sensed by the diaphragm, which begins to exert a force to compress the spring. When the diaphragm force exceeds the spring

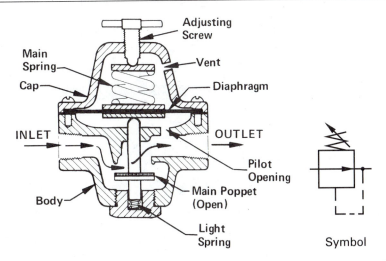

Figure 14.18 Non-relieving air pressure regulator, shown in a metering position.
Source: From *Industrial Fluid Power*, vol. 1. Published by Womack Educational Publications.

force, the diaphragm moves upward, allowing the light spring to push the poppet toward its seat. This begins to restrict the flow path, which in turn causes a pressure drop. When the downstream pressure equals the regulator setting, the poppet is fully seated, and there is no flow through the regulator.

The regulator shown in Figure 14.18 is termed a *non-vented* or a *non-relieving regulator*. This refers to the fact that, if the downstream pressure should *exceed* the regulator setting, there is no way for the regulator to act as a relief valve to relieve the excess pressure and protect the system. This problem is overcome by incorporating a relieving mechanism in the valve design.

Figure 14.19 shows a *self-relieving* or *self-vented regulator*. It is nearly identical to the unit in Figure 14.18 except that a vent hole is included in the piston or diaphragm. Up to the point at which the downstream pressure equals the pressure setting, and the poppet is fully seated, the two types work exactly the same way. If the downstream pressure increases with the poppet fully seated, the diaphragm can compress the spring more and actually lift off the poppet rod. In a self-relieving regulator, this opens the vent hole and allows air to escape through the vent port in the regulator body and relieve the downstream pressure.

Caution: An air pressure regulator is *not* a relief valve and should not be installed in place of a relief valve. While a vented regulator can protect the downstream portion of the system from overpressurization, it cannot protect the compressor side of the system. Never design or operate a system without a relief valve.

Pressure regulators may be pilot-operated. In these units, a remote pilot pressure takes the place of the spring and adjusting screw. The operating principle is the same as for the other regulators we have discussed.

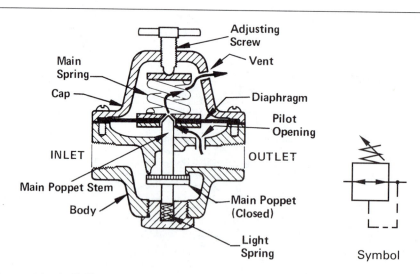

Figure 14.19 Self-relieving air pressure regulator, shown in a relieving position.
Source: From *Industrial Fluid Power*, vol. 1. Published by Womack Educational Publications.

14.11   Combined Conditioning Units

It is very common practice to combine two or three conditioning units in a single assembly. The common *combined conditioning unit* is a three-piece assembly of a filter, a regulator, and a lubricator usually referred to as an *FRL* (for filter, regulator, lubricator). Figure 14.20 is a photograph of a typical FRL. Notice that the symbol is actually a composite, with the dashed line representing the filter, the pressure gage with the arrow across it showing the regulator, and the single short line representing the lubricator. These units are always installed in the FRL order, so that the air is first filtered, then regulated, then lubricated. In systems where lubrication is not needed, the lubricator is omitted from the assembly.

Where conditioning units are used, whether individually or as assemblies, it is standard practice to dedicate one unit to each machine or branch of the circuit. This has proved to be far more effective than trying to condition an entire system with one large conditioning unit. Figure 14.21 is a schematic of the "front end" of a pneumatic system showing graphically many of the components we have discussed.

14.12   Silencers

Silencers are also called *mufflers*. As the name implies, these devices are used to reduce the noise level of the air being discharged from system components. As the air expands, it generates noise that can represent anything from a mild annoyance to a physical hazard causing hearing loss or masking audible alarms.

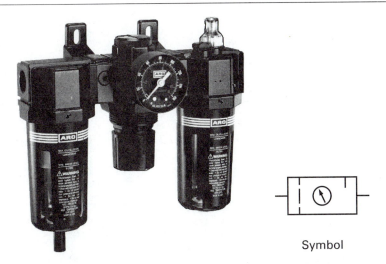

Symbol

Figure 14.20 A filter, regulator, and lubricator make up an FRL combined unit.
Source: Courtesy The ARO Corporation, Bryan, Oh.

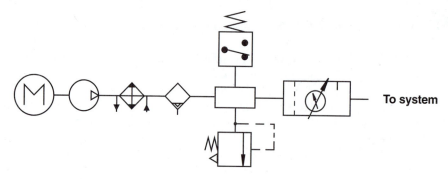

Figure 14.21 This is the basic air preparation circuit of the front end of a pneumatic system.

The continuous noise exposure limit set by OSHA is 90 dB (In some Canadian provinces the limit is 85 dB). In environments where the noise levels exceed these limits, ear protection is required. It is common practice, therefore, to either isolate noisy equipment in noise-reducing enclosures or, as is more common with pneumatic equipment, reduce the noise produced by the device. The use of silencers on exhaust ports is one method used.

As shown in Figure 14.22, a silencer may be as simple as a porous media through which the exhaust air flows, or as complicated as an acoustically tuned device that uses baffles to cancel the sound-producing vibrations. In any case,

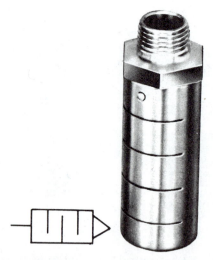

Figure 14.22 Air exhaust silencers are used to reduce the noise level of air being discharged from a system.
Source: Courtesy Parker Hannifin Corporation.

the device must represent a very low resistance to the exhaust air. Otherwise, an excessive back-pressure can be produced that can interfere with the component operation or require higher system operating pressures, which, of course, will increase the cost of operation.

14.13 Troubleshooting Tips

Symptom	Possible Cause
Compressor cycles too frequently or doesn't shut off	1. Pressure switch set too low 2. Relief valve stuck open 3. Receiver too small 4. Air leaks in system 5. Receiver full of water
Excessive moisture in system	1. Refrigeration unit not working properly 2. Desiccant dryer saturated 3. Deliquescent material depleted

14.14 Summary

Unlike the liquids used in hydraulic systems, gases are highly compressible. This compressibility means that much more attention must be paid to the thermodynamic and physical processes involved in "pumping" air than with liquids. The gas laws and the significance of adiabatic and isothermal processes must be considered.

The design of a pneumatic system requires that the compressor be sized to provide the pressures and flow rates needed to efficiently operate the tools used in the system. The sizing of the receiver (if one is used) should be considered along with the compressor in order to provide an acceptable duty cycle for the compressor.

Air preparation is critical to the life, reliability, and performance of a pneumatic system and the tools operated from it. This preparation begins with filtering the air entering the compressor. After leaving the compressor, the air is further conditioned through the use of aftercoolers, dryers, filters, and water separators, regulators, and lubricators. The importance of delivering clean, dry air to the system cannot be overemphasized.

14.15 Key Equations

Absolute pressure:
$$p_{abs} = p_{gage} + p_{atm} \tag{14.2}$$

Boyles' Law:
$$p_1 \times V_1 = p_2 \times V_2 \tag{14.5}$$

Charles' Law:
$$\frac{T_1}{T_2} = \frac{V_1}{V_2} \tag{14.6}$$

Gay Lussac's Law:
$$\frac{p_1}{T_1} = \frac{p_2}{T_2} \tag{14.7}$$

General Gas Law:

$$\frac{p_1 \times V_1}{T_1} = \frac{p_2 \times V_2}{T_2}$$
(14.8)

Actual flow rate:

$$Q_A = Q_0 \sqrt{\frac{p_A}{p_s}}$$
(14.9)

Adiabatic relationships:

$$\frac{p_2}{p_1} = \left(\frac{V_1}{V_2}\right)^k$$
(14.11)

$$\frac{p_2}{p_1} = \left(\frac{T_2}{T_1}\right)^{\frac{k}{k-1}}$$
(14.12)

$$\frac{T_2}{T_1} = \left(\frac{p_2}{p_1}\right)^{\frac{k-1}{k}}$$
(14.13)

Compressor power requirements:
U.S. Customary

$$HP = \frac{p_1 \times Q}{230} \times \left(\frac{k}{k-1}\right) \times \left[\left(\frac{p_2}{p_1}\right)^{\frac{k-1}{k}} - 1\right]$$
(14.14)

SI

$$P = \frac{p_1 \times Q}{60,000} \times \left(\frac{k}{k-1}\right) \times \left[\left(\frac{p_2}{p_1}\right)^{\frac{k-1}{k}} - 1\right]$$
(14.14a)

Receiver volume:
U.S. Customary

$$V = \frac{14.7t\,(Q_R - Q_c)}{p_1 - p_2}$$
(14.15)

SI

$$V = \frac{101t\,(Q_R - Q_c)}{p_1 - p_2}$$
(14.15a)

Review Problems

Note: The letter "S" after a problem number indicates that SI units are used.

General

1. List some advantages and disadvantages of pneumatic systems.
2. What is the primary difference between systems using gases (pneumatics) and systems using liquids (hydraulics)?
3. State Boyle's Law.
4. State Charles' Law.
5. State Gay-Lussac's Law.
6. State the General Gas Law.
7. What is the difference between positive displacement and dynamic compressors?
8. What is the difference between a centrifugal and an axial compressor?
9. List the three most common types of positive displacement compressors.
10. What is a two-stage compressor?
11. What is the function of an intercooler? Why is it necessary?
12. What is meant by scfm? What does it tell us about the incoming air?
13. What is the purpose of the receiver in a pneumatic system?
14. What does an aftercooler do?
15. What does a dryer do? Why is it needed?
16. What is the purpose of a lubricator? When is a lubricator needed?
17. What is an FRL?

18. What is the difference between gage pressure and absolute pressure?
19. What is meant by a perfect vacuum? What is the value of a perfect vacuum in psig, psia, in Hg, kPa abs, kPa gage?

Gas Laws

20. In an isothermal (constant temperature) compression, the volume of 10 ft³ of air at standard atmospheric pressure is reduced to 1 ft³. What is the pressure after compression?
21. Air is contained in a volume of 200 in³ at 150 psig. If the volume of the air is increased isothermally to 1200 in³, what is the final pressure?
22. A 4 ft³ vessel contains air at standard conditions. If the temperature of the air is increased to 100°F, what is the final pressure?
23. The pressure of air at standard conditions is increased to 30 psig while its volume is held constant. What is the final temperature?
24. Air at standard conditions is compressed from 200 ft³ to 12 ft³. The final pressure is 260 psi. What is the final temperature (°F)?
25. If the compression in Review Problem 24 were isothermal, what would the final pressure be?
26. Air at standard conditions is compressed adiabatically from 20 ft³ to 4 ft³. What is its final pressure?
27. What is the final temperature of the air in Review Problem 26?
28. As air is compressed adiabatically from standard conditions, its temperature increases to 150°F. What is its final pressure? What is its final volume if the initial volume was 400 ft³?
29. Air at standard conditions is compressed to 140 psig. Its initial volume is 250 ft³. Find its final volume and temperature for

   a. Isothermal compression
   b. Adiabatic compression

30S. Air at standard conditions is compressed from 7 m³ to 1 m³ isothermally. What is its final pressure?
31S. Air is initially at 101 kPa abs and 35°C. It is heated at constant volume to 150°C. What is its final pressure?
32S. The volume of a container of air is decreased by a factor of 10. Its initial pressure is 20 kPa gage. Its initial temperature is 25°C. Find its final temperature and pressure for:

   a. Isothermal compression
   b. Adiabatic compression

Gas Flow Rate

33. Air is flowing through a rotameter at 15 cfm. The air pressure is 90 psig. What is the flow rate in scfm?
34. The reading on a rotameter is 160 cfh. The pressure at the outlet of the meter is 30 psig. What is the flow rate in scfh?
35. Air is flowing through a flowmeter at 400 scfh. What would be the observed readings for pressures of 25, 50, and 100 psig?
36. Construct a graph that would allow you to find the actual scfm based on observed cfm readings. Use pressures of −10, −5, 0, 20, 40, 60, 80, and 100 psig. Put actual scfm on the horizontal axis.
37S. Air is flowing through a rotameter at 0.5 m³/min and 600 kPa. Calculate the flow rate in standard m³/min.
38S. A rotameter is indicating a flow of 5 m³/hr. The pressure in the rotameter is 200 kPa. Find the standard flow rate.

39S. Air is flowing at 2.5 standard m³/hr. Calculate the observed flow rate through a rotameter for pressures of 150, 300, and 500 kPa.

40S. Construct a graph that would allow you to find the actual flow rate based on observed readings of m³/min. Use pressures of -50, -25, 0, 50, 100, 300, 500, and 700 kPag.

Compressor Power

41. Determine the theoretical horsepower required for a compressor delivering 20 scfm at 125 psig.

42. A compressor in Denver, Colorado, is compressing air at 14.2 psia to 100 psig. It delivers 30 scfm. Find the theoretical horsepower required by the compressor.

43. A compressor delivers 250 cfm at 125 psi. What is the theoretical horsepower required to drive the compressor?

44. What is the horsepower required to drive a compressor that delivers 100 cfh at 150 psi? Atmospheric pressure is 14.3 psi. The overall efficiency of the compressor is 0.85.

45S. Calculate the theoretical power required for a compressor delivering 0.5 m³/min (standard) at 750 kPa.

46S. A compressor is receiving air at 90 kPa and compressing it to 800 kPa at 0.6 m³/min (standard). Find the theoretical power required by the compressor.

47S. A compressor with an overall efficiency of 0.87 is delivering 5 m³/hr (standard) of air at 650 kPa. Atmospheric pressure is 5 kPa gage. Calculate the actual power required to drive the compressor.

Receiver Sizing

48. A receiver is charged to 150 psi. It delivers 20 scfm at 100 psi. How large must it be to supply the system for 10 min with no compressor delivery?

49. A 150 ft³ receiver is charged to 130 psi. How long can it deliver 250 scfm at 100 psi with no compressor input?

50. A receiver is required to deliver 75 cfm at 50 psi for 30 min with no compressor input. It is initially charged to 120 psi. What size receiver is required?

51S. A receiver is required to deliver 2.0 m³/min (standard) at 600 kPa for 5 minutes with the compressor shut down. It is charged to 800 kPa. What size receiver is required?

52S. A 4.27 m³ receiver is charged to 750 kPa. Determine how long it can continue to deliver 1.5 m³/min (standard) at 500 kPa with the compressor shut down.

References

1. *Air Power and Control,* Design News Supplement, National Fluid Power Association.
2. Industrial Pneumatics Technology, Parker Hannifin Corporation, Cleveland, OH, 1980.
3. Ibid.

Chapter 15

Pneumatic Components, Controls, and Systems

Outline

15.1 Introduction

In Chapter 14, we discussed concepts and hardware that are peculiar to pneumatic circuits. In this chapter, we will look at components—cylinders, motors, and valves—that are comparable, though not identical, to their hydraulic counterparts. We will also look at *fluid logic*—a technique that uses air pressure as the control medium as well as the power transmission medium. Pneumatic system layout will be discussed, as well as the use of vacuum as a power transmission medium.

Objectives

When you have completed this chapter, you will be able to

- Describe and discuss the operation of pneumatic cylinders.
- Describe and discuss the operation of air motors.
- Describe and discuss the operation of pneumatic valves.
- Explain the differences between hydraulic and pneumatic components.
- Calculate the volume flow rate through flow control valves.
- Explain the concept of choking in a pneumatic valve.
- Calculate the force output of pneumatic cylinders.
- Explain the operation of common fluid logic devices.
- Explain the operation of fluid logic control circuits.
- Discuss the important considerations in pneumatic circuit design.
- Calculate the pressure losses through pneumatic system piping.
- Explain the operation of vacuum systems.

- Calculate the force capability of vacuum systems.
- Calculate the time required to achieve the required level of vacuum in a vacuum system.

15.2 Pneumatic Cylinders

Pneumatic cylinders are, in general, very similar to their hydraulic counterparts. They are available in the same basic configurations—single-acting, double-acting, and single- or double-ended—as hydraulic cylinders. Their individual parts are similar, as are their mounting options and sizes. There are some significant differences, however.

One difference is in the materials used in the cylinder structure. Since pneumatic cylinders operate at much lower pressures than hydraulic cylinders, they often can be constructed of materials other than the steels normally used for hydraulic cylinders. Aluminum is a very commonly used material. Plastics are often used in low-pressure (less than 30 psi) applications. An example of this is the *rolling diaphragm cylinder* shown in Figure 15.1. These devices are used for light loads and short-stroke (usually less than one-half inch) applications. There is no piston as in most cylinders. Instead, there is a laminated metal disk that is used to connect the rod to the rubber-like diaphragm. The diaphragm is clamped between the two end halves of the cylinder barrel or otherwise attached to the bottom half of the cylinder. When the bottom of the cylinder is pressurized, the rod extends as the diaphragm and disk are

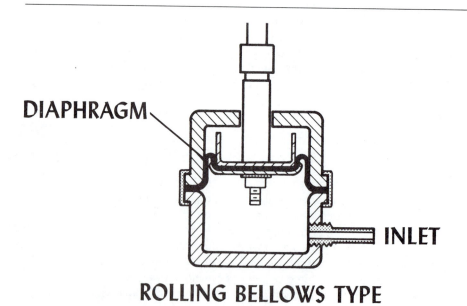

Figure 15.1 The rolling diaphragm cylinder is used for light loads and short-stroke applications.

Source: Courtesy Vega Enterprises, Inc., Decatur, Ill.

pushed upward. When the bottom is depressurized, the load forces the disk and diaphragm back down into the bottom half of the cylinder. These devices are almost always single-acting and use pressure to extend the cylinder.

Another pneumatics-peculiar cylinder is the *metal bellows device* shown in Figure 15.2. This is also a very short-stroke unit (usually less than one-half inch). It uses a pleated metal bellows in its operation. When the bellows is pressurized, the pressure inside the bellows causes it to lengthen slightly. When it is depressurized, the load forces the bellows to contract to its original length.

When additional strength or corrosion protection is required, *stainless steel cylinders* are usually used. The stainless steel, while being more expensive than other materials, reduces the likelihood of failure or performance loss that could result from water in the pneumatic system.

Nonrotating cylinders (Figure 15.3) are more common in pneumatic applications than in hydraulics. These devices are almost exclusively single-acting to extend because of the difficulty of sealing the rod end of the cylinder.

Pneumatic cylinder force is calculated using the same equation we have used previously for cylinder calculations ($F = p \times A$).

Very often you will see the term *power factor* used in cylinder manufacturers' literature. This term refers to the force output of the cylinder per psi of inlet pressure, a value that accounts for the output force lost due to seal friction. To find the output capability of the cylinder, multiply the power factor by the cylinder inlet pressure available.

Pneumatic cylinder speed is considerably more difficult to predict than for hydraulic cylinders. While cylinder speed is still a function of flow rate, the compressibility of air means that pressure also plays a major role in deter-

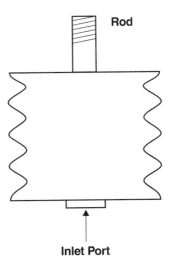

Figure 15.2 The metal bellows cylinder, like the rolling diaphragm, is used for short-stroke applications.

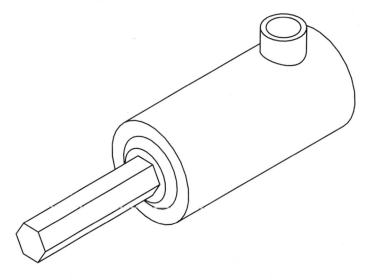

Figure 15.3 The single acting, nonrotating cylinder with hexagonal shaft is more commonly used in pneumetic applications than in hydraulics.

mining pneumatic cylinder speed. In fact, three pressures must be considered—*system (driving) pressure, back-pressure*, and the *load-induced pressure*. The driving pressure is set by the pressure regulator, while back-pressure is controlled by a meter-out flow control system.

Meter-out flow control is the preferred method of using flow control valves in pneumatic systems. The effectiveness of *meter-in flow control* is marginal to poor because of compressibility effects. The use of meter-out flow controls holds a back-pressure on the cylinder and reduces the tendency to surge or run away that is often found in meter-in applications. Even when using a meter-out system, however, calculations of speed and acceleration are uncertain.[1]

Many manufacturers offer flow control valves that mount directly into the ports of pneumatic cylinders. These devices improve speed control, because the amount of air that is between the cylinder and the valve (and which must be compressed during the control process) is very low. Such an arrangement is shown in Figure 15.4.

The discussion so far has assumed that we need to reduce the cylinder speed. Sometimes, however, we would like to increase the speed. The same problems still must be considered. In this case, a very effective technique is to reduce the back-pressure on the cylinder, usually done using a *quick exhaust* valve. This valve (which we'll discuss later in this chapter) is mounted in or very near the cylinder exhaust port. It allows air to escape from the cylinder directly to the atmosphere without flowing through the system piping and directional control valve.

In Chapter 14, we learned how to calculate the volume of air needed to operate a pneumatic cylinder. We then used this volume requirement to size the compressor for the system. Although we found the compressor size in scfm

Figure 15.4 A cylinder with flow control valves mounted in ports improves speed control.
Source: Courtesy Bimba Manufacturing Company, Monee, Ill.

or m^3/min, this is not a flow rate in the usual sense. Unless the compressor is being used to drive the cylinder directly (without a receiver in the system), the compressor size has little to do with cylinder speed.

15.3 Air Motors

Air motors are very similar to hydraulic motors in both design and operation. The most common air motor types are *vane, axial piston*, and *rotary piston*. The vane motor is the most widely used of the three.

Calculations of *air motor torque* and *power output* are the same as for hydraulic motors, although, because of the compressibility of the air, the calculations are often not as accurate as for hydraulic motors. Again, compressibility makes pressure as well as flow rate a significant consideration.

One of the most important considerations in selecting an air motor is its *starting torque*, defined as the maximum torque produced from a standstill under load at a given air pressure. Starting torque is always lower than *stall torque*, which is found by applying braking to an air motor running at full speed to determine the load required to stop the motor. Further, starting torque will vary from one start to the next, because it depends on the relative position of the moving mechanism (vanes or pistons) relative to the inlet and outlet ports. Therefore, motor manufacturers provide a minimum starting torque value as well as performance charts for their air motors. Figure 15.5 shows the performance curves for a typical air motor.

The *rated speed* of an air motor is the speed at which it produces the maxi-

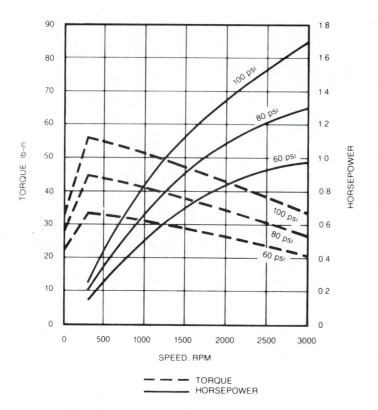

Figure 15.5 Horsepower-torque characteristics are provided by motor manufacturers. This is for a typical rotary vane air motor.
Source: Courtesy Gast Manufacturing Company, Benton Harbor, Mich.

mum power output. For most motors, the rated speed occurs at about 50 percent of the maximum speed at no load. The *rated power* of an air motor is the maximum power output with a 90 psi (about 625 kPa) line pressure.

Gast Manufacturing Corporation suggests that a rule of thumb in selecting air motors for a particular application is to choose a motor that can produce the required starting torque and power output at about two-thirds of the available air pressure.[2] Gast also suggests that the following points be kept in mind concerning air motors:

- An air motor slows down as its load increases. At the same time, its torque output increases to match the load. It will continue to provide increased torque up to stall, and it can maintain stall indefinitely without harming the motor.
- As the load is reduced, motor speed increases and torque decreases.
- When the load is increased or decreased, speed can be maintained constant by a corresponding increase or decrease in air pressure. This can be accomplished either by adjusting the pressure regulator to control inlet

pressure or by a meter-out flow control valve that affects the back-pressure.
- Because starting torque is lower than running torque, it may be necessary to provide additional pressure for starting under heavy loads.
- Air consumption increases with increased speed and pressure.

Air motors are typically very inefficient. It is common for the overall system efficiencies of pneumatic systems driving air motors to be less than 20 percent. Therefore, the applications of air motors are usually limited to those where they offer significant benefits over electric motors and where the power output requirements, weight limitations, or other considerations do not warrant the use of hydraulic motors. One important advantage of air motors is that they produce no sparks and can therefore be used in applications where electric motors would constitute a fire hazard. Other advantages include lightweight, relatively low-cost, suitability for high-temperature environments (250°F, 121°C) and the ability to be reversed instantaneously.

15.4   Directional Control Valves

As we discussed earlier, many pneumatic components are similar in construction and operation to their counterparts in hydraulic systems. While the same is true for *directional control valves*, the construction of some pneumatic valves is different from that of hydraulic valves in order to reduce internal leakage. There can also be some differences in the ways in which the valves are used. Like hydraulic directional control valves, pneumatic directional control valves come in a variety of configurations—two-way, three-way, and four-way, two-position and three-position—with a variety of actuators. Symbolically, the pneumatic valves are represented by the same symbols as the hydraulic valves.

Two-Way and Three-Way Valves

Two-way and *three-way* directional control valves are found more commonly in pneumatic applications than in hydraulic applications. While many of these valves provide the familiar control of cylinders or motors, they are also commonly used to provide pilot signals for pilot-operated devices. In the section on pneumatic controls later in this chapter, we'll see that some of these valves are used in the same way as electrical switches. They even use the same symbols as electrical switches in these applications.

Figure 15.6 shows the symbols that represent two-way and three-way valves. Notice the difference in the terminology used in describing the unactuated valve condition—*normally passing* instead of normally open and *normally not-passing* instead of normally closed.

Three-Position Valve Configurations

There are some configurations of *three-position* pneumatic valves that are not common in the hydraulic valves. Instead of having four ports, many valves have five ports that can be used in several different ways. Look, for instance,

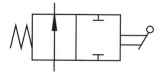

Two-way normally passing

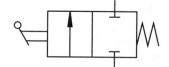

Two-way normally not-passing

Three-way normally passing

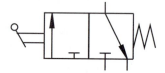

Three-way normally not-passing

Figure 15.6 Two-way and three-way valve symbols depict normally passing and normally not-passing positions.

at Figure 15.7. These symbols represent four-way three-position spools with five ports. There are several possible configurations for the center position, but the three shown here are the most common. Notice that the names given to the center positions differ from similar four-port configurations. The *blocked center* valve has all five ports blocked. In the *pressure center* configuration, the pressure port is connected to the two cylinder ports. In the *exhaust center* spool, the cylinder ports are connected to the exhaust ports directly across from them.

Each of these configurations offers a great deal of flexibility in application. The *blocked center* valve, for example, might be connected to the pressure line with the other two ports used as exhausts Figure 15.8a. This configuration would function the same way as a closed center, four-way valve, and is termed a *four-way, five-port, blocked center directional control valve*. If, however, the same valve is connected as shown in Figure 15.8b, it gives us the ability to operate a double-acting cylinder with two different pressures—say, high pressure for extending and low pressure for retracting. In this case, it becomes a *five*-way, five-port, blocked center directional control valve.

The *pressure center* also offers some interesting possibilities. When used with one double-acting cylinder, as shown in Figure 15.9a, the system performs as if the valve were a four-way, three-position valve with a regenerative center. Notice the curved dashed lines at the exhaust ports of the valve symbols. These indicate that the air is exhausted directly to the atmosphere.

Connecting the valve to two single-acting cylinders, as in Figure 15.9b, results in a completely different function. With the valve in the center position, both cylinders are extended. Shifting the spool to the right extends cylinder 1 and rectracts cylinder 2. Shifting the spool to the left retracts cylinder 1 while extending cylinder 2. There are several other possibilities as well.

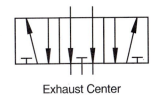

Exhaust Center

Pressure Center

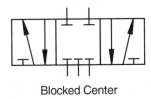

Blocked Center

Figure 15.7 Directional control valve symbols for five-port, four-way valves depict different center configurations. These are the three most common.

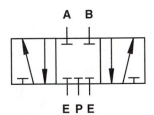

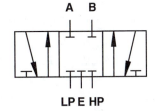

a. Four-way, five-port, blocked center.   b. Five-way, five-port, blocked center.

Figure 15.8 Options for five-port valves include a four-way and a five-way blocked center valve.

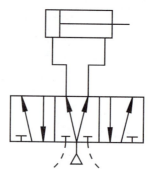

a. Five-port valve operating a double acting cylinder.

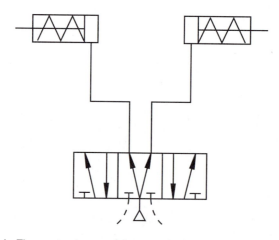

b. Five-port valve operating two single acting cylinders.

Figure 15.9 The pressure center configuration of a five-port valve can be connected to operate one double acting cylinder or two single acting cylinders.

Directional Control Value Mechanisms

The mechanisms that control flow direction may be either sliding devices or poppets. *Poppet valves* are ideal for applications using high flow rates, because a poppet can open a very large flow path with very little motion. Its short stroke allows very rapid response times when actuated by small solenoids. These are usually three-way, solenoid-actuated, pilot-operated valves. Four-way configurations are usually two three-way valves assembled to operate in opposite directions in response to the action of a single solenoid (Figure 15.10).

Sliding mechanisms come in at least five varieties—lapped spool, packed spool, packed bore, sliding plate, and rotary disk. The *lapped spool* valve is

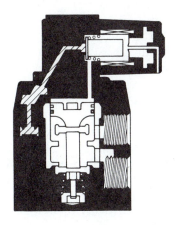

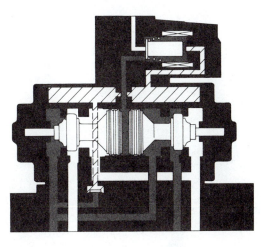

Figure 15.10 Poppet-type directional control valves are ideal for
applications using high flow rates.
Source: Courtesy of Parker Hannifin Corporation.

very similar to spool-type hydraulic valves. Because of the low viscosity of the
air, these spools must have very small radial clearances and finely machined
(lapped) surfaces to prevent significant leakage. Even so, they allow considerably more leakage than other configurations. In addition, they are highly susceptible to wear and jamming from contaminants in the air. They do offer
several advantages over other types, though, such as low operating force requirements (the leaking air acts as an air bearing) and more flow configurations
than are available in other designs.

The *packed spool* valve features grooves containing elastomeric seals. These
valves are virtually leak-proof when properly maintained. They are more tolerant of contaminants then the lapped spool valves; however, they have higher
breakout and operating force requirements and are susceptible to seal harden-

ing due to high temperature and chemical effects as well as seal swelling or hardening due to incompatibilities with lubricants. The seals also tend to adhere to the valve bore when left stationary for long periods.

A *packed bore* valve is simply a packed spool valve in reverse. A metal spool operates in a bore that has elastomeric seals between the ports. More flow path configurations are available with the packed bore than with the packed spool.

Figure 15.11 shows the action of a *sliding plate* valve. Directional control is achieved by sliding the mechanism (sometimes called a *D-slide*) to cover and uncover ports. The supply pressure on the top of the plate helps to seal the valve and prevent leakage. This same pressure, however, can mean that high forces are required to shift the valve. These valves are highly tolerant of contaminants and temperature extremes and are seldom affected by chemical compati-

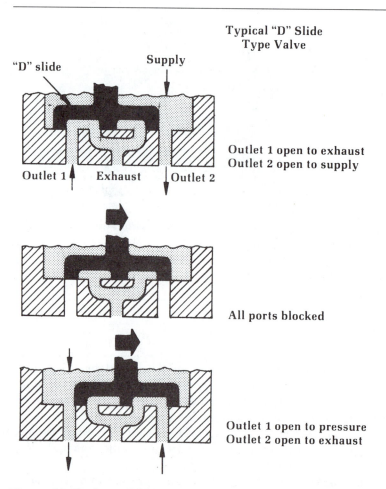

Figure 15.11 In a sliding plate valve, a D-slide covers and uncovers the ports to provide directional control.

Source: Courtesy of Parker Hannifin Corporation.

bility, lubricants, and so forth. When properly maintained (which primarily means lubricated), these valves can provide many years of trouble-free operation.

The *rotary disk* valve (Figure 15.12) is a variation of the sliding plate valve. Flow paths are selected by the positioning of the rotating disk. The port pattern of the disk determines the function of the valve in the neutral position.

15.5   Flow Control Valves

Flow control valves for pneumatic circuits are virtually identical to those used in hydraulic circuits. As in their hydraulic counterparts, *needle valves* are the most common flow control valves used for speed control in pneumatic systems. They normally include an internal *check valve* to allow free flow in one direction.

Because of the compressibility of gases, accurate speed control is difficult at best. It depends on all aspects of the forces involved, including the load itself, internal friction, and acceleration forces. Pneumatic speed controls are far more susceptible to changes in these factors than hydraulic systems. In

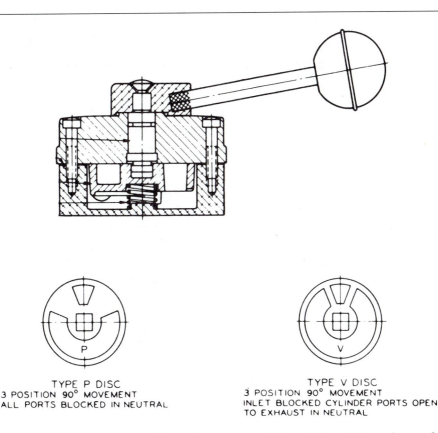

TYPE P DISC
3 POSITION 90° MOVEMENT
ALL PORTS BLOCKED IN NEUTRAL

TYPE V DISC
3 POSITION 90° MOVEMENT
INLET BLOCKED CYLINDER PORTS OPEN
TO EXHAUST IN NEUTRAL

Figure 15.12 The rotary disc valve is a variation of the sliding plate valve.
Source: Courtesy of Parker Hannifin Corporation.

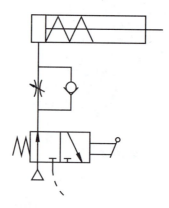

Figure 15.13 A flow control valve can provide retraction speed control of a single-acting cylinder.

fact, where high degrees of accuracy are required, hydraulic *dashpots* are often used.

As we saw previously, meter-out circuits are preferred for pneumatic speed control. Flow control valves may be installed either upstream or downstream of the directional control valve, depending on the circuit design and the functions desired. Figures 15.13 and 15.14 show two applications for single-acting cylinders. In the first circuit (Figure 15.13), the retraction speed of the cylinder is controlled, while in the second, the extension speed is controlled. Notice that the flow control valve is connected to the exhaust port in Figure 15.14 and that it includes a bypass check valve so that retraction will not be hindered.

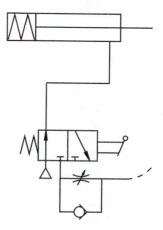

Figure 15.14 A flow control valve can provide extension speed control of a single-acting cylinder.

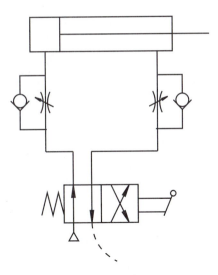

Figure 15.15 Speed control of a double-acting cylinder can be provided by two meter-out flow control valves located in the lines between the cylinder and the four-way directional control valve.

In Figure 15.15 we see a double-acting cylinder with speed control in both directions. In this case, a four-way, four-port directional control valve is used. A variation of this circuit is shown in Figure 15.16. In this circuit, the installation has been simplified somewhat in two ways. First, there is no need for the bypass check valves, because the flow control valves are installed in lines that serve only as exhaust lines. The second advantage is that the flow control valves can be installed directly in the exhaust ports of the directional control

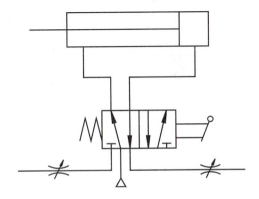

Figure 15.16 A five-port valve can simplify speed control of a double-acting cylinder as shown in this simplified installation.

valve. This eliminates some of the fittings required for the circuit in Figure 15.15.

In some applications, it may be desirable to *increase* the rate at which the air is exhausted rather than decrease it as we do with a meter-out circuit. For this purpose, we install a device called a *quick exhaust valve* (Figure 15.17) between the exhaust ports of a cylinder or motor and the directional control valve, as shown in Figure 15.18.

When the cylinder is being retracted, the pressure in the line to the rod end acts as a pilot to shuttle the quick exhaust valve to the right to allow normal air flow. Shifting the directional control valve to the left causes the cylinder to extend. Even a modest back-pressure in the exhaust line will provide sufficient pilot pressure to shift the quick exhaust valve to the left. This opens up an oversized exhaust port and allows the rod end air to escape with essentially no back-pressure. The cylinder can consequently extend much more rapidly than if the flow were forced through the lines, fittings, and directional control valve. Typical applications for quick exhaust valves are in any type of percussive tool where high extension speed of the cylinder is required to provide high impact forces (such as impact hammers and riveters).

15.6 Valve Sizing

It is important to use properly sized components in pneumatic systems. If components are oversized, unnecessary initial cost is involved. If they are undersized, excessive pressure drops will result, leading to high maintenance and operating costs. Therefore, in selecting components, always attempt to find the devices that will have the least pressure drop at the flow rate that will normally be used in the system without being oversized. Branches with different flow rates may require different size components. Usually, pressure drop information for the components we have discussed so far is obtained

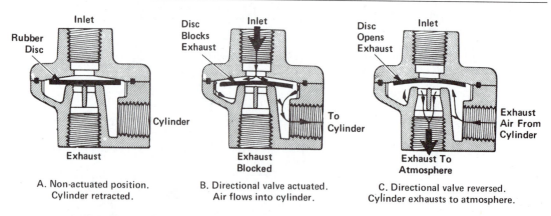

A. Non-actuated position. Cylinder retracted.

B. Directional valve actuated. Air flows into cylinder.

C. Directional valve reversed. Cylinder exhausts to atmosphere.

Figure 15.17 Quick exhaust valves increase the rate at which air is exhausted.

Source: *Industrial Fluid Power*, vol. 1. Published by Womack Educational Publication.

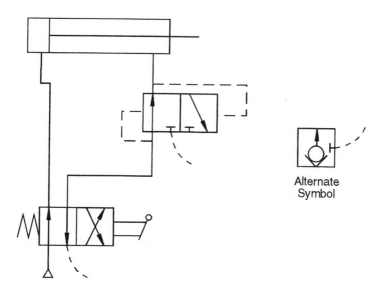

Alternate
Symbol

Figure 15.18 The quick exhaust valve is installed between the exhaust ports of the cylinder and the directional control valve.

from manufacturer's catalogs—generally in the form of graphs. The sizing of pneumatic valves is based on the *flow coefficient* (termed the C_v value) of the valve.

While discussing hydraulic flow control valves in Chapter 9, we talked about the standard orifice equation and its extension to flow control valves. That equation used a C_v term that included, among other things, the acceleration of gravity and the specific weight of water from the orifice equation as well as the valve characteristics based on the size and shape of the flow area. The C_v value for a pneumatic valve is based on the same concept; that is, it expresses the air flow through the valve at a pressure differential of 1 psi. Because of the compressibility of air, however, some additional modifications of the standard orifice equation are required. While several versions of the C_v equation are currently used, the National Fluid Power Association (NFPA) recommends Equation 15.1.

$$C_v = \frac{Q}{22.48} \sqrt{\frac{T_1 \times G}{\Delta p \times p_2}} \tag{15.1}$$

where

C_v = flow capacity coefficient.

Q = air flow rate in scfm.

T_1 = inlet temperature in °R.

G = specific gravity of air or gas ($G = 1$ for air) (see following note).

Δp = pressure drop through the valve in psi ($p_1 - p_2$).

p_2 = outlet pressure in psia.

Note: The G used in this equation is *not* the same as the Sg used in the hydraulic valve equation. The specific gravity of air at 59°F relative to water is 0.00121. The G used here is based on the specific gravity of the gas compared to air at standard conditions (14.7 psia, 59°F, 36 percent relative humidity). The constant, 22.48, accounts for this difference as well as the temperature effect on the density of the air (hence, the specific weight and specific gravity). Since the effect of relative humidity on the density of air is only about 0.6 percent over the range of 0 to 100 percent relative humidity, this effect is commonly ignored.

Example 15.1

An impact wrench uses 40 scfm of air at 90 psig and 75°F. What size valve (C_v) would be required to provide this air with a pressure drop of no more than 10 psi? Atmospheric pressure is 14.8 psia.

Solution

Equation 15.1 requires T_1 in °R and p_2 in psia.

$$T_1 = 75 + 460 = 535°R$$

$$p_2 = 90 + 14.8 = 104.8 \text{ psia}$$

Note that p_2 is the pressure *at the tool.* If we assume that the valve is near to the tool and ignore any friction losses, we can assume that this is also the pressure at the valve outlet. Putting these values into the equation, we get

$$C_v = \frac{Q}{22.48} \sqrt{\frac{T_1 \times G}{\Delta p \times p_2}}$$

$$= \frac{40 \text{ ft}^3/\text{min}}{22.48} \sqrt{\frac{535°R \times 1}{10 \text{ psi} \times 104.8 \text{ psia}}}$$

Therefore,

$$C_v = 1.27$$

Thus, any valve with a C_v of 1.27 or higher could be used.

We can rearrange Equation 15.1 to find the flow rate through a valve. Solving that equation for Q gives

$$Q = 22.48C_v \sqrt{\frac{\Delta p \times p_2}{T_1 \times G}} \tag{15.2}$$

Example 15.2

Determine the flow rate in scfm through a valve that has a flow coefficient of 2.3 if the pressure drop across the valve is 15 psi and the outlet pressure is 80 psia. The air temperature is 75°F.

Solution

The pressure is given as absolute, but we must convert the temperature to °R.

$$T = 75 + 460 = 535°R$$

Using these values in Equation 15.2 gives

$$Q = 22.48 \times 2.3 \times \sqrt{\frac{15 \text{ psia} \times 80 \text{ psia}}{535°R \times 1}}$$

$$= 77.44 \text{ scfm}$$

While this calculation seems to be fairly straightforward, there is a significant limitation to the application of Equations 15.1 and 15.2 that is related to the velocity of the gas flowing through the valve. Let's illustate the problem by discussing the partial circuit shown in the illustration for this example.

If p_1 and p_2 are equal, then Equation 15.2 tells us that there will be no flow through the flow control valve. Let's hold p_2 constant and adjust the regulator to increase p_1. As p_1 increases, air begins to flow, as predicted by Equation 15.2. In fact, the equation indicates that the higher p_1 becomes, the more flow there will be. This is true only to a certain point, though. Remember that as the flow rate increases, the velocity of the air flowing through the valve must increase also. When the velocity reaches the speed of sound (Mach 1), the valve is said to be *choked,* and no further increase in velocity is possible. This means that the flow rate cannot be increased either. The speed of sound, or *sonic velocity,* occurs in pneumatic devices when the ratio of the downstream pressure (p_2) to the upstream pressure (p_1) is 0.533; that is, $p_2/p_1 = 0.533$. Once this *critical pressure ratio* has been reached, neither increasing p_1 nor decreasing p_2 will have any effect on the flow rate. In other words, reducing p_2/p_1 to less than 0.533 will have no effect on the flow rate. This is true for all gas flow components (with the exception of a converging-diverging nozzle). When p_2/p_1 is less than the critical value, use a value for p_2 in Equation 15.2 that is equal to $0.533p_1$. This value for p_2 is termed the *critical pressure.* Remember that this will change Δp also, since $\Delta p = p_1 - p_2$.

Example 15.3

Determine the flow rate in scfm through a valve that has a C_v of 1.8. The valve is exhausting to atmosphere (14.7 psia). The pressure upstream of the valve is 50 psig and the temperature is 90°F.

Solution

Converting temperatures and pressures gives

$$p_1 = 50 + 14.7 = 64.7 \text{ psia}$$
$$p_2 = 14.7 \text{ psia}$$
$$T = 90 + 460 = 550°R$$

Before using Equation 15.2, we need to check the pressure ratio:

$$\frac{p_2}{p_1} = \frac{14.7}{64.7} = 0.227$$

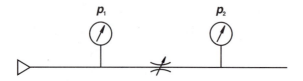

Example 15.3

This is below the critical ratio, so the valve is choked. Therefore, we need to find the critical value for p_2:

$$p_2 = 0.533p_1 = 0.533 \times 64.7 \text{ psia}$$
$$= 34.5 \text{ psia}$$

Thus,

$$\Delta p = p_1 - p_2 = 64.7 - 34.5 = 30.2 \text{ psia}$$

We now use these values of p_2 and Δp to find the flow rate:

$$Q = 22.48 \times 1.8 \times \sqrt{\frac{30.2 \text{ psia} \times 34.5 \text{ psia}}{550°\text{R} \times 1}}$$
$$= 55.7 \text{ scfm}$$

Remember, you must always check the pressure ratio to determine if the valve is choked before you attempt to calculate the flow rate through the valve.

15.7 Valve Mounting

Many standards organizations, including NFPA, ANSI, and ISO, are attempting to standardize pneumatic valve mounting dimensions in much the same way that hydraulic valve subplate mounts have been standardized. The standard that is emerging is ISO 5599. This standard has been accepted by most European valve manufacturers and has been incorporated into their valve designs. It is gradually gaining acceptance in the United States, but many manufacturers are waiting for an opportune time to make the change rather than simply scrapping all their current products and starting over.

Figure 15.19 illustrates the concept of stacking subplates to reduce the number of fittings required in a circuit. Multiple-mount manifolds are also available. This figure also shows some of the mounting interface patterns currently in use for directional control valves. Note that the ANSI and Ford interfaces shown are not interchangeable with the ISO interface.

15.8 Pneumatic Controls

One of the advantages of pneumatics is its ability to be based in applications where there is a fire hazard because motors and cylinders do not generate arcing or sparks. This attraction is further enhanced by the ability of pneumatic circuits to provide both power output and automatic control without the need for electrical switches and sensors. Of course, electrical control can be applied

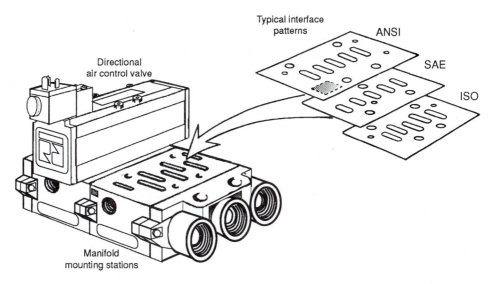

Figure 15.19 These valve-mounting interfaces look similar but are not interchangeable. Stacked subplates reduce the number of fittings required in a circuit.

Source: Courtesy Ross Operating Valve Company, Troy, Mich.

very successfully to pneumatic circuits. Solenoid valves, proportional valves, electropneumatic servovalves, and programmable logic controllers are used extensively in pneumatics. There are some hazardous applications, however, where other means of control are needed, and these applications are well served by *pneumatic controls*.

Pneumatic controls can also be attractive from a cost standpoint. A pneumatically operated machine requires that air be piped to it; however, if it were designed using pneumatic controls instead of electrical controls, there would be no need to provide electricity at the machine. This could result in a considerable cost saving when installing the machine.

There are three ways in which pneumatic control can be accomplished—pilot-operated valves with pneumatic pilot valves, pneumatic logic, and *fluidics*, a very specialized field that makes use of the phenomena of compressible fluid dynamics. Fluidics is seldom seen in common industrial applications and is beyond the scope of this text.

Pilot Valve Controls

Pilot-operated valves are positioned by directing pressurized air into the pilot ports. The most common of the pilot-operated valves are the directional control valves such as those discussed in Chapter 7. In pneumatic control systems, the pilot valves (also called *input valves*) used to direct the pilot signals to these valves are used in the same way that electrical switches are used and are indicated in circuit diagrams by symbols that are very similar to the symbols used for electrical switches. One such circuit is shown in Figure 15.20. This circuit is designed to provide a one-time automatic retraction of the cylinder after it has extended to a set position. The *START valve* here is a three-way, two-position, pushbutton-operated, spring-returned, normally not-passing directional control valve that can be represented by the symbol shown in the figure. It performs the same function as an electrical pushbutton, except that it passes air instead of electricity.

The *LIMIT valve*, which looks both symbolically and physically like an electrical limit switch, is the same type valve as the START valve, although its construction is considerably different because of its purpose in the circuit. Rather than a pushbutton, the LIMIT valve has a lever actuated by the cylinder or some other part of the machine when the cylinder has reached a predetermined position. The open circles shown at the contact points of both the START and LIMIT pilot valves indicate that the pilot line from the directional valve is exhausted when the pilot valve is not actuated. Figure 15.21 shows the symbols for some of the more common pilot valves.

To operate the system represented in Figure 15.20, the operator pushes the START pushbutton. This directs pilot pressure to the left-hand end of the directional valve, shifting it to the right. The air that was in the right-hand end of the directional valve and the air line leading to it is pushed out through the exhaust port of the LIMIT valve. When the operator releases the START

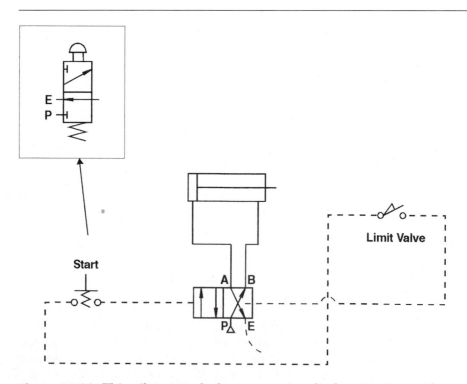

Figure 15.20 This pilot control of a pneumatic cylinder circuit provides automatic retraction.

Limit Valves (3-way)

**Normally
Not-Passing**

**Normally
Passing**

**Normally
Not-Passing
Held Passing**

**Normally
Passing
Held Not-Passing**

Limit Valves (4-way)

Pushbutton Valves (3-way)   (4-way)

**Normally
Not-Passing**

**Normally
Passing**

**Latching
Pushbutton**

Selector Valves   (4-way)

Not-Passing

Passing

Sensors (all shown normally passing, but may also be normally not-passing.)

Liquid Level   **Pressure**   **Flow**   **Temperature**

Figure 15.21 ANSI graphic symbols for pneumatic input valves and sensors.

pushbutton, the spring returns it to its unactuated position. This relieves the pressure from the left-hand end of the directional valve. That valve remains in the right-hand position, however, because there is no spring to push it back to its original position.

With the directional valve shifted to the right, air passes from P to A in the figure, causing the cylinder to extend. When the extending cylinder rod contacts

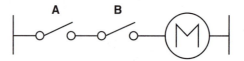

Figure 15.22 Series electrical circuit for operating an electric motor.

the LIMIT valve, that valve shifts and directs pilot pressure to the right-hand end of the directional valve. The directional valve shifts to the left, porting air from P to B to retract the cylinder automatically.

Boolean Algebra

Before we begin our discussion of pneumatic logic, it might help to take a quick look at *Boolean algebra*, which is simply a short-hand method for expressing (or analyzing) the logic (or series of events) that lead to a particular action. For instance, if we are to describe the operation of the electrical circuit represented by the ladder diagram in Figure 15.22 verbally, we might say, "If switch A is closed *and* switch B is closed, then the motor will start." Both switches must be closed in order to start the motor.

We could write a Boolean expression for this logic much more simply. When two (or more) actions are required to produce a desired result, we say that those are *series* actions. Boolean algebra expresses the idea that both actions are required by AND. Symbolically, AND is represented by a dot ($\cdot$). Since we have a series situation in Figure 15.22, we can write the Boolean equation for the circuit as

$$A \cdot B = M$$

We would say this as, "A AND B equals M."

In Figure 15.23 we have a parallel logic circuit for starting the motor. In this circuit, closing either switch will start the motor. Here, we say, "If switch A is closed *or* if switch B is closed, then the motor will start." The "or" indicates

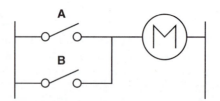

Figure 15.23 Parallel electrical circuit for operating an electric motor.

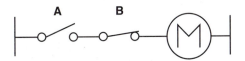

Figure 15.24 Series circuit using normally open and normally closed switches.

that we have a parallel situation. Thus, the Boolean algebra term to indicate parallel logic is OR and is represented by a plus sign (+). Accordingly,

$$A + B = M$$

would be read as, "A OR B equals M."

A third term used in Boolean algebra is NOT, meaning that an action has not taken place. NOT is represented by a dash over the letter. In Figure 15.24 we see a series circuit using one normally open switch and one normally closed switch in our motor circuit. In this arrangement, if switch A is closed and switch B is not opened, then the motor will start. The Boolean expression for this is written

$$A \cdot \overline{B} = M$$

which we read as, "A AND NOT B equals M."

Boolean algebra adheres to all of the rules of regular algebra. For instance,

$$A \cdot B = B \cdot A$$

and

$$A + B = B + A$$

There is also a set of theorems that define equalities, listed in Table 15.1.

Table 15.1 Boolean Algebra Theorems

Series (AND)	Parallel (OR)
$A \cdot A = A$	$A + A = A$
$A \cdot (A + B) = A$	$A + (A \cdot B) = A$
$(A + \overline{B}) \cdot B = A \cdot B$	$(A \cdot \overline{B}) + B = A + B$
$A \cdot B \cdot C = (A \cdot B) \cdot C$ $= A \cdot (B \cdot C)$ $= (A \cdot C) \cdot B$	$A + B + C = (A + B) + C$ $= A + (B + C)$ $= (A + C) + B$
$(A + B) \cdot (A + \overline{B}) = A$	$(A \cdot B) + (A \cdot \overline{B}) = A$
$(A + B) \cdot (A + \overline{C}) = A + (B \cdot C)$	$(A \cdot B) + (A \cdot C) = A \cdot (B + C)$
$\overline{A \cdot B \cdot C} = \overline{A} + \overline{B} + \overline{C}$	$\overline{A + B + C} = \overline{A} \cdot \overline{B} \cdot \overline{C}$

Boolean algebra is used to indicate the condition of system inputs and outputs based on binary signals, with 1 indicating that a signal is present and 0 indicating that no signal is present. Referring again to Figure 15.22, we can illustrate this concept using a *truth table*. We will say that 1 means that a signal is present (that is, that the switch is closed) and 0 means that the switch is open. We will also say that, on the output side, 1 means that we have an output, while 0 means that there is no output. The *truth table* must consider all possible combinations of the input logic (the switches, in this case).

We can start the truth table with any combination we wish, so let's say that both switches are open. This means that there is no signal from either, so $A = 0$ and $B = 0$. Since $A \cdot B = 0 \cdot 0 = 0$, this means that there is no output, so $M = 0$ also. Thus, the first line of the truth table would be

A	B	M (= A · B)
0	0	0

Our next step might be to close switch A. Now we have $A = 1$ and $B = 0$, so that $A \cdot B = 1 \cdot 0 = 0$, and $M = 0$. The truth table now expands to

A	B	M
0	0	0
1	0	0

We continue to evaluate the circuit by next considering A open and B closed, and finally both A and B closed. Our truth table fills out to show

A	B	M
0	0	0
1	0	0
0	1	0
1	1	1

This shows us that the only case in which we can have an output (the motor running) is when both switches are closed.

The truth table for Figure 15.23 would be

A	B	M
0	0	0
1	0	1
0	1	1
1	1	1

This shows us that if either switch is closed, we get an output.

The truth table for Figure 15.24 is a little more tricky, because we have the NOT B situation. This means that whatever condition we assign to B, we must use the opposite condition to determine the output state. We might write the truth table as

A	B	$\overline{\text{B}}$	$M\,(= A \cdot \overline{B})$
0	0	1	0
1	0	1	1
0	1	0	0
1	1	0	0

It is sometimes easier to think in terms of actuating or not actuating switches. Here, we might say that *actuating* (changing the position of a switch) is 1, while not actuating the switch is 0. This way, we consider the first line of the truth table to show that if switch A is not actuated (that is, not closed) and switch B is not actuated (that is, not opened), then there will be no output. This would then show us (in the second line) that the only way to get an output would be to actuate A and not actuate B.

We can also use truth tables to verify equations involving the control logic for various circuits. For example we stated earlier that

$$A + B = B + A$$

We can prove this by using a truth table; thus,

A	B	A + B	B + A
0	0	0	0
1	0	1	1
0	1	1	1
1	1	1	1

The last two columns are identical; therefore, the equation is valid. Note that the table must include *all* possible combinations of the variables (A and B, in this case) and that *every line* in the last two columns must be identical for the equality to be true.

Example 15.4

Use a truth table to prove that

$$B + (A \cdot \overline{B}) = A + B$$

Solution

For clarity of this illustration, we will add a column for $A \cdot \overline{B}$, which is normally not needed.

A	B	$A \cdot \overline{B}$	$B + (A \cdot \overline{B})$	A + B
0	0	0	0	0
1	0	1	1	1
0	1	0	1	1
1	1	0	1	1

The last two columns are identical, therefore, the equality is proved.

This technique can be very useful in analyzing complex control circuits. Often, it will show that there are more control elements than are needed. We will see in the next section that Boolean algebra also provides a simple way to describe pneumatic logic components and circuits.

Pneumatic Logic

Pneumatic logic is often referred to as *moving parts logic* or *MPL* to distinguish it from *fluidics* which deals with no moving parts. Pneumatic logic elements are devices that make use of diaphragms, disks, spools, orifices, and air chambers to control the direction of flow in a pneumatic circuit. There are several such devices, but we will look at only a few of them. Keep in mind during this discussion that the elements shown here are one manufacturer's approach. Other manufacturers accomplish the same thing using different hardware.

Since we started our Boolean algebra discussion with a series circuit, we will do the same for our logic component discussion. The devices that are used in these applications are called *elements*, so the device that performs this series function, shown in Figure 15.25, is called an *AND element*. In the element shown, if both inputs (*a* and *b*) are off, the output (*c*) is connected internally to the exhaust port. If only *a* is on, the diaphragm forces the actuator and poppet downward, closing the exhaust seat. There is still no output, however, because *c* is now connected internally to *b*, which is off. If only *b* is on, the poppet is forced against the poppet seat. There is still no output because *c* is again connected internally to the exhaust port. If both *a* and *b* are on, the

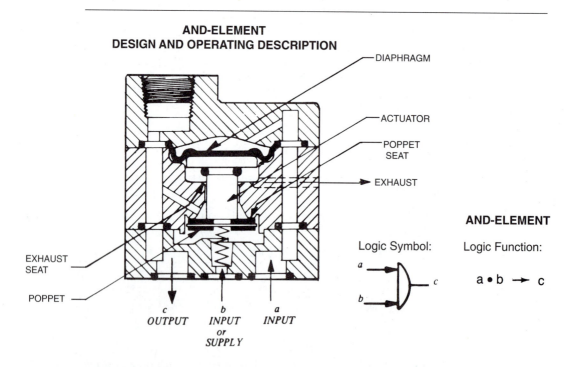

Figure 15.25 An AND logic element performs a series function.
Source: Courtesy The ARO Corporation, Bryan, Oh.

diaphragm forces the poppet down to close the exhaust port. The exhaust seat is held closed while the poppet seat is held open. This connects b to c, allowing air to flow from the input (b) to the output (c). The logic AND symbol indicates that both a and b must be pressurized to obtain an output from c.

For the element shown in this figure, the flow from b to c is 9.3 scfm at 100 psig. The opening time is 8 msec, while the closing time is 9.5 msec.

An OR element is shown in Figure 15.26. In this element pressurizing either a or b will cause flow through the device and give an output at c. The logic OR symbol indicates this situation. At 100 psig, the flow from a to c is 9.3 scfm, while the flow from b to c is 16.2 scfm. Response time of the element is 7 *msec*.

The NOT element in Figure 15.27 is designed so that an output from c will occur only when b is pressurized and a is not pressurized. When actuated, this device will pass 16.2 scfm at 100 psig. It opens in 8.5 *msec* and closes in 9 msec.

FLIP FLOP elements, such as the one shown in Figure 15.28 are memory elements that resemble pilot-operated spool valves in both their construction and operation. In the element shown here, input f has been pressurized, shifting the spool to the left and allowing flow from e to d. The spring-loaded ball will hold the spool in this position, even if f is depressurized, until input a is pressurized. At that time, the spool will shift to the right, providing a flow path from b to c. The element shown will pass 10 scfm at 100 psig. Shifting time is about 10 *msec*.

A *TIME DELAY ELEMENT*, such as the one shown in Figure 15.29, is essentially an AND element with an orifice that meters the flow of air from

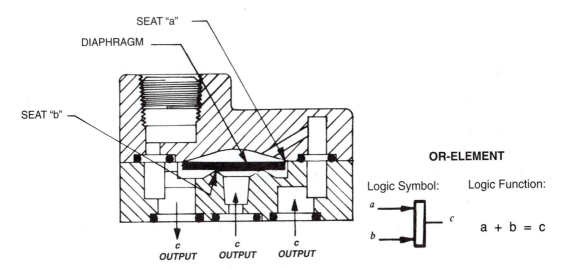

Figure 15.26 An OR logic element performs a parallel function.
Source: Courtesy The ARO Corporation, Bryan, Oh.

NOT-ELEMENT
DESIGN AND OPERATING DESCRIPTION

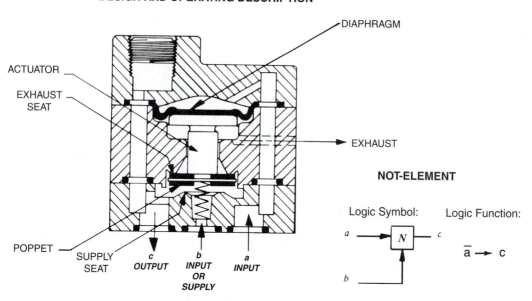

Figure 15.27 A NOT logic element provides an output (from the c port) only when b is pressurized and a is not pressurized.
Source: Courtesy The ARO Corporation, Bryan, Oh.

FLIP FLOP
DESIGN AND OPERATING DESCRIPTION

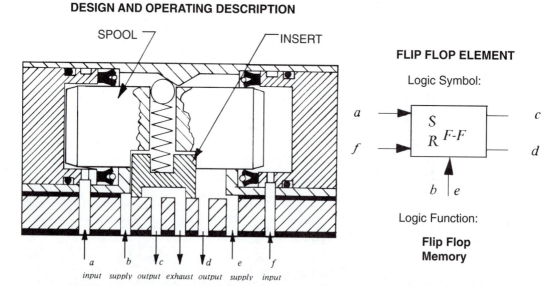

Figure 15.28 A FLIP FLOP logic element is similar to a spool valve. It is shifted by pressurizing either the a port or the f port.
Source: Courtesy The ARO Corporation, Bryan, Oh.

TIME DELAY RELAY
DESIGN AND OPERATING DESCRIPTION

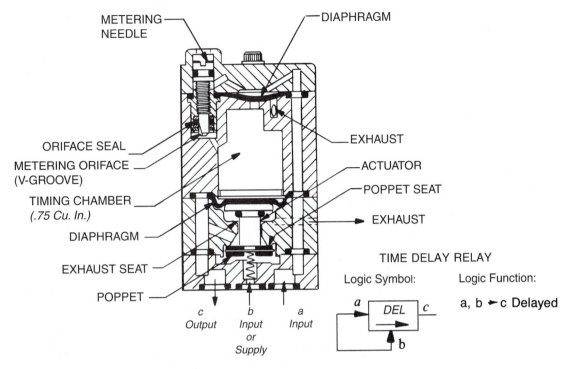

Figure 15.29 A TIME DELAY logic element requires that both a and b be pressurized in order to provide an output from c.
Source: Courtesy The ARO Corporation, Bryan, Oh.

input a into the timing chamber. When the pressure in the timing chamber reaches a preset value, the diaphragm pushes the actuator downward, forcing the poppet from its seat, and opening a flow path from b to c. Depending on the particular model used, the time delay may be fixed, or it may be adjustable over a range of 0.08 to 7.5 seconds. Additional time delay can be obtained by using a separate additional timing chamber termed an *accumulator*. The device passes 9.3 scfm at 100 psig.

A *pulse timer* is similar to the TIME DELAY element except that it opens immediately when the inlet line is pressurized, then closes when the chamber pressure reaches the preset value. The pulse duration may be fixed or adjustable from 0.08 to 7.5 seconds, depending on the model used.

Figure 15.30 is the circuit diagram for a common pneumatic logic circuit that uses several of the elements we have just discussed. This circuit is termed a *two-hand, anti-tiedown* circuit. It is a safety circuit that requires that both START pushbuttons be operated within a specified time, usually around 120 *msec* to start the system. Both pushbuttons must remain pushed or the opera-

CIRCUIT DIAGRAM

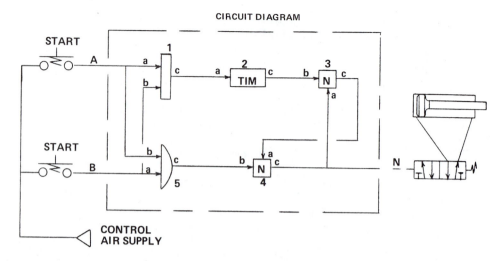

Figure 15.30 This two-hand anti-tiedown circuit requires that both START pushbuttons be operated within a specified time to start the system.
Source: Courtesy The ARO Corporation, Bryan, Oh.

tion will stop. Its purpose is to ensure that the operator's hands are safely away from the hazardous portion of the machine. The pushbuttons are separated to ensure that they cannot both be operated with one hand, one hand and elbow, etc. The time delay is set so that if the second pushbutton is not actuated within the specified time, the circuit will not operate. This prevents the operator from hitting both buttons in sequence with one hand.

Either START button can be pushed to start the sequence. The OR element (1 in the figure), needing only one input, immediately sends flow to the timer. After a preset delay, the timer opens and passes flow to the b port of NOT element (3). If the a port of NOT (3) is not pressurized, that element passes flow through port c, which pressurizes port a of the second NOT element(4). This internally directs the c port of NOT(4) to exhaust, so that there is no output to operate the directional valve. In effect, NOT (4) is closed.

To prevent NOT (4) from closing, the a port of NOT (3) must be pressurized before the timer actuates. This can occur *only* if the AND element (5) receives a signal from both pushbuttons within the timer delay period. If that occurs, then the AND element will pass flow through its c port to the b port of NOT (4). Since the a port of NOT (4) is not pressurized (because the timer has not yet actuated), flow passes through its c port. This does two things—it pressurizes the a port of NOT (3) so that it closes and remains closed, and it pressurizes the pilot port of the directional valve to shift that valve to the right and cause the cylinder to extend.

If either of the START buttons is released as the cylinder is extending, the AND element will close. This will stop the flow through NOT (4). As a result, the pilot flow to the directional valve will be lost, causing it to shift back to the left and retract the cylinder. In addition, the loss of flow from NOT (4) will depressurize the *a* port of NOT (3). This will open NOT (3) and pressurize the *a* port of NOT (4). This closes NOT (4) so that the system cannot be restarted until *both* buttons are released and reactuated.

The applications for pneumatic logic are virtually limitless. The designer's imagination and the ability to "think logic" are the only constraints. Some circuits are so frequently used that some manufacturers provide them in pre-packaged modules that are, in essence, pneumatic control computers.

15.9   System Design Considerations

When designing a pneumatic system, the major design considerations are performance and efficiency. The system must do the job it is designed to do, and it must do it efficiently. Otherwise, operating and maintenance costs might be unacceptably high.

Probably the most important factors in the overall efficiency of the system are pressure drops and air leaks. We've already touched on some aspects of pressure drop. In this section, we'll look at other aspects and at the effects of pressure drop on operating costs, then the effects of air leaks on costs.

System Piping

There are two basic plumbing layouts used for plant air distribution systems—dead-end and loop. Figure 15.31 shows a *dead-end* (or *grid*) *system*. It uses one main conduit (called simply a *main*) that begins at the receiver and terminates in a "dead end." Individual headers tap off the main to supply each tool or

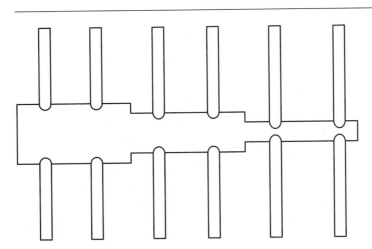

Figure 15.31 A dead-end piping system uses one main conduit that begins at the receiver and terminates in a dead end.

operating station. These systems are easy and inexpensive to install, but stations at the end of the line may have an insufficient air supply during times of heavy usage.

The *loop system* (Figure 15.32) is usually preferred because it is more efficient and more reliable. Since there is no "end of the line," the problem of air starvation is minimal. The loop system also lends itself to the use of multiple compressors to provide an extra air supply during high usage periods.

There can, of course, be a limitless number of combinations of grids with loops and loops with grids. Regardless of the configuration used, the sizing of the piping is critical to prevent excessive pressure drops.

The pressure drop in conduits is due to friction between the moving air and the inside walls of the conduit. As with liquid systems, the magnitude of the friction loss depends primarily on the velocity of the air flow. This velocity, in turn, depends on the air flow rate (Q), and the pipe diameter (d). The relationship can be calculated using the Harris equation (Equation 15.3).

$$\Delta p = \frac{c \times L \times Q^2}{3600 \times P_R \times d^5} \qquad \textbf{(15.3)}$$

where

Δp = pressure drop (psi).
L = conduit length (ft).
Q = air flow rate (scfm).
P_R = pressure ratio (pressure in conduit/atmospheric pressure).
d = conduit inside diameter (in).
c = an experimentally determined coefficient that depends on the type of conduit being used.

One of the most common types of pneumatic plumbing is Schedule 40 commer-

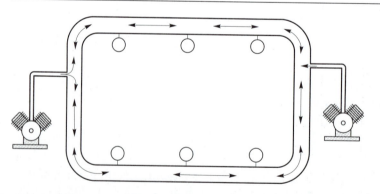

Figure 15.32 A pneumatic loop system is more efficient and reliable than a dead-end system.

Table 15.2 Values of $d^{5.31}$ for Schedule 40 Commercial
Steel Pipe

Nominal Pipe Size	d (inches)	$d^{5.31}$
$\frac{3}{8}$	0.493	0.0234
$\frac{1}{2}$	0.622	0.0804
$\frac{3}{4}$	0.824	0.3577
1	1.049	1.2892
$1\frac{1}{4}$	1.380	5.5304
$1\frac{1}{2}$	1.610	12.5384
2	2.067	47.2561
$2\frac{1}{2}$	2.469	121.4191
3	3.068	384.7707
4	4.026	1628.8448
5	5.047	5048.8738
6	6.065	14349.3284

Note: See Appendix D for values of d for various sizes and schedules of commercial steel pipe.

cial steel pipe. The value of c for this pipe has been found to be $0.1025/d^{0.31}$. Substituting this into Equation 15.3 gives

$$\Delta p = \frac{0.1025L \times Q^2}{3600\, P_R d^{5.31}} \tag{15.4}$$

The values of $d^{5.31}$ for some common Schedule 40 pipe sizes are tabulated in Table 15.2. In general use, the value of d can be found from tables such as those in Appendix D. The value of Δp can then be calculated using most hand-held calculators.

Example 15.5

Find the pressure drop thrugh 150 ft of 1½ in Schedule 40 pipe. The air flow rate through the pipe is 300 scfm and the receiver pressure is 125 psi. Atmospheric pressure is 14.7 psia.

Solution

The pressure ratio in this system is

$$P_R = \frac{139.7 \text{ psia}}{14.7 \text{ psi}} = 9.5$$

From Table 15.2, we find that $d^{5.31} = 12.5384$. Using Equation 15.4, we get

$$\Delta p = \frac{0.1025\, L \times Q^2}{3600 \times P_R \times d^{5.31}}$$

$$= \frac{0.1025 \times 150\text{ ft} \times (300\text{ ft}^3/\text{min})^2}{3600 \times 9.5 \times 12.5384}$$

$$= 3.23\text{ psi}$$

If we used 1 in Schedule 40 pipe for this application, the pressure drop would be 31.38 psi. By using 1 in instead of 1½ in pipe, we've increased the pressure drop by almost a factor of 10!

Changing the direction of the air flow—either by actually turning it or by pushing it through the passages of valves and other components—also causes a pressure drop. The easiest way to handle these losses is to consider the device—whatever it is—to be simply an equivalent length of pipe. Table 15.3 is a tabulation of the equivalent lengths of several devices for various sizes of Schedule 40 commercial steel pipe.

Table 15.3 Equivalent Lengths (in Feet) for Standard Fittings in Schedule 40 Commercial Steel Pipe

Nominal Pipe Size	45° Elbow	90° Elbow	Tees		Globe Valve*ª	Gate Valve*ª	Close Return Bend
			Run	Branch			
$\frac{1}{2}$	0.83	1.55	1.04	3.11	17.60	0.67	2.59
1	1.40	2.67	1.74	5.20	29.70	1.14	4.36
$1\frac{1}{2}$	2.14	4.02	2.68	8.10	45.50	1.74	6.70
2	2.75	5.20	3.44	10.30	59.00	2.24	8.60
3	4.10	7.70	5.10	15.40	87.00	3.34	12.80
4	5.40	10.10	6.70	20.10	114.00	4.35	16.80
5	6.70	12.60	8.40	25.20	143.00	5.50	21.00
6	8.10	15.10	10.10	30.30	172.00	6.60	25.30

ªFully opened

Example 15.6

The 1½ in piping system of Example 15.4 includes four close return bends, a globe valve, and branching flow through six tees. Determine the total pressure loss for the system.

Solution

From Table 15.3 we find that the total equivalent length for the fittings is

$$L_e = (4 \times 6.70) + 45.50 + (6 \times 8.10)$$
$$= 120.9\text{ ft}$$

We add this to the pipe length to get the total value of L for Equation 15.4.

$$\text{Total Length} = L = L_{\text{pipe}} + L_e = 150 + 120.9 = 270.9\text{ ft}$$

Our pressure drop becomes

$$\Delta p = \frac{0.1025L \times Q^2}{3600 \times P_R \times d^{5.31}}$$

$$= \frac{0.1025 \times 270.9 \times (300 \text{ ft}^3/\text{min})^2}{3600 \times 9.5 \times 12.5384}$$

$$= 5.83 \text{ psi}$$

The addition of the valve and fittings has increased the pressure drop by about 80 percent. This shows the value of keeping the number of such devices to the minimum necessary to accomplish the required task.

Piping Installations

Figure 15.33 illustrates two important aspects of pneumatic piping installations. First, piping should always be installed with a downward slope (up to 0.25 in per foot) in the direction of flow so that any moisture that gets past the dryer, etc, will migrate toward a trap to be drained off later. Second, headers (taps) for operating branches should be taken from the top of the pipe rather than from the bottom. This will help to prevent water, dirt, etc, from entering the branch piping.

The High Cost of Inefficiency

Inefficiencies in pneumatic systems stem from several sources—grossly over-sized compressors, overdrying for the application (it can happen, though not

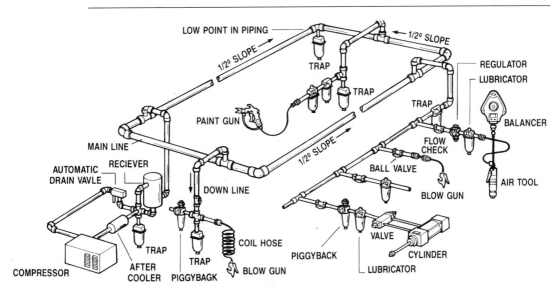

Figure 15.33 Pneumatic piping should always be installed with a downward slope in the direction of flow, and headers for operating branches should be taken from the top of the pipe rather than the bottom.
Source: Courtesy The ARO Corporation, Bryan, Oh.

often), high pressure drops in pipes and conditioners, and air leaks. Unlike inefficiencies in hydraulic systems, which show up as heat, problems in pneumatic systems usually show up as reduced work capacity of the tools and actuators.

Rated tool output depends very heavily on the pressure at the inlet to the tool. If the air pressure in lower than the rated pressure of the tool, the work capacity of the air (and, consequently, the tool) is decreased. For a system with a compressor discharge pressure of 100 psig, a pressure loss of 10 psi will result in a 14 percent loss in tool output. A 30 psi pressure drop reduces the tool output by 55 percent.[3]

How does this relate to cost? An example can help you to get a feel for this. There is a good deal of supposition in this example, but you'll get the idea. Suppose that a plant has 10 operators who make $12.00 per hour. Let's suppose that they operate tools that run continuously, perhaps in a grinding or polishing operation. If the tools are designed to operate with 100 psig at the tool, but actually have only 90 psig, the output from the 10 operators is reduced by 14 percent. This is equivalent to losing $1.68 per hour per operator, or $16.80 per hour total. If each operator works 250 days per year, 8 hours per day, the 10 operators work a total of 20,000 hours per year. Based on this, the loss in output costs $33,600 per year *in labor alone*. This doesn't account for the losses due to having fewer products to sell.

We can overcome the problem of low pressure at the tool by increasing the compressor discharge pressure. This has some costs associated with it, however. Let's examine them.

Example 15.7

A compressor delivers 300 scfm at 100 psig. The actual horsepower to drive the compressor is 76.8 HP. Find the percent increase in horsepower required if the compressor must operate at 110 psig to make up for a 10 psig line loss and deliver 100 psig to the tools. Assume an overall efficiency of 0.70.

Solution

We again apply Equation 14.14 to find the new horsepower required. In this case.

$$p_2 = 110 + 14.7 = 124.7 \text{ psia}$$

Therefore,

$$HP_T = \frac{14.7 \times 300 \times 3.5}{230} \left[\left(\frac{124.7}{14.7} \right)^{0.286} - 1 \right]$$

$$= 56.7 \text{ HP}$$

The actual horsepower, then, is

$$HP_A = \frac{HP_T}{\eta_o}$$

$$= \frac{56.7}{0.70} = 81.0$$

To find the percent increase in horsepower, we use

$$\text{Percent Increase} = \frac{HP_2 - HP_1}{HP_1} \times 100$$

$$= \frac{81.0 - 76.8}{76.8} \times 100 = 5.49\%$$

This shows that our power consumption is increased by 5.49 percent to make up that 10 psi pressure loss and keep the tools operating at their rated efficiency.

Now let's take this a step further and calculate the actual cost of the increased power requirement. We use Equation 15.5 to do this.

$$P = \frac{BHP \times 0.746 \text{ kW/HP}}{\eta_M \times PF} \tag{15.5}$$

where

$$P = \text{power in kW.}$$
$$BHP = \text{brake horsepower (the actual horsepower}$$
$$\text{output of the electric motor).}$$
$$\eta_M = \text{electric motor efficiency.}$$
$$PF = \text{power factor of the electric motor.}$$

Example 15.8

Using the information from Example 15.7, determine the annual cost to make up for the 10 psi pressure loss. Assume a motor efficiency of 0.9, a power factor of 1.0, and an electricity rate of $0.085/kWH. The system operates 2000 hours per year.

Solution

For this application, $BHP = 1.08 \times HP$. Therefore, for the 10 psig operation, we have

$$P = \frac{BHP \times 0.746 \text{ kW/HP}}{\eta_o \times PF}$$

$$= \frac{1.08 \times 76.8 \text{ HP} \times 0.746 \text{ kW/HP}}{0.9 \times 1.0}$$

$$= 68.75 \text{ kW}$$

The electricity cost at 100 psi is

$$\text{Cost} = \text{Power Consumption} \times \text{Operating Hours} \times \text{Electricity Rate}$$
$$= 68.75 \text{ kW} \times 2000 \text{ hr/yr} \times \$0.085/\text{kWH}$$
$$= \$11,688/\text{yr}$$

The *additional* cost of increasing the line pressure to 110 psi is

$$\text{Additional Cost} = 0.0549 \times \$11,688/\text{yr} = \$642/\text{yr}$$

This indicates that we could save labor costs of $33,600 and increase the annual production for an additional operating cost of only $642 per year. Thus, we would be much better off if we increased the pressure than if we continue to operate at the reduced pressure at the tool. This figure doesn't take into account the wear and tear on the system, increased cooling costs, etc., but these are not likely to exceed the labor and lost production costs.

Air leaks also contribute significantly to the operating costs, because additional air must be compressed to make up the loss. A system with sufficient leakage to be equivalent to a single 0.5 in (13 mm) hole can waste as much as 275 scfm (7.7 m³/min). This amounts to almost 12,000 scf (33,000 m³) of air per month.[4] Let's look at the cost of this leak.

Example 15.9

A compressor operating at 100 psig is required to make up 275 scfm leakage. Determine the additional operating cost per hour due to the leakage for an electricity cost of $0.085/kWH. Compressor efficiency is 0.7 and motor efficiency is 0.9. Assume $BHP = 1.08 \times HP$, and $PF = 1.0$.

Solution

Again, we apply Equation 14.14

$$HP_T = \frac{p_1 \times Q \times 3.5}{230}\left[\left(\frac{p_2}{p_1}\right)^{0.286} - 1\right]$$

$$= \frac{14.7 \times 275 \times 3.5}{230}\left[\left(\frac{114.7}{14.7}\right)^{0.286} - 1\right]$$

$$= 49.2 \text{ HP}$$

$$HP_A = \frac{HP_T}{\eta_o} = \frac{49.2}{.7} = 70.3$$

$$P = \frac{BHP \times 0.746 \text{ kW/HP}}{\eta_M \times PF}$$

$$= \frac{1.08 \times 70.3 \text{ HP} \times 0.746 \text{ kW/HP}}{0.9 \times 1.0}$$

$$= 63 \text{ kW}$$

For one hour, the cost is

$$\text{Cost} = 63 \text{ kW} \times 1 \text{ hr} \times \$0.085/\text{kWH} = \$5.36$$

Assuming that the system is never shut down, that comes to $46,910 per year, just to make up leakage! That money is totally wasted! Keep in mind that that price does not include the additional wear and tear or the additional maintenance of the equipment.

Another way to look at this loss is to base it on 1000 scfm of air compressed. When the system is leaking at 275 scfm, the compressor delivers an additional 16,500 standard cubic feet per hour at a cost of $5.36. This amounts to about $0.325 per 1000 ft^3 of air delivered.

These last examples tend to confirm some rules of thumb that you can use when you need a quick estimate of the effects of pressure losses and leaks. In a system that has a receiver pressure of 100 psig, each 1 psi pressure loss in the system increases the compressor power requirement by about 0.5 percent to maintain 100 psig in the system. Depending on the size of the system, you will pay from $0.20 to $0.35 per 1000 standard cubic feet of air. These values are not necessarily accurate, but they will get you in the ballpark.

15.10   Vacuum

To this point, we've been discussing the use of pressure above atmospheric to accomplish various tasks. There are also pneumatic applications where pressure lower than atmospheric—a *vacuum*—can be used, often with better results.

Any pressure lower than the current, local atmospheric pressure is a vacuum, and the magnitude of the pressure can be expressed in several ways—negative psig, psia, inches of mercury (in Hg) absolute, inches of mercury vacuum, kPa absolute, millimeters of mercury. Very high vacuums are measured in *torr*. A torr is 1/760 of an atmosphere (which is 14.7 psia). Beyond the torr is the *micron*, which actually means 1 μm Hg and is 0.001 torr.

For comparative purposes, vacuums are classified as rough, middle (or fine), and high. *Rough vacuum* ranges from 0 to 28 in Hg and includes most industrial systems. *Middle vacuum* is up to 1 micron. Molecular distillation, freeze drying, and coating processes fall into this category. Any vacuum higher than 1 micron is in the *high vacuum* category. Particle accelerators, mass spectrometers, electron beam welders, and electron microscopes use high vacuum.[5]

There are numerous industrial applications for vacuum, including clamping, vacuum molding, filling and sealing, transporting, and so on. The one that is of most interest as a fluid power application is lifting. Vacuum can be used to lift virtually anything that has sufficient flat surface area, from a piece of paper to sheets of steel. The job can be accomplished gently and with a great deal of finesse. Unfortunately, the capability of a vacuum system is limited by atmospheric pressure.

Figure 15.34 illustrates the action of a *vacuum cup*. The cup, which is usually rubber or an elastomer material, is placed against the flat surface. The vacuum pump evacuates the air from the volume formed between the surface and the cup. As the pressure in the volume decreases, atmospheric pressure forces the cup against the surface (or the surface against the lip of the cup, depending on the application) and holds it there as long as the vacuum exists.

The force capability of a vacuum cup depends on the vacuum and the cross-sectional area of the cup, and is found in the same way that we found cylinder force ($F = p \times A$). This is that old, familiar equation that we've used so many times before, exccept that p now refers to the *vacuum*.

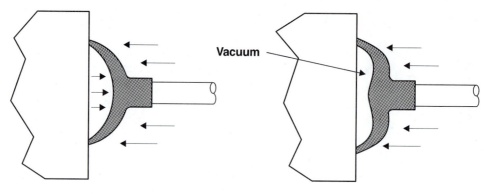

Figure 15.34 Atmospheric pressure holds evacuated suction cup (vacuum cup) against the wall.

Example 15.10

What is the maximum weight that can be lifted by a 4 in vacuum cup operating at a 10 psig vacuum?

Solution

Applying the force-pressure equation.

$$F = p \times A$$
$$= 10 \text{ lb/in}^2 \times 12.56 \text{ in}^2 = 125.6 \text{ lb}$$

Example 15.11

What is the maximum weight that can be lifted by a 10 cm vacuum cup if the atmospheric pressure is 95 kPa abs? Assume a perfect vacuum in the cup.

Solution

By assuming a perfect vacuum in the cup, we are saying that the pressure is −95 kPag.

$$F = p \times A$$
$$= 95 \text{ kPa} \times 78.5 \text{ cm}^2 = 746 \text{ N}$$

Note: There are many factors that make calculations such as these somewhat suspect in the real world. It is common practice to oversize vacuum systems by 200 to 500 percent "just to be sure."

A common vacuum application is lifting large, flat sheets such as floor covering, laminates, glass, and sheet metal. In such applications, multiple cups are arranged on a manifold (Figure 15.35) so the sheet can be supported evenly.

In designing vacuum systems, the time required to *pump down* the system is often of interest. This time is based on the initial pressure, the final pressure

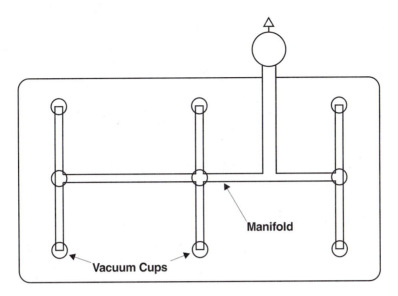

Figure 15.35 Manifolded vacuum cups are used for lifting large flat sheets.

(vacuum), the volume of the system, and the capacity of the vacuum pump at standard atmospheric pressure. It can be calculated from Equation 15.6.

$$t = \frac{V}{Q} \ln\left(\frac{p_1}{p_2}\right)$$ **(15.6)**

where

t = time (minutes).

V = volume to be pumped down (ft^3 or m^3).

Q = vacuum pump capacity (scfm or m^3/min) (standard).

p_1 = initial pressure (absolute).

p_2 = final pressure (absolute).

Example 15.12

A vacuum system with a total volume of 10 ft^3 (0.28 m^3) must be pumped down from 29.8 in.Hg to 5 in.Hg (75.7 cmHg to 12.7 cmHg). The system pump has a 5 scfm (0.142 m^3/min) capacity. How much time is required?

Solution

Using Equation 15.6,

$$t = \frac{V}{Q} \ln\left(\frac{p_1}{p_2}\right)$$

$$= \frac{10 \text{ ft}^3}{5 \text{ ft}^3/\text{min}} \ln\left(\frac{29.8}{5}\right) = 3.6 \text{ min}$$

Example 15.13

A vacuum system uses six 4 in vacuum cups to lift large steel panels weighing 200 lb. The piping system consists of 20 ft of 1 in ID header pipe and 48 ft of ¼ in ID pipe for drops. The volume of each cup is 25 in³. What capacity vacuum pump is required to pump down the system in 15 seconds? Use a safety factor of 3 when calculating the vacuum required. Assume an atmospheric pressure of 30.12 in Hg.

Solution

The total force that must be exerted by the vacuum cups is found by multiplying the actual weight by the safety factor; thus,

$$F = 200 \times 3 = 600 \text{ lb}$$

The vacuum required is found from

$$p = \frac{F}{A}$$

where A is the total area of the six vacuum cups.

$$A = 6 \times 12.56 \text{ in}^2 = 75.4 \text{ in}^2$$

Thus, the required vacuum is

$$p = \frac{F}{A} = \frac{600 \text{ lb}}{75.4 \text{ in}^2} = 7.96 \text{ psi}$$

of vacuum. Since this is the *vacuum* required, we have actually determined that a pressure that is 7.96 psi *below atmospheric pressure* is required to lift the load. Now we must convert this to an absolute pressure to use in Equation 15.6.

$$
\begin{aligned}
p_{abs} &= p_{atm} - p \\
&= 30.12 \text{ in Hg} - 7.96 \text{ psi} \times 2.04 \text{ in Hg/psi} \\
&= 13.9 \text{ in Hg}
\end{aligned}
$$

This is p_2 for Equation 15.6.

Next we need to find the total volume of the system. This includes the cups and all the piping.

$$
\begin{aligned}
V &= V_{cups} + V_{1 \text{ in}} + V_{1/2 \text{ in}} \\
&= 6 \times 25 \text{ in}^3 + 20 \text{ ft} \times 0.785 \text{ in}^2 \times 12 \text{ in/ft} \\
&\qquad\qquad + 48 \text{ ft} \times 0.196 \text{ in}^2 \times 12 \text{ in/ft} \\
&= 150 \text{ in}^3 + 188.4 \text{ in}^3 + 112.9 \text{ in}^3 \\
&= 451.3 \text{ in}^3
\end{aligned}
$$

or 0.26 ft³. Now we can use Equation 15.6 to determine the pump capacity. Solving the equation for Q gives

$$
\begin{aligned}
Q &= \frac{V}{t} \ln\left(\frac{p_1}{p_2}\right) \\
&= \frac{0.26 \text{ ft}^3}{15 \text{ sec}} \ln\left(\frac{30.12}{13.9}\right) \\
&= 0.0134 \text{ scfs} = 0.8 \text{ scfm}
\end{aligned}
$$

Thus, a vacuum pump with a capacity of 0.8 scfm will meet the time requirements.

Vacuum in a system can also be generated through the use of *venturis* in devices termed *vacuum generators* or *vacuum ejectors* (Figure 15.36). The vacuum is generated when system supply air enters the supply port, passes through the venturi section, and leaves through the exhaust port. The lower pressure generated by the flow through the venturi creates the vacuum at the vacuum port. Vacuums as high as 28 in Hg can be generated at suction flow rates up to about 1.5 scfm. Higher flow rates can be achieved, but usually at lower vacuums.

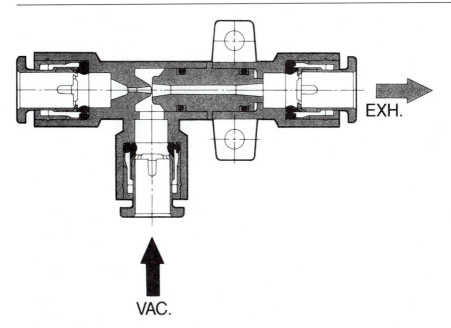

EXH.

VAC.

Figure 15.36 Vacuum can be generated through the use of a venturi in a vacuum ejector.
Source: SMC Corporation, Tokyo, Japan (in the United States: SMC Pneumatics, Inc., Indianapolis, Ind.).

Figure 15.37 shows an application of a vacuum-lifting system using a vacuum generator. This circuit is a combined vacuum/blowoff arrangement. With the directional valve in the position shown, flow is directed through the vacuum generator (shown here with an integral silencer) to create the required vacuum. Shifting the directional valve directs system pressure to the vacuum pad to break the vacuum and release the load.

15.11 Troubleshooting Tips

Because pneumatic and vacuum systems usually use atmospheric air, many of the problems encountered in these systems are the result of the dirt and moisture contained in the air.

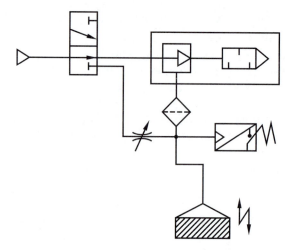

Figure 15.37 This vacuum system uses a blowoff valve.
Source: SMC Corporation, Tokyo, Japan (in the United States: SMC Pneumatics, Inc., Indianapolis, Ind.)

Symptom	Possible Cause
Air motor runs too slowly	1. Insufficient air flow or external leakage 2. Insufficient pressure 3. Flow control valve set incorrectly 4. Internal wear allowing excessive air leakage 5. Internal corrosion or rust 6. Machinery jammed 7. Misalignment 8. Choked valve somewhere in the circuit
Air motor torque too low	1. Insufficient pressure 2. High internal leakage due to wear 3. Internal resistance due to rust, corrosion, or dirt
Cylinder will not move the load or moves too slowly	1. Insufficient pressure 2. Internal leakage (wear or damaged seals) 3. Load jammed 4. Insufficient flow or external leakage
Control valves not working properly	1. Internal wear or jamming due to dirty air 2. Internal rust or corrosion due to wet air 3. Insufficient pilot pressure

Symptom	Possible Cause
Vacuum system will not hold load	1. Insufficient vacuum 2. Rough material surface 3. Suction cup damaged 4. External leaks in vacuum lines or fittings 5. Vacuum pump not working properly 6. Ejector venturi clogged
Vacuum system will not pump down in allotted time	1. Vacuum pump has internal leakage due to wear 2. Ejector venturi clogged 3. External leaks in vacuum lines or fittings

15.12 Summary

Pneumatics—as both pressure and vacuum—has a very important place in fluid power application. The applications of pneumatics, however, are limited by its relatively low power capabilities and its compressibility. Vacuum applications are limited because of the low force capabilities, but vacuum systems can provide a "gentle touch" that is not possible with pressure systems.

Properly conditioning the air is critical to the life, reliability, and performance of a pneumatic system. Both dirt and moisture must be reduced to a very low level to prevent damage to tools and, possibly, contamination of the product.

The pressure available at the tool is critical; therefore, systems should be designed to prevent excessive pressure loss. The result of these losses will be excessive cost. Either production will be lost due to tool inefficiency, or the receiver pressure must be increased to make up for the losses. Air leaks can be as costly as pressure losses, because the compressor must "work overtime" to replace the lost air.

Controlling pneumatic system operation through the use of pilot valves and pneumatic logic can offer some distinct advantages over electrical controls in many applications. Pneumatic controls eliminate the fire or explosion hazard for hazardous applications. In some cases, the initial cost of pneumatic machines can be significantly reduced because the need for running electrical lines and conduit is eliminated.

15.13 Key Equations

Valve flow capacity coefficient:

$$C_v = \frac{Q}{22.48} \sqrt{\frac{T_1 \times G}{\Delta p \times p_2}} \qquad (15.1)$$

Pressure loss in pipes:

$$\Delta p = \frac{c \times L \times Q^2}{3600 \times P_R \times d^5} \qquad (15.3)$$

Power required to drive a Compressor: $P = \dfrac{BHP \times 0.746 \text{ kW/HP}}{\eta_M \times PF}$ **(15.5)**

Vacuum system pumpdown time $t = \dfrac{V}{Q} \ln \left(\dfrac{p_1}{p_2} \right)$ **(15.6)**

References

1. Frank Yeaple, *Fluid Power Design Handbook*, 2d ed. (New York: Marcel Dekker, Inc., 1990).
2. *Air Motors Handbook*, rev. ed. (Benton Harbor, MI: Gast Manufacturing Corporation, 1986).
3. *Bulletin 150B*, Deltech Engineering, Inc., New Castle, Delaware.
4. *Industrial Pneumatics Technology* (Cleveland, OH: Parker Hannifin Corporation, 1980).
5. R. Moffat, "Putting Industrial Vacuum to Work," *Hydraulics and Pneumatics*, March 1987.

Suggested Reading

1. Bruce E. McCord, *Designing Pneumatic Control Circuits* (New York: Marcel Dekker, Inc., 1983).
2. Ralph L. Cuthbertson, *Air Logic* (Cleveland, OH: Penton Publishing, Inc., 1988).

Review Problems

Note: The letter "S" after a problem number indicates that SI units are used.

1. Explain why meter-out flow controls are preferred for pneumatic cylinders and motors.
2. Discuss the effects of load on air motor torque and speed.
3. Why is it difficult to accurately calculate the speed of pneumatic motors and cylinders?
4. What is meant by the *rated speed* of an air motor?
5. Explain the difference between starting torque and stall torque. Which is lower? Why?
6. List and describe the five different sliding mechanisms used in pneumatic directional control valves.
7. Explain the meaning of the flow capacity coefficient (C_v) for pneumatic valves.
8. Explain what is meant by the term *choked* as applied to pneumatic valves. Under what conditions does a valve become choked?
9. Discuss the advantages of pneumatic control over electrical control for pneumatic machines.
10. Explain the term *moving parts logic*.
11. Explain the difference between an AND element and an OR element.
12. Explain the basic principle of a pneumatic timer element.

13. Explain what is meant by a *two-hand, anti-tiedown* system. What is the advantage of such a system?

14. Describe and discuss the advantages and disadvantages of the two basic plant pneumatic piping layouts.

15. Explain why all drops for air supply lines should begin by going upward from the main header.

16. Explain how a vacuum can be used to lift loads.

17. Explain the operation of a vacuum ejector.

Cylinder Performance

18. A 1 in cylinder must provide a force output of 75 lb. What pressure is required?

19. A system is operating at 125 psi. Determine the maximum force output capability of a ⅜ in cylinder.

20. A cylinder is used in a system where the regulator is set at 45 psi. The cylinder has a power factor of 0.89. What is its force output capability in this system?

21S. A 3.5 cm cylinder is used to give a force output of 400 N. What pressure is required?

22S. A system is operating at 700 kPa. Calculate the force output capability of a 10 cm cylinder.

Cylinder Volume

23. A single-acting cylinder (extend) operates at 90 psi. The cylinder bore is 2 in, and its stroke is 6 in. Find the volume of air (scfm) required to cycle the cylinder at 15 cycles per minute.

24. A 1.5 in cylinder has a 0.5 in rod and a 12 in stroke. It operates at 100 psi. Find the volume of air (scfm) required to cycle the cylinder at 10 cycles per minute.

25S. A 5 cm cylinder with a 2 cm rod and a 30 cm stroke cycles 10 times per minute. Calculate the volume of air (standard cubic meters per minute) required to operate the cylinder.

Valve Flow

26. Air is flowing through a valve ($C_v = 1.7$) and exhausting into the atmosphere. The atmospheric pressure is 14.7 psia. The pressure upstream of the valve is 26 psia. The air temperature is 90°F. Calculate the flow rate through the valve.

27. An orifice is installed in a pneumatic system. The pressure downstream of the orifice is 25 psig. At what upstream pressure would maximum flow be achieved through the orifice?

28. A flow control valve in a pneumatic system is passing 250 scfm. The upstream pressure is 60 psig, and the downstream pressure is 25 psig. Is the valve choked?

29. A directional control valve in a pneumatic system has an upstream pressure of 125 psig. At what downstream pressure (critical pressure) would the valve become choked?

30. A directional control valve in a pneumatic system is operating with upstream and downstream pressures of 90 and 35 psig, respectively. What values of pressure would you use in the flow equation?

Pressure Drop in Pipelines

31. A pneumatic system operates with 50 scfm with a pressure of 100 psi. The system has 75 ft of 1 in Schedule 40 pipe. Determine the pressure drop through the pipe.

32. Develop a graph showing pressure loss versus flow rate based on pipe size. Use ½, ¾, 1, 1½, 2, and 3 in pipe.

33. A pneumatic system includes 135 ft of 1 in Schedule 40 pipe, ten 90° elbows, and two fully open globe valves. With the pressure regulator at the compressor end of the system set at 125 psi and a tool at the other end of the system using 20 scfm, find the pressure at the tool inlet.

34. A pneumatic cylinder has a 3½ in bore and 15 in stroke. It moves an 1100 lb resistance through the full stroke in 1.25 seconds. The cylinder is at the end of 70 ft of 1 in Schedule 40 pipe that includes four 90° standard elbows, three tees (runs) and one branch tee, a filter that has a 5 psi pressure drop and a lubricator that has a 3 psi pressure drop. What is the minimum regulator setting that will ensure sufficient pressure to operate the cylinder if the regulator is at the receiver outlet?

Vacuum Systems

35. A vacuum system is used to lift aluminum billets weighing 100 lb. Determine the number of 6 in vacuum cups required if the system operates at 25 in Hg vacuum. Use a safety factor of 2.

36. How long would it take a 5 scfm vacuum pump to pump a 6 ft³ tank from 29.92 in Hg to 15 in Hg vacuum?

37. The billets in Review Problem 35 are 4 × 8 ft. The vacuum cups are mounted on a vacuum manifold consisting of 22 ft of 1 in Schedule 40 pipe. The volume of each cup is 14 in³. The billets must be ready to lift in a maximum of 3 seconds after the cups contact the surface. What is the minimum flow capacity vacuum pump that can be used?

38. An aircraft altimeter system consists of 14 ft of ¼ in steel tubing (ID = 0.19 in). A vacuum unit for testing the system must be capable of simulating a 3000 ft/min rate of climb. Assuming a pressure drop of 1 in Hg/ 1000 ft, determine the flow capacity required from the vacuum pump. The tester has 8 ft of ½ in hose. Assume standard sea level pressure at the start of the test.

39S. A vacuum system uses 10 cm cups to lift sheets of wallboard weighing 90 kg. The vacuum system operates at 60 cm Hg vacuum. Determine the number of cups required. Use a safety factor of 2.

40S. A vacuum pump is evacuating a 1.7 m³ tank from standard atmospheric pressure to 45 cm Hg vacuum. The pump has a capacity of 0.11 m³/min. Determine the pump down time.

Boolean Algebra

41. Use truth tables to prove the following Boolean equalities:

     a. $A \cdot (A + B) = A$
     b. $A + (A \cdot B) = A$
     c. $(A + \overline{B}) \cdot B = A \cdot B$
     d. $\overline{A \cdot B \cdot C} = \overline{A} + \overline{B} + \overline{C}$
     e. $\overline{A + B + C} = \overline{A} \cdot \overline{B} \cdot \overline{C}$

42. Write the Boolean algebra equation for the control circuit shown here.

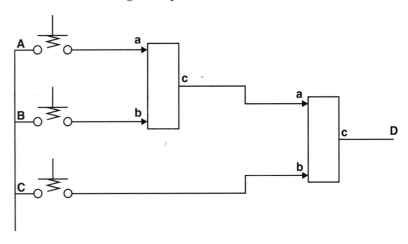

Review Problem 42

Appendix A

Fluid Power Standards

A.1 Introduction

This appendix contains a cross reference of fluid power standards including terminology, component standardization, and component and fluid testing procedures. Standards are grouped by broad categories—accumulators, bibliographies, cylinders, and so on. Within each category, specific subjects are listed in numerical order, keying on the National Fluid Power Association (NFPA) standards numbers where an NFPA standard for the subject exists.

When referring to standards and specifications, ensure that the latest revision of that standard is being used. Standards are normally reviewed (and often revised) every five years; therefore, any dated document more than five years old may have a more recent revision. The status of any standard can be ascertained from the responsible standards organization at the following addresses:

American National Standards Institute, Inc. (ANSI)
1430 Broadway
New York, NY 10018
212/868-1220

American Society for Testing and Materials (ASTM)
1916 Race Street
Philadelphia, PA 19103
215/299-5400

International Organization for Standardization (ISO)
1 Rue de Varembe
Case Postal 56
1211 Geneva 20, Switzerland
(ISO Standards are available in the United States from ANSI)

National Fluid Power Association (NFPA)
3333 N. Mayfair Road
Milwaukee, WI 53222
414/778-3344

Society of Automotive Engineers (SAE)
400 Commonwealth Drive
Warrendale, PA 15096
412/776-4841

Federal Test Methods (FTM) are available from the American Society for
Testing and Materials or from the U.S. Government Printing Office, Washington, D.C.

A.2 Cross-Reference Tables

Note that in the following tables, M demotes the use of metric (SI) dimensioning,
DIS denotes draft international standard, and DP denotes draft proposal. FTM
denotes Federal Test Method.

Table A.2.1 Accumulators

Subject	NFPA	ANSI	ISO	SAE	ASTM	FTM
Accumulator Pressure Rating	T3.4.7M	—	—	—	—	—
Procedure for Hydro-pneumatic Accumulator Use	T3.4.10M	—	—	—	—	—
Accumulator Pressure Volume Ranges, Characteristics and Identification	T.3.4.11M	B93.78M	ISO 5596	—	—	—

Table A.2.2 Bibliographies

Subject	NFPA	ANSI	ISO	SAE	ASTM	FTM
Metrication	T2.10.5	—	—	J917	—	—
Accumulators	T3.4.8	—	—	—	—	—
Hydraulic Valves	T3.5.27	—	—	—	—	—
Fluid Power Cylinders	T3.6.36	—	—	—	—	—
Tube Fittings and Conductors	T3.8.11	—	—	—	—	—
Hydraulic Pumps and Motors	T3.9.21	—	—	—	—	—
Filtration and Contamination	T3.10.12	—	—	—	—	—
Pneumatic FRLs	T3.12.9	—	—	—	—	—
Reservoirs and Power Units	T3.16.9	—	—	—	—	—
Sealing Devices	T3.19.22	—	—	—	—	—
Quick Action Couplings	T3.20.7	—	—	—	—	—

Continued

Table A.2.2 Continued

Subject	NFPA	ANSI	ISO	SAE	ASTM	FTM
Pneumatic Valves	T3.21.5	—	—	—	—	—
Fluid Power Hose, Hose Fittings, and Hose Assemblies	T3.26.1	—	—	—	—	—
Compressed Air Dryer	T3.27.4	—	—	—	—	—
Fluid Logic	T3.28.11	—	—	—	—	—

Table A.2.3 Conductors, Fittings, and Associated Hardware

Subject	NFPA	ANSI	ISO	SAE	ASTM	FTM
Metric Port Dimensions and Design	T2.8.2M	—	ISO 6149	—	—	—
Hydraulic Hose Fittings	—	—	—	J516	—	—
Hydraulic Tube Fittings	—	—	—	J514	—	—
Hyraulic Tube Fitting Test	T3.8.3M	—	—	—	—	—
Pneumatic Tube Fitting Test	T3.8.9	B93.48	—	—	—	—
4-Bolt Split Flange Dimension and Design	T3.8.19M	—	DIS 6162	J518	—	—
Hydraulic O-Ring Seals	—	—	—	J515	—	—
O-Ring Face Seal Fitting	—	—	—	J1453	—	—
O-Ring Face Seal Fitting Test	T3.8.20M	—	—	—	—	—
O.D. of Tubes and I.D. of Hoses	T3.8.22M	B93.59M	ISO 4397	J517	—	—
Nominal Pressure for Connectors and Related Components	T3.8.23M	B93.60M	ISO 4399	—	—	—
Seamless High Strength Alloy Steel Tubing	T3.8.24M	—	—	—	—	—
High-Strength Fluid Power Tubing	T3.8.25M	—	—	—	—	—
Hydraulic Hose Assembly Test Requirements	T.3.27M	B93.100M	ISO 6605	J517	—	—
Square Flange Connection Dimensions (40 to 400 bar)	T3.8.28M	—	DIS 6163	J343	—	—
Square Welded Collar Flanges Dimensions and Identification	—	—	DIS 6163	—	—	—

Continued

Table A.2.3 Continued

Subject	NFPA	ANSI	ISO	SAE	ASTM	FTM
Hydraulic Line Welded Tubing	T3.15.1M	B93.4M	—	—	—	—
Hydraulic Line Seamless Tubing	T3.15.2M	B93.11M	—	—	—	—
Tube and Fitting Assembly Pressure Rating	T3.15.8M	—	—	J1065	—	—
Quick Action Coupling Terms	T3.20.1	B93.2A	—	—	—	—
Hydraulic Quick Action Coupling Test	T3.20.2M	B93.42M	ISO 7241	—	—	—
Pneumatic Quick Action Coupling Test	T3.20.3M	B93.51M	ISO 6150	—	—	—
Quick Action Coupling Flow Pressure Drop Test	T3.20.6M	B93.64M	—	—	—	—
Quick Action Coupling Pressure Rating	T3.20.8M	—	—	—	—	—
Quick Action Coupling Dimensions and Requirements	—	B93.113M	ISO 7241	—	—	—
Agricultural Quick Action Coupling Interface Dimensions	—	—	ISO 5675	J1036	—	—
Interchangeable Industrial Quick Action Couplings	T3.20.10M	—	DIS 7471	—	—	—
Hydraulic Quick Action Coupling Surge Flow Test/ Short Duration	T3.20.11M	B93.68M	—	—	—	—
Quick Action Couplings Surge Flow Test/Long Duration	T3.20.12M	B93.69M	—	—	—	—
Hydraulic Tools Quick Action Coupling	T3.20.15	—	—	—	—	—
20 mm Quick Action Argicultural Coupling	T3.20.16M	—	—	—	—	—
Pressure Switch Terms	T3.29.1	—	—	—	—	—
Pressure Switch Pressure Rating	T3.29.2M	—	—	—	—	—
Rated Over-Range Pressure Switches	T3.29.3M	—	—	—	—	—

Continued

Table A.2.3 Continued

Subject	NFPA	ANSI	ISO	SAE	ASTM	FTM
Hydraulic Tubes/Hoses Dimensions and Designs (37° Flare/24° Flareless Fittings)	—	B93.83	ISO 8434	—	—	—
Flex-Impulse Test Procedure for Hydraulic Hose Assemblies	—	—	—	J1405	—	—
Selection, Installation, and Maintenance of Hose and Hose Assemblies	—	—	—	J1273	—	—
Power Steering Pressure Hose	—	—	—	J188	—	—
Power Steering Return Hose	—	—	—	J189	—	—
Hydraulic Brake Hose Assemblies—Non-Petroleum Base Fluids	—	—	—	J1401	—	—
Identification Codes for Fluid Conductors and Connectors	—	—	—	J846	—	—
Formed Tube Ends for Hose Connections	—	—	—	J962	—	—
Flares for Tubing	—	—	—	J533	—	—
Seamless Low-Carbon Steel Tubing Annealed for Bending and Flaring	—	—	—	J524	—	—
Welded and Cold Drawn Low-Carbon Steel Tubing Annealed for Bending and Flaring	—	—	—	J525	—	—
Welded Low-Carbon Steel Tubing	—	—	—	J526	—	—
Brazed Double-Wall Low-Carbon Steel Tubing	—	—	—	J527	—	—
Welded Flash-Controlled Low-Carbon Steel Tubing Normalized for Bending, Double Flaring, and Beading	—	—	—	J356	—	—
Automotive Pipe Fittings	—	—	—	J530	—	—

Table A.2.4 Cylinders

Subject	NFPA	ANSI	ISO	SAE	ASTM	FTM
Cylinder Bores and Piston Sizes (Inch Series)	T3.6.1	B93.3	—	—	—	—
Cylinder Dimension Code	T2.6.2M	B93.1M	ISO 6099	—	—	—
Cylinder Bore Rod Sizes	T3.6.4M	B93.8M	—	—	—	—
Cylinder Static Pressure Rating	T3.6.5M	B93.10M	—	—	—	—
Cylinder Mounting Dimensions	T3.6.7	B93.15	—	—	—	—
Cylinder Mounting Dimensions and Type Codes	—	B93.84M	ISO 6099	—	—	—
Cylinder Accessory Dimensions	T3.6.8M	B93.29M	—	—	—	—
Miniature Cylinder Dimensions	T3.6.11M	B93.34M	—	—	—	—
Cylinder Port Sizes	T3.6.17M	—	—	—	—	—
Metric Cylinder Dimensions Compact Series	T3.6.28M	—	ISO 6020	—	—	—
Tie Rod/Bolted Cylinder Pressure Rating	T3.6.29M	—	—	—	—	—
Telescopic Cylinder and Cylinders of Non-bolted End Construction	T3.6.31M	—	—	—	—	—
Large Bore Cylinder Mounting Dimensions	T3.6.32M	—	ISO 6020	—	—	—
Cylinder Bore/Rod Diameters (Metric Series)	T3.6.33M	B93.52M	ISO 3329	—	—	—
Cylinder Bore/Rod Diameters (Inch Series)	—	—	ISO 3321	—	—	—
Normal Pressure Cylinders	T3.6.35M	B93.53M	ISO 3322	—	—	—
Cylinder Buckling Strength Test	T3.6.37	—	—	—	—	—
Hydraulic Cylinders (160 bar) Medium Series	—	—	ISO 6020	—	—	—
Hydraulic Cylinders (250 bar) Mounting Dimensions	—	B93.89	ISO 6022	—	—	—
Hydraulic Single Rod Compact Cylinder Tolerances (160 bar)	—	B93.103M	ISO 8131	—	—	—

Continued

Table A.2.4 Continued

Subject	NFPA	ANSI	ISO	SAE	ASTM	FTM
Nonferrous Pneumatic Cylinder Tubes	—	B93.88M	ISO 6537	—	—	—
Piston Strokes (Metric Only)	T3.6.46M	B93.56M	ISO 4393	—	—	—
Piston Rod Threads	T3.6.48M	B93.61M	ISO 4395	—	—	—
Hydraulic Cylinder Ratios	—	B93.92M	ISO 7181	—	—	—
Piston Seal Housings Incorporating Bearing Rings	—	B93.93	ISO 6547	—	—	—
Hydraulic Cylinder Mounting Accessories (160 bar)	T3.6.51M	—	—	—	—	—
Straight Thread and Flange Port Standardization	T3.6.54M	B93.75M	—	—	—	—
Mill Cylinder Standardization	T3.6.56	B93.75M	—	—	—	—
Hydraulic Cylinder Rod End Plain Eyes	—	B93.90M	ISO 6981	—	—	—
Pneumatic Cylinder Rod End Clevis	—	B93.110M	ISO 8140	—	—	—
Hydraulic Cylinder Rod End Spherical Eyes	—	B93.91M	ISO 6982	—	—	—
Pneumatic Cylinder Rod End Spherical Eyes	—	B93.109M	ISO 8139	—	—	—
Steel Tube Specifications	—	—	ISO 4394	—	—	—
Hydraulic Cylinder Accessory Mounting Dimensions (160 and 250 bar)	—	B93.104M	ISO 8132	—	—	—
Pneumatic Cylinders (10 bar) Detachable Mounting Series	—	B93.87M	ISO 6431	—	—	—
Hydraulic Single Rod Compact Cylinders Accessory Mounting Dimensions (160 bar)	—	—	ISO 8133	—	—	—
Pneumatic Cylinders (10 bar) Integral Mounting Series	—	B93.86M	ISO 6430	—	—	—
Hydraulic Single Rod Compact Cylinder Rod End Spherical Eyes Mounting Dimensions (160 bar)	—	—	DP 8134	—	—	—
Trunnion Evaluation	T3.6.75	—	—	—	—	—

Continued

Table A.2.4 Continued

Subject	NFPA	ANSI	ISO	SAE	ASTM	FTM
Hydraulic Single Rod Cylinder Tolerances (160 and 250 bar)	—	B93.105M	ISO 8135	—	—	—
External Seal Performance	T3.6.58	—	—	—	—	—
Cushion Performance—Hydraulic	T3.6.59	—	—	—	—	—
Computer Graphics for Cylinders	T3.6.60	—	—	—	—	—
Linear Actuator Rod Seal Test	T3.19.12M	B93.62M	—	—	—	—
Cylinder Rod Wiper Seal Ingression Test	—	—	—	J1195	—	—
Hydraulic Single Rod Cylinder Port Dimensions (160 bar medium)	—	B93.106M	ISO 8136	—	—	—
Hydraulic Single Rod Cylinder Port Dimensions (250 bar medium)	—	B93.107M	ISO 8137	—	—	—
Hydraulic Single Rod Cylinder Port Dimensions (160 bar compact)	—	B93.108M	ISO 8138	—	—	—
Mounting Dimensions (160 bar) 250 mm through 500 mm	—	—	DP[a] 8141	—	—	—

Table A.2.5 Filtration and Contamination Analysis

Subject	NFPA	ANSI	ISO	SAE	ASTM	FTM
Extracting Fluid Samples	T2.9.1M	B93.19M	ISO 4021	J1227	—	—
Sample Containers	T2.9.2M	B93.20M	ISO 3722	—	—	—
Contamination Reporting	T2.9.3M	B93.30M	ISO 3938	J1165	—	—
Particle Counter Calibration	T2.9.6M	B93.28M	ISO 4402	—	—	—
Classified Test Calibration	T2.9.7M	—	—	—	—	—
Hydraulic System Cleaning	—	—	—	—	D 4174	—
Rolloff Cleanliness	T2.9.8M	B93.54M	—	J1227	—	—
Reservoir Sampling	T2.9.9M	B93.44M	—	—	—	—

Continued

Table A.2.5 Continued

Subject	NFPA	ANSI	ISO	SAE	ASTM	FTM
Particulate Count Method	T2.9.11M	—	—	—	F 312	—
In-line Particle Counting Method	T2.9.12M	B93.73M	—	—	—	—
Solid Contamination Code	T2.9.13M	—	ISO 4406	—	—	—
Fluid Contamination Gravimetric Method	T2.9.14M	—	ISO 4405	—	F 313	—
Cleanliness Levels Component Assemblies	T2.9.15M	—	—	J1227	—	—
Determination of Particulate Contaminant Levels Using Automatic Optical Particle Counters	T3.9.16	—	—	—	—	—
Counting Method Under Transmitted Light	—	—	DIS 4407	—	—	—
Counting Method Under Incident Light	—	—	DIS 4408	—	—	—
Hydraulic Filter Separator Terms	T3.10.3	—	—	—	—	—
Hydraulic Filter Separator Graphic Symbols	T3.10.4M	—	—	—	—	—
Hydraulic Filter/Separator Housing Pressure Rating	T3.10.5.1M	—	—	—	—	—
Hydraulic Filter Element End Load Test	T3.10.8.2M	B93.21M	ISO 3723	—	—	—
Hydraulic Filter Element Built-in Contaminant	T3.10.8.3M	—	—	—	—	—
Hydraulic Filter Element Integrity	T3.10.8.4M	B93.22M	ISO 2942	—	—	—
Hydraulic Filter Element Collapse/Burst Test	T2.10.8.5M	B93.25M	ISO 2941	—	—	—
Hydraulic Filter Element Compatibility Test	T3.10.8.6M	B93.23M	ISO 2943	—	—	—
Hydraulic Filter Element Flaw Fatigue Test	T3.10.8.7M	B93.24M	ISO 3724	—	—	—
Hydraulic Filter Fine Element Multipass Test	T3.10.8.8M	B93.31M	ISO 4572	—	—	—
Hydraulic Filter Element Determining Pore Size of Wire Cloth	T3.10.8.12M	B93.46M	—	—	—	—
Hydraulic Filter Element Multipass Silt Control Test	T3.10.8.16M	—	—	—	—	—
Hydraulic Filter Coarse Element Multipass Test	T3.10.8.18M	—	—	—	—	—

Continued

Table A.2.5 Continued

Subject	NFPA	ANSI	ISO	SAE	ASTM	FTM
Unsteady Flow Filter Test	T3.10.8.19M	—	—	—	—	—
Hydraulic Filter Bypass Valve Test	T3.10.9.1M	—	—	—	—	—
Hydraulic Differential Pressure Indicator Test	T3.10.9.2M	—	—	—	—	—
Filtration and Contamination Bibliography	T3.10.12	—	—	—	—	—
Hydraulic Filter Flow Rating	T3.10.14M	B93.80M	ISO 3968	—	—	—
Hydraulic Filter Requirements	—	B93.82M	ISO 7744	J931	—	—
Finite Life Hydraulic Filter Pressure Rating	T3.10.12M	—	—	—	—	—
Filter Artwork Universal Symbols	T3.10.18	—	—	—	—	—
Absorbent Filter Performance Test	T3.10.19	—	—	—	—	—
Filter Water Sensitivity Characteristics	T3.10.20	—	—	—	—	—
Air Line Filter Performance	T3.12.2	—	—	—	—	—
FRL Pressure Rating	T3.12.10	—	—	—	—	—
Air Dryer Rating and Testing	T3.27.3M	B93.45M	—	—	—	—

Table A.2.6 Fluids

Subject	NFPA	ANSI	ISO	SAE	ASTM	FTM
Synthetic Lubricants	T1.9.2	B93.50	—	—	—	—
Solid Contaminant Code for Fluids	T2.9.13M	—	ISO 4406	—	—	—
Fire-Resistant Fluids	T2.13.1M	B93.5M	ISO 6743/4	—	—	—
Fire-Resistant Fluids Trade Names	T2.13.2	—	—	—	—	—
Hydraulic Fluids Index	T2.13.3	—	—	—	—	—
Hydraulic Fluid Disposal	T2.13.4M	—	—	—	—	—
High Water Content Fluids	T2.13.5M	—	—	—	—	—
Petroleum Fluids Bulk Moduli	T2.13.7M	B93.63M	ISO 6073	—	—	—
Hydraulic Fire Resistant Fluids	T2.13.8	—	—	—	—	—
Fire-Resistant Fluids Anticorrosive Power	—	—	ISO 4404	—	—	—

Continued

Table A.2.6 Continued

Subject	NFPA	ANSI	ISO	SAE	ASTM	FTM
Hydraulic Fluid Compatibility	—	—	ISO 6072	—	—	—
Mineral Oils Characteristics	—	—	ISO 6075	—	—	—
Fire-Resistant Fluids Guidelines	—	—	ISO 7745	—	—	—
Viscosity Classification	—	—	ISO 3448	J300	D445	305
Saybolt Viscosity	—	—	—	—	—	304
Viscosity-Temperature Charts	—	—	—	—	D341	9121
Calculation of Viscosity Index	—	—	—	—	D2270	9111
Viscosity-Shear Characteristics Diesel Injector Method	—	—	—	—	D3945	—
Tapered-Plug Viscometer Method	—	—	—	—	D4741	—
Cloud Point	—	—	—	—	D97	201
Pour Point	—	—	—	—	D97	201
Pour Stability	—	—	—	—	—	203
Diluted Pour Point	—	—	—	—	—	204
Cloud Intensity at Low Temperature	—	—	—	—	—	202
Freezing Point	—	—	—	—	D1015	—
Flash Point Pensky-Martens Closed Cup	—	—	ISO 2719	—	D93	1102
Cleveland Open Cup	—	—	ISO 2592	—	D92	1103
Tag Closed Cup	—	—	—	—	D56	1101
Fire Point	—	—	ISO 2592	—	D92	1103
Effect of Evaporation on Flammability (Pipe Cleaner Test)	—	—	—	AMS 3150	—	352
High-Pressure Spray Ignition Test	—	—	—	—	—	6052
Low-Pressure Spray Ignition Test	—	—	—	AMS 3150	—	3119
Manifold Ignition Test	—	—	—	—	—	6053
Evaporation	—	—	—	—	—	323
Evaporation Loss	—	—	—	—	D972	351
Evaporation Loss-High Temperature	—	—	—	—	—	350

Continued

Table A.2.6 Continued

Subject	NFPA	ANSI	ISO	SAE	ASTM	FTM
API Gravity	—	—	—	—	D287	401
Density and Specific Gravity						
Lipkin Pycnometer	—	—	—	—	D941	402
Bingham Pycnometer	—	—	—	—	D1217	—
Specific Gravity						
(Hydrometer)	—	—	—	—	D1298	—
Emulsion Characteristics	—	—	—	—	D1401	—
Emulsifying Tendency	—	—	—	—	—	3201
Foaming Characteristics	—	—	—	—	D892	3211
Viscosity Stability at Low Temperature	—	—	—	—	D2532	307
Gelling, Crystallization and Separation	—	—	—	—	—	3458
Turbidity	—	—	—	—	—	3459
Sedimentation	—	—	—	—	D2273	3004
Lubricity Tests						
Timken Tester	—	—	—	—	D2782	6505
Falex Tester	—	—	—	—	D3233	3807
Four-Ball Tester (Extreme Pressure)	—	—	—	—	D2596	3812
Four-Ball Tester (Anti-wear)	—	—	—	—	D2266	6514
Recirculating Pump Test	—	—	—	—	D2428	—
Pump Loop Wear Test	—	—	—	—	D2271	—
Vane Pump Test	—	—	—	—	D2882	—
Load Carrying Ability						
Lubricating Oils	—	—	—	—	—	6511
Gear Lubricants	—	—	—	—	D1947	6512
Gas Turbine Lubricants	—	—	—	—	—	6509
Ryder Gear Machine	—	—	—	—	—	6508
Liquid Stability Tests Color	—	—	—	—	D1500	102
Neutralization Number by Potentiometric Titration	—	—	—	—	D664	5106
Neutralization Number by Color Indicator Titration	—	—	—	—	D974	5102
Acid Number by Semi-Micro Color Indicator Titration	—	—	—	—	D3339	5105
Carbon Residue						
Conradson	—	—	—	—	D189	5001
Ramsbottom	—	—	—	—	D524	5002
Oxidation Stability Test						
Oxidation-Corrosion Test	—	—	—	—	D4636	5307

Continued

Table A.2.6 Continued

Subject	NFPA	ANSI	ISO	SAE	ASTM	FTM
Oxidation Characteristics of Inhibited Mineral Oils	—	—	—	—	D943	5308
Thermal Stability	—	—	—	—	D2160	2508
Hydrolytic Stability (Beverage Bottle Test)	—	—	—	—	D2619	3457
Corrosiveness Tests						
Corrosion and Oxidation Stability	—	—	—	—	D4636	5308
Copper Strip Corrosion	—	—	—	—	D130	5325
Corrosion at 232°C (450°F)	—	—	—	—	—	5305
Lead Corrosion	—	—	—	—	—	5321
Rust Preventing–Steam Turbine Oils	—	—	—	—	D665	4011
Beverage Bottle Test	—	—	—	—	D2619	3457
Humidity-Type Corrosiveness Tests						
Humidity Cabinent	—	—	—	—	D1748	5329
Protection-Salt Spray	—	—	—	—	—	4001

Table A.2.7 Installations and Systems

Subject	NFPA	ANSI	ISO	SAE	ASTM	FTM
NFPA/JIC Hydraulic Systems	T2.2.1	—	ISO 4413	—	—	—
Leakage Prevention Practice	T2.24.2	—	—	—	—	—
NFPA/JIC Pneumatic Systems	T2.25.1	B93.114M	ISO 4414	—	—	—

Table A.2.8 Pumps, Motors, Power Units and Reservoirs

Subject	NFPA	ANSI	ISO	SAE	ASTM	FTM
Fluidborne Pump Noise Measurement	T2.7.2M	—	—	—	—	—
Airborne Pump Noise Levels	T2.7.4M	B93.71M	ISO 4412	—	—	—
Airborne Motor Noise Levels	T2.7.5M	B93.72M	ISO 4412	—	—	—
Pump/Motor Flange/Shaft Dimensions (Inch Series)	T3.9.2	B93.6	ISO 3019	J744	—	—
Pump Sound Test	T3.9.12M	—	—	—	—	—
Pump/Motor Terms	T3.9.13	—	—	—	—	—

Continued

Table A.2.8 Continued

Subject	NFPA	ANSI	ISO	SAE	ASTM	FTM
Motor Sound Test	T3.9.14M	—	—	—	—	—
Pump/Motor Test	—	B93.95M	ISO 4409	—	—	—
Pump Test	T3.9.17M	B93.27	—	J745	—	—
Pump Contaminant Test	T3.9.18M	—	—	—	—	—
Pressure-Compensated Pump Test	T3.9.20M	—	—	—	—	—
Hydraulic Pump/Motor Bibliography	T2.9.21	—	—	—	—	—
Pump/Motor Pressure Rating	T3.9.22	—	—	—	—	—
Motor Contaminant Test	T3.9.25M	—	—	—	—	—
Low-Speed, High-Torque Motor Flange/Shaft Dimensions	T3.9.26M	—	—	—	—	—
Pump/Motor Flange/Shaft Dimensions (Metric Series)	T3.9.27M	B93.81	ISO 3019	—	—	—
Polygon Flange Dimensions	—	—	ISO 3019	—	—	—
Pump/Motor Parameter Definitions	T3.9.29M	—	ISO 4391	—	—	—
Pump/Motor Geometric Displacements	T3.9.30M	B93.57M	ISO 3662	J745	—	—
Method of Test for Load-Sensing Pumps	T3.9.33M	—	—	—	—	—
Hydraulic Reservoir Requirements	T3.16.2M	B93.18M	—	—	—	—
Power Unit Design Practice	T3.16.2M	—	—	—	—	—
Hydraulic Power Unit Requirements	T3.16.3M	B93.41	—	—	—	—
Hydraulic Reservoir Pressure Rating	T3.16.8M	—	—	—	—	—
Hydraulic Reservoir Bibliography	T3.16.10M	—	—	—	—	—
Power Unit Installation Requirements	T3.16.10M	—	—	—	—	—
Reservoirs—Special Requirements to Accommodate HWC Fluids	T3.16.11M	—	—	—	—	—
Motor Low-Speed Characteristics	T3.9.31	—	ISO 4392	—	—	—
Motor Test Startability	T3.9.31	—	ISO 4392	J746	—	—

Table A.2.9 Sealing Devices

Subject	NFPA	ANSI	ISO	SAE	ASTM	FTM
Sealing Devices Terms	T3.19.1	—	—	—	—	—
Radial Sealing Device Dimensions	T3.19.4M	B93.76M	ISO 5597	J110	—	—
Method for Measuring Stack Heights	T3.19.5M	B93.17M	ISO 3939	—	—	—
Exclusion Devices Cavity Dimensions	T3.19.7M	B93.35M	—	—	—	—
Piston Ring Groove Dimensions	T3.19.11M	B93.36M	—	—	—	—
Sealing Device Test	T3.19.12.	B93.62M	—	—	—	—
Exclusion Device Test	T3.19.15M	—	—	J1195	—	—
Radial Compression-Type Piston Ring Groove Dimensions	T3.19.18M	B93.32M	—	—	—	—
Sealing Devices Bibliography	T3.19.22	—	—	—	—	—
Wear Ring Groove Dimensions	T3.19.23M	—	—	—	—	—
Piston Seal Dimensions and Tolerances	—	—	DP 7425	—	—	—
Plastic-faced Seal Dimensions and Tolerances	—	—	DP 7425	—	—	—
Piston Seal Housings Incorporating Bearing Rings	T3.19.26M	B93.93M	ISO 6547	—	—	—
O-Ring Dimensions Metric Series	T3.19.27M	B93.58M	ISO 3601	—	—	—
Identification—Elastometric Materials	T3.19.28M	—	—	—	—	—
Rotary Shaft Lip-Type Seals Dimensions and Tolerances	T3.19.29M	B93.98M	ISO 6194	—	—	—
O-Ring Quality Acceptance Criteria	T3.19.30	—	DP 3601	—	—	—
O-Ring Design Criteria	—	—	DP 3601	J515	—	—
O-Ring Boss Seal Size	—	—	DP 3601	—	—	—
O-Ring Pump Flanges Sizes	—	—	DP 3601	—	—	—
Fluid Compatibility with Elastomeric Materials	—	—	DIS 6072	—	—	—
Rotary Shaft Lip-Type Seals	—	—	DP 6194	—	—	—
Hydraulic Cylinder Housing for Rod Wiper Rings in Reciprocating Applications	—	B93.111M	DP 6195	—	—	—

Table A.2.10 Terminology and Symbols

Subject	NFPA	ANSI	ISO	SAE	ASTM	FTM
Fluid Power Terms	T2.1.1	B93.2	—	—	D4175	—
	T2.1.2	B89.2	DIS 5598	—	—	—
Graphic Symbols	T2.1.3M	Y32.10	ISO 1219	–	–	–
Fluid Power Diagrams	—	—	ISO 1219	—	—	—
Normal Pressures (Metric Only)	T2.1.4M	—	ISO 2944	—		
Fluid Power Mechanic-Technician-Engineer Job Responsibilities	T2.2.1	—	—	—	—	—
Preferred Metric Units	T.2.10.1M	—	—	J1322	—	—
Voluntary Metric Language Usage	T2.10.2M	—	—	—	—	—
Identifying Metric Fluid Power Components	T2.10.4M	—	—	—	—	—
Graphic Symbols for Fluidic-Devices	T3.7.2	—	—	—	—	—
Cylinder Terms	T3.6.3	—	—	—	—	—
Hose Terms	T3.8.26	—	—	J517	—	—
Pump/Motor Terms	T3.9.13	—	—	—	—	—
Hydraulic Filter/Separator Terms	T3.10.3	—	—	J1124	—	—
Drafting Practice	—	Y14.17	—	—	—	—
FRL Terms	T3.12.7	—	—	—	—	—
Sealing Devices Terms	T3.19.1	—	—	—	—	—
Quick Action Couplings Terms	T3.20.1	B93.2A	—	—	—	—
Fluid Logic Diagramming	T3.28.9	—	—	—	—	—
Standard Reference Atmosphere	—	—	DP 8778	—	—	—

Table A.2.11 Valves

Subject	NFPA	ANSI	ISO	SAE	ASTM	FTM
Hydraulc Valve Interfaces	T3.5.1M	B93.7M	ISO 4401	—	—	—
Fluid Power Valves Connection Symbols	T3.5.2M	B93.9M	—	—	—	—
Hydraulic Valves Marking	T3.5.2M	—	—	—	—	—
Hydraulic Valve Interfaces (315 bar)	T3.5.9M	B93.40M	—	—	—	—
Hydraulic Valve Metering Test	T3.5.14M	B93.66M	—	—	—	—

Continued

Table A.2.11 Continued

Subject	NFPA	ANSI	ISO	SAE	ASTM	FTM
Hydraulic Valve Leakage Test	T3.5.15M	B93.112	—	J1235	—	—
Hydraulic Flow Control Valve Test	T3.5.16M	—	—	—	—	—
Hydraulic Relief Valve Testing	T3.5.24M	—	—	—	—	—
Hydraulic Valve Pressure Rating	T3.5.26M	—	—	—	—	—
Hydraulic Valve Differential-Flow Characteristics	T3.5.28M	B93.49M	—	J1117	—	—
Hydraulic Valve Electrical Connector	T3.5.29M	B93.55M	—	—	—	—
Hydraulic Direction Valve Response Test	T3.5.30M	—	—	—	—	—
Mounting Cavities for Hydraulic Cartridge Valve	T3.5.31M	—	DP 7789	—	—	—
Electrical Plug Connector Characteristics and Requirements	T3.5.32M	B93.94	ISO 4400	—	—	—
Cylinder Actuator Mounted Valve Dimensions	T3.5.33M	B93.77M	—	—	—	—
Valve Mounting Surfaces Code for Identification	—	—	ISO 5783	—	—	—
Hydraulic Valve Interfaces Antirotation Device	T3.5.35M	—	—	—	—	—
Hydraulic Valve Interface Auxillary Port	T3.5.36M	—	—	—	—	—
Modular Stack Valve Dimensions	—	B93.102	ISO 7790	—	—	—
Hydraulic Valves Controlling Flow and Pressure—Methods of Test	—	—	ISO 6403	—	—	—
Hydraulic Pressure Control Valve Mounting Surfaces	—	—	ISO 5781	—	—	—
Hydraulic Flow Control Valve Mounting Surfaces	—	—	ISO 6263	—	—	—
Hydraulic Pressure Relief Valve Mounting Surfaces	—	—	ISO 6264	—	—	—
2-Pin Electrical Connector Characteristics and Requirements	—	—	ISO 6952	—	—	—
Hydraulic Valve Pressure Differential Flow Characteristics	—	—	ISO 4411	—	—	—

Continued

Table A.2.11 Continued

Subject	NFPA	ANSI	ISO	SAE	ASTM	FTM
Hydraulic Servovalve Test Methods	—	B93.99	ISO 6404	—	—	—
2-Port Hydraulic Slip-in Cartridge Valves	T3.5.45M	—	DIS 7368	—	—	—
Relief Valve Interfaces	T3.5.46M	—	—	—	—	—
Air Line Regulator Requirements	T3.12.3	B93.13	—	—	—	—
Pneumatic Plug Dimensions	T3.21.3	—	—	—	—	—
Pneumatic Valve Interface	T3.21.1	B93.33	—	—	—	—
Pneumatic Flow Rating Test	T3.20.14	—	—	—	—	—
Pneumatic Valve Pressure Rating	T3.21.4	—	—	—	—	—
Pneumatic Valve Electrical Connector	T3.21.6	—	—	—	—	—
Pneumatic Valve Interface Surface	T3.21.7	—	—	—	—	—
Pneumatic Valve Response Test	T3.21.8	—	—	—	—	—
Pneumatic Valve Marking	T3.21.15	—	—	—	—	—
Fluidic Performance Data Presentation	T3.28.3	B93.14	—	—	—	—

ISO Graphic Symbols

The graphic symbols presented in this appendix are taken from ISO 1219-1976. The ISO symbology has been used throughout this textbook and is preferred because it is understood internationally. These symbols are very similar to the ANSI and NFPA symbols, which are used extensively in the United States, and the German DIN symbols.

fluid power graphic symbols

0   Introduction

In fluid power systems, power is transmitted and controlled through a fluid (liquid or gas) under pressure within a circuit.

Graphic symbols are used in diagrams of hydraulic and pneumatic equipment and accessories for fluid power transmission.

1   Scope and field of application

This International Standard establishes principles for the use of symbols and specifies the symbols to be used in diagrams of hydraulic and pneumatic transmission systems and components.

The use of these symbols does not preclude the use of other symbols commonly used for pipework in other technical fields.

2   Reference

ISO 5598, *Fluid power - Vocabulary*[1]
1)   In preparation

3   Definitions

For definitions of terms used, see ISO 5598.

4   Identification statement
(Reference to this International Standard)

Use the following statement in test reports, catalogues and sales literature when electing to comply with this International Standard:

"Graphic symbols shown in accordance with ISO 1219, *Fluid power systems and components - Graphic symbols.*"

5   General basic and functional symbols

Symbols for hydraulic and pneumatic equipment and accessories are functional and consist of one or more basic symbols and in general of one or more functional symbols. The symbols are neither to scale nor in general orientated in any particular direction. The relative size of symbols in combination should correspond approximately to those in clauses 11 and 12.

DESCRIPTION		SYMBOL	APPLICATION
5.1	**BASIC SYMBOLS**		
5.1.1	**Line**	2)	
5.1.1.1	— continuous		flow lines
5.1.1.2	— long dashes	$L > 10E$	
5.1.1.3	— short dashes	$L < 5E$	
5.1.1.4	— double	$D < 5E$	mechanical connections (shafts, levers, piston-rods)
5.1.1.5	— long chain thin (optional use)		Enclosure for several components assembled in one unit
5.1.2	**Circle, semi-circle**		
5.1.2.1			As a rule, energy conversion units (pump, compressor, motor)
5.1.2.2			Measuring instruments
5.1.2.3			Non-return valve, rotary connection, etc
5.1.2.4			Mechanical link, roller, etc
5.1.2.5			Semi-rotary actuator

2) L = Length of dash, E = Thickness of line, D = Space between lines

DESCRIPTION		SYMBOL	APPLICATION
5.1.3	Square, rectangle		As a rule, control valves (valve) except for non-return valves
5.1.4	Diamond		Conditioning apparatus (filter, separator, lubricator, heat exchanger)
5.1.5	Miscellaneous symbols	3)	
5.1.5.1		$d \approx 5E$	Flow line connection
5.1.5.2			Spring
5.1.5.3			Restriction:
5.1.5.3.1			— affected by viscosity
5.1.5.3.2			— unaffected by viscosity
5.2	FUNCTIONAL SYMBOLS		
5.2.1	Triangle:		The direction of flow and the nature of the fluid
5.2.1.1	— solid	▼	Hydraulic flow
5.2.1.2	— in outline only	▽	Pneumatic flow or exhaust to atmosphere
5.2.2	Arrow		Indication of:
5.2.2.1			— direction
5.2.2.2			— direction of rotation
5.2.2.3			— path and direction of flow through valves.
			For regulating apparatus as in 7.4 both representations, with or without a tail to the end of the arrow, are used without distinction
			As a general rule the line perpendicular to the head of the arrow indicates that when the arrow moves, the interior path always remains connected to the corresponding exterior path
5.2.3	Sloping arrow		Indication of the possibility of a regulation or a progressive variability

3) E = Thickness of line

fluid power graphic symbols

	DESCRIPTION	SYMBOL	USE OF THE EQUIPMENT OR EXPLANATION OF THE SYMBOL	
6.1	**PUMPS AND COMPRESSORS**		To convert mechanical energy into hydraulic or pneumatic energy.	
6.1.1	**Fixed capacity hydraulic pump:**			
6.1.1.1	—with one direction of flow			
6.1.1.2	—with two directions of flow			
6.1.2	**Variable capacity hydraulic pump:**			
6.1.2.1	—with one direction of flow		The symbol is a combination of 6.1.1.1 and 5.2.3 (sloping arrow)	
6.1.2.2	—with two directions of flow		The symbol is a combination of 6.1.1.2 and 5.2.3 (sloping arrow)	
6.1.3	**Fixed capacity compressor (always one direction of flow)**			

	DESCRIPTION	SYMBOL	USE OF THE EQUIPMENT OR EXPLANATION OF THE SYMBOL	
6.2	**MOTORS**		To convert hydraulic or pneumatic energy into rotary mechanical energy	
6.2.1	**Fixed capacity hydraulic motor:**			
6.2.1.1	—with one direction of flow			SPRING
6.2.1.2	—with two directions of flow			
6.2.2	**Variable capacity hydraulic motor:**			
6.2.2.1	—with one direction of flow		The symbol is a combination of 6.2.1.1 and 5.2.3 (sloping arrow)	
6.2.2.2	—with two directions of flow		The symbol is a combination of 6.2.1.2 and 5.2.3 (sloping arrow)	
6.2.3	**Fixed capacity pneumatic motor:**			
6.2.3.1	—with one direction of flow			
6.2.3.2	—with two directions of flow			
6.2.4	**Variable capacity pneumatic motor:**			
6.2.4.1	—with one direction of flow		The symbol is a combination of 6.2.3.1 and 5.2.3 (sloping arrow)	
6.2.4.2	—with two directions of flow		The symbol is a combination of 6.2.3.2 and 5.2.3 (sloping arrow)	
6.2.5	**Oscillating motor:**			
6.2.5.1	—hydraulic			
6.2.5.2	—pneumatic			

fluid power graphic symbols

	DESCRIPTION	SYMBOL		USE OF THE EQUIPMENT OR EXPLANATION OF THE SYMBOL	
6.3	PUMP/MOTOR UNITS			Unit with two functions, either as pump or as rotary motor	
6.3.1	Fixed capacity pump/motor unit:				
6.3.1.1	— with reversal of the direction of flow			Functioning as pump or motor according to direction of flow	
6.3.1.2	— with one single direction of flow			Functioning as pump or motor without change of direction of flow	
6.3.1.3	— with two directions of flow			Functioning as pump or motor with either direction of flow	
6.3.2	Variable capacity pump/motor unit:				
6.3.2.1	— with reversal of the direction of flow			The symbol is a combination of 6.3.1.1 and 5.2.3 (sloping arrow)	
6.3.2.2	— with one single direction of flow			The symbol is a combination of 6.3.1.2 and 5.2.3 (sloping arrow)	
6.3.2.3	— with two directions of flow			The symbol is a combination of 6.3.1.3 and 5.2.3 (sloping arrow)	
6.4	VARIABLE SPEED DRIVE UNITS			Torque converter. Pump and/or motor are of variable capacity. Remote drives, see 12.2	
6.5	CYLINDERS			Equipment to convert hydraulic or pneumatic energy into linear energy	
6.5.1	Single acting cylinder:	Detailed	Simplified	Cylinder in which the fluid pressure always acts in one and the same direction (on the forward stroke)	
6.5.1.1	— returned by an unspecified force			General symbol when the method of return is not specified	
6.5.1.2	— returned by spring			Combination of the general symbols 6.5.1.1 and 5.1.5.2 (spring)	

	DESCRIPTION	SYMBOL		USE OF THE EQUIPMENT OR EXPLANATION OF THE SYMBOL	
6.5.2	**Double acting cylinder:**			Cylinder in which the fluid pressure operates alternately in both directions (forward and backward strokes)	
6.5.2.1	— with single piston rod				
6.5.2.2	— with double-ended piston rod				
6.5.3	**Differential cylinder**			The action is dependent on the difference between the effective areas on each side of the piston	
6.5.4	**Cylinder with cushion:**				
6.5.4.1	— with single fixed cushion			Cylinder incorporating fixed cushion acting in one direction only	
6.5.4.2	— with double fixed cushion			Cylinder with fixed cushion acting in both directions	
6.5.4.3	— with single adjustable cushion			The symbol is a combination of 6.5.4.1 and 5.2.3 (sloping arrow)	
6.5.4.4	—with double adjustable cushion			The symbol is a combination of 6.5.4.2 and 5.2.3 (sloping arrow)	
6.5.5	**Telescopic cylinder:**				
6.5.5.1	— single acting			The fluid pressure always acts in one and the same direction (on the forward stroke)	
6.5.5.2	— double acting			The fluid pressure operates alternately in both directions (forward and backward strokes)	

fluid power graphic symbols

DESCRIPTION	SYMBOL		USE OF THE EQUIPMENT OR EXPLANATION OF THE SYMBOL	
6.6 **PRESSURE INTENSIFIERS:**	Detailed	Simplified	Equipment transforming a pressure x into a higher pressure y	
6.6.1 — for one type of fluid			E.g. a pneumatic pressure x is transformed into a higher pneumatic pressure y	
6.6.2 — for two types of fluid			E.g. a pneumatic pressure x is transformed into a higher hydraulic pressure y	
6.7 **AIR-OIL ACTUATOR**			Equipment transforming a pneumatic pressure into a substantially equal hydraulic pressure or vice versa	
7 **CONTROL VALVES**				
7.1 **METHOD OF REPRE-SENTATION OF VALVES (EXCEPT 7.3 AND 7.6)**			Made up of one or more squares 5.1.3 and arrows In circuit diagrams hydraulic and pneumatic units are normally shown in the unoperated condition	
7.1.1 One single square			Indicates unit for controlling flow or pressure, having in operation an infinite number of possible positions between its end positions so as to vary the conditions of flow across one or more of its ports, thus ensuring the chosen pressure and/or flow with regard to the operating conditions of the circuit	
7.1.2 Two or more squares			Indicate a directional control valve having as many distinct positions as there are squares. The pipe connections are normally represented as representing the unoperated condition (see 7.1). The operating positions are deduced by imagining the boxes to be displaced so that the pipe connections correspond with the ports of the box in question	

	DESCRIPTION	SYMBOL	USE OF THE EQUIPMENT OR EXPLANATION OF THE SYMBOL	
7.1.3	Simplified symbol for valves in cases of multiple repetition	[3]	The number refers to a note on the diagram in which the symbol for the valve is given in full	
7.2	**DIRECTIONAL CONTROL VALVES**		Units providing for the opening (fully or restricted) or the closing of one or more flow paths (represented by several squares)	
7.2.1	**Flow paths:**		Square containing interior lines	
7.2.1.1	—one flow path			
7.2.1.2	—two closed ports			
7.2.1.3	—two flow paths			
7.2.1.4	—two flow paths and one closed port			
7.2.1.5	—two flow paths with cross connection			
7.2.1.6	—one flow path in a by-pass position, two closed ports			
7.2.2	**Non-throttling directional control valve**		The unit provides distinct circuit conditions each depicted by a square	
7.2.2.1			Basic symbol for 2-position directional control valve	
7.2.2.2			Basic symbol for 3-position directional control valve	
7.2.2.3			A transitory but significant condition between two distinct positions is optionally represented by a square with dashed ends

A basic symbol for a directional control valve with two distinct positions and one transitory intermediate condition | |

fluid power graphic symbols

	DESCRIPTION	SYMBOL	USE OF THE EQUIPMENT OR EXPLANATION OF THE SYMBOL	
7.2.2.4	**Designation:** The first figure in the designation shows the number of ports (excluding pilot ports) and the second figure the number of distinct positions			
7.2.2.5	**Directional control valve 2/2:**		Directional control valve with 2 ports and 2 distinct positions	
7.2.2.5.1	—with manual control			
7.2.2.5.2	—controlled by pressure operating against a return spring (e.g., on air unloading valve)			
7.2.2.6	**Directional control valve 3/2:**		Directional control valve with 3 ports and 2 distinct positions	
7.2.2.6.1	—controlled by pressure in both directions			
7.2.2.6.2	—controlled by solenoid with return spring		Indicating an intermediate condition (see 7.2.2.3)	
7.2.2.7	**Directional control valve 4/2:**	Detailed	Directional control valve with 4 ports and 2 distinct positions	
7.2.2.7.1	—controlled by pressure in both directions by means of a pilot valve (with a single solenoid and spring return)	Simplified		
7.2.2.8	**Directional control valve 5/2:**		Directional control valve with 5 ports and 2 distinct positions	
7.2.2.8.1	—controlled by pressure in both directions			

	DESCRIPTION	SYMBOL	USE OF THE EQUIPMENT OR EXPLANATION OF THE SYMBOL	
7.2.3	**Throttling directional control**		The unit has 2 extreme positions and an infinite number of intermediate conditions with varying degrees of throttling All the symbols have parallel lines along the length of the boxes. For valves with mechanical feedback see 9.3	
7.2.3.1			Showing the extreme positions	
7.2.3.2			Showing the extreme positions and a central (neutral) position	
7.2.3.3	— with 2 ports (one throttling orifice)		For example: Tracer valve plunger operated against a return spring	
7.2.3.4	— with 3 ports (two throttling orifices)		For example: Directional control valve controlled by pressure against a return spring	
7.2.3.5	— with 4 ports (four throttling orifices)		For example: Tracer valve, plunger operated against a return spring	
7.2.4	**Electro-hydraulic servo valve:** **Electro-pneumatic servo valve:**		A unit which accepts an analogue electrical signal and provides a similar analogue fluid power output	Torque motor Spool T B P A T Torque motor armature
7.2.4.1	— single-stage		—with direct operation	
7.2.4.2	— two-stage with mechanical feedback		— with indirect pilot operation	
7.2.4.3	— two-stage with hydraulic feedback		— with indirect pilot operation	

fluid power graphic symbols

	DESCRIPTION	SYMBOL	USE OF THE EQUIPMENT OR EXPLANATION OF THE SYMBOL	
7.3	**NON-RETURN VALVES, SHUTTLE VALVE, RAPID EXHAUST VALVE**		Valves which allow free flow in one direction only	
7.3.1	**Non-return valve**			
7.3.1.1	— free		Opens if the inlet pressure is higher than the outlet pressure	
7.3.1.2	— spring loaded		Opens if the inlet pressure is greater than the outlet pressure plus the spring pressure	
7.3.1.3	—pilot controlled		As 7.3.1.1 but by pilot control it is possible to prevent	
7.3.1.3.1	— a pilot signal closes the valve			
7.3.1.3.2	— a pilot signal opens the valve			
7.3.1.4	— with restriction		Unit allowing free flow in one direction but restricted flow in the other	
7.3.2	**Shuttle valve**		The inlet port connected to the higher pressure is automatically connected to the outlet port while the other inlet port is closed	
7.3.3	**Rapid exhaust valve**		When the inlet port is unloaded the outlet port is freely exhausted	
7.4	**PRESSURE CONTROL VALVES**		Units ensuring the control of pressure. Represented by one single square as in 7.1.1 with one arrow (the tail to the arrow may be placed at the end of the arrow). For interior controlling conditions see 9.2.4.3	
7.4.1	**Pressure control valve:**		General symbols	
7.4.1.1	— 1 throttling orifice normally closed			

	DESCRIPTION	SYMBOL	USE OF THE EQUIPMENT OR EXPLANATION OF THE SYMBOL	
7.4.1.2	— 1 throttling orifice normally open			
7.4.1.3	— 2 throttling orifices, normally closed			
7.4.2	Pressure relief valve (safety valve):		Inlet pressure is controlled by opening the exhuast port to the reservoir or to atmosphere against an opposing force (for example a spring)	
7.4.2.1	— with remote pilot control		The pressure at the inlet port is limited as in 7.4.2 or to that corresponding to the setting of a pilot control	
7.4.3	Proportional pressure relief		Inlet pressure is limited to a value proportional to the pilot pressure (see 9.2.4.1.3)	
7.4.4.	Sequence valve		When the inlet pressure overcomes the opposing force of the spring, the valve opens permitting flow from the outlet port	
7.4.5	Pressure regulator or reducing valve (reducer of pressure):		A unit which, with a variable inlet pressure, gives substantially constant output pressure provided that the inlet pressure remains higher than the required outlet pressure	
7.4.5.1	— without relief port			
7.4.5.2	— without relief port with remote control		As in 7.4.5.1 but the outlet pressure is dependent on the control pressure	
7.4.5.3	— with relief port			

fluid power graphic symbols

	DESCRIPTION	SYMBOL		USE OF THE EQUIPMENT OR EXPLANATION OF THE SYMBOL	
7.4.5.4	— with relief port, with remote control			As in 7.4.5.3, but the outlet pressure is dependent on the control pressure	
7.4.6	**Differential pressure regulator**			The outlet pressure is reduced by a fixed amount with respect to the inlet pressure	
7.4.7	**Proportional pressure regulator**			The outlet pressure is reduced by a fixed ratio with respect to the inlet pressure (see 9.2.4.1.3)	
7.5	**FLOW CONTROL VALVES**			Units ensuring control of flow excepting 7.5.3 positions and method of representation as 7.4	
7.5.1	**Throttle valve:**			Simplified symbol (Does not indicate the control method or the state of the valve)	
7.5.1.1	—with manual control			Detailed symbol (indicates the control method of the state of the valve)	
7.5.1.2	— with mechanical control against a return spring (braking valve)				
7.5.2	**Flow control valve:**	Detailed	Simplified	Variations in inlet pressure do not affect the rate of flow	
7.5.2.1	— with fixed output				
7.5.2.2	— with fixed output and relief port to reservoir			As 7.5.2.1 but with relief for excess flow	

	DESCRIPTION	SYMBOL	USE OF THE EQUIPMENT OR EXPLANATION OF THE SYMBOL	
7.5.2.3	— with variable output		As 7.5.2.1 but with arrow 5.2.3 added to the symbol of restriction	Control chamber — Inlet — Tank — Outlet — Vent connection
7.5.2.4	— with variable output and relief port to reservoir		As 7.5.2.3 but with relief for excess flow	
7.5.3	**Flow dividing valve**		The flow is divided into two flows in a fixed ratio substantially independent of pressure variations	
7.6	**SHUT-OFF VALVE**		Simplified symbol	
8. ENERGY TRANSMISSION AND CONDITIONING				
8.1	**SOURCES OF ENERGY**			
8.1.1	**Pressure source**		Simplified general symbol	
8.1.1.1	**Hydraulic pressure source**		Symbols to be used when the nature of the source should be indicated	
8.1.1.2	**Pneumatic pressure source**			
8.1.2	**Electric motor**		Symbol 113 in IEC Publication 117.2	
8.1.3	**Heat engine**			
8.2	**FLOW LINES AND CONNECTIONS**			
8.2.1	**Flow line:**			
8.2.1.1	— working line, return line and feed line			
8.2.1.2	— pilot control line			

fluid power graphic symbols

	DESCRIPTION	SYMBOL	USE OF THE EQUIPMENT OR EXPLANATION OF THE SYMBOL	
8.2.1.3	— drain or bleed line			
8.2.1.4	— flexible pipe		Flexible hose, usually connecting moving parts	
8.2.1.5	— electric line			
8.2.2	**Pipeline junction**			
8.2.3	**Crossed Pipelines**		not connected	
8.2.4	**Air bleed**			
8.2.5	**Exhaust port:**			
8.2.5.1	— plain with no provision for connection			
8.2.5.2	— threaded for connection			
8.2.6	**Power take-off:**		On equipment or lines, for energy take-off or measurement	
8.2.6.1	— plugged			
8.2.6.2	— with take-off line			
8.2.7	**Quick-acting coupling:**			
8.2.7.1	— connected, without mechanically opened non-return valve			
8.2.7.2	— connected, with mechanically opened non-return valves			
8.2.7.3	— uncoupled, with open end			
8.2.7.4	— uncoupled, closed by free non-return valve (see 7.3.1.1)			

	DESCRIPTION	SYMBOL	USE OF THE EQUIPMENT OR EXPLANATION OF THE SYMBOL	
8.2.8	Rotary connection:		Line junction allowing angular movement in service	
8.2.8.1	— one way			
8.2.8.2	— three way			
8.2.9	Silencer			
8.3	RESERVOIRS			
8.3.1	Reservoir open to atmosphere:			
8.3.1.1	— with inlet pipe above fluid level			
8.3.1.2	— with inlet pipe below fluid level			
8.3.1.3	— with a header line			
8.3.2	Pressurized reservoir			
8.4	ACCUMULATORS		The fluid is maintained under pressure by a spring, weight or compressed gas (air, nitrogen, etc.)	Air or gas →
8.5	FILTERS, WATER TRAPS, LUBRICATORS AND MISCELLANEOUS APPARATUS			
8.5.1	Filter or strainer			Bowl — Filter element
8.5.2	Water trap:			

fluid power graphic symbols

DESCRIPTION	SYMBOL	USE OF THE EQUIPMENT OR EXPLANATION OF THE SYMBOL	
8.5.2.1 — with manual control			
8.5.2.2 — automatically drained			
8.5.3 **Filter with water trap:**			
8.5.3.1 — with manual control		Combination of 8.5.1 and 8.5.2.1	
8.5.3.2 — automatically drained		Combination of 8.5.1 and 8.5.2.2	Float
8.5.4 **Air dryer**		A unit drying air (for example, by chemical means)	Desiccant
8.5.5 **Lubricator**		Small quantities of oil are added to the air passing through the unit, in order to lubricate equipment receiving the air	Inlet air Lubricated air
8.5.6 **Conditioning unit**		Consisting of filter, pressure regulator, pressure gauge and lubricator	
8.5.61 — Detailed symbol			
8.5.6.2 — Simplified symbol			
8.6 **HEAT EXCHANGERS**		Apparatus for heating or cooling the circulating fluid	

	DESCRIPTION	SYMBOL	USE OF THE EQUIPMENT OR EXPLANATION OF THE SYMBOL	
8.6.1	**Temperature controller**		The fluid temperature is maintained between two predetermined values. The arrows indicate that heat may be either introduced or dissipated	
8.6.2	**Cooler**		The arrows in the diamond indicate the extraction of heat	
8.6.2.1			— without representation of the flow lines of the coolant	
8.6.2.2			— indicating the flow lines of the coolant	
8.6.3	**Heater**		The arrows in the diamond indicate the introduction of heat	
9. **CONTROL MECHANISMS**				
9.1	**Mechanical components**			
9.1.1	**Rotating shaft:**		The arrow indicates rotation	
9.1.1.1	— in one direction			
9.1.1.2	— in either direction			
9.1.2	**Detent**		A device for maintaining a given position	
9.1.3	**Locking device**	*	* The symbol for unlocking control is inserted in the square	
9.1.4	**Over-center device**		Prevents the mechanism stopping in a dead center position	
9.1.5	**Pivoting devices:**			
9.1.5.1	— simple			

fluid power graphic symbols

DESCRIPTION		SYMBOL	USE OF THE EQUIPMENT OR EXPLANATION OF THE SYMBOL	
9.1.5.2	— with traversing lever			
9.1.5.3	— with fixed fulcrum			
9.2	CONTROL METHODS		The symbols representing control methods are incorporated in the symbol of the controlled apparatus, to which they should be adjacent. For apparatus with several squares the actuation of the control makes effective the square adjacent to it.	
9.2.1	Muscular control:		General symbol (without indication of control type)	
9.2.1.1	— by pushbutton			
9.2.1.2	— by lever			
9.2.1.3	— by pedal			
9.2.2	Mechanical control:			
9.2.2.1	— by plunger or tracer			
9.2.2.2	— by spring			
9.2.2.3	— by roller			
9.2.2.4	— by roller, operating in one direction only			
9.2.3	Electrical control:			
9.2.3.1	— by solenoid:			
9.2.3.1.1			— with one winding	
9.2.3.1.2			— with two windings operating in opposite directions	

	DESCRIPTION	SYMBOL	USE OF THE EQUIPMENT OR EXPLANATION OF THE SYMBOL	
9.2.3.1.3			— with two windings operating in a variable way progressively, operating in opposite direction	
9.2.3.2	— by electric motor			
9.2.4	**Control by application or release of pressure**			
9.2.4.1	Direct acting control:			
9.2.4.1.1	— by application of pressure			
9.2.4.1.2	— by release of pressure			
9.2.4.1.3	— by different control areas		In the symbol the larger rectangle represents the larger control area, i.e., the priority phase	
9.2.4.2	Indirect control, pilot actuated:		General symbol for pilot directional control valve	
9.2.4.2.1	— by application of pressure			
9.2.4.2.2	— by release of pressure			
9.2.4.3	Interior control paths		The control paths are inside the unit	
9.2.5	**Combined control:**			
9.2.5.1	— by solenoid and pilot directional valve		The pilot directional valve is actuated by the solenoid	
9.2.5.2	— by solenoid or pilot directional valve		Either may actuate the control independently	

fluid power graphic symbols

	DESCRIPTION	SYMBOL	USE OF THE EQUIPMENT OR EXPLANATION OF THE SYMBOL	
9.3	Mechanical feedback	1) 2) 1) Controlled apparatus 2) Control apparatus	The mechanical connection of a control apparatus moving part to a controlled apparatus moving part is represented by the symbol 5.1.1.4 which joins the two parts connected. (For examples see 11.1.2 and 12.1.1)	
10. SUPPLEMENTARY EQUIPMENT				
10.1 MEASURING INSTRUMENTS				
10.1.1 10.1.1.1	Pressure measurement: — pressure gauge		The point on the circle at which the connection joins the symbol is immaterial	 Bourdon tube
10.1.2 10.1.2.1	Temperature measurement: — Thermometer		The point on the circle at which the connection joins the symbol is immaterial	
10.1.3 10.1.3.1 10.1.3.2	Measurement of flow: — Flow meter — Integrating flow			
10.2 10.2.1	OTHER APPARATUS Pressure electric switch			High pressure adjustment Low pressure adjustment

Typical Graphic Symbols for Electrical Diagrams

This appendix contains the symbols for the more commonly used electrical switches and sensors for controlling electrohydraulic valves. The symbols are shown as they would appear in standard electrical ladder diagrams. They represent the function of the device only. There is no attempt to represent the physical construction of the device or to imply the location or means of connecting the device into the circuit.

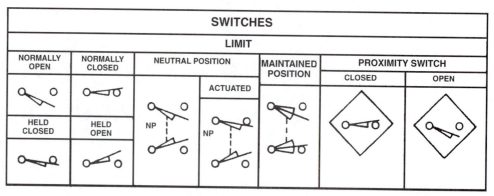

SWITCHES

LIMIT

NORMALLY OPEN	NORMALLY CLOSED	NEUTRAL POSITION		MAINTAINED POSITION	PROXIMITY SWITCH	
			ACTUATED		CLOSED	OPEN

| HELD CLOSED | HELD OPEN | NP | NP | | | |

	LIQUID LEVEL		VACUUM & PRESSURE		TEMPERATURE		FLOW (AIR, WATER ETC.)	
NORMALLY OPEN	NORMALLY CLOSED	NORMALLY OPEN	NORMALLY CLOSED	NORMALLY OPEN	NORMALLY CLOSED	NORMALLY OPEN	NORMALLY CLOSED

FOOT		TOGGLE			
NORMALLY OPEN	NORMALLY CLOSED	SPST*	SPDT**	SPDT (CENTER OFF)	DPST***
		NORMALLY OPEN			
		NORMALLY CLOSED			

*SINGLE POLE, SINGLE THROW   **SINGLE POLE, DOUBLE THROW   ***DOUBLE POLE, SINGLE THROW

PUSHBUTTONS			CONNECTIONS, ETC.					
SINGLE CIRCUIT	DOUBLE CIRCUIT		MAINTAINED CONTACT	CONDUCTORS		GROUND	CHASSIS OR FRAME NOT NECESSARILY GROUNDED	PLUG AND RECP.
		MUSHROOM HEAD		NOT CONNECTED	CONNECTED			
NORMALLY OPEN								
NORMALLY CLOSED								

CONTACTS						
TIME DELAY AFTER COIL				RELAY, ETC.		THERMAL OVER-LOAD
ENERGIZED		DE-ENERGIZED		NORMALLY OPEN	NORMALLY CLOSED	
NORMALLY OPEN	NORMALLY CLOSED	NORMALLY OPEN	NORMALLY CLOSED			

COILS					
RELAYS, TIMERS, ETC.		SOLENOIDS, BRAKES, ETC.			THERMAL OVERLOAD ELEMENT
	GENERAL	2-POSITION HYDRAULIC	3-POSITION PNEUMATIC	2-POSITION LUBRICATION	

Standard Steel Pipe Sizes

Nominal Pipe Size (in)	Outside Diameter	Wall Thickness		Inside Diameter		Flow Area	
		Sched 40	Sched 80	Sched 40	Sched 80	Sched 40	Sched 80
U.S. Customary	(in)	(in)	(in)	(in)	(in)	(in^2)	(in^2)
⅛	0.41	0.07	0.10	0.27	0.22	0.057	0.036
¼	0.54	0.09	0.12	0.36	0.30	0.104	0.072
⅜	0.68	0.09	0.13	0.49	0.42	0.191	0.140
½	0.84	0.11	0.15	0.62	0.55	0.304	0.234
¾	1.05	0.11	0.15	0.82	0.74	0.533	0.432
1	1.32	0.13	0.18	1.05	0.96	0.864	0.719
1¼	1.66	0.14	0.19	1.38	1.28	1.495	1.282
1½	1.90	0.15	0.20	1.61	1.50	2.035	1.766
2	2.38	0.15	0.22	2.07	1.94	3.354	2.951
2½	2.88	0.20	0.28	2.47	2.32	4.785	4.236
3	3.50	0.22	0.30	3.07	2.90	7.389	6.602
3½	4.00	0.23	0.32	3.55	3.36	9.882	8.883
4	4.50	0.24	0.34	4.03	3.83	12.724	11.491

Nominal Pipe Size (in)	Outside Diameter	Wall Thickness		Inside Diameter		Flow Area	
		Sched 40	Sched 80	Sched 40	Sched 80	Sched 40	Sched 80
SI Units	(cm)	(cm)	(cm)	(cm)	(cm)	(cm^2)	(cm^2)
⅛	1.03	0.17	0.24	0.68	0.55	0.366	0.234
¼	1.37	0.22	0.30	0.92	0.77	0.671	0.462
⅜	1.71	0.23	0.32	1.25	1.07	1.231	0.906
½	2.13	0.28	0.37	1.58	1.39	1.959	1.510
¾	2.67	0.29	0.39	2.09	1.88	3.439	2.788
1	3.34	0.34	0.45	2.66	2.43	5.573	4.638
1¼	4.22	0.36	0.49	3.51	3.25	9.645	8.272
1½	4.83	0.37	0.51	4.09	3.81	13.128	11.395
2	6.03	0.39	0.55	5.25	4.93	21.638	19.041
2½	7.30	0.52	0.70	6.27	5.90	30.873	27.330
3	8.89	0.55	0.76	7.79	7.37	47.670	42.592
3½	10.16	0.57	0.81	9.01	8.54	63.754	57.312
4	11.43	0.60	0.86	10.23	9.72	82.089	74.136

Dimensions of Steel Tubing

Outside Diameter		Wall Thickness		Inside Diameter	
in	mm	in	mm	in	mm
1/8	3.18	0.028	0.711	0.069	1.758
		0.032	0.813	0.061	1.554
		0.035	0.889	0.055	1.402
3/16	4.76	0.032	0.813	0.123	3.134
		0.035	0.889	0.117	2.982
1/4	6.35	0.035	0.889	0.180	4.572
		0.049	1.245	0.250	6.350
		0.065	1.651	0.120	3.048
5/16	7.94	0.035	0.889	0.243	6.162
		0.049	1.245	0.215	5.451
		0.065	1.651	0.183	4.638
3/8	9.53	0.035	0.889	0.305	7.752
		0.049	1.245	0.277	7.041
		0.065	1.651	0.245	6.228
1/2	12.70	0.035	0.889	0.430	10.922
		0.049	1.245	0.402	10.211
		0.065	1.651	0.370	9.398
		0.083	2.108	0.334	8.484
5/8	15.88	0.035	0.889	0.555	14.102
		0.049	1.245	0.527	13.391
		0.065	1.651	0.495	12.578
		0.083	2.108	0.459	11.664
3/4	19.05	0.049	1.245	0.652	16.561
		0.065	1.651	0.620	15.748
		0.083	2.108	0.584	14.834
		0.109	2.769	0.532	13.513
7/8	22.23	0.049	1.245	0.777	19.741
		0.065	1.651	0.745	18.928
		0.083	2.108	0.709	18.014
		0.109	2.769	0.657	16.693

Continued

Outside Diameter		Wall Thickness		Inside Diameter	
in	mm	in	mm	in	mm
1	25.40	0.049	1.245	0.902	22.911
		0.065	1.651	0.870	22.098
		0.083	2.108	0.834	21.184
		0.109	2.769	0.782	19.863
1¼	31.75	0.049	1.245	1.152	29.261
		0.065	1.651	1.120	28.448
		0.083	2.108	1.084	27.534
		0.109	2.769	1.032	26.213
1½	38.10	0.065	1.651	1.370	34.798
		0.083	2.108	1.334	33.884
		0.109	2.769	1.282	32.563
1¾	44.45	0.065	1.651	1.620	41.148
		0.083	2.108	1.584	40.234
		0.109	2.769	1.532	38.913
		0.134	3.404	1.482	37.643
2	50.80	0.065	1.651	1.870	47.498
		0.083	2.108	1.834	46.584
		0.109	2.769	1.782	45.263
		0.134	3.404	1.732	43.993

ISO Viscosity Grades

VG	cSt (40°C, 104°F)	m²/s	ft²/sec
2	2	2×10^{-6}	2.15×10^{-5}
3	3	3×10^{-6}	3.23×10^{-5}
5	5	5×10^{-6}	5.38×10^{-5}
7	7	7×10^{-6}	7.53×10^{-5}
10	10	1×10^{-5}	0.000108
15	15	1.5×10^{-5}	0.000161
22	22	2.2×10^{-5}	0.000237
32	32	3.2×10^{-5}	0.000344
46	46	4.6×10^{-5}	0.000495
68	68	6.8×10^{-5}	0.000732
100	100	0.00010	0.001076
150	150	0.00015	0.001614
220	220	0.00022	0.002367
320	320	0.00032	0.003443
460	460	0.00046	0.004950
680	680	0.00068	0.007317
1000	1000	0.00100	0.010760
1500	1500	0.00150	0.016140

Answers to Selected Problems

Chapter 2

13. 400 psi
15. 15,000 lb
17. 1592 psi
19S. 499.6 kPa
21. a. 63.7 psi
    b. 63.7 psi
    c. 63.7 psi
    d. 3200 lb
23. 0.5 in
25. a. 160 strokes
    b. 20
    c. 2.24 in
27S. 3127 N
29. 10,000 ft lb/sec
31. 0.746 kW
33. 160 lb
35. 59.7 ft lb
39. 1071 psi
41S. 2546 kPa, 3.33 kW
43S. a. 16.67 kW
     b. 88.4 N m
     c. 16.67 kW
45. 244.8 gpm
47. 36.8 ft/sec
49. Q = 195.7 gpm
    v(2.5) = 12.8 ft/sec
    v(3) = 8.9 ft/sec
51S. 88.4 Lpm
53S. 6.8 m/s, 2.7 m/s
55S. 353.5 Lpm

Chapter 3

25. 1.69 slug/ft^3, 54.29 lb/ft^3
27. 0.85
29. 396.1 lb

31. 155, 600 in^3, 90 ft^3, 673.5 gal
33. a. 966 lb
    b. 468 lb
35. 22.4 slug/ft^3
37. 1.86 slug/ft^3, 59.9 lb/ft^3
39. 0.941, 1.83 slug/ft^3
41. 532.8 gal
43S. 9.03 kN/m^3, 920 kg/m^3
45S. 11.04 kN/m^3, 1.125
47S. 3000 kg/m^3, 29.43 kN/m^3, 3
49. -3.33%
51S. -7.14%
53. 50

Chapter 4

13. 8.5 HP
15. a. 14.6 HP
    b. 1.28 in^3
    c. 0.97
17. 0.92, 0.87
19. 0.855
21. a. 91%
    b. 91%
    c. 86%
23S. 54.95 kW
25S. 0.926
27. 21.67 gpm
29. Q_T = 12.18 gpm
    Q_A = 10.11 gpm
31S. Q_T = 50 Lpm
     Q_A = 40 Lpm
33. 33.18 HP
35. 9.45 HP, 7.1 kW
37. 6.38 HP, 4.76 kW
39. 173 kWH, $14.75
41. 0.305 year or 3.65 months

Chapter 5
11. 14 HP
13. 11.1 HP
15. 328.25 ftlb
17S. 15.7 kW
19. 0.833
21. 23.52 HP
23. 0.87
25. 8.8 gpm
27. 870 rpm
29. 60.45 ftlb
31S. 12 kW
33S. 34 Lpm
35. 825 rpm
37. N_T = 888.5 rpm
    N_A = 808.5 rpm
39. 15.6 gpm
41S. 2000 rpm
43S. 24.7 gpm
45. 1890 inlb
47. 837.76 psi
49. 1091 psi
51S. 220 Nm
53S. 1.9 MPa
55. 111.94 in³/rev
57. 29.17 HP, 21.73 kW, 57%

Chapter 6
17. 6000 lb
19. F_{ext} = 14,140 lb
    F_{ret} = 6240 lb
21S. 14 kN
23. 566 psi
25. p_{ext} = 955 psi
    p_{ret} = 1274 psi
27. 5556 psi
29S. p_{ext} = 2546 kPa
    p_{ret} = 1516 kPa
31. 0.817 ft/sec
33. 0.681 ft/sec
35. 22.04 gpm
37. 12.25 gpm
39. 11.25 gpm (on retraction)
41S. v_{ext} = 0.085 m/s
    v_{ret} = 0.113 m/s
43S. 21.6 Lpm
45. v_{ext} = 0.511 ft/sec
    v_{ret} = 0.595 ft/sec
    F_{ext} = 18,840 lb
    F_{ret} = 16,190 lb

47. 955 psi, 19.57 gpm, 10.91 HP
49S. v_{ext} = 12.7 cm/s
    v_{ret} = 16.98 cm/s
    F_{ext} = 157 kN
    F_{ret} = 117.8 kN
51. 31.53 HP
53. A_{PISTON} = 2 A_{ROD}
55. p_{accel} = 76.1 psi
    p_{ss} = 71.7 psi
57. a. 13.6 HP
    b. 16.5 gpm, 1414.4 psi
    c. 13.6 HP
    d. 14.9 HP
    e. 17.1 HP
    f. 79%

Chapter 7
11. 1.4 HP
13S. 0.42 kW
15. 9.23° F
17S. 0.42 kW, 0.125° C

Chapter 8
15. 23.34 HP
17S. 1.33 kW
19. 50 lb
21. 875 psi
23S. 52.5 N
25S. 140 Pa
27. 3180 Btu/min
29. 10.55° F
31S. 0.357° C
35. Q = 91.89 gpm
    p_{crack} = 1937 psi
    HGR = 6941 Btu/min
    T = 15.95° F

Chapter 9
11. 9.81 gpm
13. 53.6 psi
15. 3.68 psi
17S. 36.8 Lpm
19S. 435.75 kPa
21S. 61.98 kPa
23. 6.43 ft/s
25. a. 3000, 6000, 0
    b. 3000, 7500, 0
27S. 0.6° C/1000 kPa

Chapter 10
11. 0.333 in/V
13. 2016 rpm

Chapter 11
11. 15
13. d
15. 9.38 gpm
17. Extension: $Q_{out} = 3$ gpm
Retraction: $Q_{out} = 12$ gpm

Chapter 12
11. 1 in Sched 40
13. 3/4 in Sched 40
15. 0.13 in
17. 2831
19. 53,011
21. 27.5 psi
23S. 51.39 kPa

Chapter 13
13. 480 ft^2
15S. 4500 W
17. a. 162.2 in^3
 b. 7.07 in^3
19. a. 87.5 in^3
 b. 215.5 in^3
 c. 286.5 in^3
21S. a. 1.12 L
 b. 3.02 L
 c. 3.77 L

23. a. 19 min
 b. 7.13 ft^2

Chapter 14
21. 27.45 psia
23. 1118° F
25. 245 psia
27. 528° F
29. a. 59° F, 23.76 ft^3
 b. 557° F, 46.54 ft^3
31S. 37.7 kPa (gage)
33. 40 scfm
35. 243.4 cfh, 190.7 cfh, 143.2 cfh
37S. 1.32 scmm
39S. 1.59 m^3/h, 1.25 m^3/h, 1.02 m^3/h
41. 4.05 HP
43. 483.1 HP
45S. 0.021 kW
49. 1.22 min
51S. 10.1 m^3

Chapter 15
19. 13.8 lb
21S. 415.8 kPa
23. 1.16 scfm
25S. 8.06 m^3/min
27. 50.4 psig
29. 59.8 psig
31. 0.531 psi
33. 124.79psi
35. 1 cup
37. 5.05 scfm
39. 2 cups

Index